Electronic Packaging Materials Science III

MATERIALS RESEARCH SOCIETY SYMPOSIUM PROCEEDINGS   VOLUME 108

Electronic Packaging Materials Science III

Symposium held November 30-December 4, 1987,
Boston, Massachusetts, U.S.A.

EDITORS:

Ralph Jaccodine
Lehigh University, Bethlehem, Pennsylvania, U.S.A.

Kenneth A. Jackson
AT&T Bell Laboratories, Murray Hill, New Jersey, U.S.A.

Robert C. Sundahl
Allied Signal, Des Plaines, IL, U.S.A.
   (Presently at Intel Corporation, Chandler, Arizona, U.S.A.)

MRS MATERIALS RESEARCH SOCIETY
Pittsburgh, Pennsylvania

This work was supported by the Office of Naval Research under Grant Number
N00014-87-G-0234. The United States Government has a royalty-free license
throughout the world in all copyrightable material contained herein.

Published by:

Materials Research Society
9800 McKnight Road, Suite 327
Pittsburgh, Pennsylvania 15237
Telephone (412) 367-3003

Library of Congress Cataloging in Publication Data

Electronic packaging materials science III.

(Materials Research Society symposium proceedings, ISSN 0272-9172 ; v. 108)
Bibliography: p.
1. Electronic packaging—Congresses.   2. Electronic packaging—Materials—
Congresses.   I. Jaccodine, Ralph.   II. Jackson, Kenneth A.   III. Sundahl, Robert C.
IV. Series: Materials Research Society symposium proceedings ; v. 108.

TK7870.E218   1988   621.381'046   88-9395
ISBN 0-931837-76-6

Manufactured in the United States of America

Manufactured by Publishers Choice Book Mfg. Co.
Mars, Pennsylvania 16046

Contents

*Invited Paper

PART III: INTERCONNECT TECHNOLOGY

PART IV: POLYMER INTERFACES AND ADHESION

*Invited Paper

*Invited Paper

*Invited Paper

*Invited Paper

Financial support for the symposium was provided by:

Advanced Micro Devices
Allied-Signal Corporation
Amoco Corporation
Dow Chemical USA
IBM Corporation
MCC
Office of Naval Research

Preface

This volume is the Proceedings of the third in a series of symposia on Electronic Packaging Materials Science, held during the Fall Meeting of the Materials Research Society on November 30 to December 4, 1987 in Boston, Massachusetts. A total of 77 invited and contributed papers were presented at the symposium. The symposium included overviews as well as papers which addressed levels of packaging. An important component of the symposium was two sessions which were organized jointly with the symposium on Polymer Surfaces, Interfaces and Adhesion. Papers from these joint sessions are included in this volume.

The packaging and interconnection technologies for electronic systems are becoming an increasingly important part of these systems in terms of their cost, functionality, and complexity. This trend is being driven by system trends toward higher density, higher speed, and higher thermal dissipation. The nature of these trends, and their impact on the demands placed on the materials were covered in a plenary session.

This theme was continued in a session devoted to the specific materials problems associated with the emerging packaging technology of multichip modules. This packaging approach was first introduced into main frame computers and is now proliferating into other applications. The complex demands on the hybrid materials systems involving layers of polymers, metals oxide films and ceramics is a major component of these contributions.

The joint symposium on Polymer Interfaces and Adhesion expanded on specific issues associated with the mechanical and chemical interactions occurring at polymer-substrate interfaces. New processes for depositing, etching and promoting interfacial adhesion are described.

Particular materials issues associated with interconnection on the IC, from the IC to its package and between ICs are covered in sessions dealing with these subjects. Finally, a special session dealt with issues particular to the packaging of GaAs and optoelectronic circuits.

As the symposium co-chairs, we are indebted to the important contributions made by the participants and, in particular, those who took the extra time and effort to prepare and revise their manuscripts for these proceedings. The organizing committee played a particularly important role in this symposium in the organization of the symposium, the running of the sessions, and the considerable effort involved in assembling and reviewing the manuscripts for this volume. Without the dedicated contributions of B.G. Bagley (BellCore), Allyson J. Beuhler (Amoco), Gordon Buchi (Ciba-Geigy), Paul Ho (IBM), Edwin Lillie (MCC), and Sherri Wen (IBM), this symposium and these Proceedings would not have been possible.

MATERIALS RESEARCH SOCIETY SYMPOSIUM PROCEEDINGS

ISSN 0272 - 9172

Volume 71—Materials Issues in Silicon Integrated Circuit Processing, M. Wittmer, J. Stimmell, M. Strathman, 1986, ISBN 0-931837-37-5

Volume 72—Electronic Packaging Materials Science II, K. A. Jackson, R. C. Pohanka, D. R. Uhlmann, D. R. Ulrich, 1986, ISBN 0-931837-38-3

Volume 73—Better Ceramics Through Chemistry II, C. J. Brinker, D. E. Clark, D. R. Ulrich, 1986, ISBN 0-931837-39-1

Volume 74—Beam-Solid Interactions and Transient Processes, M. O. Thompson, S. T. Picraux, J. S. Williams, 1987, ISBN 0-931837-40-5

Volume 75—Photon, Beam and Plasma Stimulated Chemical Processes at Surfaces, V. M. Donnelly, I. P. Herman, M. Hirose, 1987, ISBN 0-931837-41-3

Volume 76—Science and Technology of Microfabrication, R. E. Howard, E. L. Hu, S. Namba, S. Pang, 1987, ISBN 0-931837-42-1

Volume 77—Interfaces, Superlattices, and Thin Films, J. D. Dow, I. K. Schuller, 1987, ISBN 0-931837-56-1

Volume 78—Advances in Structural Ceramics, P. F. Becher, M. V. Swain, S. Sōmiya, 1987, ISBN 0-931837-43-X

Volume 79—Scattering, Deformation and Fracture in Polymers, G. D. Wignall, B. Crist, T. P. Russell, E. L. Thomas, 1987, ISBN 0-931837-44-8

Volume 80—Science and Technology of Rapidly Quenched Alloys, M. Tenhover, W. L. Johnson, L. E. Tanner, 1987, ISBN 0-931837-45-6

Volume 81—High-Temperature Ordered Intermetallic Alloys, II, N. S. Stoloff, C. C. Koch, C. T. Liu, O. Izumi, 1987, ISBN 0-931837-46-4

Volume 82—Characterization of Defects in Materials, R. W. Siegel, J. R. Weertman, R. Sinclair, 1987, ISBN 0-931837-47-2

Volume 83—Physical and Chemical Properties of Thin Metal Overlayers and Alloy Surfaces, D. M. Zehner, D. W. Goodman, 1987, ISBN 0-931837-48-0

Volume 84—Scientific Basis for Nuclear Waste Management X, J. K. Bates, W. B. Seefeldt, 1987, ISBN 0-931837-49-9

Volume 85—Microstructural Development During the Hydration of Cement, L. Struble, P. Brown, 1987, ISBN 0-931837-50-2

Volume 86—Fly Ash and Coal Conversion By-Products Characterization, Utilization and Disposal III, G. J. McCarthy, F. P. Glasser, D. M. Roy, S. Diamond, 1987, ISBN 0-931837-51-0

Volume 87—Materials Processing in the Reduced Gravity Environment of Space, R. H. Doremus, P. C. Nordine, 1987, ISBN 0-931837-52-9

Volume 88—Optical Fiber Materials and Properties, S. R. Nagel, J. W. Fleming, G. Sigel, D. A. Thompson, 1987, ISBN 0-931837-53-7

Volume 89—Diluted Magnetic (Semimagnetic) Semiconductors, R. L. Aggarwal, J. K. Furdyna, S. von Molnar, 1987, ISBN 0-931837-54-5

Volume 90—Materials for Infrared Detectors and Sources, R. F. C. Farrow, J. F. Schetzina, J. T. Cheung, 1987, ISBN 0-931837-55-3

Volume 91—Heteroepitaxy on Silicon II, J. C. C. Fan, J. M. Phillips, B.-Y. Tsaur, 1987, ISBN 0 931837-58-8

Volume 92—Rapid Thermal Processing of Electronic Materials, S. R. Wilson, R. A. Powell, D. E. Davies, 1987, ISBN 0-931837-59-6

Tungsten and Other Refractory Metals for VLSI Applications, R. S. Blewer, 1986; ISSN: 0886-7860; ISBN: 0-931837-32-4

Tungsten and Other Refractory Metals for VLSI Applications II, E.K. Broadbent, 1987; ISSN: 0886-7860; ISBN: 0-931837-66-9

Ternary and Multinary Compounds, S. Deb, A. Zunger, 1987; ISBN:0-931837-57-x

Tungsten and Other Refractory Metals for VLSI Applications III, Victor A. Wells, 1988, ISSN 0886-7860; ISBN 0-931837-84-7

Atomic and Molecular Processing of Electronic and Ceramic Materials: Preparation, Characterization and Properties, Ilhan A. Aksay, Gary L. McVay, Thomas G. Stoebe, 1988, ISBN 0-931837-85-5

PART I

Electronic Packaging Materials

DEVICE LIMITATIONS

ROBERT W. KEYES
IBM T. J. Watson Research Laboratory, P. O. Box 218, Yorktown Heights, NY 10598

ABSTRACT

Packaging technology must deal with the inexorable trend of semiconductor technology towards higher levels of integration. Extrapolation of present trends suggests that chips with 100 million devices will be produced by the end of the present century. The ability of technology to miniaturize pin-outs will limit the utilization of all of these devices for purposes other than memory. This limitation plus problems of supplying power and removing heat means that chips for high-performance large systems, where the demand for pins follows a well known rule, will probably be limited to levels of integration less than 100,000. A model of large system wiring shows that large increases in the density of wires in system packages and in the rate at which heat can be removed will be needed.

Less severe limitations apply to low cost applications. No large increase in power per chip can be anticipated. However, more powerful microprocessors will become available and will need increased amounts of input-output capability.

INTEGRATION

The success of semiconductor technology is a result of its ability to continuously and rapidly reduce the cost of providing electronic functions. The chip is the basic unit that is handled, sold, bought, and packaged into systems. Since the invention of the integrated circuit around 1960 steadily decreasing costs have been achieved by increases in the level of integration, the number of components (and of other circuit entities such as bits of memory or logic gates) on a single chip. The history of the maximum level of integration of chips in commercial products and a forecast of its future is shown in Fig. 1 [1]. The trend shown is driven by an inexorable economic force, the augmentation of human capabilities by powerful computational facilities [2], and will continue to the limits of physical laws.

The maximum integration is generally achieved in memory chips, where a regular structure simplifies designs. Microprocessor chips have also reached high levels of integration. Chips for use in the highest performance computers, however, have been restricted to many fewer components than the maximum used in memories and microprocessors. The reasons for this are not unrelated to packaging concerns, and will be discussed below.

The packaging of chips for large mainframe computers is a more complex subject than packaging other chips. A system contains hundreds of chips. Obtaining the highest performance from the machine requires minimizing the distances that signals travel. Space is extremely valuable and the chips must be placed in closest proximity to one another. The restrictions on

4

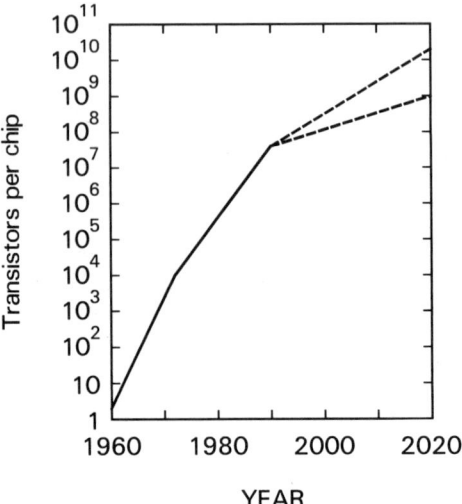

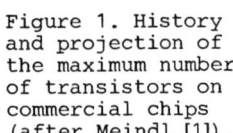

Figure 1. History
and projection of
the maximum number
of transistors on
commercial chips
(after Meindl [1]).

available space make packaging of large computers a difficult
technology with many facets, and we concentrate on it here.
It is also, however, a subject about which it is difficult to
generalize, as there is no unanimity among system designers as
to the best approach to packaging large, fast machines. Thus,
there is a much greater variability among the packaging technologies of different systems than among their chip technologies.

CONNECTIONS

     The number of components on a chip does not directly interact
with the package. The package communicates electrically with
the chip by means of "pin-outs" that connect lands on a chip
to package wiring. The number of pin-outs on a chip is more
relevant to the packaging of the chip than the number of devices
that it contains. Some of the pin-outs are used to supply
electrical power to the chip and others are used to receive and
send information.
     Memory chips, although reaching the highest levels of inte-
gration, do not need many pin-outs. Even one megabit memory
chips are provided with only about twenty pin-outs [3]. A modest
amount of information is read from a memory chip at each access,
and even this can be multiplexed and transmitted through a
single pin-out. The power requirements of highly integrated
memory chips are small, and electrical power can be supplied
through a few terminals.
     Microprocessors need more intensive communication with the
outside world. As the number of transistors on a microprocessor
chip has grown, longer words and wider data paths have demanded
increasing numbers of pin-outs. There has been a rough pro-
portionality between word length and number of pin-outs (al-

though the former moves in steps of a factor two while the latter is an almost continuous variable). The relation between number of transistors and of word length to number of pin-outs for some microprocessor chips is shown in Fig. 2. It can be anticipated that 64-bit microprocessors will eventually appear and will call for something like 250 pin-outs.

The most intensive demand for off-chip communication is encountered in gate arrays in large computers. These chips are part of systems containing hundreds of chips and perhaps a million logic gates. The results of logic operations that are performed on them are often needed in a distant part of the system, and many pin-outs are necessary to supply the demand for rapid transmission of information throughout the computer. The need for many pin-outs in logic chips that are part of a large system was recognized long ago [4], and is described by a relation known as Rents Rule. In fact the rule is not applicable only to chips, but to multichip modules, boards, or any partition of a large computing system. It has the form

$$P = BN^s \tag{1}$$

Here N is the number of logic gates on a chip or other partition and P is the number of signal connections that must be made to it. The general form of this rule is supported by a large number of empirical studies [5].

It is certainly possible to fabricate a chip with fewer connections than the number given by equation (1), but the basis of (1) is that it is then not possible to utilize most of the circuits on the chip as a part of the system. Both pin-outs and circuits contribute to the cost of a system, and (1) is a compromise aimed at obtaining a certain functional capability at minimal cost. Hardware designers differ as to the optimal way to make the compromises involved. Thus, it is not surprising that different authors disagree as to the best values for the parameters in equation (1). The exponent s is variously given in the range 1/2 to 2/3 and the factor B is about 3 or 4.

It is noteworthy that the number of pin-outs grows more slowly than the number of circuits on the chip. It appears,

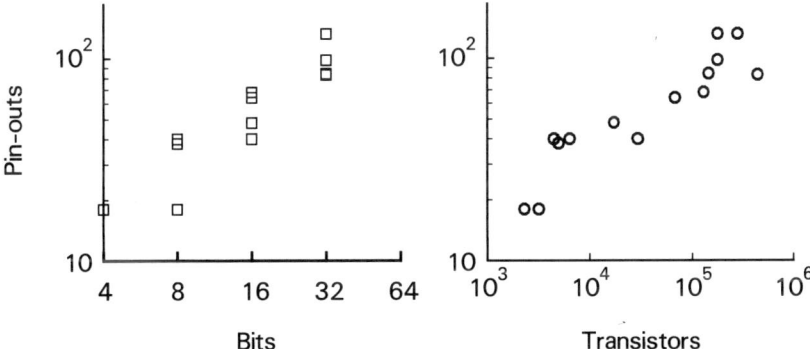

Figure 2. The dependence of the number of pin-outs on a microprocessor chip on the parameters of the chip.

as might reasonably be expected, that signal destinations are found on the chip of origin more frequently as the level of integration increases. This fact has also been explicitly noted [6]: the fraction of logic steps in which the result is transmitted to a different chip decreases with increasing level of integration. The reduction of the fraction of off-chip signal paths provides a motivation in addition to cost reduction for increasing integration in large machines: the off-chip paths are long and cause longer delays than paths on a chip; longer series of logic operations on a single chip aid performance. Indeed, this fact is responsible for a substantial part of the improvement in computer performance in the last decades. Interconnections have been moved from packages to chips, where they are shorter, in addition to being less expensive and more reliable. Paths that pass from one chip to another at low integration can be completed on a chip at high integration [6,7].

In spite of the economic force and the performance advantage pressing towards high levels of integration, the number of components on chips in large machines lags far behind the maximum capability shown in Fig. 1. One reason for this is the large number of pin-outs. Higher levels of integration are achieved by miniaturization of features on the chip; the size of the chip may increase little, if at all. The increased number of connections to the chip must be accommodated without a corresponding increase in area, and constant miniaturization of the connections is needed. It is difficult for technology to provide as many as 400 pin-outs on a chip, a number that can support less than 10,000 logic gates according to Rents rule.

The problem of connections between one electronic package level and the next does not end at the meeting of the chip and the module. The multichip modules must be mounted on and connected to large boards. The boards contain the wiring that connects the modules to one another. In accord with Rents rule, very large numbers of connectors are needed on the module. For example, the IBM Thermal Conduction Module has 1800 pins to connect to the next higher level of package [8]. Another somewhat similarly packaged computer uses over 2000 connections to the multichip module [10]. These connections are constructed with a quite different technology than the connections to the chip, as they must be easily separable for purposes of assembly and testing, and for replacement in the field to allow engineering changes and repairs. They are much larger than the connections between the chip and the module. The area per connection in contemporary large computers is about 3×10^{-3} cm^2 on chips and 0.06 cm^2 for multichip modules to boards.

WIRING

The large number of connections to the chips and to higher package levels must be joined by a dense network of wires. For example, the IBM Thermal Conduction Module, a ceramic substrate on which 100 chips can be mounted in an area of less than 100 cm^2, can contain over 100 m of wire [8]. Each square centimeter of the substrate contains well over 100 centimeters of wire. The wiring has a pseudo-random character; wire channels are provided and wires placed in them to interconnect the chips to implement the logical structure of the computer. As levels of integration increase, more and more logic is placed on such multichip modules and increasing wire densities will be required in the package to support them. A similar trend is evident on chips; chips now contain of the order of 1000 cm of wire per cm^2 of chip area. The high wire densities on chips and packages are achieved by fabricating the wiring in many layers.

The trend to higher wire density can be expected to continue. In fact, greater wire density is associated with increasing levels of integration through (1). This can be illustrated by a simple model. Let M chips be placed on a multichip module. Then the total number of signal pin-outs on the module is MBN^s. If these pin-outs are joined in pairs, there are $(1/2)MBN^s$ wires. The wires have an average length that is some fraction f of the linear dimension of the module. Measured in units of the chip-to-chip distance (the chip "pitch"), the side of the module is $M^{1/2}$, and the average wire length can be written $fM^{1/2}$ Multiplying these terms together, the total amount of wire on a module (in pitches) is

$$L_{tot} = (1/2)fM^{3/2}BN^s \qquad (2)$$

The amount of wire that must be supplied under each chip is

$$L_{tot}/M = (1/2)fM^{1/2}BN^s \qquad (3)$$

It is desirable in the interest of speed that chips be placed as closely together as possible. The provision of the large amount of interconnecting wire is one of the factors that limits the space that must be allotted to each chip. Let wires be placed in K layers. Also assume that wire channels have a width W. The channel width W must include space for vias that allow one layer to be connected to another, so that W is considerably greater than the actual width of the wire itself. Further, because of the pseudo-random character of the wiring not all of the channel can be used, the utilization of the available channel space is likely to be only 50% [8]. Additional notation to describe these influences will not be introduced, they are regarded as included implicitly in W and in the average wire length.

If each chip occupies an area A on the module, then the length of wire needed to traverse one chip site is the chip pitch, $A^{1/2}$, and the area that it occupies is $WA^{1/2}$. The area of all of the wiring needed under a chip, now found by using (3), must equal the total area available for wiring, KA.

$$KA = (1/2)fM^{1/2}BN^sWA^{1/2} \qquad (4)$$

Solving (4) for A yields

$$A = (fB/2K)^2MN^{2s}W^2 \qquad (5)$$

The area per chip required for the wiring is inversely proportional to the square of the number of layers. For example, using numbers somewhat representative of the IBM TCM [8], if f = 1/2, K = 16, M = 100, W = 0.05 cm, and N = 500, and assuming that B = 3 and s = 0.6, A turns out to be 0.95 cm^2. Of course, A must also be greater than the area of the chip plus additional space needed for assembly.

The need for more layers of wiring is seen most clearly by considering the resistance of the wire. Let $\rho_\square$ be the sheet resistance of the wire. (Again, a factor to account for the fact that the the wire is narrower than the channel appears here and is assumed to be absorbed in $\rho_\square$.) Then, since the length of the average wire is $fM^{1/2}A^{1/2}$, its resistance is

$$R = fM^{1/2}A^{1/2}\rho_\square/W \tag{6}$$

Taking A from (5) gives

$$R = f^2 MBN^s \rho_\square/2K \tag{7}$$

The resistance is inversely proportional to K. Note also that increasing W does not directly impact R; W must be determined by limitations of the fabrication process and by reduction of the distance between chips, $A^{1/2}$, equation (5).

Equations such as (5) and (7) are often more useful when solved for K and used to estimate the number of wire layers needed from other design criteria, for example, the acceptable resistance and module area per chip.

POWER SUPPLY AND HEAT REMOVAL

The quest for high speed has prevented the power dissipation per circuit in gate arrays for large machines from decreasing very much as the level of integration of the chips has increased. There is a trade-off between power and delay; the highest speed demands that the gates be operated at high power. Many pin-outs are needed to carry the electrical power on to the chips. The package must include conductors that distribute the power throughout the system. The power supply wiring is simpler in the sense that it does not have the pseudo-random character of the signal network. Electrical resistance in the power supply leads must be carefully controlled, however. A module may be powered by 100 A of current, which must reach the chips with less than 5% voltage loss, for example. The number of wire levels in the module used to deliver current at the supply voltage and return it through ground is not much different from the number needed for signal transmission.

An aspect of the chip that poses a serious problem for the package is the intermittent delivery of large amounts of current to the signal lines by driving circuits on the chip [11]. Each such event changes the current drawn from the power supply by the chip by tens of mA. Sometimes many driver circuits switch simultaneously, causing a sudden large change in the current drawn from the power supply, a change of a sizable fraction of an Ampere. The inductance of the power supply leads then produces a significant change of supply voltage at the chip. For example, if a current change of 0.25 A in 250 ps is to produce a voltage transient of less than 0.1 V, the inductance must be less than 0.1 nH. This is approximately the inductance of only 100 micrometers of a wire. The power supply transient can be controlled with capacitive filters, but the package must then provide the capacitor [11].

The electrical power delivered to the chip is dissipated to heat, and another demand made on the package by high-speed gate array chips is the removal of large amounts of heat. Today a power of 10 Watts may be produced on a gate array chip. Such high heat currents are carried away from the chip by special cooling structures that transfer the heat to air or another fluid that carries it out of the machine and eventually out of the building. Chip temperatures must be kept below some maximum value; only a certain finite temperature difference is available as a driving force for the heat current. The limited heat transfer coefficients at interfaces require that area be provided to allow heat to be transferred from one material to another with an acceptable temperature drop. Apertures through which fluids flow must be large enough to permit the flow to take place at pressures that can be provided by inexpensive and quiet apparatus. All of these restrictions limit the rate of heat removal to a few Watts per square centimeter of module area and constitute a lower limit to the area per chip. Fairly complex structures are needed to achieve even these rates.

A novel approach to cooling by passing water through channels etched into the back of chips has a potential to greatly relieve the heat removal limitations [9]. Although 1000 Watts have been removed from a single chip by this method, the limit on density of heat removal in a closely packed assembly of chips will depend upon ingenuity in miniaturizing the plumbing that carries the water to and away from the chips.

Relief from the heat transfer limitations may also be found in recent attempts to build large mainframe computers from CMOS (Complementary Metal-Oxide-Semiconductor) transistor circuits. CMOS circuitry dissipates much less power than high-speed bipolar circuitry. Operation at the temperature of liquid nitrogen is necessary to achieve the required high performance.

SUMMARY

The aim of packaging for high speed is the reduction of the distance between chips, or, equivalently, the area per chip. The area that must be reserved on the module for each chip is limited by several things: the area of the chip itself plus a peripheral area needed for handling the chip, the area of the cooling apparatus that carries heat away from the chip, the area of the connections between the chip and the multichip module, the area of the wiring in the module, and the area of the connections between the module and the next higher level of packaging. These limits are summarized in the following equations (8)-(12) and Figures 3 and 4.

Call the factor by which the area of the chip must be increased to permit mechanical access β. Then

$$A > \beta A_{chp} \qquad (8)$$

If cooling technology is described by a parameter Q, the number of watts that can be removed from a square centimeter of module, and the power dissipated on the chip is taken to be proportional to the number of logic gates on the chip, power = Np, then

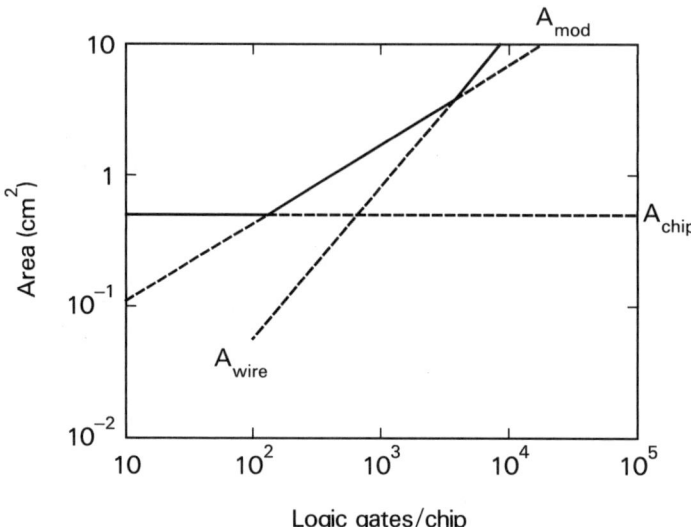

Figure 3. An example of the limits of equations (8) to (12). The parameters have the values β = 2, A_{chp} = 0.25 cm², p = 3 mw, Q = 4 W/cm², B = 3, s = 0.6, a_p = 0.002 cm² , M = 64, f = 0.5, W = 0.05 cm, K = 20, a_M = 0.05 cm² .

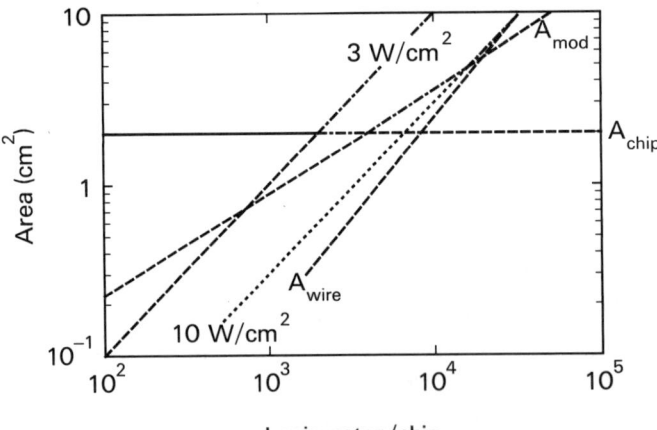

Figure 4. Another example of the limits of equations (8) to (12). The parameters have the values β = 2, A_{chp} = 1 cm² , p = 3 mw, K = 20, B = 3, s = 0.6, a_p = 0.002 cm² , M = 64, f = 0.5, W = 0.02 cm, a_M = 0.025 cm² . The cases Q = 3 W/cm² and Q = 10 W/cm² are shown.

$$A > Np/Q \tag{9}$$

Let a_p be the area required per connection to the chip. The area of the chip must be at least Na_p and A is limited by

$$A > \beta BN^s a_p \tag{10}$$

From (5)

$$A > M(fBN^sW/2K)^2 \tag{11}$$

Let the area per connection from the module to the next higher package level be a_M. The area per chip is at least the area of these pins divided by the number of chips.

$$A > B(MN)^s a_M/M \tag{12}$$

The example of Fig. 3 uses numbers typical of 1985 packaging technology. At the lowest levels of integration the chip packing density is limited by the area of the chip and the space needed around it. Above about 100 gates per chip over 600 pins are required to connect the module to the next level, and the space used by these pins limits the area. When the number of gates per chip exceeds 4000 the area is limited by the space occupied by the wiring in the module (with 20 layers of wire). The example of Fig. 4 shows how the rate of heat removal can limit the the area per chip. Narrower wire channels in the module and larger chips intended for higher levels of integration are assumed. At 3 W/cm^2 cooling limits the chip spacing above about 2000 gates per chip. If the cooling density is increased to 10 W/cm^2, which is technologically feasible, the cooling limit is reduced to a value close to that set by the wiring and the module pins, and a more balanced design is achieved.

Let us finally consider a very advanced packaging technology, in which wiring approaches the characteristics found on today's chips. Assume that $W = 10~\mu m$ and that 4 layers with this dimension can be fabricated. To achieve high performance, assume further that the area per chip must be less than 2 cm^2. The time for a signal to travel four chip pitches at half of the speed of light with this spacing is 380 ps, significant in a future very high performance system. According to (5) 2830 pin-outs, corresponding to a chip with 91,000 gates, can then be accomodated. Packaging such chips would strain all aspects of technology. The area of the module available for each connection to the chip is only 7×10^{-4} cm^2. The chip would probably dissipate about 25 W, requiring major advances in cooling methods. Thus, we conclude that it will be extrenely difficult to package chips with a level of integration as high as 10^5 into large high-performance systems.

REFERENCES

1. J. Meindl, IEEE Trans. Electr. Dev. ED-31, 1555-63 (1984).

2. M. E. Jones, W. C. Holton, and R. Stratton, Proc. IEEE 72, 1380-1409 (1982).

3. S. Asai, Proc. IEEE 74, 1623-35 (1986).

4. B. S. Landman and R. L. Russo, IEEE Trans. Computers EC-20, 1469-79 (1971).

5. For example, T. Chiba, IEEE Trans. Computers C-27, 319-25 (1978); C. T. Goddard, IEEE Trans. Components, Hybrids, Manuf. Technol. CHMT-2, 367-71 (1979); T. S. Steele, Proc. Int. Microelectronics Conference (Anaheim 1981), pp. 13-22.

6. D. Balderes and M. L. White, Proc. 35th Electronic Components Conference, 351-55 (1985).

7. R. T. Evans, Proc. 34th Electronic Components Conference, 374-78 (1984).

8. A. Blodgett, Scientific American 121 (1), 86-96 (1983).

9. D. B. Tuckerman and R. F. W. Pease, IEEE Electron Device Lett. EDL-2, 126-29 (1981).

10. T. Watari and H. Murano, Proc. 35th Electronic Components Conference, 192-98 (1985).

11. C. W. Ho, et al, IBM Journal Res. Dev. 26, 286-296 (1982).

MATERIALS LIMITATIONS IN THE HIGHER ELECTRONIC PACKAGING LEVELS

C.A. NEUGEBAUER
General Electric Corporate Research & Development
P.O. Box 8, KWC-1605, Schenectady, New York 12301

INTRODUCTION

Microminiaturization is desirable for reasons of lower cost, lower volume and weight, higher performance, and higher reliability. Microminiaturization is best achieved by downscaling the feature sizes on the chip itself, and this indeed continues to be the major driving force in the electronics revolution. It is normally cheaper, weighs less, is faster, and is more reliable to put as much of a system's interconnection effort on the chip itself (i.e., packaging level 0). Costs increase steeply for interconnections on the higher levels.

However, while much greater functionality on a chip has become possible with the downscaling of the feature sizes, it is nevertheless not always possible to put the entire electronic system on one single chip. This is partly because modern electronic systems require an increasingly greater functionality, and partly because processing yield sets a finite limit to the maximum feasible chip size which can be processed. Indeed, while the functionality on a chip has increased by many of magnitude over the last 20 years, the chip size itself has increased much less rapidly (Fig. 1). The maximum feasible chip size has increased at about the same rate as the maximum wafer diameter. It is thus clear that most systems require one or more chips to be interconnected on a hybrid or 2nd level package, albeit at a much higher cost per interconnection than the intraconnections on the chip itself.

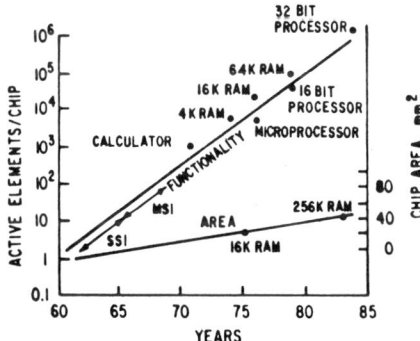

Figure 1 Historical increase in functional
density per chip and increase in
chip size.

This report concentrates on the materials requirements anticipated for future packaging of electronic components beyond the 1st level package. This includes hybrid assembly of bare chips (level 1.5), printed wiring board (PWB) assembly, including surface mount (level 2), and higher levels (connectors, mother boards, and cables). In addition, high performance digital VLSI logic packaging only is addressed, to the exclusion of memory, analog, and power circuitry (except power supplies). This is not to imply that the latter do not require a packaging effort utilizing new materials solutions for the future, but only that they are paced by the progress in high speed, high pin count VLSI chips.

The drive toward the downscaling of the minimum feature size in VLSI has lower cost as the principal motivation, but it also has direct conse-quences on the required packaging. It is thus critical to be able to predict what the VLSI chip of the future (5-10 years) is. Most predictions describe the future digital VLSI chip approximately as shown in Table I.

TABLE I - Future VLSI Chip Drivers

- Maximum feasible die size up to 2.5 cm on an edge

- 100 ps rise time and setting time

- 1000 I/O's

- 256 I/O's switching simultaneously

- Power dissipation as high as 10-100w

- 1.5V power supply voltage

- Immune to hostile environment

- 3D structures

FUTURE VLSI CHIP DRIVERS

The maximum economically feasible die size is not likely to exceed 1 inch (2.5cm) in the near future, although there have been abortive attempts at much larger chips. Most of the functional density increase will continue to be achieved by downscaling.

The performance capability of digital chips has continued to increase over the years, and this trend will continue, although not necessarily for the same reasons. Downscaling results in active devices with larger tran-sconductance, which can therefore charge an internal RC node faster, thus decreasing the propagation delay. As chips get larger and the feature size drops, this speed improvement becomes smaller, however, due to the increasing resistance of narrow aluminum conductors and the high sidewall capacitance between neighboring conductors. This was pointed out by Saraswat and Mahamadi [1]. However, means to decrease the line resistance will be found, such as superconductors, or cooling Al to liquid nitrogen temperatures, which will guarantee a continued increased performance of future VLSI chips, as shown in Fig. 2.

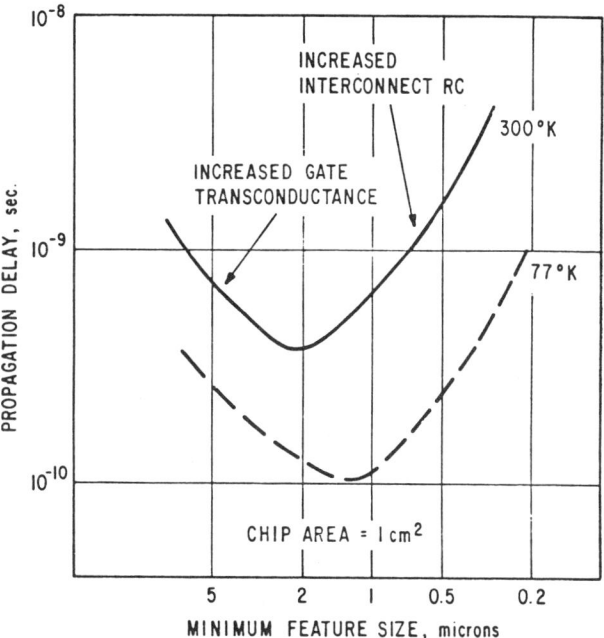

Figure 2   Propagation delay in large VLSI chips as
a function of downscaling, showing that the
increasing RC delay of the interconnections
dominate at small dimensions [1].

The number of I/O's in large digital chips depends principally on the circuit. In the worst case it is given by Rent's rule, such as in gate arrays. As long as the feature size was of the order of 2 to 3 μm or more, the traditional 0.008" pitch for the I/O pads resulted in a sufficiently close I/O pad spacing around the chip periphery as not to be size limiting. Instead, the actual circuitry dictated the chip size. This 0.008" pitch provided a reasonably comfortable fit with the inner lead pitch of most ceramic packages, at 0.010 more more.

The situation is reversed when the feature size drops to 1 μm and below. In order not to be I/O bound, the chip must have an I/O pad periphery of less than 0.008" pitch, such as .004" or below.

A large number of I/O's also implies a large number of simultaneously switching I/O's. A large number of simultaneous switching I/O's at high frequency then imply a very high di/dt. This is further exaggerated by the low power supply voltage in chips with small feature size, dictated by the breakdown voltage, which also scales downward. This means that the power for the chip must be supplied mostly as current in the IV product. It should be noted that while the power per gate does scale downward, the total power per chip does not, and in fact, will increase as performance increases, to a large extent due to switching power in which the capacitance in the interconnection network is constantly charged and discharged (CV^2f power). This not only requires power and ground conduits of very high electrical conductivity, (and low inductance) but also puts extraordinary requirements on the decoupling capacitors to suppress the L $\Delta I/\Delta t$ voltage spikes. A very high dielectric constant material is thus required in very close proximity to the chip power pins, and, indeed near the low dielectric constant insulator which normally separate the signal and power/ground planes.

One of the most severe materials requirements will be imposed by the formidable cooling effort which must be applied to future electronic assemblies. This is due to a greatly increased dissipated power density, and the chip contributes to it. The reason for a higher power density (W/cm^2 of Si) in future digital chips is due to:

- Higher switching speeds, thus resulting in increased CV^2f power dissipation.

- Although the transistors are smaller, more of the chip area is covered with active devices, each contributing to power dissipation.

The power supply voltage in future VLSI chips will be of the order of 1.5V because of the required signal to noise ratio, it is not likely that the threshold voltage of the VLSI chips will drop much below 0.5V, leaving a signal swing of only 1.0V. As mentioned above, since the power dissipation will continue to increase, the low supply voltage thus means the current density will be very high.

Future chips will be more adequately passivated, as part of the wafer process making them more immune to hostile environment. The impetus is given by the plastic package which is how most commercial chips are packaged, and whose first line of defense is a chip which was already "bullet proofed" in the wafer fab.

Finally, there will appear an increasingly larger number of chips which have 3-D construction. In simples cases, this may consist of 2 or more layers of diffused or implanted devices, or one or more chips stacked one on top of another [2].

FUTURE ELECTRONIC SYSTEMS DRIVERS

The principal electronic system drives will continue to be

- Performance

- Electrical noise control

- Packing density

- Reliability

- Manufacturability

The maximum possible clock rates is the measure for performance. The maximum clock rate is given by the signal rise time and the settling time required before a new state can be unambiguously recognized. The signal rise times, which depend on the chip I/0's, can be as small as 100 ps or less. On the other hand, the settling time depends on charging times (RC, L/R, etc.), transmission line delays (speed of light, dielectric constant, etc.), noise budgets (crosstalk, reflections), etc. Many simultaneously switching I/O's will, at short rise times, cause a very large current to rise quickly in the power supply lead, leading to a large possible voltage overshoot (L $\Delta I/\Delta t$), which can cause breakdown on the chip unless suppressed. Lossless lines give a minimum signal degradation due to IR drops, thus the highest possible conductance is desirable for the signal lines between chips.

Performance and noise budgets go hand in hand. In general, performance suffers if the noise tolerance is low. The permissible noise is higher the higher the threshold voltage, and the higher the power supply voltage. However, the power supply voltage must scale down with the feature size. A typical total noise budget from all sources might be 0.5V.

Packing density, measured in smaller volume and weight, is another systems requirement. This is obvious for airborne and space applications, where weight/volume has a pay load penalty associated with it, but is also an important consideration for ground based electronic systems where the support costs are often proportional to the systems size (such as the cost of warehousing of spare parts). For commerical systems it is often required that the electronics are a small fraction of the size of the overall system size or that it be tightly packed because of overall size constraints (computers). While the principal driving force for increased functional density certainly has been downscaling of the chip feature sizes, suitable packaging can also substantially reduce the size, such as use of bare chip and squeezing the various packaging levels together (3D) as closely as possible.

Future systems will increasingly require higher reliability. Reliability problems in packaging have been principally due to TCE interfacial differences between silicon and common packaging materials, thermal issues such as too high a thermal resistance and too low a heat transfer coefficient to the ultimate heat sink (normally air), and environmental attack. Using bare chip forgoes the first line of defense against environmental attack, such as corrosion, and thus put further demands on the chip passivation.

Finally, future packaging technology must be manufacturable, inspite of the demands encountered above. This not only is an important factor for cost and availability, but also strongly influences reliability. Problems in the manufacturability of future electronic systems are:

- Large die attach control

- Repair capability

- Demountability

- Solder control

- Lead control

- Fine lead spacings

- Inspectability

- Low resistance metallurgical joints.

Other additional systems requirements include freedom from EMI and lack of susceptibility to EMP, upset, total dose γ etc., and other radiation related events.

MATERIALS LIMITATIONS IN THE HIGHER PACKAGING LEVELS

Packaging materials limitations are apparent in the present state of the technology brought about by both, the future VLSI chip capabilities, and the increased future systems requirements. These are summarized in Tables II, III, and IV for various packaging levels, and are discussed below. The approach taken considers the impact of each chip driver on the various systems requirements.

1. Performance

The very short rise times and settling times of which future chips are capable will require low resistivity signal lines imbedded in a low dielectric constant medium. This is not only to reduce RC charging times, but also to reduce the speed of light delay. it is apparent that to achieve the lowest feasible value, dielectric foams may have to be considered. Optical connections, such as glass fibers or light pipes, are also subject to the dielectric constant limitation. Propagation of light through air or vacuum would be optimum.

In order to preserve high systems performance in face of very high pin counts in VLSI chips, it is necessary to place the chips very closely together, to reduce the signal delay between them, and still manage a very high density of interconnections. This requires a very fine interconnection grid. A conductor pitch of 1 mil would be desirable. A pitch finer than that would be undesirable because the conductor crossection would become too small and sidewall capacitance too large. Superconducting signal lines could, of course, go to still higher density, provided the critical current density is not exceeded.

Two orthogonal conductor levels should be sufficient to interconnect all devices. This requires that internal vias (as opposed to plated-through- holes) be used for the interconnections between layers, and that the dimensions of the vias do not require a coarsening of the conductor pitch. Further, groundplanes should be not much further removed from the conductors than the conductor pitch itself, to reduce crosstalk. This means very thin dielectric layers and precise thickness control, and requiring planarization steps in the fabrication process. A stiffener may have to be provided to back up printed circuit boards to provide sufficient dimensional stability.

The mounting of pin grid array packages on a printed wiring boards poses a performance problem because of the pin inductance and absence of shielding, and requires careful placement of ground and power pins.

High power density must be achieved in submicron technologies at ever decreasing power supply voltages. Unfortunately, digital power supplies become more and more inefficient at lower output voltages, because the diode threshold voltage of the power output rectifiers is a constant 0-3V. Use of synchronous rectifiers can reduce the threshold voltage to near zero, depending on device dimensions. Hower, they do require the lowest possible IR drops in the power supply package itself, such as solder bump contacts to the cathode, and use of thick copper straps.

Finally, 3-D chip construction, if it results in thicker chips or chip composites, will result in greater distances between substrates or boards, which detracts from performance.

TABLE II

MATERIALS LIMITATIONS IN LEVEL 1.5 (BARE CHIP HYBRIDS)

Systems Drivers / VLSI Chip Drivers	Performance	Electrical Noise Control	Packing Density	Reliability	Manufacturability
Large Die Size			High H Low R_θ	TCE compatibility of die/substrate, fatigue	Void control & inspectability of die attach.
100ps rise/settling times	High λ conductors. Low κ dielectrics. Optical fibers.	Impedance control planearizable multilayer interconnects, terminations.	Distributed power supplies.		Demountable 3-D construction (tight spacing).
1000 I/O's	1 mil pitch. 1:1 aspect ratio.		Staked vias	TCE compatibility.	Damage control in repair. Solder control, outer lead bonding materials for TAB, engineering change pad periphery, planarizibility.
256 simultaneously switching I/O's		High κ decap's closely located near chip.	High κ decap's.		
High Power density	Low IR drops in signal & power lines.		No heat spreading possible.	If liquid cooling: - corrosion - charge transport	
1.5V Power supply	Low forward drops in power supplies	Tight E,L,C control.	High efficiency power supplies. Power by light.		Low resistance joining materials/processes.
3-D construction	Shielded vertical conductors			Corrosivity of high λ materials, radiation hardness. Hiarchy of sealing materials, adhesion & purity control of plastics, out gassing, higher Tg's, non-hermetic enclosure.	
Other		EMI, EMP, Radiation upset (γ).			Materials to resist vibration, g-forces.

20

TABLE III

MATERIALS LIMITATIONS IN LEVEL 2 (ASSEMBLY OF PRE-PACKAGED CHIPS)

System Drivers / VLSI Chip Drivers	Performance	Electrical Noise Control	Packing Density	Reliability	Manufacturability
Large Die Size	Flatness Requirement		High H, Low R_θ	TCE compatibility of package to board, fatigue, barrel cracking.	
100ps rise/settling time	Low κ boards. <.013" hole sizes. Thinner boards, optical fibers.	Impedance Control	Distributed power supplies.		Demountable 3-D construction.
1000 I/O's	Internal vias. Multilayer, 1 mil pitch, mounting of pin-grids.		Blind vias		Solder control, damage control in repair, engineering change pad periphery, .010" lead pitch, lead fragility & control, lead solderability.
256 simultaneously switching I/O's		High κ decap's closely located near chip.	High κ decap's		
High Power density	Low IR drops in conductors.		Heat Spreading in 1st level package only, thermal vias, heat flow through board, along board. High efficiency power supplies, power by light.	If liquid cooling: – corrosion – board stability – charge transport	
1.5V Power supply	Low IR drops in power supplies.				
3-D construction	Tight board-to-board spacings.				
Other		EMI, EMP, radiation upset.		Corrosivity of Cu, board coatings. Deterioration in sliding contact surface.	

TABLE IV
MATERIALS LIMITATIONS IN LEVEL 3 AND HIGHER (BACK PLANES, CONNECTORS, CABLES)

System Drivers / VLSI Chip Drivers	Performance	Electrical Noise Control	Packaging Density	Reliability	Manufacturability
Large Die Size				TCE compatibility board to connectors, fatigue	
100ps rise/ settling time	Optical fibers	Impedance Control			
1000 I/O's			Blind vias	Reliable demountable contacts	Tighter connector pitch, 3-D construction, demountability
256 simultaneously switching I/O's					
Higher Power density			Lateral thermal conductivity, high efficiency heat exchangers, plumbing, radiators	If liquid cooling: – corrosion – compatibility – charge transport	
1.5V Power Supply					
3-D Construction					
Other				Corrosion of Cu, protective coating deterioration of demountible contacts	

2. Electrical Noise Control

Adequate electrical noise control requires that unwanted voltage transients are kept to a minimum. Unwanted voltage transients can occur due to
the following: resistive line loss, capacitive and inductive coupling, skin
effect, signal dispersion, signal reflection, and fan-out. These have a
well defined relationship to line length and crossection, proximity to other
lines, parallel and crossed, T-junctions, cross junctions, ground planes and
terminations, and the dielectric constant of the insulting matrix.

Impedance control of the signal lines is easier if each signal layer is
planerizable, using filled vias which can be stacked, and well defined
geometric relationships between the conductor runs themselves, as well as
the ground and power planes. Thus, all the elements of a particular conductor layer are to be located in one plane, with only vias making the transitions between planes. Normally this is easiest with filled vias, as opposed
to open ones. Control of the thickness of the conductor layers is crucial
to assure equal thickness throughout. When the layers are built up by a wet
plating process, preferential growth at asperities due to spurious high
current density sites must be suppressed, or overgrowth must subsequently be
removed. Processes such as mechanical grinding, if applied to this step,
require extraordinary flatness control over the entire substrate. Similarly, the dielectric layer must fill the holes and crevices in the preceding layer, without leaving hills and valleys. Depending on whether it is
applied by spinning on, stacking of foils, or deposition, the planarization
process presents different challenges. Thus, in the spin-on process of curable polymeric liquids of suitable viscosity, a planarizibility in excess of
80% is difficult to achieve at present.

Another aspect of impedance control is the shape of the conductor
sidewall. While perfectly vertical sidewalls are normally envisioned, and
are even realizable under certain conditions, other shapes are more common,
such as trapezoidal crossections, depending on how the sidewall is defined,
i.e., trench filling, subtractive etching, lift-off, etc.

One of the most damaging voltage transients which requires suppression is due to high current requirements of simultaneously switching I/O's. This noise spike is commonly suppressed by use of decoupling capacitors located close to the power pins on the VLSI chip. This capacitance is typically required to be in the 10,000 pF or higher range to be effective, particularly if shared between several power pins or even chips. A high dielectric constant material is required here to preserve area; however, at very fast risetimes it must be ascertained that the propagation delay in the capacitor electrodes themselves, which also depends on κ, does not make a large portion of the decoupling capacitor ineffective in the L $\Delta I/\Delta t$ suppression.

Suceptibility to electromagnetic pulses (EMP) and the broadcasting of electromagnetic radiation (EMI) present materials challenges, principally in shielding.

3. Packaging Density

Perhaps the greatest limitation to achieving higher packing density in the face of higher performance and decreased power supply voltages is the increasing difficulty to deliver and remove power to the circuit. Low voltage, high current power delivery requires that the power supply be highly efficient, and that it be distributed. The efficiency of conventional power supplies decreases as the output voltage decreases, because the forward biased diode potential of the Schottky output rectifier. This voltage, which must be reached before any current flows in the output circuit, becomes a greater and greater percentage of the output voltage as the output voltage itself decreases. To restore greater efficiency, the Schottky diodes should be replaced with power MOS devices operating as synchronous rectifiers, so that the IR drop associated with their on-resistance can be reduced to arbitrary low levels by increasing their size. This requires extraordinarily low resistance contacts.

Distributed supplies are needed in order to reduce the length of conductors which must carry high current, and replacing them with high voltage conductors. The idea is to distribute power over 48V or higher voltage lines, and locally transform it to 1.5V by DC to DC converters, distributed amongst the VLSI chips in a ratio of perhaps 1:12. This also requires that the DC to DC converters be quite small. The principal obstacle to reducing power supply size is the size of the magnetics. This can be done by going to higher frequencies. Screen printable magnetic materials for the fabrication of flat transformers would be highly desirable.

Finally, powering by light has recently been proposed. Power removal will require extraordinary measures. Previously heat spreading was available, allowing use of relatively high thermal resistance materials. Thus, mounting the chip in a ceramic chip carrier, which in turn was mounted on a printed wiring board, resulted in reasonably low thermal resistances from the chip to the cooling air because the heat generated in the chip quickly spread in the ceramic, which in turn presented a much larger surface area toward the surroundings. Even if the thermal resistivity of these surroundings was high, such as for printed wiring boards, the contact area is larger, thus giving a relatively low overall thermal resistance.

Future 1st level packages will be much smaller, if they exist at all. Heat spreading will thus be less available to reduce the thermal resistance between the chip and the cooling medium. It will therefore become necessary to apply lower thermal resistance materials and construction techniques. Composite materials, in which thermal conductors such as copper can be mixed with good electrical insulators such as polymers, might give rise to materials exhibiting better thermal conductivity without significantly sacrificing electrical resistance. Recent progress in the low cost deposition of diamond promise to make it practical as substrate material.

In addition to the thermal conductivity, it is the heat transfer coefficient to the cooling medium which determines the overall power removal capability. Maximizing contact area, such as in heat fins, maximizing the emissivity for heat transfer by radiation, and the use of thermal greases and thermal vias can improve the heat transfer coefficient.

Liquid cooling will become necessary for high power densities. The most prominent cooling liquids are water, fluorocarbons, and liquid nitrogen.

Finally, the use of stacked vias in the construction of the upper level interconnection network gives higher packing density, as does the use of higher dielectric constant decoupling capacitors, because of their greater volume efficiency.

4. Reliability

Primary among the reliability issues is the TCE compatibility between the die and the various packaging materials with which it comes into contact, and between the packaging materials themselves. The well known solder fatigue problem in surface mounted leadless chip carriers on printed wiring boards becomes still greater when higher pin count chips are involved. Use of constraining layers, such as Kevlar or Invar, with FR4 boards, while successfully reducing the X-Y direction expansion coefficient of the composite, result in a corresponding increase in the Z direction expansion coefficient, and can give rise to barrel cracking in the plated through holes. Materials improvements required fall into the following categories: Materials with expansion coefficients closer to that of silicon, such as AlN or diamond, ductile copper for compliance, and fatigue free solders and fatigue free solder joint construction. Falling into a similar category is the deterioration of mechanical or sliding contacts. Needed here are materials which do not corrode and gall, and yet are sufficiently compliant to give a low resistance contact under repeated connect/disconnect cycling.

Corrosion is the major issue in liquid cooling. This is particularly true if the liquid comes in direct contact with high thermal conductivity materials, such as copper. Often the reaction is slow and damage is not visible for years. This is due to the slow build-up of corrosive reaction products due to the decomposition of the cooling fluid. Thus, electrolysis of water produces oxygen, and an electrical discharge through fluorocarbon vapor due to charge build-up gives fluorine or chlorine. These can then react with copper, aluminum, solder, and the other common package construction materials to weaken metallurgical bonds or cause electrical short circuits.

Another cooling liquid materials issue is electrical charge transport, which is possible under conditions of misting, where fine droplets can transport charge from one region of the package to another, causing charge build up and eventual discharge and damage. A possible solution is to increase the electrical conductivity of the cooling liquid slightly to provide a leakage path to prevent charge build up, but not so high as to interfere with circuit operation.

A hierarchy of melting points or softening temperatures for adhesives, either metallic or polymeric, is normally required in package construction. This is because reflow must be avoided in sequential processing steps, thus, the lowest temperature adhesives are used last. A particular problem is the glass transition temeperature of polymers, where large changes in thermal expansion coefficient may occur, causing high mechanical stresses. Finally, outgassing of plastics, if contained in a hermetic enclosure, can give rise to failure due to reaction with the other construction materials. Water is the most common gaseous impurity.

Hermetic enclosures exact a heavy price in reduced packing density. It is likely that, with the rapid progress in low temperature CVD techniques, a method of encapsulating the chip with an inorganic coating can be found which will hermetically seal the chip completely against the environment, particularly the corrosive influence of water, thus making the hermetic enclosure redundant from a reliability point of view.

5. Manufacturability

Manufacturability or producibility not only reflect the cost of a particular packaging process but also the likelihood of it ever reaching a high level of maturity. Process maturity in turn is related to reliability.

Voids in adhesives used in packaging are a common manufacturing problem. They are due to trapped gas or the outgassing of the adhesive itself. If this cannot be prevented in the adhesive material itself, it is at least necessary that the void content is easily inspectable. Present techniques, such as X-ray or ultrasonic scanning, add greatly to the manufacturing cost.

Conventional 3-D construction consists of stacking boards or substrates at specified spacings, and interconnections between the boards by means of a motherboard. This requires long conductor lines to and from the boards to the motherboard, and is incompatible with the required systems performance and connectivity. Conductor vias interconnecting boards or substrates in the Z direction, not only at the periphery, but also in the interior, are highly desirable for a stacked construction. Further, the spacing between boards in the stack should be kept as small as is consistent with the required access of the cooling medium, either gas or liquid. An additional requirement for these interconnections is that they be readily demountable for easy access for testing, fault isolation, and repair[3]. Attempts in this direction are indicated in the report by Grinberg et al. Using demountable spring contacts between wafers. One of the requirements for stacked construction is the use of throughholes through the boards or substrates.

As the lead pitch of packages decreases to .020" and below, solder control for lead attachment becomes critical. This is also true for outer lead bonding of TABed chips. More accurate solder thickness control on the bonding pads is required, as well as better solder cream constituents and means of dispensing it. Shorts between pads are common due to solder spill-over, spitting, or dust particles under non-cleanroom assembly conditions.

Robustness toward damage by repair is a key manufacturability requirement. Usually this requires easy demountability and desoldering of previously mounted packages or components and their replacement without affecting the quality of the neighboring parts. Key for repairability is the ability to make engineering changes, such as the engineering change pad[4] periphery around each chip used in the IBM Thermal Conduction Module. Unfortunately, this is usually done at the expense of the packaging density. A means must be found to make engineering changes compatible with a density of silicon greater than 25%.

Lead control is equally important in fine lead pitch package attachment, because of their fragility and propensity to bend or crimp. Stiffer lead materials, complete elimination of flexible leads, and a positive means of controlling their location in the attachment step are desirable.

Joining processes in the higher packaging levels often use high resistivity joints, such as Kovar, crimped leads, solder, and slidable or compression contacts, all of which can give rise to high resistance. Finally, it is in these levels that materials are particularly prone to damage by vibration and g-forces.

REFERENCES

[1] K.C. Saraswat and F. Mahammadi, IEEE J. Sol. St. Circuits, SC-17, 275 (1982).

[2] F.R. Rossi and W. Straehle, Proc. NEPSCON East 87 201 (1987).

[3] J. Grinberg, G.R. Nudd, and R.D. Etchells, IEEE Trans. Computers, C - 33, 69 (1984).

[4] C.W. Ho, D.A. Chance, C.H. Bajorek, and R.E. Acosta, IBM J. Res. Develop. 26, 286 (1982).

SUPERCONDUCTING INTERCONNECTIONS IN FUTURE HIGH PERFORMANCE SYSTEMS

R. C. FRYE, AT&T Bell Laboratories, Murray Hill, NJ 07974

ABSTRACT

New, high temperature superconducting materials could eventually be used for interconnections in electronic systems. Such interconnections would undoubtedly cost more to implement than conventional ones, so the most likely applications would be for complex, high-speed systems that could benefit from the performance advantages of a resistance-free interconnecting medium. The problem with conventional conductors in these systems is that the resistance of wires increases quadratically as dimensions are scaled down. The most important advantage offered by superconductors is that they are not linked to this scaling rule. Their principal limitation is the maximum current density that they will support and this determines the range of applications for which they are superior to conventional conductors. An analysis will be presented which examines the relative advantages of superconductors for different critical current densities, wire dimensions and system sizes.

If their critical current densities are adequate, and if they can statisfy a number of processing criteria, then superconductors could find useful applications in a number of high performance electronic systems. The most likely applications will be those demanding very high interconnection densities. Several of these systems will be discussed.

INTRODUCTION

Recent materials breakthroughs are changing many of our notions about applications for superconductors. Superconductivity can now be achieved at temperatures that are compatible with semiconductor device operation, raising a number of interesting possibilities for device interconnection. Efforts currently underway to fabricate high T_c films are too numerous to catalog, and the rapid pace of improvement would quickly make any attempt to do so absolete. It appears likely, given the scope of the effort in this area, that good quality films will soon be available, and that many of the problems in processing these materials into useful interconnecting structures will be solved.

It is probable that superconducting interconnections will cost more to implement than conventional ones, so the important issues to examine are the benefits of a resistance-free interconnecting medium. This paper will examine these benefits as they relate to semiconductor electronic devices and, more importantly, within the context of all of the peripheral interconnection technology that has been built up around these devices. It is important to recognize that when we remove a limiting constraint from a system - in this case the constraint being line resistance - the limits of performance will be dictated by some other element in the system. Transmitting a signal from one point in a circuit to another may involve sending it through many different interconnecting media including chips, packages, circuit boards and even wires. Each of these different media imposes its own unique constraints on the performance of a system, and the expected improvement that can be obtained from superconductors depends on where they find use within this interconnecting framework.

We will examine interconnections at three different levels - printed wiring boards, thin-film hybrids and integrated circuits. These levels represent a progression from very large to very small structures, and from low to high wiring density. One feature that they all have in common is that they are planar interconnection technologies. It is probably safe to say that any application for superconductors in this area will be for high performance applications, which generally means high speed. Planar interconnections have been thoroughly studied at high frequencies, and if we expect to realistically weigh the benefits of superconducting technology, we must first consider the limitations that are imposed by the finite conductivity of conventional lines.

PROPERTIES OF LOSSY TRANSMISSION LINES

Figure 1 shows a conventional microstrip transmission line structure. It consists of a simple, rectangular cross-section line which is run over a uniform ground plane with an intervening dielectric. This basic structure is common to the three interconnection technologies that we will be considering. At low frequencies and low values of resistance, such a line will behave like a simple wire i.e. there will be a single,* well defined voltage along the entire length of the line. At higher frequencies, the finite speed of signal propagation becomes important. If we were suddenly to apply a voltage at one end of the wire, it would not appear instantly at the other. Viewed on a fast enough time scale, it can not be said that the entire length of the wire is at any particular potential.

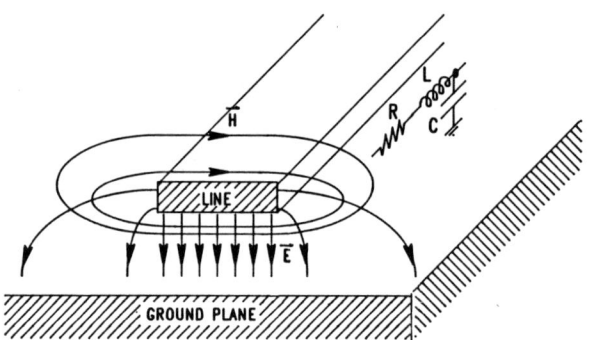

FIGURE 1: Microstrip transmission line structure.

In this electrodynamic regime, the line is characterized by its resistance, inductance and capacitance per unit length, R, L, and C. The resistance per unit length is given by

$$R = \frac{\rho}{A} \tag{1}$$

where ρ is the metal resistivity and A is the line cross-sectional area. The inductance is a measure of the total magnetic flux per unit current that is induced between the two conductors. Similarly, capacitance is a measure of the total electric flux (or charge) per unit potential that originates on one conductor and terminates on the other. If the line is very wide, and lies close to the ground plane so that the fringing fields are negligible, then L and C are approximately given by

$$L = \frac{\mu d}{w} \tag{2}$$

and

$$C = \frac{\epsilon w}{d} \tag{3}$$

where μ and ϵ are the permeability and permittivity of the dielectric between the lines, w is the linewidth and d is the separation between the line and the ground plane.

Typical microstrip transmission lines do not satisfy the conditions of negligible fringing fields, so Equations 2 and 3 are not generally (or even usually) true. These expressions, however, demonstrate a very important principle. Resistance per unit length in conventional lines is determined by the cross sectional area of the wire. Inductance and capacitance per unit length, on the other hand, are determined by the aspect ratio of the line - its width, height and separation from the ground plane. If we scale down all of the dimensions of a line by some scaling factor X, then the capacitance and inductance per unit length are unchanged, but the resistance per unit length increases by a factor of X^2.

To be general, the model circuit shown in Figure 1 should also include a conductance in parallel with the capacitance to represent dielectric loss, but these losses can usually be neglected for the kind of transmission lengths that we will be considering.

The model for a transmission line consists of a sequence of RLC circuits like the one in Figure 1, connected in series, each representing an infinitesimal length of the line. The equations describing propagation of the voltage, V, and current, I, along the line in the z direction are[1]

$$\frac{\partial V}{\partial z} = -RI - L\frac{\partial I}{\partial t} \tag{4}$$

and

$$\frac{\partial I}{\partial z} = -C\frac{\partial V}{\partial t} \tag{5}$$

If we decouple the above wave equations in the sinousoidal steady state, the solutions are

$$V = V_+ e^{i(\omega t - kz)} + V_- e^{i(\omega t + kz)} \tag{6}$$

and

$$I = I_+ e^{i(\omega t - kz)} + I_- e^{i(\omega t + kz)} \tag{7}$$

where

$$k = \omega\sqrt{LC}(1 - \frac{iR}{\omega L})^{1/2}. \tag{8}$$

Here, ω is the angular frequency ($\omega = 2\pi f$) and k is the propagation coefficient. Both the voltage and current solutions consist of forward and reverse traveling waves. The amount of each component is determined by the boundary conditions on the lines.

The expression for the propagation constant, k, illustrates some of the problems caused by loss in a transmission line. If R is non-zero then k is complex. This means that the exponentials in equations 6 and 7 have a negative real, component, i.e. the signal gets attenuated as it travels along the line. A second effect is in the phase velocity, v_p, given by

$$v_p = \frac{\omega}{\text{Re}[k]} = \frac{1}{\sqrt{LC}} \left[\frac{2}{1+(1+R^2/\omega^2L^2)^{1/2}} \right]^{1/2} \tag{9}$$

For no resistance, this becomes

$$v_p = \frac{1}{\sqrt{LC}} = \frac{1}{\sqrt{\mu\,\epsilon}}, \tag{10}$$

which is just the speed of light through the dielectric medium. If R is non-zero, the velocity becomes frequency dependent. In addition to attenuation, resistance causes dispersion.

Returning to the wave equation, it can also be shown that

$$\frac{V+}{I+} = -\frac{V-}{I-} = Z_o \tag{11}$$

where Z_o, the characteristic impedance of the line, is given by

$$Z_o = \frac{\sqrt{L}}{C}\left(1 - \frac{iR}{\omega L}\right)^{1/2} \tag{12}$$

The significance of the characteristic impedance can be seen by considering the reflection coefficient. If we terminate the end of a line of length ℓ with a general load impedance Z_L , then the reflection coefficient is given by

$$\Gamma = \frac{V-}{V+} = e^{-2ik\ell}\frac{Z_L - Z_o}{Z_L + Z_o} \tag{13}$$

In other words, if the load termination is an impedance equal to the characteristic impedance, the backward travelling reflected wave on the line vanishes. When R is very small, Z_o is a constant real number.

$$Z_o = \sqrt{\frac{L}{C}} \tag{14}$$

However, for non-zero R, because of the peculiar frequency dependence of Equation 12, it is not possible to match the impedance of the line at all frequencies.

Notice once again the important difference in the way R scales with wire size, as compared to L and C. For an ideal conductor, the characteristic impedance, Z_o, and the velocity, v_p, are both scale independent, since they depend only on L and C.

TECHNOLOGICAL COMPARISONS

In order to make some kind of comparisons among various interconnection approaches, we need to define a standard line geometry. Figure 2 shows the

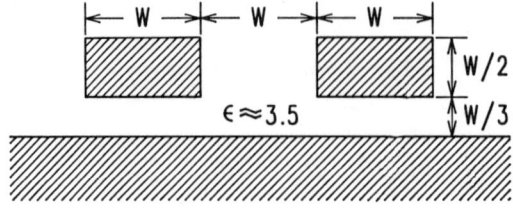

FIGURE 2: Example line geometry.

structure that we will use. This particular structure results in a characteristic impedance of about 50 Ω. The final results are not very sensitive to the choice of geometry provided that extreme shapes are avoided. The purpose of choosing a particular geometry like this is that we can now relate a number of factors. The wiring pitch, i.e. number of wires across a unit length of substrate, is

$$Pitch = 2/W \tag{15}$$

Equation 1 becomes

$$R = 2\rho/W^2 \tag{16}$$

So the resistance per unit length can be directly related to the wiring pitch.

At high frequencies, skin effect begins to increase the resistance of conventional lines. The skin depth is

$$\delta = \sqrt{\frac{\rho}{\pi f \mu_o}} \tag{17}$$

For $\delta > W/2$, the resistance per unit length is a constant. However, at high frequencies R increases, since only a thickness δ actually carries the current.

Superconducting transmission lines do show nonzero loss at finite excitation frequencies.[2,3] These losses arise from the presence of normal conduction electrons alongside the superconducting pairs. At nonzero frequencies, the finite voltage drop along the inductive component of the superconducting line will cause dissipation in the normal component. The amount of loss depends on the temperature and frequency. For the line sizes and frequencies that we will consider, however, these losses are insignificant.

The basic problem that we wish to consider is illustrated in Figure 3. We have a line of some appropriate length being driven by an ideal driver having a resistance R_S, and being terminated by a load impedance Z_L. There are two useful approaches to selecting the terminating impedance. If Z_L is a resistor of value $\sqrt{L/C}$, and R_S is small, then the signal will be perfectly matched at the load, and no undesired reflections will interfere with the signal. An alternative approach is to

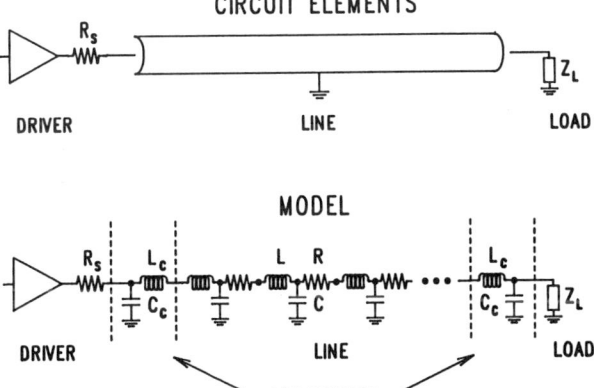

FIGURE 3: Circuit elements and model including parasitic elements at the line terminations.

make Z_L the high impedance input of a receiver circuit, and make $R_S = \sqrt{L/C}$. In this approach, the signal reflects once at the load, but the reflection gets damped at the source and can not return to the load a second time to cause problems. This second approach is often preferable, since it does not require static power dissipation to the load once the output has been charged up. This advantage is particularly important for very high wiring density applications in which heat removal can be a problem.

The model, shown in the lower half of Figure 3, accounts not only for the distributed resistance, inductance and capacitance of the line, but also includes a parasitic series inductance and shunt capacitance at each of the connections between the line and its driver or load. For most kinds of electronic applications, these parasitic elements are of little importance. However, for multi-gigabit device operation, they can dominate system performance.[4] It is also important to point out that in this model we have assumed the existence of an ideal voltage driver that has a constant output resistance over the entire frequency range of interest. Such drivers do not generally exist. Moreover, the output stages of chips usually have a resistance that is voltage dependent. Often the pull-up and pull-down resistances may be quite different. Trying to model a more realistic driver, however, would only obscure the primary issue of this analysis which concerns the performance of the interconnections. We, will therefore, assume that we have at our disposal drivers with arbitrarily good characteristics.

Table 1 lists the parameters used to evaluate each of the three technologies. For printed wiring boards, the connector parasitics mostly arise from the need to use a package between the line and a chip. The values of L_C and C_C were chosen to be typical of good, high performace packages.[5] Package inductance, in particular, has been recognized as one of the important limits to system performance.[6]

For thin-film hybrid technology, these parameters are less well defined because the technology is not as mature and a variety of chip attachment and

PARAMETERS

TECHNOLOGY	W (um)	L_C (nH)	C_C (pF)	MAXIMUM LENGTH (cm)	METAL
PRINTED WIRING BOARD	50–200	2	1	50	Cu
THIN FILM HYBRID	10–50	0.1	0.25	10	Cu
INTEGRATED CIRCUIT	2–10	0	0	1	Al

COMMON PARAMETERS

Z_0, R_s: 50 Ω	Z_L: .01pF
L: 3.1nH/cm	C: 1.2pF/cm

TABLE 1: Values of the model parameters used for three different interconnection technologies.

interconnection techniques are being employed.[7-9] Some of the possible techniques range from tape-automated bonding,[10] which has a reported lead inductance of around 1.2nH, to direct solder attachment, which can give an inductance as low as 20pH[9]. Intermediate values of inductance have been obtained for Josephsen junction packaging structures.[11] For the calculations in this paper, we will use a value of 0.1nH, which is weighted toward the high performance end of the spectrum. The shunt capacitance of 0.25 pF corresponds to a typical 100μm x 100μm pad on a chip.

In integrated circuits, the interconnection line from driver to load does not go through any intermediate connector, so the parasitics are essentially absent. For the purposes of calculating line resistance, we will use the conductivity of aluminum instead of copper since its use in IC's is much more common.

The maximum length of the lines in each of these three technologies is also a very important parameter. The longest line will have the poorest performance within each of the technologies, and will therefore determine the system constraint.

Figure 4 shows the magnitude and phase of the frequency response for 50 cm long interconnections on a printed wiring board. In these calculations and in the ones to follow, the inherent propagation delay has been substracted from the phase to make it easier to see the relative phase shifts at the output. The top trace in this figure is the response of a superconducting line (R=0), and the remaining curves show the effects of resistance in lines of varying width, or wiring density. The ripples in the response, which are particularly evident in the superconducting line, are a result of signal reflections resonating inside the line. Ideally, the 50Ω source resistance would eliminate these resonances, but at high frequencies the small connector inductance and capacitance contribute to the terminating impedances, causing mismatch and signal reflection.

It is interesting to note that in all cases, superconducting and normal, the signal dispersion in these lines, evidenced by the departure of the phase angle from zero, is more or less equivalent. Most of the dispersion in this system comes from the packages. At this particular level of technology superconductors will not reduce the

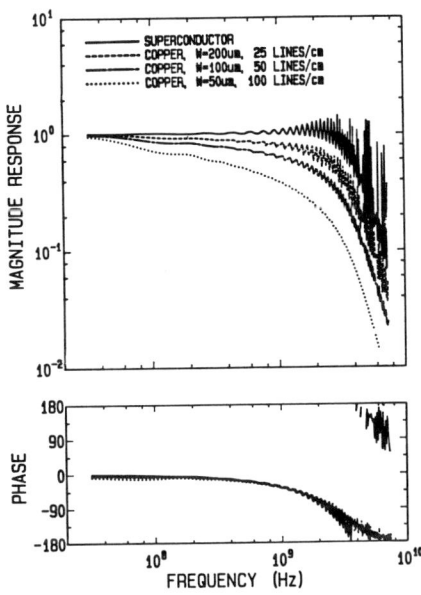

FIGURE 4: Magnitude and phase of the frequecies response for 50cm lines, printed wiring board technology.

signal dispersion. The improvements in bandwidth for a superconductor compared with low density (i.e. large linewidth) copper lines are marginal. Note, however, that as we reduce the linewidth in conventional lines, the magnitude of the response falls off at lower and lower frequencies. An important advantage of superconductors is that they do not experience the quadratic rise in loss as their dimensions are reduced.

Similar calculations for 10cm long lines on thin-film hybrid structures are shown in Figure 5. Because these lines are shorter, their resonances are shifted up to higher frequencies. Just as was the case for printed wiring boards, the benefits for thin-film hybrids at their lowest level of interconnection density are marginal. Even for the low values of connector parasitics offered by this technology, the dispersion is still dominated by the connectors.

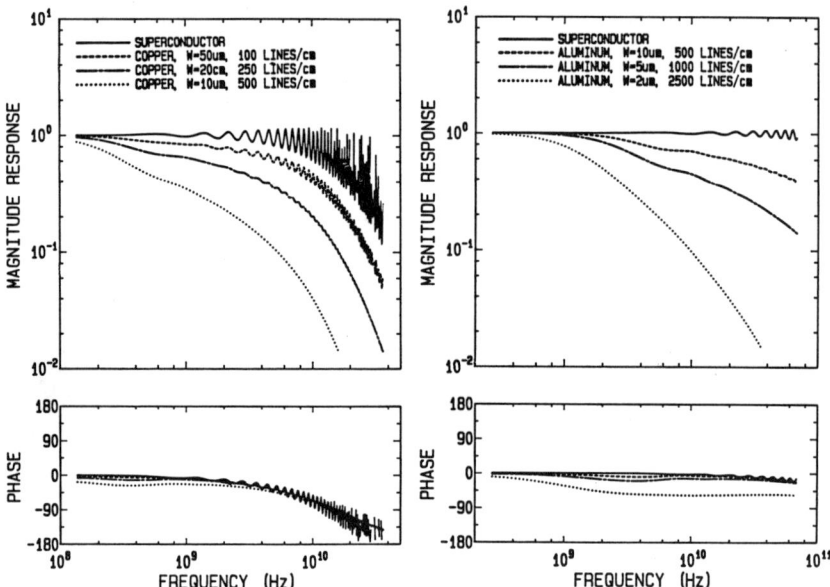

FIGURE 5: Magnitude and phase of the frequency response for 10cm lines, thin-film hybrid technology.

FIGURE 6: Magnitude and phase of the frequency for 1cm lines, integrated circuit technology.

The results for lines on integrated circuits are shown in Figure 6. In this case there are no connector parasitics to limit the transmission properties. The only non-ideality in the superconducting line is the load of .01p F (basically the input capacitance of a small transistor) that begins to cause a small amount of reflection above 10GHZ. At this scale, superconducting lines offer more substantial benefits because the ordinary metal lines constitute one of the most important limits to the performance of the system. These benefits become significant for multigigahertz bandwith applications.

DISCUSSION AND ANALYSIS

Like most technological comparisons, these results are necessarily dependent on the accuracy of the parameters that enter into the model. Rather than draw specific conclusions from the previous examples, let us instead consider the trends that they suggest. If we focus on one group of characteristics, it is clear that for conventional metallic conductors, within the confines of a particular technology, there is a limit to the dimensions of the wires below which resistance becomes a problem. The question to ask is at what point does this happen? The answer is central to transmission line theory. We can consider the effects of resistance to be small as long as

$$R\ell \ll \sqrt{\frac{L}{C}}.$$

or

$$R\ell \ll Z_o \tag{18}$$

In other words, as long as the total resistance of a line is small compared to the characteristic impedance then the effects of that resistance are small.

Compared with other properties of interconnecting structures, characteristic impedance is a parameter over which we have relatively little control. This is partly because of its insensitivity to changes in scale. Realistic microstrip transmission line geometries exhibit impedances ranging from 20Ω to about 150Ω. The value of 50Ω chosen for the examples represents a fairly typical, midrange impedance. By contrast, the resistance per unit length in these same examples varied by four orders of magnitude.

Figure 7 shows the relationship between wire pitch and line length for nearly loss free behavior. These curves assume 50Ω characteristic impedance lines. Actually, the onset of lossy behavior is difficult to define precisely. These curves are for $R\ell = 5\Omega$, or ten percent of the characteristic impedance. The curves for copper and aluminum films change slope for wire pitches below 500/cm to account for the fact that $5\mu m$ is probably a realistic upper limit for film deposition. Thicker films like those on printed wiring boards, are often plated or laminated. A curve for lead-tin solder has also been included, since it is often used on printed wiring boards. It can be made quite thick, but as the figure shows, it is not nearly as good as copper for low loss lines. Finally, an additional curve has been included for aluminum at $77°K$, since this may be used in conjunction with high T_c superconductors.

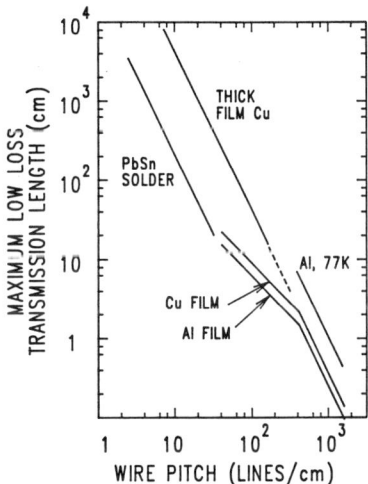

FIGURE 7: Maximum low-loss transmission distance vs. wire pitch for various metals.

The curves in Figure 7 represent the limits of conventional metalic interconnections in high performance applications. If we operate at some point above these lines then we can obtain more interconnection density or greater transmission length only at the expense of decreased bandwidth. The applications that are most likely to be impacted by superconducting interconnections will be those that demand both high wiring density and high speed, which is not generally possible to achieve with conventional interconnections.

Although the resistance of superconducting lines is insensitive to scale changes, the current density flowing down the line increases as we reduce the wire size. Superconductors are only capable of supporting a limited current density. The peak current flowing in a signal line, for the kind of high frequency operation that we have been considering, is simply

$$Ipeak \sim V_{sig}/Z_o \qquad (19)$$

Where V_{sig} is the voltage amplitude and Z_o, as before, is the characteristic impedance. The value of V_{sig} is a very technology dependent issue. CMOS circuits, for example, currently use 5 volt logic levels, but will probably soon be using 3 volts. High speed bipolar circuits, like ECL, use 800 mV signals. Future, higher speed technologies would benefit greatly by reducing the signal levels even further. Figure 8 shows the relationship between peak current densities and wire pitch for our example geometry, using the typical existing signal levels. It is important to keep in mind, however, that these particular curves are very dependent on the output levels of the semiconductor devices.

If we consider Figures 7 and 8 together, we can determine the critical current density above which a superconducting line has a significant advantage over a conventional one. If we are, for example, making a 1cm long connection using copper, we are limited to a wiring pitch of about 500 lines/cm. A superconductor will give us a better wiring density if it can support more current density. This gives us the break even limits for superconductors vs. copper shown in Figure 9. For very long conducting paths, superconductors are better because they do

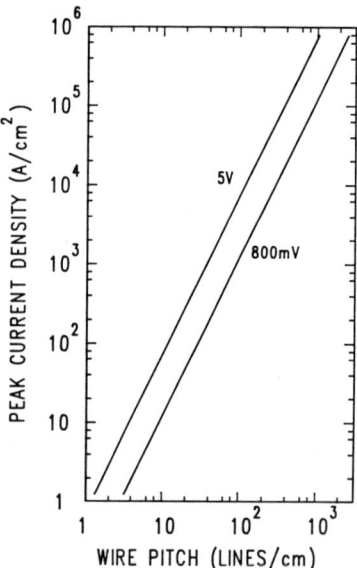

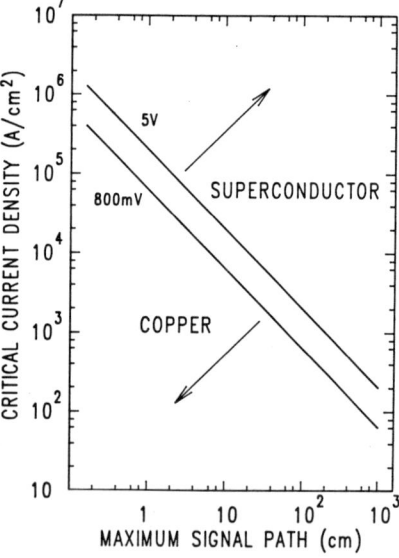

FIGURE 8: Peak current density on 50.ohm impedance transmission lines vs. wire pitch.

FIGURE 9: Break-even limits for copper vs. superconducting interconnections as a function of signal length and superconductor critical current density.

not accumulate resistance as they are made longer. For very short paths, metal lines are less of a problem, so the wiring density that we demand of the superconductor has to be greater before we would consider replacing the wire. For lower voltages of operation, superconductors become more and more attractive. If, however, we compare them with aluminum at 77°K, the curves in Figure 8 would be shifted upwards by about a factor of 4.

Note that small structures, like integrated circuits, make the most stringent demands on a superconductor, whereas larger structures are relatively relaxed. Unfortunately, as our examples showed, it is probably the smallest structures that stand to benefit most from a superconducting technology.

POSSIBLE SYSTEM APPLICATIONS

If the current density requirements can be met, superconducting interconnections show the most promise for systems that demand both high bandwidth and high interconnection density. These two issues are not unrelated. Signals in transmission lines typically travel at about half the speed of light. Computational systems are approaching the point where this constrains the system size.

Consider, for example, a personal computer operating at 10Mbit/sec. This system has 100 nsec in which to change its electrical state and communicate this change to its various parts. If we budget half of this time for propagation, then we find that the important parts of the system must fit in a space with dimensions on the order of 10 meters. This is not a very challenging task.

One decade higher in speed, at 100 Mbit/sec, we are in the realm of the supercomputer. Now, we have only 10 nsec and the system must be built with dimensions of about 1 meter. This is a more difficult task, but it can clearly be done.

If we ever build systems that work at 1Gbit/sec, then they must be made in a very small space having 10 cm dimensions. On this kind of scale, interconnection density becomes a very valuable commodity. Power dissipation is also very important, because it is generating heat in a small volume. Superconductors may be the only possible way to build such systems.

High bandwidth telecommunications switching is another example of an application that might benefit from superconducting interconnections. In this particular case, wiring density, rather that speed, is the primary consideration. Optical fibers can carry many channels of high frequency data. The task of demultiplexing these channels, re-routing them, and then multiplexing them into new fibers requires an enormous fan-out of wires. As we have seen, systems that are physically very large are usually not well suited to high frequency applications. Superconductors, by reducing the size of such systems, could have a beneficial impact on their performance.

One final application, although it is really a subsystem rather than a system, is the heavily loaded line. Such lines are common in computer back-planes and shared clock lines on chips. By adding multiple, capacitive taps on a clock line, we effectively make a transmission line with a very large capacitance per unit length and, consequently, a very low characteristic impedance. Such a line is very intolerant of any appreciable resistance.

SUMMARY

Interconnections between semiconductor electronic devices could possibly be improved by using new, high Tc superconductors. Potential benefits are high bandwidth signal transmission at high interconnection densities. Whether or not these benefits can, in fact, be realized will depend in part on the current density that these new materials will support. Very small structures, like integrated circuits, make the most stringent demands on the current density and, as a consequence, will be the most difficult to implement.

REFERENCES

[1] T. C. Edwards, "Foundations for Microstrip Circuit Design" John Wiley and Sons, New York (1981).

[2] W. D. McCaa and N. S. Nahman, J. Appl. Phys. 39 2592 (1968).

[3] W. D. McCaa and N. S. Nahman, J. Appl. Phys. 40 2098 (1969).

[4] A. J. Rainal to be published.

[5] C. J. Stranghan and B. M. MacDonald, IEEE Proc. 35th Electronic Components Conf. 361 (1985).

[6] A. J. Rainal, AT&T Bell Laboratories Tech. J. 63 177 (1984).

[7] E. T. Lewis, IEEE Trans. Components, Hybrids and Manuf. Tech. CHMT-2 441 (1979).

[8] C. W. Ho, D. A. Chance, C. H. Bajorek, and R. E. Acosta, IBM J. Res. Develop. 26 286 (1982).

[9] K. L. Tai, these proceedings.

[10] D. Herrell and D. Carey IEEE Trans. Components, Hybrids and Manuf. Tech. CHMT-10 199 (1987).

[11] K. Sato, H. Yoshikiyo, K. Aoki and F. Ohiro IEEE Trans. Components, Hybrids and Manuf. Tech. CHMT-9 145 (1986).

IDENTIFICATION OF THE FAILURE MECHANISM OF A THIN FILM ON A THICK
SUBSTRATE BY MEANS OF SYNCHROTRON X-RAY TOPOGRAPHY COMBINED
WITH TRANSMISSION ELECTRON MICROSCOPY

D. GOYAL*, W. NG*, A.H. KING* AND J.C. BILELLO**
*Department of Materials Science and Engineering, State University of New York,
Stony Brook NY 11794-2275.
**School of Engineering and Computer Science, California State University,
Fullerton CA 92634.

ABSTRACT

We have used synchrotron x-ray topographic techniques to study the stresses
in thin films formed upon silicon substrates either by evaporation or sputtering.
It is found that the film stress generally decreases with increasing film thickness
for evaporated films, but film delamination occurs at a well defined film thickness. Transmission electron microscope studies have been performed on the same
specimens in order to reveal what mechanisms are involved with the delamination
of the films

INTRODUCTION

Large internal stresses can be induced when thin films are formed by
deposition from the vapor phase onto a substrate. Particularly for very thin
films, the stresses can exceed the bulk yield strength of the film material, but do
not seem to cause plastic deformation because the small dimensions of the
specimens prevent dislocation multiplication from occuring. The stresses, however,
may lead to delamination of the film from the substrate and thereby cause
catastrophic failure of the film substrate system. Such failures occur both for
compressively loaded films, where they are theoretically impossible (1), and for
films under tensile stress. We will show that the stress in a tensile-loaded film
decreases rapidly with increasing film thickness, and yet the failure of the system
occurs when a well defined thickness is exceeded. The occurrence of this type of
failure is therefore something of a puzzle, irrespective of whether the film is
loaded in tension or compression. We have demonstrated in another paper (2) that
the stress in a thin film is not homogeneous and this may have some effect upon
the mechanism of the failures.

In this paper we report on a series of experiments that were undertaken to
measure the average film stresses developed during growth of thin films by
evaporation and sputtering, and were linked to transmission electron microscope
observations of the film-substrate system in order to elucidate the failure
mechanism.

EXPERIMENTAL

Thin films were prepared on silicon wafer substrates using techniques
described elsewhere (2). Both sputtering and evaporation were used as deposition
techniques. After the formation of the films, the stresses present in them were
determined by the method of determining the radius of curvature of the substrate,
which was achieved using an x-ray diffraction experiment. The curvature of the
substrate ensures that only a small portion of it can satisfy Braggs' Law at any
one time, if a highly collimated and well monochromatized beam is used. These
conditions are obtained at the topography facility at the National Synchrotron
Light Source at Brookhaven National Laboratory. The shift of the diffracting area
associated with a small rotation of the crystal provides an accurate means of
determining the radius of curvature. Following the determination of the stresses,
TEM specimens were prepared from the wafers, as previously described (2).

Mat. Res. Soc. Symp. Proc. Vol. 108. ©1988 Materials Research Society

RESULTS

The variation of the average film stress with film thickness is given for sputtered niobium and tungsten on (111) silicon substrates in Fig.1. Comparable results for evaporated nickel are shown in Fig.2. It is found that the stresses are larger for (100) oriented than for (111) oriented wafers. For sputtered films, the stress is initially tensile in the plane of the film, but becomes compressive as the film thickness increases, in most cases. For evaporated films, the stress is always tensile, but it decreases rapidly with increasing film thickness.

A distinctive phenomenon for all of our specimens has been the failure of the films, which occurs when a well defined "critical thickness" is reached. The thickness at which failure typically occurs is between 100 and 160 nm, under the growth conditions that we have used, with the exact value depending upon the film material and the substrate orientation. For compressively loaded (sputtered) films, the failure took the form of a wavy wrinkle in the film, as shown in Fig.3, corresponding to a region of delamination. When these wrinkles are formed at low densities, they tend to be straight and parallel to the <110> directions of the substrate. Failure of the tensile (evaporated) films occurs by the formation of cracks in the films.

A TEM image of a substrate with a film still attached is shown in Fig.4, where a crack in the film is seen to be associated with extensive deformation of the substrate. Fig.5 shows a case where the substrate exhibits slip traces like those observed in metals with strong oxide films, although no failure of the film is observed to be associated with the deformation. It would appear that the contrast arises because of the surface constraint imposed by the adherent film. In Fig.6, a substrate is seen without the film which has completely delaminated. In this case, a long array of cracks can be seen, bounded by an array of dislocations which appear to have been emitted from the crack tips.

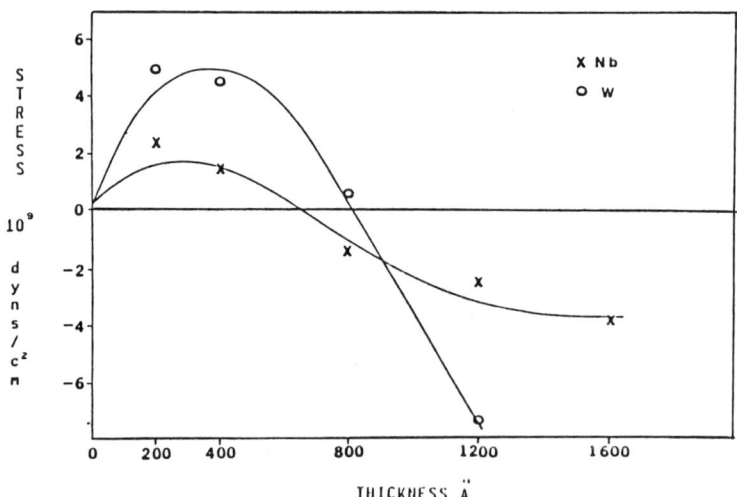

Fig.1. Average stress in a thin film on a substrate as a function of film thickness, for the case of sputtered niobium and tungsten films on (111) silicon substrates.

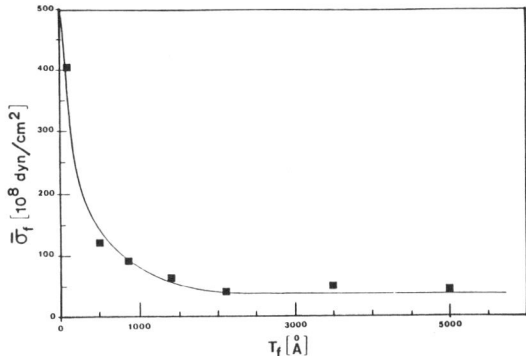

Fig.2. Average stress in a thin film on a substrate as a function of film thickness, for the case of evaporated nickel films on (111) silicon substrates.

Fig.3. Scanning electron micrograph of wavy wrinkles formed in a 160nm film of tungsten on a (111) silicon substrate.

Fig.4. (Above) Bright field TEM image of a 5nm chromium film on a (111) silicon substrate. A crack in the film is associated with severe distortion of the substrate.

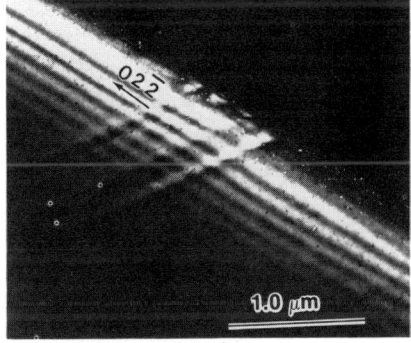

Fig.5. (Left) Dark field TEM image of a 5nm chromium film on a (111) silicon substrate. Slip traces parallel to <110> directions in the substrate are clearly observed, although there is no apparent failure of the film.

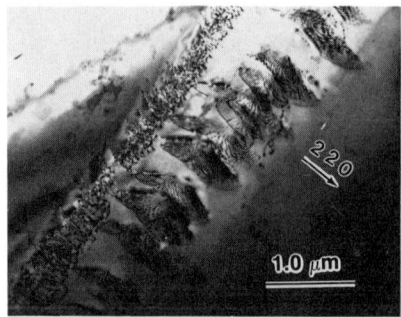

Fig.6. Bright field TEM image of a (100) silicon substrate which had a 500nm chromium film deposited on it. An array of cracks is observed, and an array of dislocations which emanate from the crack tips lies parallel to the array of cracks.

DISCUSSION

It is the general experience of the materials scientist that mechanical failure of any form occurs when some critical stress is exceeded. It is difficult, therefore, to rationalize the failure of thin films on substrates at some well defined thickness, where the stress is decreasing with increasing thickness. On the other hand, the stress in the substrate, immediately below the coating, increases with increasing film thickness, so it is reasonable to anticipate that failure occurs first in this region.

In our experiments, the thinning of the substrate to the point of electron transparency further increases the stress in the silicon and increases the density of film failures. Without the occurence of this effect, it would be nearly impossible to observe the delaminations, given the small sampling area available to the electron microscopist. The effect shown in Fig.5 clearly demonstrates that slip occurs in the substrate before the delamination of the films occurs: it is the continued adherence of the thin film that causes the bending of lattice planes necessary to create the contrast. The bending occurs because the surface slip step cannot be fully relaxed because it is constrained by the film. Fig.4 provides evidence that severe deformation of the substrate is associated with film failure, and Fig.6 indicates that the type of substrate damage that might be expected includes the formation of cracks, which in this case form long straight arrays.

In general, the damage is of a form that creates a step on the substrate surface, and we infer that this step provides the necessary incipient buckle and stress concentration form the delamination of a compressive film, or causes the shear of a tensile film thus precipitating its failure.

CONCLUSIONS

The failure of film-on-substrate systems appears to be associated with damage to the substrate rather than to the film or the interface. This would indicate that improved adherence might be obtained if the substrate material can be toughened.

ACKNOWLEDGMENT

This work was supported by the Army Research Office, under contract number DAAG2984K0168.

REFERENCES

1. D.J. Srolovitz and M.P. Anderson, Acta Met. 32, 1089 (1984).
2. D. Goyal, A.H. King and J.C. Bilello, this volume.

MEASUREMENT OF DIE STRESS FROM PACKAGING
AND EFFECTS OF THERMAL CYCLING

MURRAY J. ROBINSON, CLIFF TSAY, MATTHEW BUYNOSKI, AND
RAJENDRA PENDSE
National Semiconductor, 2900 Semiconductor Drive, Santa
Clara, CA 95051

ABSTRACT

During the processing and packaging of silicon
integrated circuits there are many sources of die stress.
Probably the least understood and controlled is the stress
introduced during the die attach and molding processes. The
largest packaging stresses are due to the mismatch of the
thermal expansion coefficients between the leadframe and the
die,and the die and the molding material. We report results
obtained by monitoring the resistance changes of implanted,
p-type resistors, undergoing various heat treatments with a
matrix of mold and leadframe materials.

Introduction

Die stress has been well studied [1] and can be
qualitatively understood and modeled using finite element
simulation programs. However, the changes in the stress
during and after thermal treatment of die are much more
complex and have not received as much attention in the
literature. Shallow surface devices, such as bifet
transistors and implanted resistors are particularly
susceptible to this stress. Changes in the stress and
therefore the silicon resistivity can create severe
reliability problems for silicon ICs. For this reason we
studied the changes in resistance of implanted resistors
which received different packaging processes. We then
compared the resistance changes before and after subjecting
the packaged die to various heat treatments.

Since the die stress is proportional to the temperature
coefficient of expansion (TCE) mismatch between the die and
the adjacent materials then we expected a larger resistance
shift from the materials with the largest mismatch to silicon
(TCE=2.5 ppm/C).

Experimental

For this study we used an array of 4 resistors forming
the edges of a square. The resistors (width/length=10)
measured about 14 kohms at room temperature. They were p-
type, implant resistors with a junction depth of
approximately 0.3 μms. The die size was 31 mil x 36 mil.
The die were packaged in a 14 pin DIP. Two types of
commercially available mold compounds, a standard B8 (TCE=24
ppm/C) and a lower stress Sumitomo 6300 (TCE=17 ppm/C) were
used on both copper (TCE=17 ppm/C) and Alloy 42 (58% Fe:42%
Ni, TCE=4 ppm/C) leadframes. We also tried a B8/copper
leadframe leg using a conformal die coating (concoat) on the
die surface to reduce the contact between the die and the

molding material.

The resistors were Kelvin probed using a high accuracy, constant current source (HP4141B) and an integrating voltmeter (HP706A). Since the temperature coefficient of resistance for these resistors is about 0.3% per degree centigrade the largest error in the measurement came from determining the exact temperature. To minimize this error we used an oven accurate to 0.14C, which gave us a resistance error of approximately 0.03%.

We used three heat treatments for each leg: a 25C to -55C to 25C half-cycle, 500 -65C to 150C cycles and a further 500 -65C to 150C hot/cold cycles. The half-cycle heat treatment consisted of, 90 minutes at 25C followed by 90 minutes at -55C then 90 minutes at 25C. We measured the resistors for the final 30 minutes at each temperature to check for any drift in the resistance. We then plotted the difference in resistance between the first and second 25C measurement (figure 1).

We then thermal cycled these parts 500 times using the -55C/165C cycle (1/2 hour per cycle) and measured the difference in resistance with respect to zero cycles (figure 2). A final 500 cycles was followed by another delta resistance measurement (figure 3).

Results

The resistance shift through a single "half-cycle" (figure 1) clearly shows that the high stress assembly process causes the largest resistance shift. The lowest shift is for the Alloy 42 leadframe with low stress mold compound. However, the matrix utilizing a concoat also displayed a low shift. These results are all consistent with stress being the most likely cause of resistance shift. This data also indicates that the dominant factor controlling the shift through a single half-cycle is the leadframe material. However, after 500 -55/165C thermal cycles the difference in leadframe material is less significant compared to the molding material, this becomes more significant after 1000 cycles. The benefits obtained using a concoat are lost after 1000 cycles compared with the zero and 500 cycle data. The 1000 cycle data also displays a difference in shift between resistors 1 and 3, and 2 and 4. Further experiments with die of different aspect ratios showed that this was due to the non-unity aspect ratio of the die with the longer side displaying the larger shifts. This has been previously reported in the literature [1].

Conclusions

We have shown that the packaging process introduces a resistance shift which is consistent with a stress arising from the expansion coefficient mismatch between the various components. We have also shown that the post-assembly changes in the shift are highly dependent on the post-assembly thermal treatment. The packaging stress effects are well known, the variation of this stress with thermal treatment is neither well documented nor understood. These effects need more study since they are a major reliability problem.

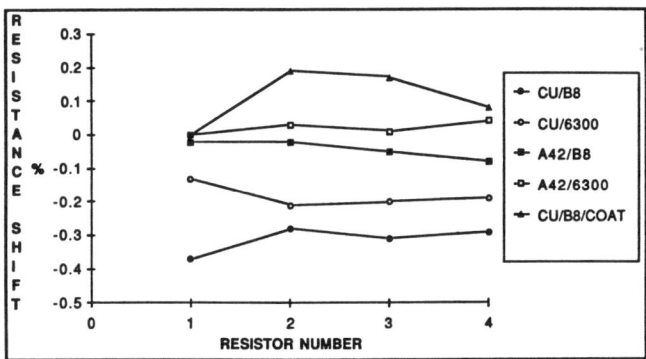

FIG. 1 HALF CYCLE

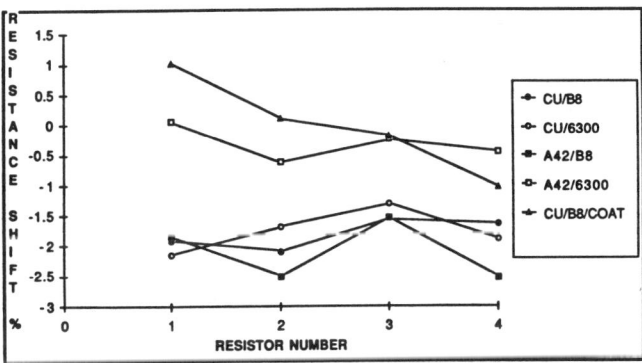

FIG. 2 500 TEMP CYCLES

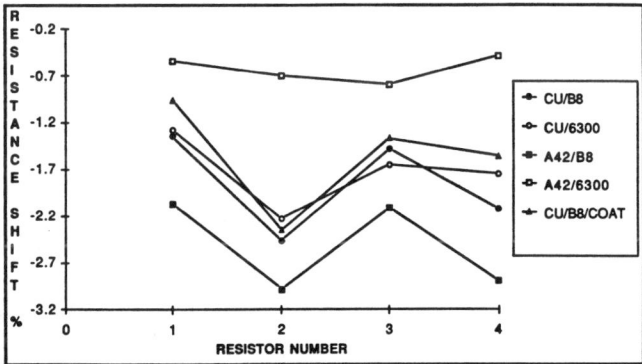

FIG. 3 1000 TEMP CYCLES

References

[1]   Walter H. Schroen, James L. Spencer, John A. Bryan, Robert D. Cleveland, Terry D. Metzgor, and Darvin R. Edwards; "Reliability Test and Stress in Plastic Integrated Circuits", 19th Annual Proceddings, International Reliability Physics Symposium, 1981, IEEE Catalog No. CH1619-6, pp. 81-87.

THERMAL STABILITY OF AMORPHOUS Ni-Nb THIN FILMS
FOR USE AS DIFFUSION BARRIERS

S.N. Farrens, J.H. Perepezko, Univ. of Wisc., 1500 Johnson Dr., Madison, WI 53706, B.L. Doyle and S.R. Lee, Sandia Nat'l. Labs, Albuquerque, NM 87185-5800

Abstract

The interdiffusion and crystallization reactions between amorphous Ni-Nb alloy films and Si substrates and several overlayer metals have been monitored by x-ray diffraction and high resolution Rutherford backscattering spectroscopy. Free standing amorphous thin films of Ni-Nb alloys crystallize in one hour at temperatures between 600-625 °C and show little dependence of the crystallization temperature, T_x, on composition over the range from 30-80 at.% Ni. However, in films that are sputter deposited onto Si substrates T_x tends to increase with increasing Nb composition. $Ni_{60}Nb_{40}$ samples without overlayers crystallize at 650-700 °C. Enhancement of the thermal stability to 700-750 °C is achieved with a Nb overlayer. In contrast, a Ni overlayer can reduce T_x to 450 °C. At the film/substrate interface silicide formation reactions with Ni from the film contribute to a destabilization of the amorphous alloy. The modification of T_x with Ni, Nb, and other overlayers appears to be related to changes in the reaction kinetics associated with penetration of the overlayer into the film.

1. Introduction

The unique structural characteristics and physical properties of amorphous metals are attractive in a number of applications involving diffusion barriers, primary metallizations, and resistance to electromigration damage. With most electronic applications the amorphous metal is in contact with, and can interact with crystalline material at the substrate or overlayer interface. Previous work shows that under these conditions the inherent thermal stability of an amorphous film to crystallization can undergo a significant modification.[1] It is also clear that the examination of a single amorphous alloy composition with a single substrate and overlayer is incomplete and can be misleading in judging the optimum performance of an amorphous alloy metallization. As a result the present work has been focused on a more comprehensive evaluation of amorphous Ni-Nb alloy thin films including a simultaneous investigation of substrate and overlayer interactions.

2. Experimental

Dual cathode magnetron sputtering allowed successive deposition of the amorphous layer (.75-1.0μm) and the polycrystalline metal overlayer (15-20nm) without breaking vacuum. Silver overlayers were deposited in a separate evaporation system. Following deposition, anneals were done in a tube furnace at base pressures of 10^{-7} Torr. The Rutherford backscattering analysis was done using two geometries. A high resolution geometry featuring a glancing angle of incidence was used for near surface analysis. This enabled us to determine overlayer thickness changes as small as 25 Å through precise measurements of variation in the overlayer energy width. A standard normal incidence geometry was used to investigate the interactions with the substrate.

The RBS samples were isothermally annealed in a geometric series for 1,2,4,...96 hours at temperatures of 400-550 °C in 50 °C increments. At temperatures of 600-750 °C one hour isothermal anneals were done as well as in situ isochronal anneals at 5, 10, 20, and 40 minutes.

3. Results

The onset of crystallization temperature, T_c, for these experiments is defined as the temperature at which the "first" indication of crystallites becomes apparent in an x-ray pattern following a one hour anneal. The one hour crystallization temperature, T_x, is defined as the lowest temperature at which the x-ray pattern shows sharp crystalline features. Figure 1 shows the x-ray patterns for the $Ni_{75}Nb_{25}$ sample with a Nb overlayer which was isothermally annealed up to 600 °C. Note: T_c=500 °C and T_x=550 °C for this sample. The same composition sample without an overlayer has a T_x of 500 °C. Similar overlayer effects have been observed in past work on amorphous Ni-Nb alloys and also with amorphous W-Si alloys.[1,2]

Table I reports the T_x results obtained from the x-ray data. The T_x increased with decreasing Ni content for all cases considered. Also, T_x is enhanced for samples with Nb overlayers while some overlayers have little effect and Ni overlayers can actually decrease the thermal stability of the amorphous layer.[3]

Table I. Crystallization Temperature, T_x, of Amorphous Ni-Nb films on Silicon

Alloy	Metallic Overlayer					
	None	Nb	Mo	Ag	Cu	Ni
$Ni_{60}Nb_{40}$	700 °C	700 °C	700 °C	700 °C	650 °C	450 °C
$Ni_{65}Nb_{35}$	700 °C	700 °C	700 °C	650 °C	650 °C	
$Ni_{70}Nb_{30}$		650 °C	600 °C		600 °C	
$Ni_{75}Nb_{25}$	500 °C	550 °C	500 °C	500 °C	550 °C	

RBS analysis of samples annealed at or above the T_c showed rapid diffusion and indications of compound formation at both the substrate and overlayer interface. The samples annealed below T_c showed extremely sluggish diffusion typical of amorphous materials. The diffusion coefficients of Nb from the overlayer into the fully amorphous alloys ranged from 8.7×10^{-20} to 2.7×10^{-19} cm^2/sec. These values for the diffusion coefficient are in agreement with work done by previous workers with amorphous Ni-Nb alloys.[4,5]

4. Analysis and Discussion

The diffraction patterns initially observed during crystallization are complicated by metastable Ni-Nb phases as well as equilibrium Ni-Nb phases. Preliminary analysis of the x-ray patterns indicate that the amorphous material crystallizes into a mixture of metastable Ni-Nb compounds and several nickel silicides. Above T_c the RBS analysis indicates that Si has diffused to the surface of the samples and x-ray work suggests that the layer may contain Nb_3Si. Further analysis will be done to confirm these findings. These results are in agreement with other reports of mixed phase formation upon the crystallization of amorphous Ni-Nb alloys and will be discussed elsewhere.[6]

The T_x of free standing amorphous Ni-Nb films has been reported by others as 600-615 °C with little dependence on the composition.[4] As shown in figure 2, the T_x of samples both with and without Nb overlayers exhibit a linear increase in T_x with decreasing Ni content. Comparison with the Ni-Nb equilibrium phase diagram shows an approximately constant undercooling below the liquidus for crystallization. This suggests that a critical level of driving free energy is required to initiate the interdiffusion

Figure 1.
X-ray Patterns for Amorphous
$Ni_{75}Nb_{25}$ Films with Nb Overlayers

Figure 2.
One Hour Crystallization Temperatures
of Amorphous Ni-Nb Films

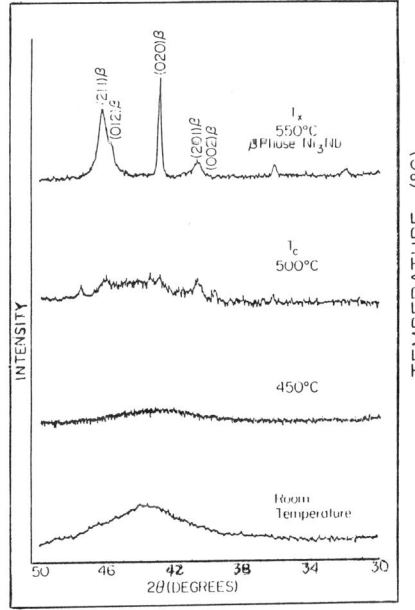

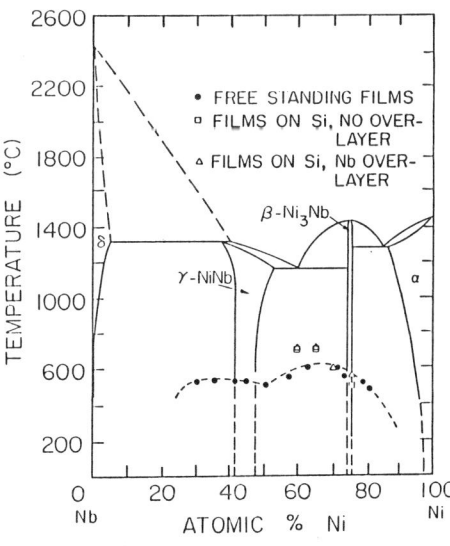

reactions involved in crystallization. Moreover, the increase in T_x in the samples without overlayers compared to freestanding films may require consideration of the difference in preparation techniques. The samples prepared for these experiments were most likely under a compressive stress. Preliminary examination of similar films has shown that the stresses are as high as 7×10^9 dynes/cm^2.[2]

The increase in T_x for the thin film samples on Si as compared to free standing films may be due to kinetic competition between the phases attempting to nucleate and grow. It is proposed that a competition between the silicide phases and the Ni-Nb phases for Ni or Nb may be responsible in part for the added increase in T_x. For instance, while NiSi may be favored to form at 400 °C in bulk samples, local concentration gradients and slow diffusion rates in films may delay NiSi formation until higher temperatures are reached. Once these higher temperatures are reached, $Nb_x Si_{1-x}$ may also exhibit a strong driving force for nucleation. Another study of amorphous alloys involving a refractory component has also observed similar behavior.[7] In all of these cases it appears that the formation of a refractory silicide governs the kinetics. Further analysis is presently underway to evaluate the validity of this argument as well as the role of local interfacial equilibrium at the reaction interfaces.

5. Conclusion

The T_x of amorphous Ni-Nb films is dependent on composition over the composition range of 60-75 at% nickel. The 60 at% Ni samples exhibited the highest T_x of 700 °C. The T_x can be modified significantly by selected overlayers. The diffusion rates of Nb into the amorphous Ni-Nb films were determined as a function of temperature and composition. Diffusion coefficients in the range of 8.7×10^{-20} to 2.4×10^{-19} cm^2/sec were measured.

6. Acknowledgements

We would like to acknowledge support by the U.S. Department of Energy under contract number DE-AC04-76DP00789 (Sandia) and DE-FG02-84ER45069 (UW-Madison).

7. References

[1] E.A. Dobisz, D.B. Aaron, K.J. Guo, J.H. Perepezko, R.E. Thomas, and J.D. Wiley, J. Non Crystl. Sol., 61&62, 901, (1984).
[2] R.E. Thomas, Ph.D. Thesis, Univ. of Wisc., Madison, WI, 75, (1987).
[3] L.E. Collins, N.J. Grant, and J.B. Van der Sande, J. Mat. Sci., 18, 804, (1983).
[4] B.L. Doyle, P.S. Peercy, R.E. Thomas, J.H. Perepezko, and J.D. Wiley, Thin Sol. Films, 104, 69, (1983).
[5] D. Akhtar and R.K.K. Misra, Scripta Metall., 19, 603, (1985).
[6] E.G. Reinke and O.T. Inal, Mat. Sci. Engr., 57, 223, (1983).
[7] L.S. Hung, F.W. Saris, S.Q. Wang, and J.W. Mayer, J. Appl. Phys., 59, 2416, (1986).

MULTILAYER ANALYSIS BY FOCUSED MeV ION BEAM

M. TAKAI, A. KINOMURA, M. IZUMI, K. MATSUNAGA, K. INOUE, K. GAMO,
M. SATOU*, AND S. NAMBA
Faculty of Engineering Science, Osaka University, Toyonaka, Osaka 560, Japan
*Government Industrial Research Institute Osaka, Ikeda, Osaka 563, Japan

ABSTRACT

A high-energy (MeV) helium ion beam has been focused down to 1 µm by a combination of piezo-driven objective slits and a magnetic quadrupole doublet. Rutherford backscattering (RBS) mapping techniques using focused MeV ion beams were, for the first time, applied to multilayered structures of metals, isolated with insulators, representing a test structure for multilayered wiring or interconnections of integrated circuits to nondestructively analyze the imperfection of the structures.

INTRODUCTION

There has been increasing interest in analyzing multilayered structures of metals, isolated with insulating layers, for semiconductor wiring or interconnections, in which more than one million of gates or interconnections of devices are fabricated. Three dimensional analysis to nondestructively check those structures is necessary to develop reliable process technologies. SEM or Auger electron spectroscopy (AES), for example, suffers charge-up effects for such structures or cannot nondestructively provide depth information. RBS spectroscopy, on the other hand, nondestructively provides depth information with less lateral resolution due to the beam spot size of 0.5 - 1 mm [1-3], which limits the applicability of this technique to the recent integrated circuits having a feature size of a few microns down to submicron.

Microbeam lines with a beam spot diameter less than 3 µm with MeV ions have been developed by several groups to accomplish three dimensional analysis [4-10]. Two methods have been used to realize a microbeam. One is to insert a small aperture in a beam line, through which only a small part of the beam passes onto a target[6]. The other is to combine an objective aperture and focusing lenses in a beam line, where an imaginary ion source defined by the objective aperture is demagnified with focusing lenses onto a target[4,5,7-10]. The obtained beam spot, therefore, is the projected image of the objective aperture. Although the latter is much more complicated than the former, higher beam intensity with smaller beam spot size can be obtained by the latter.

In this study, a microbeam line with 1.5 MeV helium ions for micro RBS and PIXE measurements was realized at the Government Industrial Research Institute Osaka by inserting piezo-driven objective slits and a magnetic quadrupole doublet in the beam line . RBS mapping techniques with a minimum beam spot size of 1.0 µm x 1.2 µm were, for the first time, applied to multilayered structures of metals, isolated with insulators, to nondestructively analyze three dimensional structures.

EXPERIMENTAL PROCEDURES

Figure 1 shows the schematic diagram of the beam line, consisting of objective slits, subsidiary slits, scanning coils, and a magnetic quadrupole doublet. The objective aperture is formed by two pairs of collimator slits which are driven by piezo-elements mounted on micrometers. The slit width is controlled by piezo-elements and micrometers in a range of up to 30 and

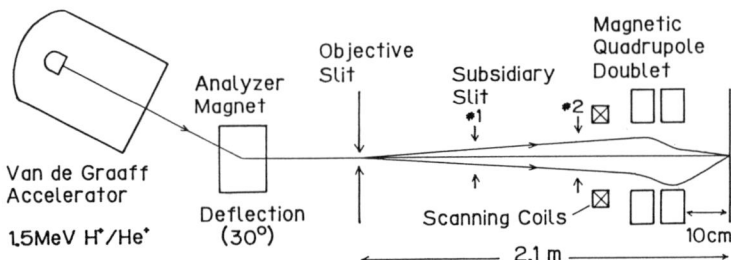

Fig. 1 A schematic diagram of the focused ion beam line at Osaka

1000 μm, respectively. The slit is made of tantalum and the design of the slit edges is similar to that of the Heidelberg system [5], to decrease the forward scattering effects at the slit edge. Subsidiary slits are used to eliminate scattered beams.

The focusing lenses for demagnifying the beam are a magnetic quadrupole doublet with a pole gap and length of 5 and 45 mm, respectively, yielding a maximum field gradient of 4.9 kG/cm. The combination of the objective slits and the quadrupole doublet gives rise to horizontal and vertical demagnification factors of 1/4.7 and 1/21.9, respectively, since the effective distances between the slits and the doublet and between the doublet and the target are 1846 and 135 mm.

Two pairs of scanning coils are attached to the lens system which can deflect beams horizontally by 150 μm and vertically by 50 μm at the target surface for 1.5 MeV helium ions with a maximum scanning frequency of 100 Hz. This scanning system is used to get a mapping image or to locate a beam on micro areas.

The minimum beam spot size obtained by this system is 1.0 μm x 1.2 μm with a current of 50 pA [10].

The target chamber has an optical microscope and three detectors: a photo–multiplier–tube coupled with a scintillator for secondary electrons, a surface–barrier Si solid–state–detector (SSD) with an energy resolution of 15 keV for RBS and a pure Ge SSD for PIXE. The optical microscope, monitoring the target with a glancing angle, uses fluorescence from a quartz plate to align the beam or to roughly check the focusing of the beam. The Si SSD with a sensitive–area of 25 mm^2 is located 30 mm away from the target with an angle of 45 degrees normal to the beam axis. The solid angle is about 28 mstr.

Figure 2 shows the data acquisition system for RBS, PIXE and secondary electrons. A 16–bit microcomputer controls the beam scanning and collects the signals from three detectors. 12–bit D/A or A/D converters with settling or conversion times of 20 – 25 μs are used for generating scanning signals or collecting secondary electron yields and target ion currents. All of the measurements can, thus, be controlled by software. Secondary electron and RBS mapping data are stored in computer memories with 2700 pixels (90 x 30). Each of the intensities (secondary electron) or counts (RBS) stored in the corresponding memories is displayed in the monitor by different colors or symbols in 6 or 12 steps for mapping images. The minimum data acquisition time for one frame of secondary electron or RBS mapping images is 9 seconds. This time duration is limited by data handling time in the computer. Detailed description on the microbeam line was published elsewhere [8–10].

18 nm thick gold grating patterns with a periodicity of 20 μm, isolated

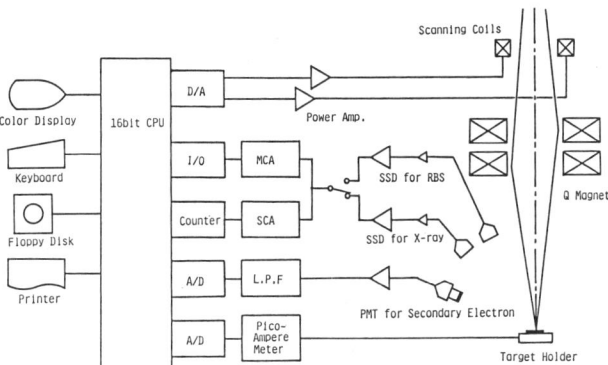

Fig. 2 A schematic diagram of the detecting system

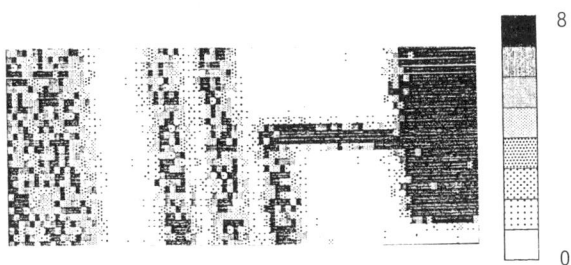

Fig. 3 RBS mapping of molybdenum silicide gate and
interconnection metals

with SiO$_2$ layers and delineated on a silicon substrate, were used to obtain
a test structure of multilayered wiring up to the 4th layer.

A single channel analyzer (SCA) was used to determine energy windows
for RBS mapping.   The beam spot size of 3 μm x 3 μm was used for RBS
mapping to get good statistics in counts.

RESULTS AND DISCUSSION

Figure 3 shows the RBS image mapping for molybdenum silicide gate and
interconnection metals with a thickness of 20 nm on an insulator (10 nm
thick SiO$_2$) in an integrated circuit. The energy window for RBS signals is
adjusted to molybdenum signals by a single channel analyzer.   The RBS image
clearly indicates the gate and the connecting part with a 3 micron rule. It
takes about 1 hour to obtain an RBS mapping image with a beam current of 100
pA. The total ion dose for this RBS mapping was 2.9 x 10^{16}/cm^2.

A test structure for multilayered wiring was analyzed by both secondary
electron and RBS mapping.   Figure 4 shows the schematic diagram of a
demonstrated multilayered structure of metals (gold layers) isolated with
insulating layers of SiO$_2$ and the corresponding RBS spectrum obtained by a
defocused beam of several hundred microns. The secondary electron mapping
image is also shown. The four gold-grating patterns are offset by 45

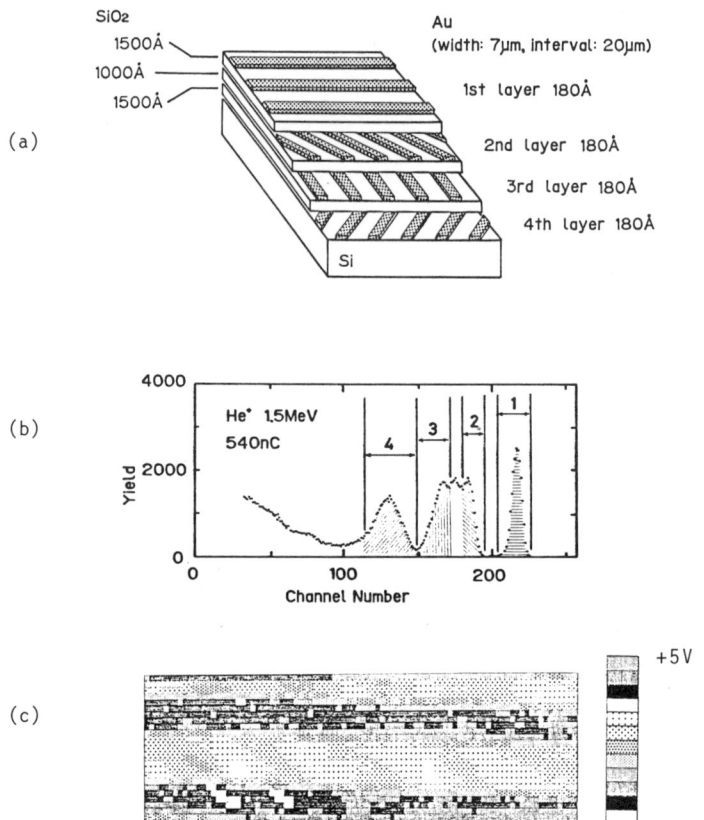

Fig. 4 A schematic illustration of the multilayered structure (a),
corresponding RBS spectrum (b), and secondary electron mapping (c).

degrees with respect to each other to correctly identify each layer by RBS
mapping images.   Three major peaks observed in the RBS spectrum correspond
to the first, the overlapping of the second and the third, and the fourth
gold layers, respectively. The energy windows for RBS mapping were adjusted
at the channel number indicated in the spectrum.   The secondary electron
mapping image in Fig. 4c was, at first, collected to position the beam and
to adjust the focusing parameters of the beam optics.   Since the secondary
electron intensity depends not only on the atomic number of an element but
also on the surface geometry, the image of the second gold layer weakly
appears due to the surface geometry of the insulating layer.
      Figure 5 compares RBS mapping images of the sample.   The RBS mapping
image of the first gold layer clearly indicates the grating pattern without
influence of the surface geometry of the insulating layer.   The second layer

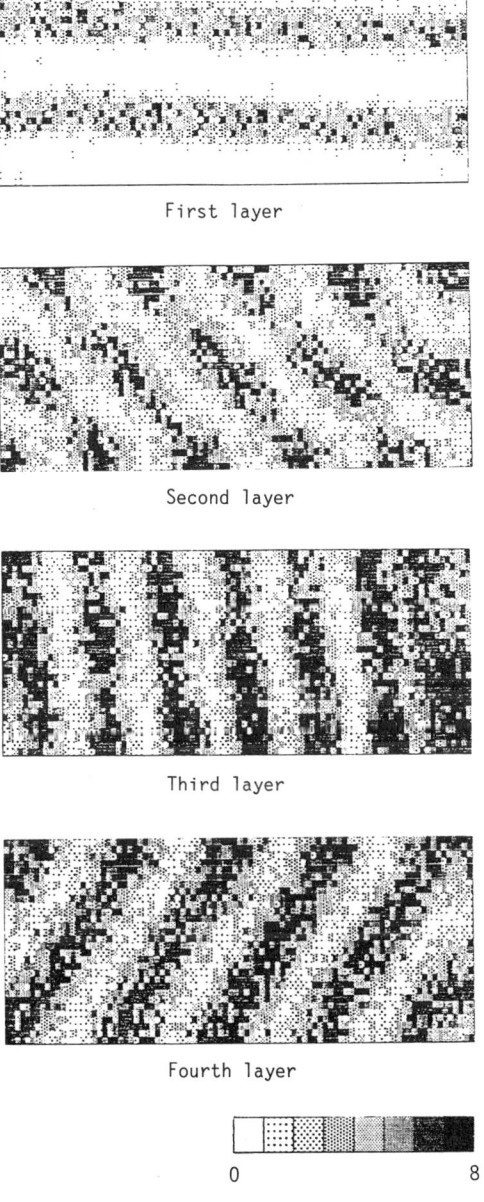

First layer

Second layer

Third layer

Fourth layer

0   8

Fig. 5 A comparison of RBS mapping images for the multilayered structure as shown in Fig. 4a

mapping, however, reflects the structure of the first gold layer because the energy loss of the incident or scattered helium depend on the upper layer structure. It should be noted that the process imperfection of gold grating at the right hand side of the third layer was clearly resolved in the mapping image. Although the gold grating of the fourth layer was also resolved, the line width became somewhat broader than those of the upper layers. This is due to the difference in the energy window during mapping, where the window was adjusted wider than those for the upper layers as in Fig. 4b to get better statistics.

CONCLUSIONS

RBS mapping techniques using focused MeV ion beams were, for the first time, applied to nondestructively analyze multilayered structures of metals isolated with insulating layers, representing multilayered wiring or interconnection structures of integrated circuits, which could not be inspected by AES or SEM.

ACKNOWLEDGEMENTS

The microbeam line at the Government Industrial Research Institute Osaka is an adaptation and modification of the Heidelberg system. The authors are indebted to K. Traxel for his advice on the construction of the beam line in the initial stage of this study.

REFERENCES

1. M. Takai, K. Gamo, K. Masuda and S. Namba: Japan. J. Appl. Phys. 12, 1926 (1973)
2. M. Takai, K. Gamo, K. Masuda and S. Namba: Japan. J. Appl. Phys. 14, 1935 (1975)
3. Backscattering Spectrometry, edited by W.K. Chu, J.W. Mayer and M.A. Nicolet (Academic Press, New York, 1978)
4. G.J.F. Legge: Nucl. Instrum. and Methods. 197, 243 (1982)
5. R. Nobiling: Nucl. Instrum. and Methods. 218, 197 (1983)
6. P. Bayerl and P. Eichinger: Nucl. Instrum. and Methods. 149, 663 (1978)
7. H. Kneis, B. Martin, R. Nobiling, B. Povh and K. Traxel: Nucl. Instrum. and Methods. 197, 79 (1982)
8. M. Takai, K. Matsunaga, K. Inoue, M. Izumi, K. Gamo, M. Satou, and S. Namba, Japan. J. Appl. Phys. 26, L550 (1987)
9. M. Takai, A. Kinomura, K. Inoue, K. Matsunaga, M. Izumi, K. Gamo, S. Namba, and M. Satou, in Proc. of the 1st Intern. Conf. on Nuclear Microprobe Technology and Applications, edited by G.W. Grime and F. Watt (to be published in Nucl. Instrum. and Methods 1988)
10.A. Kinomura, M. Takai, K. Inoue, K. Matsunaga, M. Izumi, T. Matsuo, K. Gamo, S. Namba, and M. Satou, in Proc. of the 12th Intern. Conf. on Atomic Collisions in Solids, edited by F. Fujimoto (to be published in Nucl. Instrum. and Methods 1988)

PART II

Multichip Modules

SMALL DIMENSIONS
or
LOW DIELECTRIC CONSTANT

The Competing Approaches to High Density Interconnect

John W. (Jack) Balde, Senior Consultant
Interconnection Decision Consulting
Flemington, New Jersey 08822

EXTENDED ABSTRACT

Cost Considerations

Ten years ago, the conventional wisdom as cited by Rex Rice and others was that interconnect wiring on a silicon chip was much less expensive than interconnections on a ceramic hybrid, a printed circuit board, or cable interconnect. That led to a major emphasis on increasing the size and complexity of the silicon chip, with the other interconnect media left for the overflow or leftovers that could not be placed on the chip.

A major change of thinking was triggered by Knausenberger and Schaper of AT&T (1), with the realization that costs normalized per inch of wire length were nearly identical for all forms of interconnect. Literally an inch of interconnection circuit costs the same whether that circuit was on silicon or on ceramic, whether that circuit was on a printed circuit board or in cable.

If the only important criteria is the length of the interconnect, then a system or a board of the smallest size and area for a given circuit will have the shortest path lengths and the lowest cost. The dominant criteria is the area of the interconnection medium that carries the active silicon.

Speed Considerations

If cost was one consideration, certainly speed of the circuit performance was the other. The industry yardstick has traditionally been cost/speed; one looked for the most thruput for the lowest cost. Here the conventional wisdom was to look for the lowest line capacitance for the interconnection line, and with that lowest capacitance to achieve the highest propagation velocity. The standard wisdom was to lower the capacitance, and thereby increase the speed of the __same size__ board.

That wisdom was negated by the work of Evan Davidson (2) and others who correctly pointed out that a transmission line of too high an impedance was prone to crosstalk and noise problems. A line with too low an impedance suffered from other problems, not the least of which was excessive power consumption. Changing the impedance from the 50 to 70 ohm region was unprofitable, because the problems created were difficult ones.

The problem then was to increase density without changing the capacitance and characteristic impedance.

The Fine Line Approach

With a dielectric constant of 4 to 5, and a desired characteristic impedance or about 50, the cross-section of the line capacitance is approximately "square". By this I mean that the width of the capacitor in cross-section is the same as the thickness of the dielectric. If one wants to increase the density, one has to decrease the thickness of the stripes and lower the thickness simultaneously. In printed circuit board materials, with constant dimensions of the reinforcing fiberglass, it is difficult to reduce the thickness. Furthermore, increasing the density by putting lines closer together produces disabling crosstalk from one line to another.

The low dielectric constant insulation material approach

It is much better to use an insulating material of lower dielectric constant on which to deposit the lines, thereby reducing the thickness of the dielectric. Then the circuit density is doubled, and the area reduced to 1/4, with not only cost benefit but also the speed benefit that could have been achieved wastefully by changing the propagation delay.

This matter is discussed in a paper present at IPC World, and reprinted here with permission. (3)

Which dielectric is the best

The highest density interconnection is now being realized in multichip packaging. Silicon-on-silicon can achieve densities that wafer scale could not, because there is no wasted area for real or possible dead cells. As density increases, wattage goes up, and it is useful to consider better substrates than silicon - Aluminum Nitride or Silicon Carbide being the ones of most current interest. Alternatively, low dielectric constant ceramics can be used.

But the low dielectric constant ceramics have very poor thermal conductivity, making them only suitable for packaging constructions like the IBM TCM and the similar NEC and Mitsubishi approaches. Where the heat must flow from the die to the substrate, one must use better conducting materials. Furthermore, the high dielectric constant of the better heat conducting materials requires heroic measures such as the waffled ground planes of the TCM to reduce the capacitance. It is better to use organic dielectrics to carry the circuits, and better to use copper as the conductor so that fine conductors need not be lossy.

The best organic dielectrics

Polyimide has been the choice for the organic dielectric to support the copper interconnection wiring. But it is a poor dielectric because it absorbs water and then has a dielectric constant of over 5. New materials such as the Dow Chemical and Celanese Hoechst materials of low water absorbtion and dielectric constants about 2.5. The resultant achievable circuit density is then 2:1 in linear dimensions, and 1/4 the area of the circuit with a higher dielectric constant material.

The new Multichipping

The major efforts in lowering the cost and increasing the speed of circuits is through the use of copper circuits on organic dielectrics with $\epsilon = 2.5$, all deposited on a good thermal substrate such as Silicon Carbide. Furthermore, the density increases are sufficient to produce significant increase in operating speed and sufficient reductions in area and cost that it is not necessary to use full conversion to ASIC and gate array chips. Rather, for cost performance systems for which time to market is an important criterion, it is possible to mix vlsi, ssi and msi chips; to mix TAB, wirebond, and soldered interconnects; and to achieve very respectable speed and density.

This is the new emerging technology, and the Rockwell paper at SAMPE tells it as well as any. (4) It will be of major importance in the computer and analog circuit designs of the 90's

John W. Balde
12/1/87

Bibliography

(1) Interconnection Costs of Various Substrates — The Myth of Cheap Wire W.H. Knausenberger amd L. W. Schaper, IEEE Transactions on Components, Hybrids and Manufacturing Technology, Volume 7, Pg 261, September 1984

(2) Electrical Design of a High Speed Computer Packaging System, E. E. Davidson, IEEE CHMT Volume 6, pg 272, September 1983

(3) Low Dielectric Constant — the Substrate of the Future, J. W. Balde and G. W. Messner, Printed Circuit World Convention IV, Paper WCIV-59, June 1987

(4) Advanced Packaging Concepts — Microelectronics Multiple Chip Modules Utilizing Silicon Substrates, A. A. Evans and J.K.Hagge,   1st International SAMPE Electronics Conference Proceedings, pg 37, June 1987

LOW DIELECTRIC CONSTANT - THE SUBSTRATE OF THE FUTURE

John Balde, IDC Corp and George Messner, PCK Inc.

If a group of system designers were asked to list the ways to increase the high speed performance of a system, the list would be sure to include lowering the dielectric constant. Lower dielectric constant would either be first, or at least very high on the list. If you asked for changes that would increase the density of packaging and reduce the size and cost, lowering the dielectric constant would not even be mentioned. Yet decreasing the dielectric constant is one of the most effective ways to increase the circuit line density and thus reduce the circuit board area. Consider the electrical characteristics of a semiconductor circuit:

The operating speed of a computer is determined by the sum of the circuit delays. Of these, a major delay is the time for a pulse to travel the length of a circuit board to another gate. The faster the velocity of propagation, the faster the computer can operate.

The propagation velocity for ECL and other bipolar circuits is:

$$v = 1/ \sqrt{LC} \qquad \therefore \quad v \approx 1/ \sqrt{C}$$

and the propagation delay = $1/v \approx \sqrt{C}$

If the dielectric constant is reduced from 5.5, a typical value for epoxy glass boards, to 2.8, a possible value with new board materials, the velocity would increase by $\sqrt{2}$, and the system speed proportionally.

If the speed of the system was previously satisfactory, with a reduction of the dielectric constant the linear dimensions of the board could be increased in each direction by 40 %, and the board doubled in area. This is an important result, because it is much more cost efficient to provide fewer larger boards than many small ones, and in fact many computers can be reduced to only one logic board.

For CMOS circuits, the effect is even more important. The output of a CMOS circuit cannot be loaded down as could the ECL circuit. A low resistance load would drain the output capacitor of the CMOS circuit, requiring additional current draw from the supply voltage that would make the CMOS circuit consume power continuously for all circuits. The power and heat load would be intolerably excessive.

CMOS circuits, therefore, are not terminated. The reflections are kept manageable by making the path lengths short, so that the reflected pulses occur harmlessly during the rise time. Lower dielectric constant can shorten the effective electrical length of the leads, and help with the reflection management. The lower dielectric constant affects the delay by reducing the skew or charging delay caused by the line capacitance that prevents signal buildup to the signal value. Each element of a circuit trace has a capacitance to the ground plane, and the traveling pulse cannot charge the next segment of the line till the previous segment has received enough electrons to charge up the line capacitance of that segment. Thus for CMOS circuits the time constant of the line will be proportional to the line capacitance and to the dielectric constant.

Thus, for CMOS circuits, the propagation delay (or the time for signal buildup) depends directly on C ! On average, CMOS circuits benefit from reduced dielectric constant more than ECL circuits do, and the limiting distance can be doubled for the same electrical performance by changing the ϵ of the substrate from 5.5 to 2.8. For a CMOS circuit, therefore, lowering the dielectric constant can reduce the delay directly, and reducing the dielectric constant to 1/2 of the original value can increase the size of a circuit board to 4 times the previous area if the operating speed of the system does not need to be increased.

All of this is standard system engineering. Alternatively, if the operating speed needs to be increased, a lowered dielectric constant, for the same size board, can increase the speed by $\sqrt{2}$ (ECL) or by a factor of 2 (CMOS). If the lower dielectric constant organic printed wiring boards are used instead of ceramic interconnection substrates, the comparison graph of Figure 1 indicates the advantages in system performance to be obtained using substrates of lower dielectric constant.

As shown in Figure 1, a system operating frequency of 500 MHz is desired (a system cycle time of 2 nanoseconds), the maximum board path length for a system using 1 gigahertz switching devices can only be 4 inches in ceramic, 6 inches in epoxy-glass, and 8 inches in 2.8 ϵ material for ECL circuits. For CMOS the usable distance may be 12 inches ! That is a tremendous board size advantage. If the present operating speed is satisfactory, the board size can be doubled or quadrupled without paying a penalty in operating thruput.

POWER SAVINGS ARE SIGNIFICANT

The operating power of a system is also affected by the change of dielectric constant. Again let us consider the performance of ECL semiconductor logic. At the low impedance of ECL or other bipolar circuits (TTL, E^2L) the signal line from one chip to another is terminated in a resistor of a value identical to the characteristic impedance of the line.

This value is related to the line characteristics by the expression:

$$Z = \sqrt{L/C}$$

where L is the line inductance, and C is the capacitance of the line. The power requirements of the circuit, and of the board with all it's circuits, is determined by this resistance load on each circuit.

By simple substitution in the equation for power:

$$W = E^2 / R, \quad R = Z = \sqrt{L/C}, \quad \therefore \quad W = E^2 / \sqrt{L/C}$$

it is apparent that the power dissipated is determined by the dielectric constant.

Eliminating the constant terms:

$$W \approx \sqrt{C}$$

Lower the capacitance and you lower the power consumption requirements. For large computers, a logic board might dissipate 300 to 1500 watts. Cutting the power consumption by $1/\sqrt{2}$ or 30 % is an important consideration, as it is in small portable computers which need to avoid high speed air or liquid cooling. Clearly, lowering the dielectric constant to reduce power is an important consideration .

All the above discussion assumed constant circuit board trace and thickness dimensions, and lowered circuit capacitance and an increase in the characteristic impedance. There can be problems with major changes in the characteristic impedance 'Z' of the line, however. Control of crosstalk is dependent on the impedance, but so is the pickup of external interfering circuit noise. In a most useful discussion of the importance of these effects, Evan Davidson of IBM concluded that optimum circuit performance occurs when 'Z' is in the region from 40 to 120 ohms, with the lowest practical value 40 and the highest 105. (Figure 2) If a circuit was designed at 70 ohms, a doubling of its 'Z' to 140 ohms would introduce serious electrical performance problems. The improvement in electrical speed performance might not be the only important consideration.

USING THE LOWER DIELECTRIC CONSTANT TO REDUCE LAMINATE THICKNESS

Another alternative might be much more cost effective. If the thickness of the board is decreased by the amount of the dielectric constant reduction, other interesting possibilities emerge. Consider the fields around the circuit traces. If the traces are located too close together, the field around an active line is picked up by a nearest quiet line. Typical spacings for high performance systems might be 3 mil (75 micron) conductor line width and 9 mil (225) micron spaces. A closer conductor distance is not possible without excessive crosstalk. If the dielectric constant is reduced, however, the ground plane can be moved closer to the signal traces till the same capacitance and characteristic impedance is obtained. (Figure 3) The field spread, however, will then be so reduced that the crosstalk becomes much lower.

Another way of looking at this is to consider that the crosstalk is determined in great part

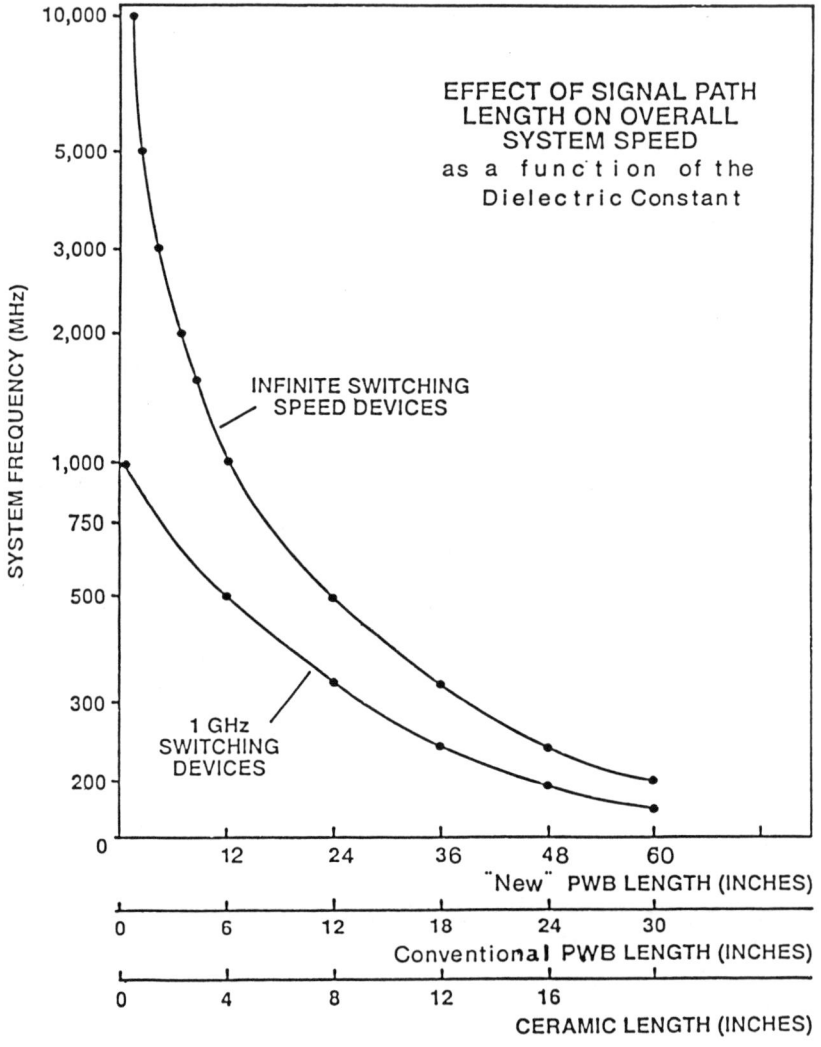

Figure 1   Effect of Signal Path Length on Overall System Speed
(Showing the effects of changing board dielectric constant)

IEEE TRANSACTIONS ON COMPONENTS, HYBRIDS. AND MANUFACTURING TECHNOLOGY, VOL. CHMT-6, NO. 3, SEPTEMBER 1983

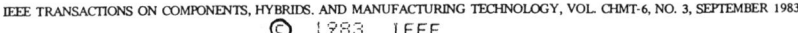

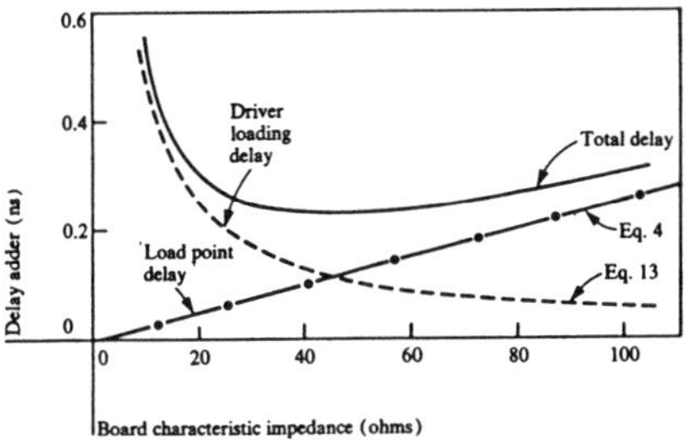

Delay adder versus characteristic impedance for discrete net, with $C_D = 5$ pF, $T_R = 1$ ns, and $Z_D = 10$ Ω.

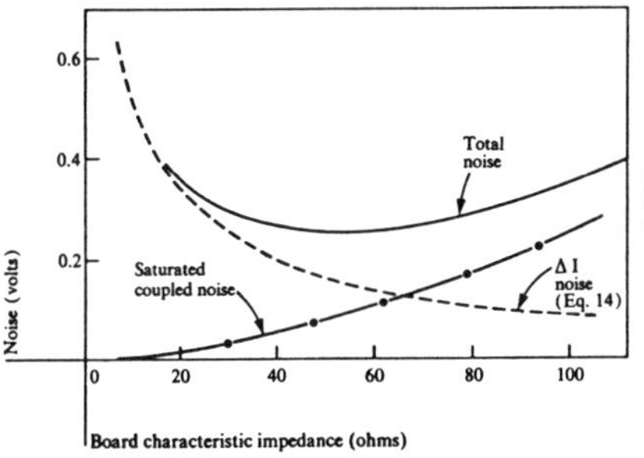

Noise amplitude versus characteristic impedance with two active lines and one quiet line. Parameter values are $L_{eff} = 0.5$ nH, $N = 20$, $V_D = 1$ V, $T_R = 1$ ns, $Z_D = 10$ Ω, $W = 100$ μm, $s = 400$ μm, $t = 25$ μm, and $\epsilon_r = 4.0$.

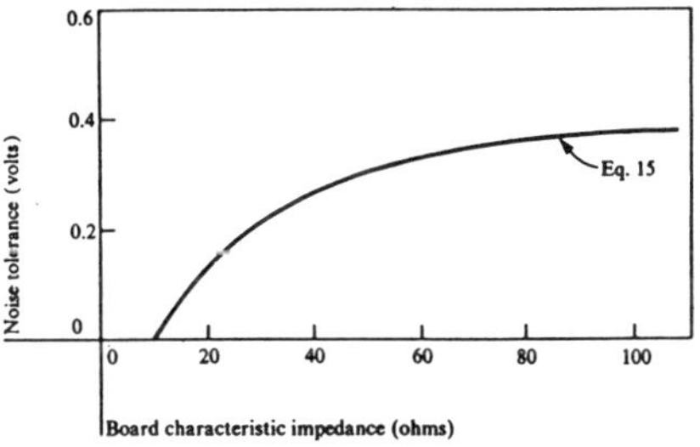

Noise tolerance versus characteristic impedance with $V_D = 1$ V, $Z_D = 10$ Ω, $Z_0' = 50$ Ω, and $NT' = 0.3$ V.

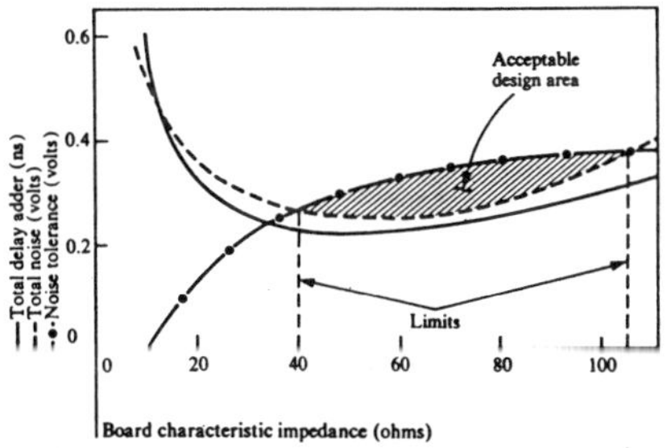

Design space for characteristic impedance.

Figure 2   The dependence of Electrical Noise and delay on the characteristic impedance
(Four charts from the Evan Davidson paper on High Speed Computer Packaging (Ref 1)

by the capacitance between the driven and the quiet line. Half the capacitance, and you halve the crosstalk. The field considerations are more appropriate, however, because the crosstalk is also determined by the magnetic field, which is also reduced by the closer proximity of the ground plane, and the closer approach of the return flowing currents to the outgoing signal lines.

Lowering the crosstalk from the originally satisfactory design value is rarely needed. The previous circuit crosstalk was usually chosen to provide satisfactory performance in the system. Rather than lower the crosstalk, it is better to reduce the line spacing, bringing the nearest trace closer to the signal line.

The amount of the reduction is directly proportional to the reduction in the dielectric constant. Reducing the dielectric thickness and bringing the traces closer together can nearly double the possible wiring density of a circuit board. True, only the space between the holes will see the increased density, but that was the only space available for wiring in the first place.

Consider the advantages. If the use of surface mount packages required 6 or 8 signal layer for interconnection, the board with reduced channel spacing can do the job with 4 layers – at a cost that is much lower than half, because printed wiring board costs are related to the square of the number of layers. If the number of layers is assumed constant, surface attach and low dielectric constant boards can be half the size of their predecessors, with half the propagation delay and half their long trace capacitance. More importantly, half the area of a circuit board can reduce the system size and lower the system costs, which are determined by the enclosed volume of the electronics.

WHAT ARE THE CHOICES FOR LOW DIELECTRIC CONSTANT ?

Unreinforced dielectrics are one possibility. Polyimide, if it is deposited on a metal substrate, can offer a dielectric constant as low as 2.8. There are problems, one of which can be quite serious. Polyimide absorbs water, as does epoxy glass, and that changes the dielectric constant. Figure 4 shows the amount of this interaction of the dielectric constant with included moisture for epoxy-glass. This may be a problem for electrical characteristics, but it is more of a production problem, particularly for unsealed multichip substrates. If curface mount packages are to be surface soldered to a substrate with included water, the presence of moisture can be a disaster. Included water turning to steam in the temperatures of a vapor phase or infra-red reflow soldering operation can destroy the assembly. Of course the solution is to dry the board before soldering, but that is costly and time consuming, and does not solve the problem of repairs and changes.

An alternate to the use of polyimide is a new resin from Dow Chemical called Quatrex. It has lower moisture absorbtion and a more consistent low dielectric constant as shown in figure 4a. If the material is not reinforced with included fibers, but built up on a metal substrate, this can also be used for fully additive board manufacture. This technique, like polyimide on metal, is better suited to small boards and single chip or hybrid packages. This is an important market for electronic substrates, but this solution does not address the large substrate applications; boards 16 inches square or larger.

Reinforced Teflons are another choice. Even though glass fibers have a higher dielectric, the low dielectric constant of Teflon can produce a composite with an ϵ of 2.8. Glass reinforced Teflons have been in use for some time for microwave applications, with many manufacturing problems. Low melting point has made soldering virtually impossible, and the softening of the Teflon has caused the conductor locations to move or "swim" to other positions during lamination of a multilayer board. No – the glass reinforced Teflons were not acceptable for widespread application.

There is a new alternative. The Rogers Corporation has introduced a new board composite RO 2800. (Figure 5) It uses PTFE Teflon as the matrix, but uses a combination of ceramic filler and chopped short glass fibers to provide the rigidity and stiffening that the older Teflon board technology lacked. The shortness of the fibers, and the low dielectric constant of the ceramic filler keeps the dielectric constant low, yet the board is able to withstand soldering temperatures. Like all the Teflons, there is little water absorption, and the composite is a low loss material.

There is still one disadvantage – there always is. The adhesion of the copper requires a

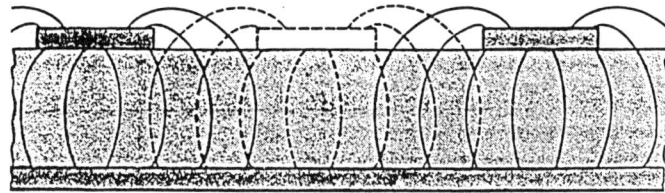

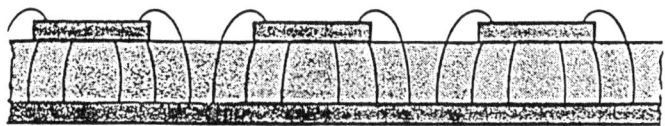

Figure 3   The compression of the field around a signal conductor for
          a reduced dielectric thickness, suggesting the ability to
          reduce line spacing and achieve the same crosstalk.

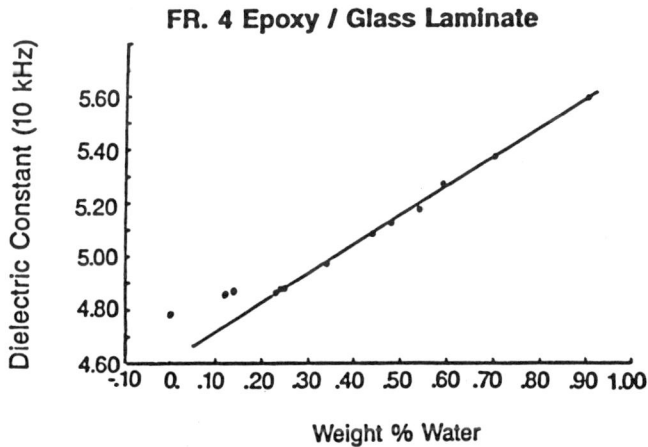

Figure 4 - The change in dielectric constant with
          absorbed water for FR4 epoxy-glass

Experimental Low Dielectric Resin / Glass Laminate

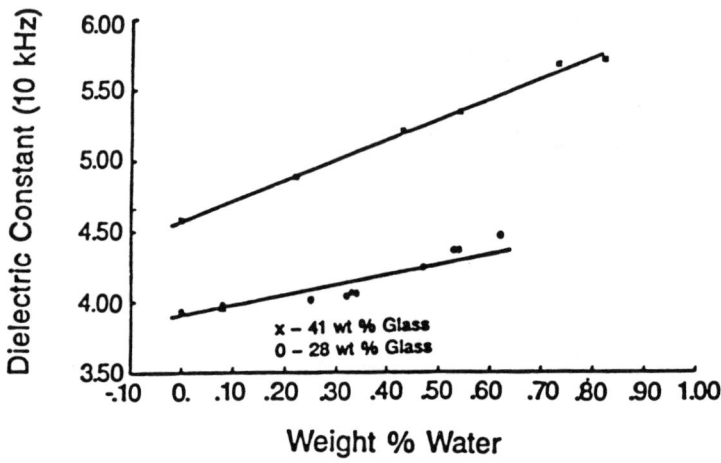

x – 41 wt % Glass
0 – 28 wt % Glass

Dielectric Constant (10 kHz) (y-axis)

Weight % Water (x-axis)

Figure 4A The dependence of dielectric constant on absorbed water for epoxy-glass and Quatrex
Polycynate-glass laminates (Ref 2)

RO2800™ Fluoropolymer Composite

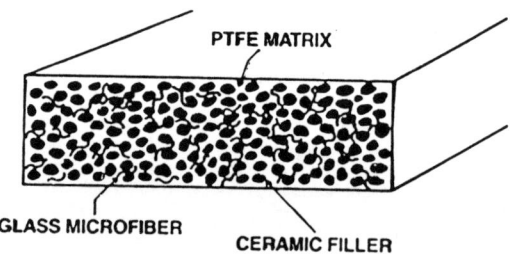

PTFE MATRIX

GLASS MICROFIBER   CERAMIC FILLER

- **Highly** Filled
- Reinforced with Randomly
  Oriented Microfibers

Figure 5   Rogers RO 2800 Fluoropolymer Composite, showing the mix of ceramic and glass microfiber
filler materials (Ref 3)

surface etch using sodium napthalene etchant – not the usual printed board shop material. Not a really big problem, but one of accepting the need to go to new materials to use this board.

Standard epoxy and Teflon reinforcement is the next new choice. The Gore corporation has been marketing stretch expanded PTFE Teflon for some time, and it can be used as an alternative reinforcement to the glass fibers of the conventional epoxy-glass boards. Once again the composite is 2.8, and the material can be laminated, processed, drilled and soldered just like a conventional epoxy glass board. There are some slight differences, but this is a material much like the stuff that is familiar. The benefits of the lower propagation delay on the higher density have been reported by Gore in recent papers and presentations (Ref 5).

Gore has come up with an interesting utilization of this new board material. The signal laminates can be retained in epoxy glass, because the capacitance through a single layer only affect the very small areas of the perpendicular trace crossovers – so small that the capacitance is unimportant. The capacitance that counts is the capacitance to ground (or the AC ground of the power planes), and that capacitance is determined by the dielectric constant of the "B" stage pre-preg used to assemble the circuit laminated into a multilayer board. Use the Gore Tex reinforced material in the pre-preg layers (Figure 6), and conventional epoxy glass for the signal, power-ground and-pad layers, and all the operations with the copper circuit paths are being performed in conventional materials. As long as the pre-preg drills and plates so that the via holes work, the board is capable of conventional processing.

A great idea. One that is applicable to the Rogers ceramic-chopped fiber PTFE Teflon also. Probably a clear winner in the low dielectric constant board race for expanded area or lower paropagation delay, but not yet for the reduced thickness increased density application. The thickness cannot be reduced very much because there must be enough resin around the fairly thick fibers. It is still one of the winners, except for the possible entry of another new material.

Liquid Crystal Resins, notably PBT (Poly benz-bisthiazole) offers another solution. It has the low dielectric constant both in the matrix and in the reinforcing fibers – because the reinforcing fibers are made of PBT also ! (Figure 7) If the formulations and the processing is properly controlled, the PBT liquid crystal material forms a network of microscopic fibers that can stiffen and reinforce the material without paying a dielectric constant penalty. These fibers tend to have a majority orientation in one direction, so the first films were reinforced in only one direction. Foster-Miller had worked on the process, and can control the formation of the fibers so that they are directionally oriented through a range of axes, with the top layers of the film orthogonal to the bottom. Furthermore, they can control the dielectric constant and the temperature coefficient of expansion. Because the fibers are so small, that is a material that can provide the strength of reinforced films in thicknesses as little as 1/2 mil !

The control of the TCE is the most interesting. The use of bonded copper-Invar-copper metal plates to printed circuit board materials has recently become of great concern. If the metal provides expansion restraint in the X-Y directions, the inherent volume expansion of the dielectric laminate material in X-Y is forced to create additional Z axis expansion. Epoxy glass boards were never wonderful in the Z axis direction – values of 55 to 85 ppm/$^\circ$C are typical for boards with 16 TCE in the laminate planes. The lower expansions of the Rogers material of 24 in the Z direction is a great help, but trying to reduce the X-Y expansion to 8 or 9 makes that 24 expansion great enough to crack via barrels. Controlling the X-Y expansion to 9 and having a 24 TCE in the Z axis direction can look like salvation to the military manufacturer worried about solder joints and via cracking.

It too has a problem. It is not yet commercial, and the ability to control the TCE and the expansion tightly have to be demonstrated in production, even though the development work is promising.

CHOOSING THE BEST

Clearly, one has to make a choice. For commercial applications where leaded packages like the PLCC and the QUADS , "J" leads or Gull wing, or butt"I" leads are most interested in board TCE's of 16 to reduce the strain on the copper circuit conductors. After all, having the ability to go to smaller and more closely spaced lines for large sized boards means long traces that must match the

70

DUAL STRIPLINE - MIXED DIELECTRIC

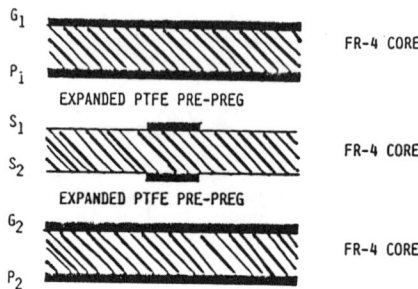

Figure 6   Gore-Ply pre-pregs to reduce Capacitance to ground only (Ref 4)

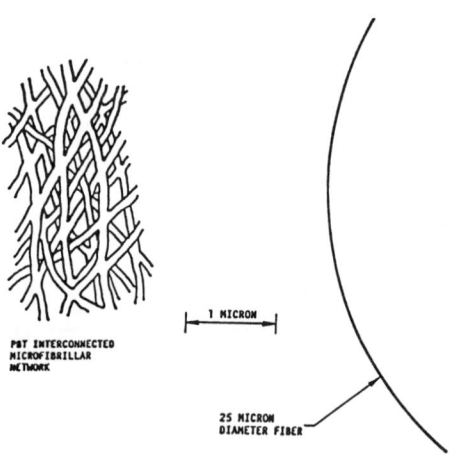

Figure 7   The fiber structure and dimensions of the PBT liquid cristal self-reinforcing polymer material (Ref 5)

TCE to copper, or the traces will snap. The lower TCE's's can be left to the military that is still in the leadless chip carrier technology. With the rapid introduction of leaded ceramic packages, and the promise of military acceptance of gel protected plastic packaging, the leadless package may no longer be necessary. But if you are using leadless now, and can't wait, rapid development of the liquid crystal board would be wonderful.

If you are into small boards, hybrids, and multichip packages, polyimide or the alternative, Quatrex material on a plastic substrate is available now. Solve the moisture problem or use Quatrex, and these boards are available now.

If, however, larger boards are needed, the Rogers and the Gore solutions look particularly good. Both make satisfactory laminates, and both claim good soldering a processing rigidity. There are processing differences, and other handling characteristics that distinguish one from the other, and the electricals have different high frequency effects. Do the homework, look at the choices, and pick one.

Why make that suggestion ? Because changing the dielectric constant is the easiest way to increase the circuit density – far easier to do reproducibly than decreasing the line width and spacing. The system costs reduce, either because the density is greater, or because the lower dielectric constant is used to increase the speed and lower the power consumption. Do a little of each – sometimes only a slight increase in density can add another trace between vias, and reduce the number of signal layers. After all, many times the last pair of layers are not fully populated but carry only overflow wires. Similarly, even a little increase in the characteristic impedance can reduce power and permit a slightly larger board for high speed systems. That also reduces cost and improves efficiency.

The coming availability of low dielectric board materials can make for significant improvement in system design, performance and cost.

LIST OF REFERENCES

1. Electrical Design of a High Speed Computer Packaging System, E.E. Daviscon, IBM, IEEE CHMT Transactions, Volume 6, Number 3, page 280

2. Design Improvements In PWB Electrical Performance via Thermoset Resin Technology, M. E. Lyssy, M.P. Kubisiak, and P. D. Aldritch, Dow Chemical, Proceedings of the Sixth Internationalk Electronics Packaging Conference, page 109

3. Electrical and Mechanical Characteristics of Low Dielectric Constant Printed Wiring Boards, D. J. Arthur, Rogers, IPC Spring Conference, 1986

4. High Performance Low Dielectric Constant Substrate, D. Johnson, Gore and Hirosuzuki and Telsuro Umebayshi, Junkosho, IEEE Computer Packaging Workshop, Oiso Japan, January 1986

5. PBT Liquid Crystal plastic for self-reinforcing printed Wiring Board Use, R. Lusignea, Foster-Miller, IEEE Computer Packaging meeting, December 1986, New York

RECENT ADVANCES IN THIN FILM MULTILAYER INTERCONNECT TECHNOLOGY FOR IC PACKAGING

RONALD J. JENSEN

Honeywell Sensors and Signal Processing Laboratory, 10701 Lyndale Avenue South, Bloomington, MN 55420

ABSTRACT

A high-performance packaging technology being developed at Honeywell and a number of other companies uses thin-film processes to pattern high-density interconnections in multiple layers of a high-conductivity conductor (e.g., copper) and a polymer dielectric, primarily polyimide. This paper describes the physical characteristics and unique advantages of this thin film multilayer (TFML) interconnect technology; it then summarizes the results of recent work done at Honeywell in processing TFML structures, assessing the stability and reliability of the materials system, and fabricating test vehicles and demonstration packages.

INTRODUCTION

The high circuit densities and fast switching speeds of advanced integrated circuits (ICs) have created a need for new technologies to interconnect and package these devices. Fine-line, multilayer interconnections are required to accommodate the large number of signal input/outputs on highly integrated circuits. Interconnections must be short and have well-controlled electrical characteristics in order to propagate high speed signals with minimal delay and distortion. Finally, the interconnect structures must be incorporated into a package that can provide power to the ICs, dissipate heat, and protect the ICs from the environment.

A high-density interconnect technology that is being actively developed throughout the electronics, computer, and IC packaging industries is based on multiple layers of a thin film conductor such as gold, aluminum, or copper and a polymer dielectric [1-8]. Polyimides are the most widely used dielectric material because of their high stability, their processability in solution form, and their low dielectric constant (typically 3.5). The low dielectric constant results in high signal propagation velocity, low crosstalk, and low interconnect capacitance for reduced power consumption and signal risetime degradation. High-density multilayer interconnect structures are fabricated using thin film deposition, etching and photolithographic processes similar to those used in patterning on-chip interconnections. This thin film multilayer (TFML) technology thus provides a combination of high-density interconnection and high-speed performance.

The TFML technology is also highly flexible in that it can be applied to a wide range of substrates, interconnect geometries, and performance requirements. It is being widely developed for multichip packaging in high-performance computer systems [4], and for military systems employing VHSIC (Very High Speed IC) technology [10]. TFML interconnections also have unique advantages for the packaging of high-speed GaAs ICs. Finally, the TFML materials system and process technology can be extended to even finer geometries required for wafer scale integration [5], or to high-density optical interconnections using thin-film waveguides [9].

This paper will describe the physical characteristics of the TFML technology, and then summarize recent advances at Honeywell in processing TFML structures, assessing the stability and reliability of the materials system, and fabricating test vehicles and demonstration packages. Several recent review papers [1-3, 10, 11] describe the Honeywell TFML technology in greater depth; the reader is referred to these for more details.

THIN FILM MULTILAYER PACKAGING APPROACH

TFML interconnect structures can be fabricated on a variety of substrates and incorporated into a variety of package designs. Figure 1 shows three packaging approaches using TFML interconnections. In the first approach, the TFML interconnections are patterned on a multilayer co-fired ceramic substrate. This substrate may contain internal metal layers for power and ground distribution, a grid array of pins for connecting the package to a printed wiring board (PWB), metallized strips to provide thermal contact to the PWB, and a metallized ring around the perimeter for attachment of a seal ring and lid for hermetic sealing. In the second approach in Figure 1, the TFML interconnections are patterned on a blank metal or ceramic substrate which is then mounted into a hermetically-sealable package such as a metal flatpack with perimeter leads. In the third approach, TFML structures are fabricated on larger substrates to create high-density board-level interconnections between single or multi-chip packages.

In all of these packaging approaches, the high-density interconnections are patterned in TFML structures of copper (Cu) conductor and polyimide (PI) dielectric. Figure 2 shows a typical cross-section consisting of five metal layers: two layers of signal lines sandwiched between ground or voltage planes, and a top metal layer for chip attachment and bonding. This places the signal lines in an offset stripline configuration with controlled characteristic impedance. Table I shows typical dimensions and electrical characteristics of these offset striplines. For typical 50-ohm signal lines, the conductor lines are 25 μm wide and 5 μm thick and the dielectric thickness is 15 μm between signal layers and 21 μm between signal lines and reference planes. The high conductivity of Cu results in a low resistance of 1.3 ohm/cm for the narrow-cross-section conductor lines, while the low dielectric constant of PI results in a low capacitance of 1.2 pf/cm for the 50-ohm lines. The pitch between signal lines is typically 75-125 μm; the minimum pitch will be dictated by yield constraints and crosstalk considerations.

The chips are attached to pads on the surface of the polyimide, and electrically bonded to the TFML interconnections by wire bonding (Al or Au wire) or by tape-automated-bonding (TAB) with a solder bond. An array of copper vias may be used to conduct heat through the polyimide layers to the substrate.

TABLE I

Typical TFML Interconnect Characteristics

Signal Line Dimensions

Conductor Thickness	5 μm
Linewidth	25 μm (.001")
Line Pitch	75-125 μm (.005")
Dielectric Thickness	15-25 μm
Via Diameter	25-35 μm

Electrical Characteristics (Offset Stripline)

Propagation Delay (Lossless)	62 ps/cm
Characteristic Impedance	50 ohm
Resistance	1.26 ohm/cm
Capacitance	1.2 pf/cm
Inductance	3.1 nH/cm
Max. Backward Crosstalk	-40 dB

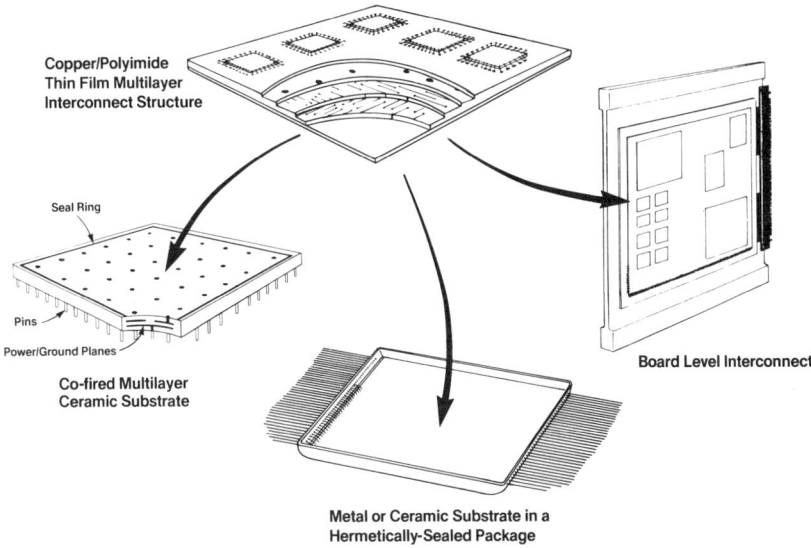

Figure 1.   Different packaging schemes incorporating thin film multilayer (TFML) interconnections.

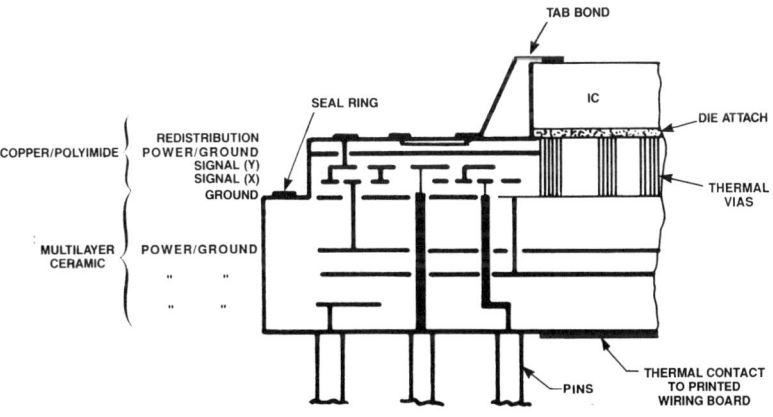

Figure 2.   Cross-section of TFML interconnections on a pinned multilayer ceramic substrate.

PROCESSING

The TFML interconnect geometry presents a number of unique challenges for thin film processing. The relatively thick conductor and dielectric films, which are required for low interconnect resistance and capacitance, require long processing times and can create significant strain energy due to the thermal expansion mismatch between PI and substrates. The high aspect ratios of conductor lines and PI vias require anisotropic patterning processes, and produce large topographies that must be planarized. A wide variety of substrate materials and sizes may be used to support the TFML structures, and each substrate presents unique problems in terms of surface roughness, flatness, mechanical and thermal stability, and compatibility with TFML process equipment and reactants. Finally, the large substrate size places severe demands on process yield; very low defect densities and/or in-process test and repair techniques are required in order to achieve acceptable process yields.

A variety of processes have been used to fabricate TFML structures, including both subtractive [1, 2, 7] and additive [5, 8] approaches. In the subtractive process developed at Honeywell [1, 2], the copper conductor layers are deposited by sputtering and patterned by photolithography and wet etching, and the polyimide is deposited by spin or spray coating and patterned by reactive ion etching. These processes have been used to fabricate five metal layers on three-inch square substrates, with the signal line dimensions given in Table I. TFML structures have been patterned on a variety of substrates, including tape-cast ceramic, co-fired ceramic (pinned and unpinned), and metals including aluminum, copper, molybdenum, copper-clad Invar, and porcelainized steel.

Alternative processes based on direct photopatterning or laser writing offer the potential for reducing the cost and turnaround time and improving the yield of TFML processes. Direct photopatterning of photosensitive polyimide [2, 12] eliminates several process steps required in conventional wet or dry etching of PI. Direct-write laser-based processes such as laser ablation of polyimide [13], laser-assisted etching of metals [14], and deposition of conductor lines by laser-assisted CVD or plating [15] eliminate the need for photolithographic masks, and thus permit quick-turnaround programmable patterning of interconnections. The laser processes can also be used to repair faults in TFML conductor lines.

RELIABILITY STUDIES

A number of material stability and reliability issues are of concern for a package containing an organic material such as PI. These issues need to be addressed at three levels: (1) the stability of PI alone, including the effects of PI chemistry, cure conditions, and environmental stress (e.g., temperature, humidity, radiation) on the electrical, thermal, mechanical, and outgassing properties of PI, (2) the stability of metal/PI multilayer structures, including the effects of materials, processes, and environmental stress on interlayer adhesion, internal stress, corrosion, and the electrical characteristics of the interconnects, and (3) the reliability of assembled and sealed packages containing Cu/PI interconnects, including their bond reliability, hermeticity, moisture content, and performance under MIL-STD tests.

Initial reliability studies at Honeywell have been directed at the most critical potential failures of TFML structures [1, 2, 11, 16]. Although generalizations must be treated carefully, the following statements summarize the essential findings from these studies: (1) The mechanical and electrical properties of PI are unaffected by total gamma radiation doses of up to 10^8 rad (Si); thus PI has excellent radiation resistance in addition to thermal stability; (2) The tensile strength of PI is five times greater than the internal tensile stress in PI films caused by thermal expansion mismatch between PI and low-thermal-expansion substrates (specifically Si); (3) By chosing the correct interfacial materials (e.g., Cr or TiW adhesion metals or amino-silane adhesion promoters on ceramic), PI adheres strongly to substrates and internal metal layers, and the adhesion does not degrade with humidity, gamma radiation, thermal shock or temperature cycling; (4) Thermal shock and cycling do not affect the continuity of vias; (5) Humidity has a significant effect on the dielectric properties of PI, and it has the greatest potential for degrading adhesion and via continuity and causing corrosion of copper. For these reasons, hermetic sealing of TFML interconnections is advisable; and (6) Packages containing fully-cured PI can be baked and hermetically sealed to establish internal moistures levels of less than 3500 ppm.

TECHNOLOGY DEMONSTRATION

A number of test vehicles and multichip packages have been fabricated at Honeywell to evaluate the feasibility and performance of TFML interconnections. Multichip ECL ring oscillator test vehicles have been used to determine the dynamic electrical characteristics of TFML interconnections [1, 17]; results have been in good agreement with lumped-element simulations of a lossy transmission line. A thermal test vehicle was used to test alternative methods of conducting heat from the chip through the TFML layers to the substrate [10]. An array of staggered copper-metallized vias provided a thermal impedance of less than 1°C/W through the polyimide layers.

The most advanced test vehicle fabricated at Honeywell, shown in Figure 3, is designed to test bipolar driver and receiver circuits on ICs with submicron features sizes [10]. The test vehicle is fabricated on a 3" x 3" pinned co-fired ceramic substrate and contains sites for 15 ICs. The five layers of TFML interconnections are patterned to the design dimensions given in Table I. The interconnect geometries and electrical characteristics of these interconnections were measured and compared with design and calculated values [10, 17].

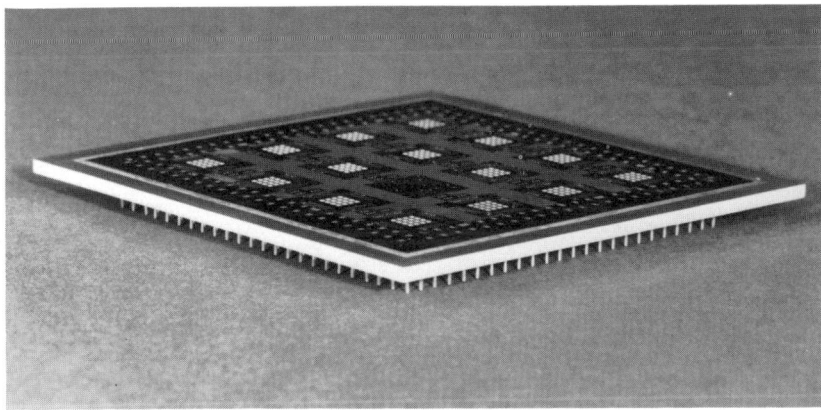

Figure 3.   Multichip test vehicle for submicron bipolar ICs, containing five metal layers of TFML interconnections on a 3" x 3" pinned co-fired ceramic substrate.

Functional packages have also been fabricated at Honeywell to demonstrate the use of TFML technology in a variety of system applications. For high-performance computer applications, a microprocessor module containing nine bipolar gate array ICs was designed and fabricated [2]. The ICs are TAB-bonded to an 80 mm-square substrate with six metal layers of Cu/PI TFML interconnections and voltage planes. Another demonstration package, designed for an image processing application, contains 18 chips on a 2.25 inch-square substrate mounted in a hermetically-sealable metal flatpack [18]. This multichip package provides a 60% area reduction over the current PWB with surface-mounted components.

Finally, a single-chip package has been developed to demonstrate the use of TFML technology for packaging high-speed, high I/O digital GaAs ICs. The package accomodates a single GaAs chip with 200 I/Os. The signal lines are designed for data rates of up to 3 Gbit/s, with thin film resistors patterned very close to the chip to minimize reflections. The low dielectric constant of polyimide and the high-density patterning of TFML technology permit high-speed signal lines to be closely spaced with acceptable levels of crosstalk. This GaAs single-chip package will be described in more detail in a forthcoming publication.

CONCLUSIONS

Thin film multilayer (TFML) structures of Cu conductor and PI dielectric can provide a combination of high-density interconnection and high-speed performance for IC packaging. A variety of processes can be used to pattern TFML structures; the key issue for large substrates is yield. Reliability studies have shown that properly- processed TFML structures are stable over a wide range of environmental conditions; however, moisture has a significant effect on the dielectric properties of PI. It is therefore advisable to hermetically seal TFML packages, and moisture levels of <3500 ppm have been demonstrated. The feasibility and performance of TFML interconnections have been demonstrated at Honeywell through the design and fabrication of a variety of test vehicles and functional single- and multi-chip packages.

REFERENCES

1. R. J. Jensen, J. P. Cummings, H. Vora, IEEE Trans. Components, Hybrids and Manuf. Technol. CHMT-7, 383-393 (1984).

2. R. J. Jensen in Polymers for High Technology: Electronics and Photonics, edited by M. J. Bowden and R. S. Turner, ACS Symposium Series 346, (American Chemical Society, Washington, D.C., 1987) pp. 466-483.

3. R. J. Jensen, Proc. ASM Third Conference on Electronic Packaging (ASM International, Metals Park, OH, 1987) pp. 25-31.

4. T. Watari, H. Murano, IEEE Trans. Components, Hybrids, Manuf. Technol. CHMT-7, 462-467 (1985).

5. J. F. MacDonald, A. J. Steckl, C. A. Neugebauer, R. O. Carlson, A. S. Bergendahl, J. Vac. Sci. Technol. A4, 3127-3138 (1986).

6. C. W. Ho., in VLSI Electronics: Microstructure Science, edited by N. G. Einspruch (Academic, New York, 1982) Vol. 5, Chapter 3.

7. H. Tsunetsugu, A. Takagi, K. Moriya, Int. J. Hybrid Microelectronics 8, 21-26 (1985).

8. H. Takasago, M. Takada, K. Adachi, A. Endo, K. Yamada, T. Makita, E. Gofuku, Y. Onishi, Proc. 36th Electronic Components Conf., 1986, 481-487.

9. C. T. Sullivan, M. C. Roth, T. Budzynski, Proc. SPIE, O-E/Fibers, August, 1987.

10. C. J. Speerschneider, F. J. Belcourt, R. J. Jensen, J. M. Smeby, Proc. VHSIC Packaging Conference, 1987, 131-143.

11. R. J. Jensen, R. B. Douglas, J. M. Smeby, T. J. Moravec, Proc. VHSIC Packaging Conference, 1987, 193-205.

12. A. S. Deutsch, R. Schulz, Proc. Sixth Int. Elec. Packaging Conf., 1986, 331-339.

13. J. T. C. Yeh, J. Vac. Sci. Technol. $\underline{A4}$, 653 (1986).

14. C. R. Moylan, T. J. Chuang, Extended Abstr. ECS Fall Meeting, 1987, Abstr. 436.

15. T. H. Baum, J. Electrochem. Soc. $\underline{134}$, 2616 (1987).

16. R. B. Douglas, J. M. Smeby, Proc. 37th Electronic Components Conference, 1987, 197-201.

17. F. J. Belcourt, T. A. Lane, R. J. Jensen, Proc. 37th Electronic Components Conference, 1987, 614-622.

18. D. J. Kompelein, T. J. Moravec, M. DeFlumere, Proc. Int. Symp. Microelectronics, 1986, 749-757.

THIN FILM WIRING FOR INTEGRATED ELECTRONIC PACKAGES

SHEREE H. WEN AND JUNGIHL I. KIM
IBM Thomas J. Watson Research Center, Yorktown Height, NY 10598

ABSTRACT

Among the available packaging technologies, thin film wiring is most suitable for high density package applications. This paper discusses the materials choices for the thin film wiring and some possible fabrication processes. The Cu- polymer or Cu- glass system are most promising from both performance and manufacturability perspectives. The interface adhesion and high inter layer interconnection resistance are critical problems for this structure. The use of a Cr, Cu blend structure as an interface layer will solve this problem. The microstructure, interface chemistry and the adhesion mechanisms of this blend layer are discussed.

INTRODUCTION

A good integrated electronic package structure should provide high performance, adequate I/O, and high reliability as well as manufacturability. Thin film wiring provides high density wiring which minimizes the distance between active circuits and the complexity of multilayer provides higher I/O count.

In this paper we will describe the choices of materials and processes for thin film wiring and major problems encountered in thin film wiring. Then we will focus on interface adhesion between dielectrics and conductors, as well as the via resistance between layers of interconnections.

Materials and Choices

To achieve ultimate performance, one would like to have the highest propagation speed and the lowest distortion of the signal waveform during transmission.

Since the propagation delay Tpd is the function of the insulator dielectric constant,

$$Tpd = 0.0333\sqrt{Er}\ l$$

where Er is the dielectric constant of the insulator and l is the wiring length. It is preferable to use low dielectric constant insulators.

The signal distortion depends on the loss in the conductor lines and the noise picked up from the surrounding lines or circuits. All these depend on the impedance of the conductor, dielectric constant, inductance and capacitance of the materials used as well as the circuit design. [1-5]

It is clear that Cu, Ag, Au have the lowest resistance and Cu further has the benefit of lower cost. Among the commonly available insulator materials,the dielectric constant of Glass (3.9) and polyimide (3.5) are considerably less than the others, such as alumina(8.5-9.5),Silicon nitride (7.5), alumina nitride (8.9), epoxy-glass(5) and etc..

In addition to the performance requirements, the material choices need to provide compatibility between insulators and conductors to ensure a manufacturable and reliable structure.

The thermal coefficent of expansion (TCE) mismatch between insulators and conductors often causes stress or even cracking during or after manufacturing process. The TCE mismatch between the multilayer substrate and the Si or GaAs chip may also cause problems in the solder joint

,substrate cracking, or even cause threshold current and voltage shift in the device [6]. A low TCE, soft, high strength and ductile material will provide crack resistance. Alumina has TCE value 6-6.5 x 10^{-6} which matches GaAs very well but not Si. Polyimide has very high TCE value (66 x 10^{-6}). The paper by Numata et.al. in this book will discuss new low TCE polymers for thin film wiring applications.

Based on performance,a low dielectric constant insulator such as polymer or glass is the best insulator choice. Cu is the best conductor choice.[7,8]. A multilayer fine line structure based on these materials must be fabricated through a manufacturing process which is compatible with the materials set.

Fabrication Processes and Critical Issues

There are several different ways to make an integrated, multilayer structure:

Co-fired ceramics:
Sheets of ceramics that are made of ceramic powders with binders. Holes are punched into the sheets to make interconnection vias. Metal paste is screened on the sheets to make conductor patterns. These sheets are then laminated under pressure to make a multilayer structure. The whole laminate is then put into a furnace to burn out the binders and sinter into the final product. Due to the tolerance of the process, the grain sizes of the ceramic and metal powers, and the resistivity of the conductor, the conductor lines are usually in the millimeter or sub millimeter range which prevents high wiring density. Furthermore, except for a few glass ceramics which are still under development, the dielectric constant of most of available and processable ceramics are quite high. This also prevents very high density structures from being built.

Epoxy-Glass Printed Circuit Board (PCB):
Insulator layers are formed by dipping the woven glass into an epoxy bath. Circuit patterns are formed by lithographic patterning followed by seeding and plating of copper. Interlayer and intralayer connections are formed by plating of laser or mechanical drilled holes. Multilayer structures are formed by lamination. The roughness of woven glass prevents high density thin film wiring. The size of the plated through-hole is also limited by the mechanical drill bit. The alignment accuracy and the size of the board also limit the potential wiring density to the sub millimeter range.

Molded Printed Circuit Board(PCB):
The dielectric layer is fabricated by molding plastic powder. The circuit patterns can be generated by molding grooves and holes. The metal lines can be deposited by silk screen or sputtering through a mask. Metal plating followed by lithographic process is also being used to define the circuit patterns. Double-sided signal layer circuit boards have been demonstrated. Multilayer molded PCB is still under development. The accuracy of the molded grooves, holes, and the silk screen mask, also limits the density of the patterning.

Polymer, Glass-Cu Multilayer Structure:
Polyimide-Cu or glass-Cu modules provide the most desirable electrical properties, such as low dielectric constant and small transmission line resistance. Furthermore, polyimide and glass also have good characteristics for fine line processing. The microstructures of the polymer , glass and thin film copper are fine and uniform, The materials are patternable with lithographic imaging and followed by RIE or chemical wet etching process. The polyimide or glass may be sprayed or spun onto the substrate. Metal layers may be deposited by e-beam evaporation, resistance heating

evaporation, plating, or plasma assisted sputtering. Micron or submicron dimensions are achievable with these processes.

To have a reliable thin film wiring structure, adhesion between layers of dielectrics and metals, and via resistance for inter layer connections, are critical. Good adhesion and low via resistance are required even after temperature and humidity cycling test.

Interface Adhesion and Via Resistance

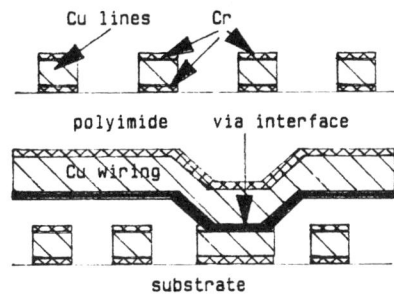

Figure 1 shows a sample multilayer thin film wiring structure with polyimide as dielectric and copper as conductor. If one uses pure copper, the thin film lines very often lift from the substrates during the process or during temperature and humidity cycles. Therefore, a layer of Cr is introduced to promote adhesion. Since Cr is highly oxidizable, a layer of Cr oxide is formed in the via between the two conductor layers. This produces high via resistance which causes difficulties in inter layer interconnections particularly during temperature and humidity test.

Fig 1. A thin film wiring structure

The problem can be solved by use of a layer of Cr/Cu blend to replace the Cr layer. In this case, high adhesion strength >50g/mm and low via resistance <0.12mΩ were obtained.

Two processes have been used to fabricate Cr/Cu blend layers: dual e-beam evaporation and resistance heating flash evaporation.

The dual e-beam evaporation process employs two individually controlled electron guns for Cr and Cu evaporation. The co-evaporation starts by opening the two shutters simultaneously after the power of the electron gun reaches predetermined steady states.

The resistance heating flash evaporation is accomplished by a high power, short pulse which vaporizes Cr and Cu in seconds. The source material is millimeter sized chunks of pure Cr and Cu.

Nature of the interface

Pure Cu has a face centered cubic (FCC) structure and pure Cr has a body centered cubic (BCC) structure. The resistance heating flash evaporated structures show the co-existance of FCC and BCC structure through the whole compositional range. Figure 2 shows the transmission electron diffraction patterns for 25% Cr, 50% Cr, 75% Cr and Cu alloys. Figure 3a,b,c, are TEM micrographs showing the changes of microstructure due to compositional changes. The 75% Cr, 25% Cu has about 150Å grain sizes (Figure 3a). The 50% Cr, 50% Cu consists of ~100Å Cr and ~150Å Cu grains (Figure 3b). The 25% Cr 75% Cu consists of ~350Å Cu grains and 50Å Cr crystals (Figure 3c). The dual e-beam coevaporated structure shows a metastable phase around 50% Cu Cr mixture. A detailed study is described in [9].

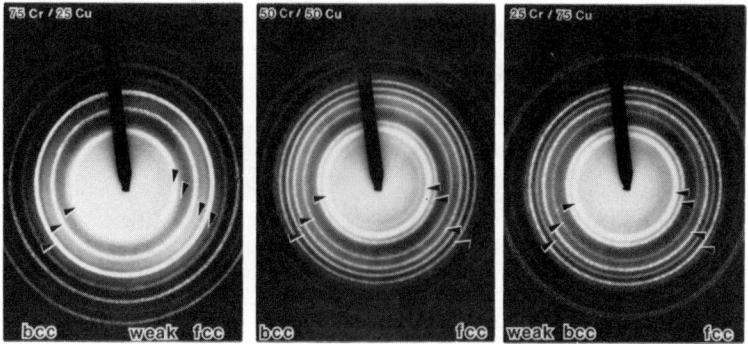

Fig. 2   TEM electron diffraction pattern of Cr/Cu
blend film

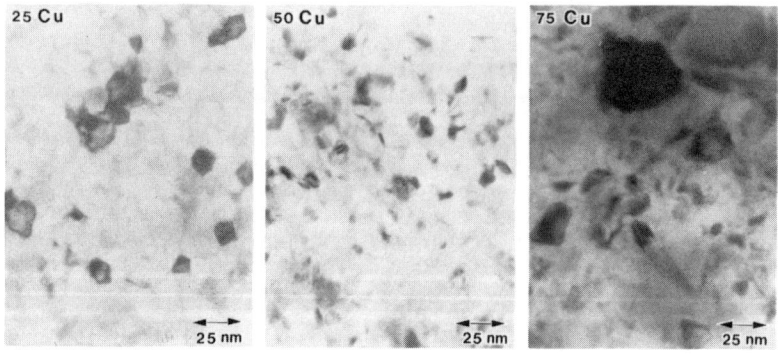

Fig. 3   TEM bright field image of Cr/Cu blend film

Figure 4 shows a TEM lattice image of the interface between the microcrystalline Cr/Cu films and polyimides. The fringes are atomic planes inside the Cr and Cu. The interface is rugged in microscale and shows intimate contact between this composite metal film and polyimide. In fact, by using ESCA and optical microscopy, we found the adhesion of Cr Cu blend film to polyimide is so good that the separation always occurs either inside the polyimide or between the polyimide or the ceramic substrates. A key implication of the result is that the adhesion of Cr Cu to PI is stronger than the PI to itself. Figure 5a,b,c,d shows the fractured surface on PI after the metal film has been peeled off. Figure 5a shows very little change in the film surface if Cr is not present. Figure 5b shows a deep trench inside the PI for the e-beam co-evaporated blend interface. Figure 5c shows the marks created by flashed thin Cr interface layer and the Figure 5d shows the marks created by flashed blend layer. This data clearly demonstrates that the interaction occurred between PI and metal creates a ditch when Cr is present at the interface.

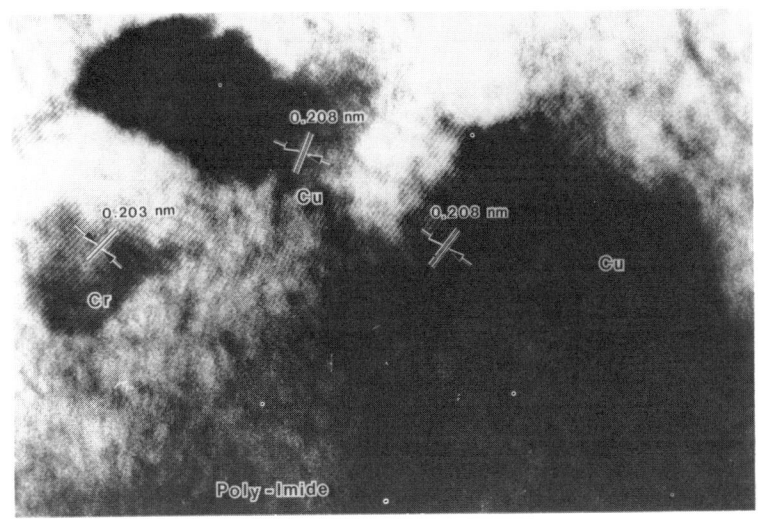

Fig. 4   TEM high resolution atomic plan lattice image
           of Cr/Cu - PI interface

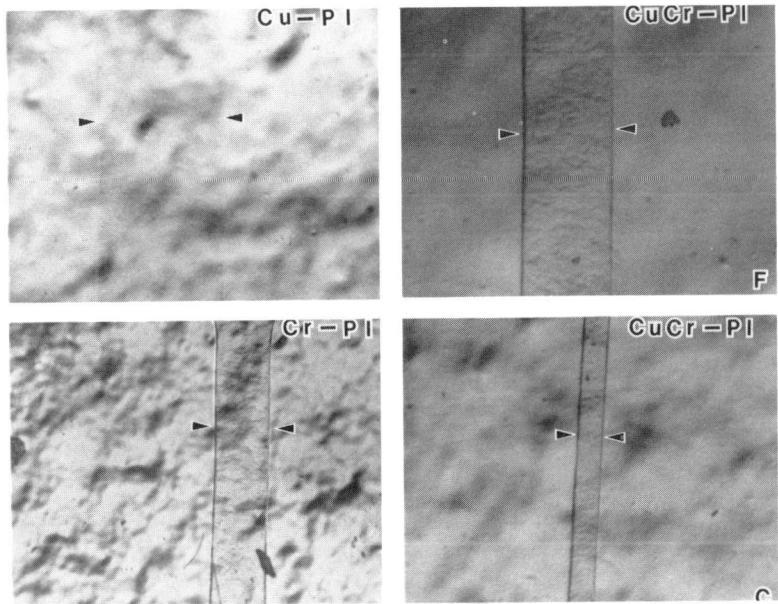

Fig. 5   Optical microscopy of the polyimide surface
           after metal lines have been peeled off

The above observations indicate that the separation occurred inside the PI immediately below the metal films. Thus the peel strength values given above do not represent the Cr Cu - PI adhesion and the limit is determined by the polyimide itself.

Figure 6 shows the ESCA survey spectrum on the surface of the underside of the metal after peeling of the PI. More than 80% of the area is covered by C,N, and O. These three elements are constituents of polyimides. The C 1s spectrum was analyzed in detail by high resolution small spot ESCA (Figure 7). The three identified peaks show the PI functional groups. The peak height ratios are not exactly the same as for fresh cured PI; this may indicate some kind of reaction takes place between metal and PI.

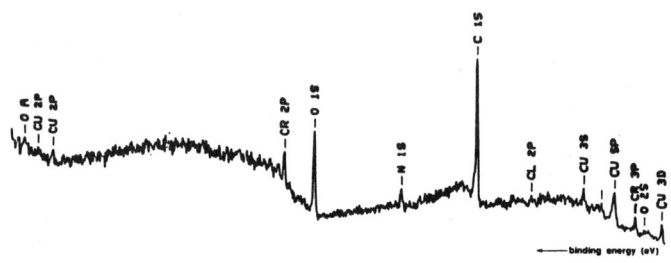

Fig. 6   ESCA spectrum of the underside of the peeled off
          metal film

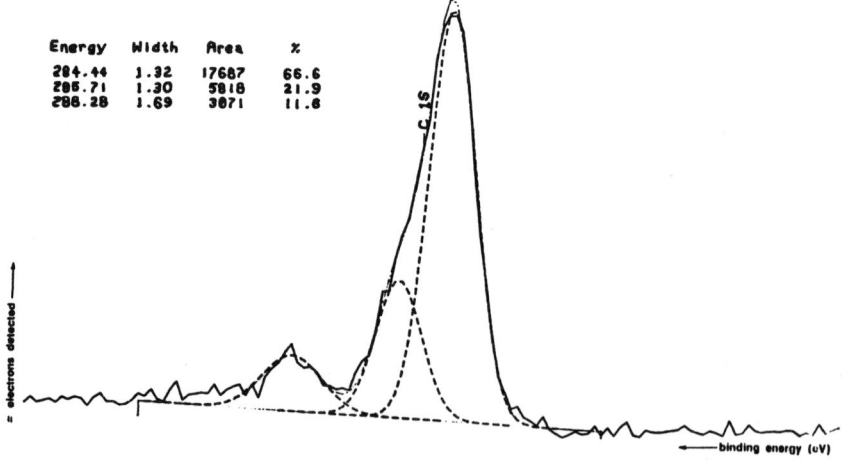

Fig. 7   High resolution ESCA spectrum of the carbon 1S peak

CONCLUSION

The best candidates for thin film wiring dielectrics are ploymers or glass due to the fine microstructure, low dielectric constant, and ease of manufacture. The best choice for conductor lines is Cu, which has low resistivity, easy processing and etchability. For the interface layer, we recommend Cr/Cu blend film. This provide good adhesion ($> 50g/mm$), good via resistance ($<12m\Omega$) and good reliability.

REFERENCES

1. S. Wen, Advances in Electronic Materials, edited by B.W. Wessels and G.Y. Chin. (American Society for Metals, 1984), p. 263.
2. C.W. Ho, D.A. Chance, C.H. Bajorek and R.E. Acosta, IBM Journal of R & D, 26, 286. (1982)
3. C.J. Bartlett, J.M. Segelken, N.A. Teneketeyer, IEEE Electronic Component Conference Proceeding 37, 518. (1987)
4. C.J. Stanghan and B.M. MacDonald, IEEE Trans. Components, Hybrids, and Manufacturing Technology, Vol CHMT-8, No. 4, 468. (1985)
5. H. Tsunetsugu, A. Takagi, and K. Moriya, Int. J. Hybrid Micro- electronics, 8, 21. (1985)
6. M. Kanamori, H. Ono, T. Furutsaka, J. Matsui, IEEE Electron Device Letters, Vol. EDL-8, No. 5, 228. (1987)
7. T.A. Lane, F.J. Belcourt, and R.J. Jensen, IEEE Electronics Components Conference Proceeding, 614. (1987)
8. R.J. Jensen, J.P. Cummings, and H. Vora, IEEE Trans. Components, Hybrids, and Manufacturing Tech., CHMT-7, 384. (1984)
9. J.I. Kim, S. Wen, and D. Yee, Journal of Vacuum Science & Technology, July-August, 1988.

MICROSTRUCTURE AND PHYSICAL PROPERTIES OF NEW SIC MATERIALS WITH HIGH THERMAL CONDUCTIVITY

S. S. SHINOZAKI*, J. HANGAS* AND K. MAEDA**
*Ford Motor Company, Research Staff, Dearborn, Michigan 48121
**Hitachi Ltd., Hitachi Research Laboratory, Hitachi, Japan

ABSTRACT

Silicon carbide materials with BeO addition (2 wt%) have the unique properties of high electrical resistivity, high thermal conductivity, and a thermal expansion coefficient close to that of silicon. The materials have been used as a chip carrier material for high power LSI packaging. Microstructures were correlated by means of analytical electron microscopy (AEM), with the physical properties. Generally, BeO particles are evenly distributed mostly at triple points and the grain growth is anisotropic and many grains are elongated with an aspect ratio of 2 or 3. The average grain size is measured to be around 5μm and the morphology is typically thick tabular.

AEM analysis has shown that large middle section of each SiC grain is mostly 6H polytype with a few or no stacking faults. On both sides of the 6H polytype, sheaths are formed, which consist of a large number of extremely thin 4H or other polytype lamellae. Along grain boundaries, no second phase formation is observed with a few exception of Be_2C and impurity transition metal compounds lamellar formation.

The results indicate that direct and clean contacts between 6H lamellae and BeO grain or other 6H lamellae form a path of high thermal conductivity. On the other hand, complex network of thin disordered (4H rich) lamellae doped with BeO forms the electrically high resistive path.

INTRODUCTION

New silicon carbide (SiC) materials hot-pressed with BeO (2 wt%) have higher thermal conductivity than metallic aluminum, a high electrical resistivity comparable to other ceramic materials, and a thermal expansion coefficient close to that of silicon. These materials have been used as a chip carrier for high power LSI packaging [1,2]. Generally, physical properties in polycrystalline SiC can be changed over a relatively wide range by changing sintering conditions and the type of sintering aids. The properties are mostly determined by formation of second phase zones along grain boundaries and elemental doping level in SiC grains by the sintering aids. For this aspect, it has been reported that the type of doping affects polytype distribution in SiC grains, which has somewhat different electronic states, and that the effect is different, depending on whether starting powder is α-SiC (with hexagonal or rhomboheral symmetry) or β-SiC (FCC) [3]. Thus, the microstructure of the new SiC materials has been characterized by means of high structural resolution and microbeam chemical analysis to correlate the microstructure with the physical properties, in order to optimize the materials to their fullest extent.

So far as the effects of sintering aids on polytypes are concerned, there have been several typical cases reported. Such

Mat. Res. Soc. Symp. Proc. Vol. 108. ©1988 Materials Research Society

as, when β-SiC is sintered with boron (B) and carbon (C), and is heat-treated in vacuum, β-SiC transforms predominantly to the 6H polytype. When the same material is subsequently heat-treated in a metal rich environment such as aluminum or silicon, the predominant phase is 4H polytype [4]. When β-SiC is sintered with B, C, and metallic aluminum (Al), the β-SiC is transformed to 4H through an intermediate phase of 15R [5]. In these cases, the phase transformation is always initiated by nucleation and propagation of partial dislocations to form complex lamellae of these polytypes and thus the disordered lamellar region often includes low level of sintering aids, such as Al. In this investigation of the SiC-BeO system, a complex lamellar formation of the polytypes during phase transformation and a doping of BeO in these regions was observed. It is proposed that these lamellae may contribute to formation of the high electrically resistive zone. Furthermore, single crystal of SiC and BeO possess excellent thermal conductivity as long as the materials are high purity. The thermal conductivity of the materials is therefore dependent mainly on grain boundary area, chemistry and the density of lattice defects in both the SiC and BeO grains.

In this report, a possible formation mechanism of a zone of high electrical resistivity will be explained to correlate the unique properties to the microstructure of the new SiC materials.

MATERIALS AND ANALYTICAL TECHNIQUES

The SiC-BeO materials were prepared at Hitachi Research Laboratory and the details of the materials preparation have been reported elsewhere [6]. High purity α-SiC powder (mostly 6H) was mixed with a binder and extremely fine BeO powder. The mixture was hot-pressed at 2050°C for one hour under a pressure of 30MPa to form a large disc of around 70 mm in diameter and 1.5 mm in thickness. These discs displayed a variation in electrical resistivity across their diameter. AEM specimens were cut from zones of high electrical resistance of around 10^{13} Ω-cm and relatively low resistance of around 10^{10} Ω-cm. Additional specimens were prepared in the same conditions, except that fine powder of β-SiC was used. After each piece was ground to a thickness of 100 μm, discs of 3mm in diameter were cut ultrasonically. Discs were further thinned by means of mechanical microthinning and argon ion beam milling to perforation.

Microstructural and microbeam chemical analyses were carried out using Siemens EM 102 CTEM and JEOL 2000-FX AEM, equipped with LINK LZ5 light element energy dispersive x-ray (EDX) spectrometer. For analysis of elemental Be, Gatan electron energy loss spectrometer (EELS) was used. Polytype distribution in SiC grains was analyzed by means of structural lattice imaging and x-ray diffraction analyses.

RESULTS AND DISCUSSION

When β-SiC powder is sintered or hot-pressed, β - α phase transformation is initiated with thin lamellar formation of α-SiC polytypes along a $\{111\}_\beta$ plane, even at a relatively low temperature of around 1850°C [7]. On the other hand, polytypic transformation in α-SiC (6H) is an extremely slow process, irrespective of the type of sintering aids. This means that β-

4H phase transformation in SiC-BeO materials occurs at a temperature of around 2050°C within 1 hour through a disordered state, but, for example, 6H - 4H transformation occurs at a higher temperature over 2100°C for over 12 hours. Figure 1 shows an example of α phase lamellar formation in β-SiC grains with an insert of a selected area electron diffraction pattern, which shows strongly disordered 4H pattern (complex mixture with heavily twinned β-phase and other polytypes, measured by structural lattice image spacings in the figure). On the other hand, when α-SiC powder is hot-pressed such as SiC-BeO materials, overall microstructure within each SiC grain is nearly the same, as shown in Figure 2. However, the polytype distribution within each SiC grain is completely different, when compared to Figure 1 (Figure 3). Here, the center of the grain is entirely 6H (1 - 2 μm), outside of which are completely disordered lamellae. The disordered sheaths are always formed on both sides of the 6H lamella. This result may be evidence that the newly grown region on the 6H α-SiC grains during hot-pressing is β-SiC, which subsequently transforms to α-SiC. The effect of the sintering

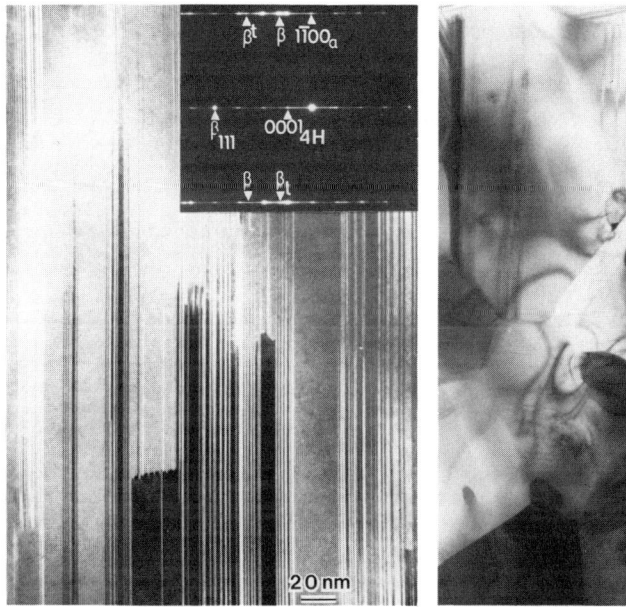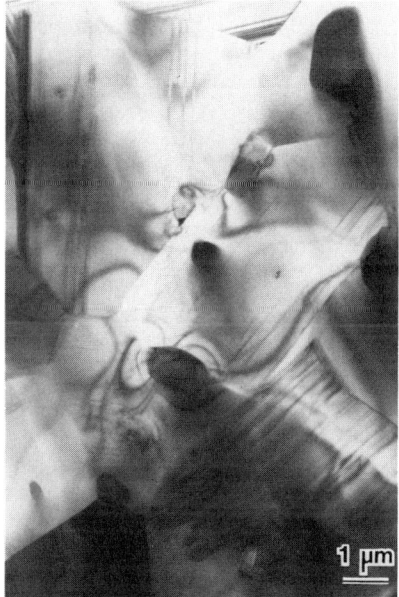

Figure 1: Extensive α phase (mostly 4H) lamellar formation in β-SiC matrix, with an insert of electron diffraction pattern of 4H and β phase mixture.

Figure 2: Overall microstructure of a hot-pressed α-SiC material, showing homogeneous distribution of BeO particles.

aid is indeed different on β-SiC and α-SiC regions. Formation of the sheaths in sintered α-SiC materials has been reported elsewhere [8]. In the present case of SiC-BeO materials, dissolved BeO or Be element was not detected in the disordered sheaths, mainly due to the present capability of EELS and EDX spectroscopies. However, similar mechanisms of 4H formation in β-SiC with other sintering aids, suggest that the disordered sheaths are also doped with Be or BeO in the SiC-BeO materials, creating a network of thin zones of high electrical resistivity.

Figure 3: Structural lattice image of an interface between 6H lamella and disordered sheath within a single SiC grain.

In addition, because BeO itself possesses high thermal conductivity and high electrical resistivity, significant contributions of the BeO grains to microstructures in SiC-BeO materials cannot be ignored, which are summarized below: (1) BeO is well dispersed homogeneously mostly in triple points with a few particles entrapped in SiC grains during hot-pressing process. (2) BeO prevents exaggerated grain growth. (3) BeO grains are normally single crystalline and make direct contact with SiC grains without any second phase (Fig. 4(a,b)). On the other hand, deterioration of the physical properties is associated with : (a) fine cracks formed within BeO grains, due to phase transformation at around 2100°C (Fig. 4(c)), (b) Thin second phase zone observed in SiC-BeO grain boundaries, when the materials were hot-pressed at 2200°C (Fig. 4(d)). Thus, it seems that the effectiveness of BeO as a sintering aid is reduced by the BeO phase transformation.

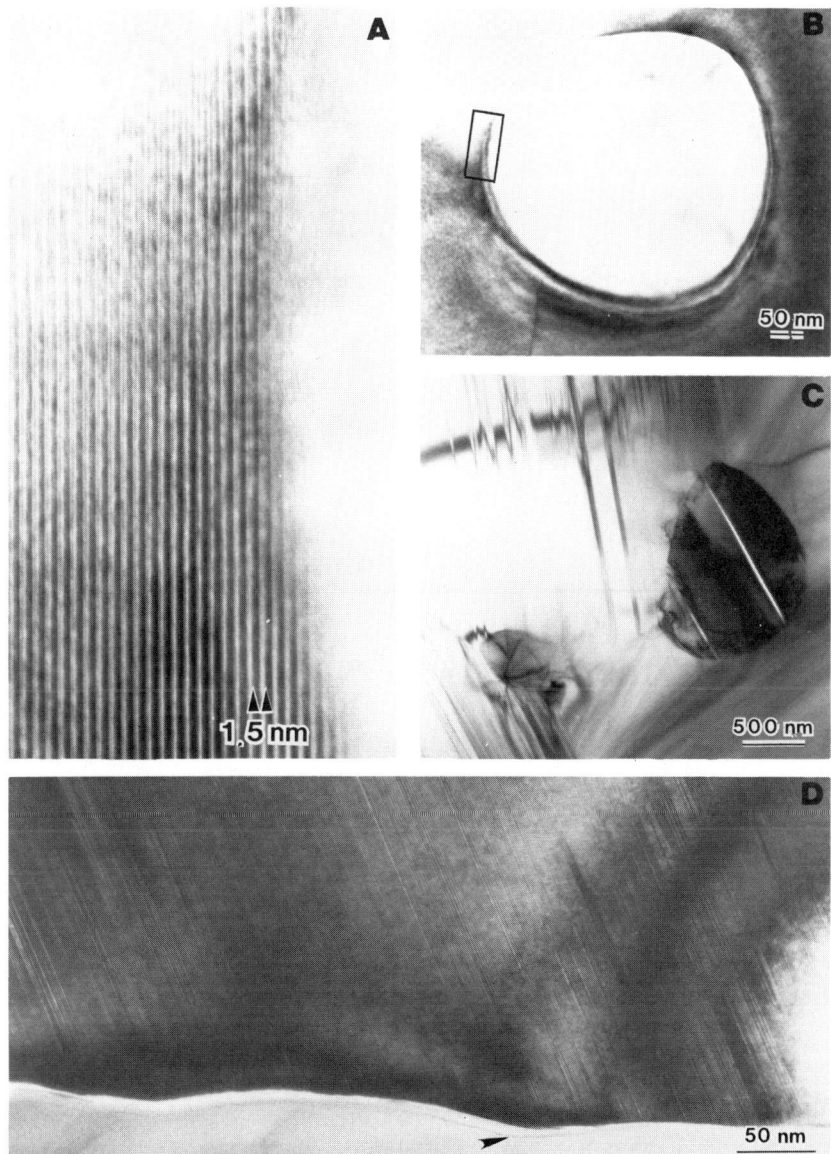

Figure 4: (a) TEM micrograph shows an excellent contact between BeO grain trapped in a SiC grain, and (b) structural lattice image shows the contact without strain. (c) Fine crack formation within a BeO grain, formed by BeO phase transformation. (d) Thin second phase formation in SiC-BeO grain boundary in a SiC-BeO material hot-pressed at 2200°C.

From these results of microstructural analysis, the extensive capability of analytical electron microscopy (AEM) to correlate the microstructures and the physical properties of the SiC-BeO materials has been demonstrated. Here, some interesting conclusions can be summarized, as follows:

(1) The hot-pressing process was effective to trap the BeO grains mostly at triple points for an excellent dispersion of the BeO grains and for formation of adequate thick tabular grain shape with an aspect ratio of 2 - 3.

(2) In order to achieve the excellent physical properties, the newly grown β-SiC needs to be transformed to 4H polytype and to have an excellent basal plane growth to achieve good contact between grains. This is provided by hot-pressing process using a large graphite die, which helps to maintain a high temperature for a reasonable period of time.

(3) Sintering temperature control is a critical factor, since phase transformation of BeO occurs and decomposition of SiC is reported around 2100°C. The physical properties may be easily deteriorated by the formation of second phases and microcracks.

REFERENCES

1. M. Ura and O. Asai, Development of SiC Ceramics Having High Thermal Conductivity and Electrical Resistivity, F. C. Report, 1, No. 4, Japan Fineceramics Association, 1982

2. K. Maeda, T. Miyoshi, Y. Takeda, K. Nakamura, S. Ogihara, and M. Ura, in Additives and Interfaces in Electronic Ceramics, edited by M. F. Yan and A. H. Heuer (Advances in Ceramics, Vol.7, 1984), pp.260.

3. S. Shinozaki and K. R. Kinsman, in Processing of Crystalline Ceramics, edited by Hayne Palmour III, R. F. Davis and T. M. Hare (Mat. Sci. Res. Vol. 11, 1978) pp. 641

4. S. Shinozaki, J. Hangas and K. Maeda, presented at the Silicon Carbide Symposium, Columbus, Ohio, 1987

5. S. Shinozaki, R. M. Williams, B. N. Juterbock, W. T. Donlon, J. Hangas and C. R. Peters, ACS Ceramic Bulletin, 64, 1389 (1985)

6. Y. Takeda, K. Usami, K. Nakamura, S. Ogihara and M. Ura, in Additives and Interfaces in Electronic Ceramics, edited by M. F. Yan and A. H. Heuer (Advances in Ceramics, Vol. 7, 1984), pp 253

7. A. Soeta, K. Maeda, and Y. Suzuki, Japan Ceramic Sac., Ann. Meeting Bulletin, 5119 (1985)

8. K. R. Kinsman and S. Shinozaki, Scripta Met., 12, 517 (1978)

REACTION MECHANISMS IN ORGANIC BINDER REMOVAL DURING CERAMIC PROCESSING: PMMA/CORDIERITE AS A PROTOTYPE SYSTEM

William E. Farneth*, Ralph H. Staley*, and
Theodore Budzichowski*
*E. I. du Pont de Nemours & Co., Central Research and
Development Department, Experimental Station, E356/B37,
Wilmington, DE 19898.

INTRODUCTION

As the performance requirements of ceramic components are refined, the demands placed on the materials used in their processing also become more stringent. For example, in advanced ceramic packaging applications, it is increasingly important that the organic polymers used as binders be capable of degradation and removal at low temperatures and in non-oxidizing atmospheres. This paper examines the chemistry that can occur during two fundamental stages of the binder removal process, (1) the breakdown of the C-C backbone of the polymer during pyrolysis, and (2) the interaction of the small molecules that are produced during pyrolysis with ceramic surfaces. The work has focused on acrylate and methacrylate systems, both because this is an important family of polymers for applications in tape casting, and because there is a relatively extensive literature on their pyrolysis chemistry.

EXPERIMENTAL

A microbalance/temperature-programmed desorption technique was employed as the central tool in this work. This instrumentation was used in two different, complimentary ways. Details of the apparatus have been published elsewhere.[1] The polymer samples were either commercial materials, i.e. Aldrich Chem. Co., PMMA # 18,224, or prepared from monomer by group transfer polymerization.[2] The precordierite powder was obtained from Specialty Glass (SP-980). The TiO_2 was a Du Pont material (#R-900).

In order to study the thermal decomposition of the polymer, samples of pure polymer or polymer/ceramic composites were loaded onto the pan of the microbalance enclosed in a high vacuum chamber. The system was evacuated to a background pressure of $<1 \times 10^{-7}$ torr by pumping at room temperature. After pumpdown, samples were heated in vacuum at variable heating rates to 420°C and held at that temperature for varying periods of time. Mass changes during the pumpdown, heating, and high temperature hold were followed continuously. The mass spectrum of the evolving gas stream was monitored simultaneously. The effect of changes in the ambient atmosphere on the degradation rate and product distribution may also be examined using this apparatus. Oxygen, for example, could be added to the chamber either after heating in vacuum or during the ramp. In order to study the chemical reactivity of the small molecule fragments produced during pyrolysis with the ceramic surface, a powder of a cordierite precursor was loaded onto the sample pan. After pumpdown and pretreatment of the powder, the sample was exposed to the vapor of the organic molecules of interest for times,

temperatures and partial pressures that varied from experiment
to experiment. Mass changes due to chemisorption could be
determined, and the thermal reactivity of the chemisorbed
species could be examined by a subsequent heating ramp using
both the balance and the mass spectrometer to follow the
changes.

RESULTS and DISCUSSION.

The decomposition mechanism of the pure polymer serves as
a starting point for these studies. It has been shown that
PMMA degrades by a process in which the backbone unzips
leading to monomeric methylmethacrylate as the principal
reaction product.[3] However, even in the absence of ceramic
powders and even for polymers like PMMA that degrade
relatively cleanly along a single reaction pathway, the
mechanism of polymer degradation can be complex. For example,
for PMMA, one observes that small structural changes can lead
to large differences in the effective decomposition
temperature in vacuum. In figure 1 the vacuum TGA curves for
several PMMA samples are compared. While in all cases, the
polymer degrades completely after ramping to 400°C and holding
at that temperature, there are significant differences in the
temperature at which mass loss occurs. The samples shown in
figure 1 would all be identified as "pure" PMMA materials, but
they differ from one another in two ways, molecular weight
distribution and end group functionality. These kinds of
variations in PMMA degradation rates have been previously
described. For example, it was shown that samples prepared by
anionic polymerization show a single TGA peak at ~370°C under
a certain set of conditions. However, samples prepared by

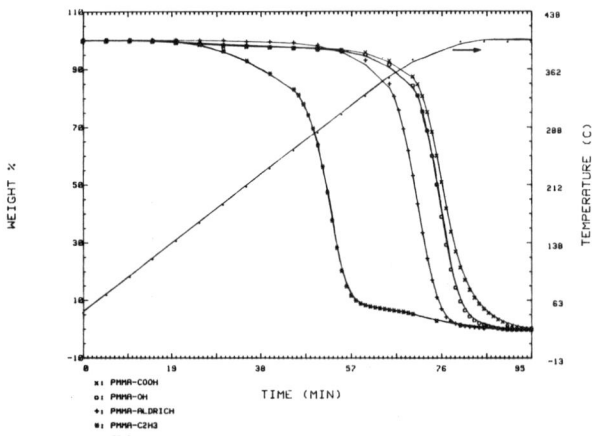

Figure 1. % weight loss versus time during a linear ramp from
25°C to 400°C in vacuum. Each curve represents different
PMMAs that vary in chain terminating group and molecular
weight.

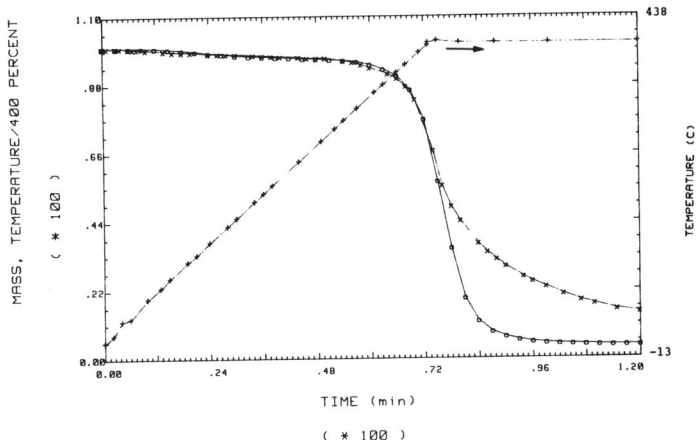

Figure 2. % Organic remaining versus time during a linear ramp to 400°C in vacuum for 5 weight % PMMA on calcined and uncalcined TiO$_2$ rutile.

radical polymerization show, in addition, peaks at ~150°C and ~270°C.[4] These differences result from specific "weak links" in the polymer backbone that can act as initiation sites for depolymerization. From the standpoint of binder applications these results demonstrate the importance of recognizing that PMMA represents a family of polymers that may have very different degradation behavior based on their preparation histories.

When polymers are thermally decomposed in the presence of ceramic powders, both rates and reaction product distributions can be markedly different from the polymer alone. In the case of PMMA, the water content of the ceramic is an important variable. Figure 2 shows weight loss curves for polymer degradation in 5% PMMA/TiO$_2$ composites. These composites were prepared by drying a slurry of the two materials that had been made up in methylethyl ketone. Calcining the rutile powder at 750°C overnight before making up the slurry had a pronounced effect on the efficiency of polymer removal. The symbols x represent the degradation of the polymer on the uncalcined ceramic. The symbols o show the calcined samples. We believe that hydrolysis of the ester side chains produces acrylic acid groups which bind tightly to the surface. In support of this hypothesis, we have shown that (under these conditions) polyacrylic acid cannot be removed completely from a similar composite with either calcined or uncalcined powder.

Once volatile fragments have been created, then removal requires diffusion of these fragments to vapor interfaces in competition with chemical reactions at surfaces. There is no question that chemisorption can occur during this stage, for the types of small molecules that are generally produced in PMMA degradation, and the types of surfaces that are generally present on the ceramic powders. The heterogeneous catalysis literature, for example, shows that alumina surfaces build up

"coke" deposits after relatively short periods of time in hydrocarbon atmospheres at several hundred degrees Centrigrade.[5] The plausibility of this type of "coking" mechanism for carbon formation during binder removal has been tested by exposing ceramic precursor powders to vapors of small organic molecules at various temperatures. With certain, but not all combinations of small molecule and ceramic powder, the growth of non-volatile carbon deposits can be observed. An example of this type of experiment is shown in Figure 3. Figure 3 is a representation of the changes in sample mass (SP-980 powder ~100mg.) with time over four stages of an experimental cycle.
These four stages are:
<u>t0-t1</u> exposure of the sample to one torr of methylmethacrylate vapor at 200°C. Mass gain is observed due to chemisorption and physisorption on the surfaces of the powder.
<u>t1-t2</u> pumpdown to $<10^{-7}$ torr still at 200°C Mass loss is observed due to removal of the weakly bound material. A substantial mass of strongly bound residue remains.
<u>t2-t3</u> heating from 200°C to 400°C at $<10^{-7}$ torr Mass loss is observed as some of the residue decomposes and is pumped away. Some material cannot be removed even by prolonged heating at 400°C.
<u>t3-t4</u> addition of 500 torr of oxygen to the chamber with the temperature still at 400°C. Mass falls as oxygen accelerates the residue removal through new oxidative chemical mechanisms. In 2 hours under these conditions, the sample has returned to its original mass.

It is clear that some of the behavior of the polymer/ceramic composites can be mimicked by these dosing experiments. This suggests that an important mechanism for carbon retention may be oligomerization initiated by strong chemisorption of small fragment molecules on the inorganic surfaces.

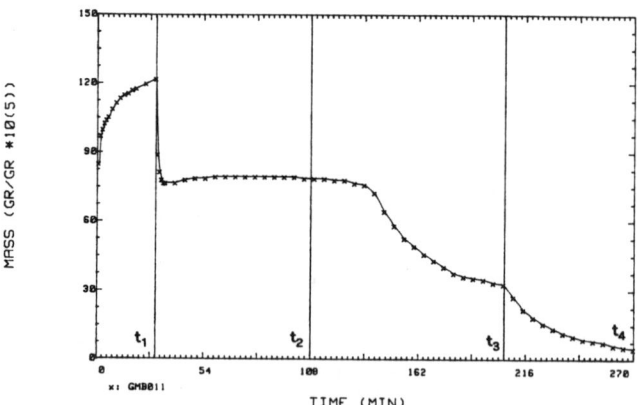

Figure 3. Mass changes during chemisorption/desorption cycle of methyl methacrylate on cordierite SP-980.

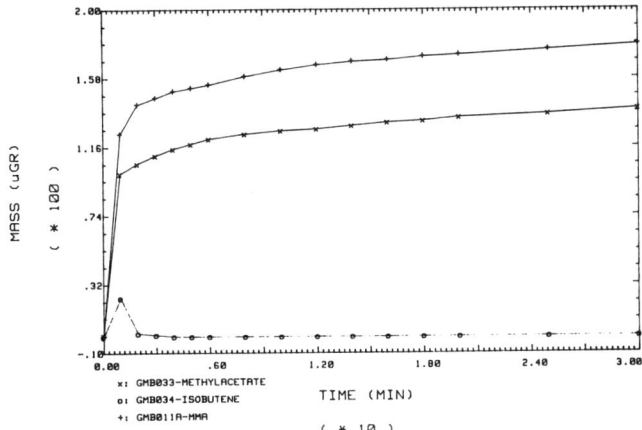

Figure 4. Mass changes during chemisorption of methyl methacrylate, +, methyl acetate, x, and isobutene, o, on cordierite SP-980 at 200°C.

An interesting extension of this approach is shown in figure 4. Again, mass increases on SP-980 are followed with time of exposure to three different organic molecules at 1 torr and 200°C. Methylmethacrylate has two reactive functional units, an ester group and an olefin unit, either or both of which could conceivably be involved in the surface chemistry. Figure 4 compares the chemisorption behavior of methylmethacrylate (MMA), methyl acetate, which contains the ester function but not the olefinic function, and isobutene which has the olefin but not the ester. The difference between the two monofunctional adsorbates is dramatic. Methyl acetate behaves similarly to MMA, a rapid uptake followed by a slow growth. Isobutene on the other hand shows no surface reaction under these conditions. The clear implication is that it is the ester functionality and not the olefin that is responsible for anchoring the carbonaceous deposit to the surface. To the extent that these experiments can be used to understand carbon build up during polymer degradation, then polymers that decompose to olefins rather than carboxylate-containing species might be good candidates for low residue binder materials.

[1] W.E. Farneth, F. Ohuchi, R.H. Staley, U. Chowdhry, and A.W. Sleight, J. Phys. Chem. 89, 2493 (1985).
[2] obtained from G. Cohen, personal communication.
[3] C. David in "Comprehensive Chemical Kinetics Vol. 14 Degradation of Polymers", C.H.. Bamford and C.F.N. Tipper, eds., Elsevier Pub. Co., Amsterdam (1975), p. 53.
[4] T. Kashiwagi, A. Inaba, J. Brown, K. Hatada, T. Kitayama, and E. Masuda, Macromolecules 19, 2160 (1986).
[5] M. Guisnet, P. Magoux and C. Canaft in "Chemical Reactions in Organic and Inorganic Constrained Systems", R. Setton (ed.), D. Reidel Pub. Co., New York, (1986), p. 131 and references therein.

COMPOSITES FOR ELECTRONIC SUBSTRATE APPLICATIONS

R. GERHARDT
Center for Ceramics Research, Rutgers University, P.O. Box 909, Piscataway, NJ 08855

ABSTRACT

The need for low dielectric constant, high thermal conductivity, matched thermal expansion and co-processability in electronic substrates is reviewed. Since no single phase material is able to satisfy all the requirements, a microscopic composite approach is proposed. Recent experimental evidence supporting the concept is also presented.

INTRODUCTION

The property requirements for electronic substrates are becoming more stringent as the frequency of operation gets higher and the number of circuits per unit area increases[1,2]. No single material possesses all the desired properties (see Table I) such as low dielectric constant (k<5), high thermal conductivity (K>100 W/mK), good thermal expansion match to Si or GaAs($\sim$ 3 ppm/^{o}C), high resistivity, fair mechanical strength ($\sim$ 40MPa) and co-processability with high conductivity metals. The first requirement (i.e. low k) has been met quite adequately by the usage of polymers whose low atomic mass guarantee a low dielectric constant but also an equally low thermal conductivity. For chip-on-board applications this has been the right way to go. On the other hand, multilayer structures have required the use of Al_2O_3 whose thermal conductivity is one to two orders of magnitude better than that of polyimide but this gain has been obtained at the expense of the dielectric constant which is 3 times that of polyimide. However, in order to be able to dissipate up to 100 W/cm^2 [3], even higher thermal conductivities are desired. Nitrides and carbides are excellent in thermal conductivity but the temperatures needed to process them are prohibitively high (>1600°C) and incompatible with metal co-firing. In addition, the impurities added to improve densification sometimes render them semiconducting. Other materials such as BeO present a toxicity problem. Nevertheless, engineering ingenuity has led to the

Table I
Properties of Single Phase Materials

MATERIAL	k (25°C, 10^6 Hz)	K (R.T. - 300°C)	α (ppm/^{o}C)
Air	1	< 0.1	--
Polyimide	3 - 4	0.2	> 50
Amorphous SiO_2	3.8 - 4.1	2	1
BN (hexagonal)	4.1	42	3.7
Quartz	4.6	2	10
Cordierite	4.9	4	1
Diamond	5.7	650 - 2000	3.5
BN (cubic)	5.8	950 - 1300	4
Si_3N_4	6.0	33	3
BeO	6.5	250	8
Mullite	6.6	7	5
AlN	8 - 9	100 - 320	4 - 5
Al_2O_3	9 - 10	20 - 35	7 - 9
SiC	~40	270	4

Mat. Res. Soc. Symp. Proc. Vol. 108. ©1988 Materials Research Society

102

development of the TCM (thermal conduction module) by IBM[4] and layered substrates[5,6] consisting of ceramic or polymer layers on metal substrates to solve the problem of removing the heat generated by the IC's (integrated circuits). These solutions are, however, not always feasible to use such as in the case of embedded active devices and multilayer set ups. Therefore, a new solution involving the combination of two or more materials on a microscopic scale in order to obtain the right properties is proposed here. A review of the combination rules is also presented.

COMPOSITE APPROACH

Before describing the equations relating property changes with volume fraction of added second phase, it is desirable to define the different types of composites possible. There are particulate and fiber composites, layered composites and porous composites. Fig. 1 illustrates the three simplest possible geometrical arrangement for two phases. They have been classified according to what an electric field would see in its path. The parallel model would predict the behavior of composites with open pores as well as any composites with the 2 phases stacked parallel to the electric field. Therefore, continuous fiber composites would fit under this category if the fibers are perpendicular to the substrate surface. The series arrangement, on the other hand would require the fibers to be parallel to the substrate surface and hence, perpendicular to the electric field. Commercial layered products combining a polymer top layer and a highly thermally conductive bottom layer[6] or a ceramic layer on top of a metal [5] also fit into this category. The third arrangement which I have called a 0-3 composite [7] could be represented by closed pore materials. Any composite consisting of a continuous phase (connected in 3-dim) that contains an embedded second phase (not connected with itself) would fall in this category. This includes particulate and chopped fiber composites as well. These three arrangements provide the elements necessary to arrive at any other possible configurations.

PHASE GEOMETRY

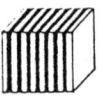 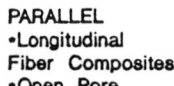 

PARALLEL
•Longitudinal Fiber Composites
•Open Pore Materials

SERIES
•Layered Composites
•Transverse Fiber Composites

0-3
•Particulate Composites
•Closed Pore Materials

Fig. 1 A schematic of the three simplest geometrical arrangements of two phases.

The models that can predict the properties are relatively simple especially in the case of the dielectric constant, k, and the thermal conductivity, K. The combination of properties will always fall between the bounds determined by the series and parallel models. The equation for a parallel arrangement is given by

$$P = V_1 P_1 + V_2 P_2 \qquad (1)$$

where P is the property such as k or K, and V_1 and V_2 are the volume fractions of the respective component phases. For a series arrangement the model becomes

$$\frac{1}{P} = \frac{V_1}{P_1} + \frac{V_2}{P_2} \tag{2}$$

In the case of a 0-3 composite Maxwell's model is the most appropriate

$$P = \frac{2(1-V_1)P_2 + (1 + 2V_1) P_1}{(2+V_1)P_2 + (1 - V_1) P_1} [P_2] \tag{3}$$

Thermal expansion, α ,on the other hand incorporates mechanical interactions and is not as simple as the k or K cases. The bulk moduli and the shear moduli of the component phases play an important role in determining the coefficient of thermal expansion of composites. Predictions of resistivity and mechanical strength are also quite straight forward but will not be considered here due to space constraints. Besides, the three properties discussed above are the most crucial in achieving an all-around desirable substrate material. For a more lengthy discussion of composite properties the reader is referred to reference 8.

Fig. 2 shows a schematic of the composite k or K properties for the parallel, series and 0-3 arrangement. Figs. 3 and 4 display the composite properties for several property ratios. Note that the higher the difference in properties the more evident a percolation threshold becomes, especially in the series case. It may be added that the microstructure of a composite generally will not be the same as the ratio of the two phases changes, therefore, no single model can be expected to fit the whole composition range.

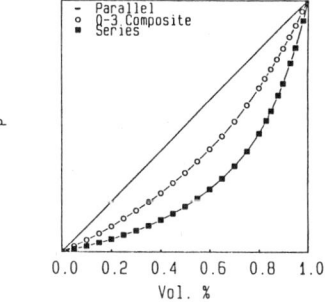

Fig. 2 A schematic for the composite property P, vs the volume fraction of added phase.

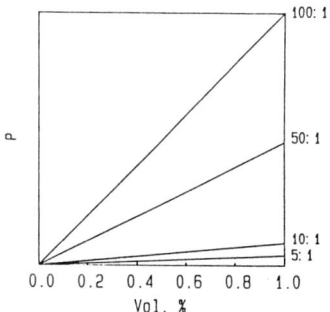

Fig. 3 Calculated composite properties for the parallel arrangement at four different P_2/P_1 ratios.

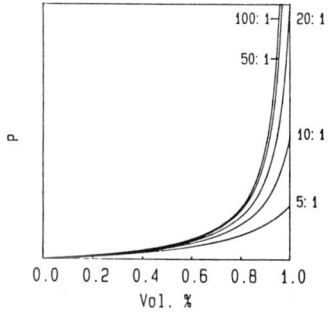

Fig. 4 Calculated composite properties for the series arrangement at 5 different P_2/P_1 ratios.

RECENT EXPERIMENTAL EVIDENCE

Probably the most studied composite material has been the glass alumina system[9]. This material was originally studied for its excellent mechanical properties but more recently new glass compositions with lower k's have been made by Kyocera[10]. They have shown that it is possible to cofire Cu with this system but its thermal properties still leave a lot to be desired. Research on cordierite and mullite systems[11,12] has shown that one can obtain a tailored coefficient of thermal expansion to match that of the chips but their k values are not as low as one might wish while thermal conductivity remains relatively low as well.

A lot of activity has been evident in the nitride area[13], particularly on AlN. However, the most promising single phase material is now cubic BN. Hirano et al.[14] demonstrated that BN could be made to have a k of 5.1 and a K of 950W/mK. These properties make nitrides extremely attractive but their high sintering temperatures still limit their applicability.

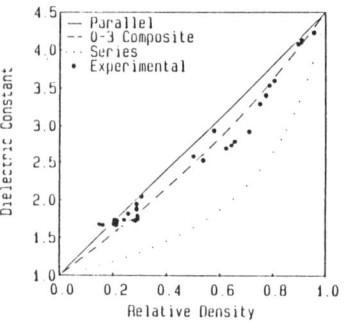

Fig. 5 Dielectric constant vs relative density for bulk porous silica at room temperature for 50 kHz.

Recently some emphasis has been put on the development of porous ceramics[15], particularly porous silica, and k values as low as 1.6 have been obtained[16] but their thermal conductivities are still dismally low. Nevertheless, the potential for porous ceramics has been demonstrated[17] and applications where speed is the most important consideration could benefit from the ability to tailor the dielectric constant by varying the porosity content. This can be seen in our own work in Fig. 5 where the dielectric constant varies from 1.6 up to 4.3 just by varying the porosity from 85% down to nearly zero. The decrease in k with porosity is linear up to 50% volume fraction and then tapers off possibly due to pore interconnection[16]. One can envision adding thermally conducting fillers into these porous ceramics to improve their thermal properties. This has in fact been done successfully for other systems. Fig. 6 presents thermal conductivity data for silicon carbide whisker reinforced mullite. Russell et al.[18] showed that the greatest change in thermal conductivity took place in the range between 2-20% of filler phase. Since the whiskers were assumed to be aligned perpendicular to the hot pressing direction the series model of Fig. 4 applies. Agreement of the experimental data with a thermal conductivity ratio of 5 is quite good for the rice hull whiskers while data for VLS whiskers is best fit by an infinite ratio. The dielectric properties of this system have not as yet been measured since it was

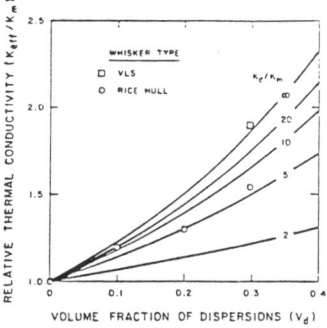

Fig. 6

Comparison of relative changes in thermal conductivity at room temperature of silicon carbide whisker reinforced mullite with those predicted by the series model (reprinted by permission of the American Ceramic Society, Ref. 18).

studied for its potential as a structural material.

An extensive amount of work has been carried out on polymer materials for many years because of their low k's. They remain the material of choice in many applications due to the ease of fabrication and low cost. However, heat dissipation is still a problem. Button et al. [19] from DuPont have taken the lead in making composites based on polymers and recently made alumina fiber/polyimide composites with improved thermal properties without sacrificing the dielectric constant by cleverly manipulating the microstructure. The values of K with alumina additions are still low, however they have proposed using AlN as a filler which should improve the thermal properties even further.

CONCLUDING REMARKS

It is clear that enough evidence is available to show that a composite approach is viable. Certainly layered composites can still be exploited but a mixing of phases on the microscopic scale will prove to be more versatile in the end. Polymers and polymer composites will still continue to keep a large portion of the market but we must not overlook the opportunities in glass ceramics systems. This is especially true if embedded active devices become the norm. Polymers would not be able to withstand cofiring. At Rutgers we are currently working on glass:BN composites[20] and expect to look at many other composites with potential for use in electronic substrate applications in the near future. I think the time is ripe to stop thinking of composites as materials purely for structural applications but also with potential for electronic applications.

REFERENCES

1. Albert J. Blodgett, Jr., Scientific American, July 1983, pp. 86-96.
2. Bernard Schwartz, Am. Ceram. Soc. Bull., 63[4] 577-581 (1984).
3. The estimated requirement as per Neugebager, Paper 11.3 ,MRS Fall 1987 Meeting in Boston.
4. A.J. Blodgett, Jr. and D.R. Barbour, IBM J. Res. Develop., 26[1] 30-36 (1982).
5. Commercially available Ferro-ECA CCMS and Texas Instruments CERCIC, presented at ISHM Meeting Sept.1987 in Minneapolis and at Electronics Division ACerS Meeting Oct. 1987 in Denver by V.N. Shukla, A. Amin and S. Hingorany.
6. Review of work at Honeywell, and Boeing at MRS Fall 1987 Meeting given by J. Balde in Paper 12.1.
7. R.E. Newnham, and L. E. Cross, Mater. Res. Bull., 13[5] 525-536 (1978).
8. D.K. Hale, J. Mat. Sci., 11, 2105-2141 (1976).
9. Y. Shimada, K. Utsumi, M. Suzuki, H. Takamizawa, M. Nitta and S. Yano, in Proc. 33rd Electronic Components Conf., pp.314-319 (1983).
10. H. Emura, K. Onituka and H. Manuyama, to be published in Advances in Ceramics as Proc. of Int'l. Conference on Ceramic Substrates and Packages (ICCSP) held in Denver, Oct. 1987, ed. Man Yan, The American Ceramic Society.
11. Bernhard H. Mussler and Merrill W. Shafer, Am. Ceram. Soc. Bull., 63[5] 705-710 (1984).
12. R. Anderson, R. Gerhardt, J.B. Wachtman,Jr. and D.G. Onn, to be published in Advances in Ceramics as ICCSP, Denver, 1987.
13. As evidenced by 15 papers presented in Denver at Electronics ACerS Meeting, Oct. 1987 and 4 papers presented in Boston At MRS Fall Meeting, Dec. 1987.
14. S. Hirano, N. Fuji and S. Naka, to be published in Advances in Ceramics, ICCSP, Denver, 1987.
15. There were 4 papers on this subject at the recent ACerS Denver Meeting.
16. W. Cao, R. Gerhardt and J.B. Wachtman,Jr., to be published in Advances in Ceramics (ICCSP), Denver, 1987.
17. W.A. Yarbrough, T.R. Gururaja and L.E. Cross, Am. Ceram. Soc. Bull., 66[4] 692-98 (1987).
18. L.M. Russell, L.F. Johnson, D.P.H. Hasselman and R. Ruh, J. Am. Ceram. Soc.,

70[10] C226-C229 (1987).
19. D.P. Button, B.A. Yost, J.D. Bolt, R.H. French, M.J. Kletter, W.Y. Hsu, H.M. Zhang, R.E. Giedd and D.G. Onn, to be published in Advances in Ceramics (ICCSP), Denver, 1987.
20. S. Clark and R. Gerhardt, to be presented at the 90th Annual ACerS Meeting, Cincinnati in May 1988.

THE MICROSTRUCTURE AND CHEMISTRY OF THE REACTION
BETWEEN Ti AND α-Al$_2$O$_3$

J.H. Selverian[*], M. Bortz[**], F.S. Ohuchi[**] and M.R. Notis[*]

[*]Lehigh University, Department of Materials Science and Engineering, Bethlehem, PA 18015

[**]E.I. du Pont de Nemours and Co., Central Research and Development Department, Experimental Station, Wilmington, DE 19898

ABSTRACT

The Ti/Al$_2$O$_3$ system is being considered for advanced aerospace applications and is important in microelectronics where Ti is used for metallization on Al$_2$O$_3$ substrates. We have studied the intrinsic reactivity of Ti with the R-plane (10$\bar{1}$2) surface of α-Al$_2$O$_3$ by x-ray photoemission spectroscopy (XPS) and transmission electron microscopy (TEM) techniques. XPS results indicate that a for deposition of several monolayers of Ti at 25 °C, the Ti reduces the Al$_2$O$_3$ surface to produce Ti-O bonds. At 1000 °C the Ti reduces the Al$_2$O$_3$ to produce Ti-O and Ti-Al bonds. Cross-section TEM specimens were prepared from 200 nm thick Ti films deposited on Al$_2$O$_3$ at 25 and 800 °C to observe the interface region. In samples deposited at 25 and 800 °C, without further annealing, no reaction zone could be seen in the TEM. Ordered Ti$_3$Al (78 ± 5 Ti, 16 ± 3 Al and 6 ± 3 O wt pct) was observed in a sample deposited at 25 °C and annealed at 800 °C for 2 hours.

INTRODUCTION

Lefakis et al [1] studied reactions between 100 nm thick films of oxidized Al and monolayer coverages of Ti with Auger electron spectroscopy (AES) and ultraviolet electron spectroscopy (UPS). At room temperature, Ti was found to reduce Al$_2$O$_3$ to produce a 1 nm thick layer which was identified as a mixture of TiO$_2$ and Al in accordance with the reaction proposed by Lofton and Swartz [2].

Chaug, Chou and Kim [3] deposited Ti onto bulk crystals of SiO$_2$ and Al$_2$O$_3$ and studied the chemical bonding changes with XPS. Their conclusions concerning Al$_2$O$_3$ as a substrate included observing a small shift in the Ti(2p) peaks interpreted as the formation of a TiO$_x$ (1 < x < 1.5) compound. Also, no shift was observed in the Al(2p) peak and therefore no Al-O bonds were broken. However, evidence for local rearrangement of the bulk Al$_2$O$_3$ structure at the interface was seen. The Ti-O bonds that form at the interface act to "bridge" across the interface and join the bulk Ti and bulk Al$_2$O$_3$ lattices.

EXPERIMENTAL PROCEDURE

Al$_2$O$_3$ single crystals in the 10$\bar{1}$2 orientation from Union Carbide were used in both the XPS and TEM investigations. For the XPS studies Al$_2$O$_3$ crystals were mounted on a heating stage in an ion pumped ultra-high vacuum (UHV) chamber, with base pressure better than 5 x 10^{-10} torr. XPS was used to probe changes in the Al(2p), O(1s) and Ti(2p) core levels with monochromatized Al K$_\alpha$ x-rays and a hemispherical energy analyzer with resolution of ~ 0.9 eV. Absolute binding energies were not obtained due to charging problems associated with the Al$_2$O$_3$ substrates. All samples were prepared in the analysis chamber to minimize contamination. Substrates were sputter cleaned with 1.5 keV Ar$^+$ for 10 minutes followed by annealing to 1000 °C for 5 minutes. Titanium was deposited by heating Ti ribbon until the pressure in the chamber had risen to ~9 x 10⁻⁹ torr.

Titanium films for the TEM study were 200 nm thick and were deposited and annealed in the system described above. A Philips 430 TEM was operated at 150 and 250 keV with a windowless EDAX x-ray energy detector for chemical analysis. All of the reported compositions were corrected for x-ray absorption.

RESULTS AND DISCUSSION

XPS Investigation

Initially Ti was deposited on Al₂O₃ at 25 °C. At very low, submonolayer, coverages no reaction was observed, Figure 1a. At higher coverages a metallic Al(2p) peak appears on the low binding energy side of the oxide Al(2p) peak and a reduction of the Al-O bonds by Ti occured. Metallic Ti was not seen, only oxidized Ti was seen in the Ti(2p) peak. Figure 2a describes the interfacial reaction in this sample, Ti + Al-O = Ti-O + Al. When Ti was deposited on Al₂O₃ at 1000 °C two types of oxygen bonds were seen, Figure 1b, indicating that Ti substitutes for Al with Ti-O and Al-O bonds forming, Figure 2b. Figure 1c shows that the Al(2p) peak also is split into two peaks, indicating Al-O and Ti-Al bonds and the Ti(2p) peak was split into an oxidized and a metallic peak. These findings are in qualitative agreement with those of Lefakis et al [1].

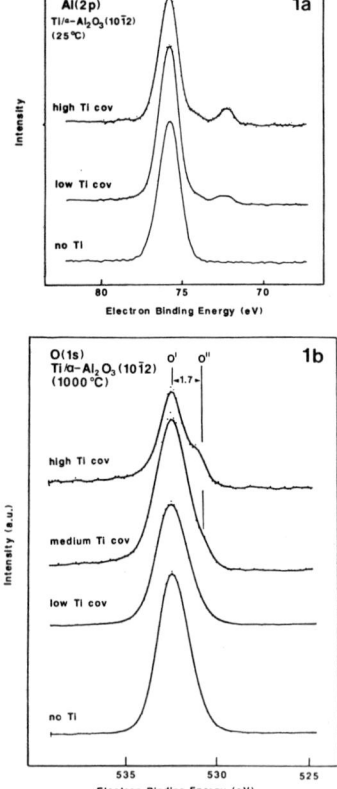

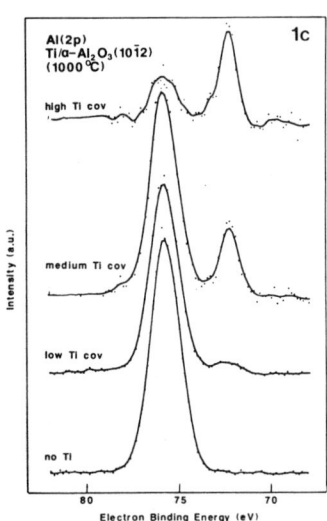

Figure 1. a) The Al(2p) peak for Ti deposited at 25 °C shows the appearence of a metallic Al peak on the lower binding energy side of the Al oxide peak. b) The O(1s) peak for Ti deposited at 1000 °C is split into two peaks, indicating Al-O and Ti-O bonds. c) The Al(2p) for Ti deposited at 1000 °C shows a metallic Al peak indicating Ti-Al bonds.

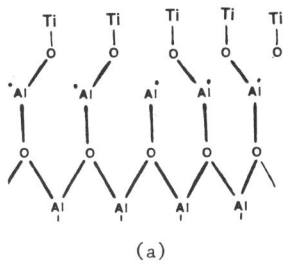 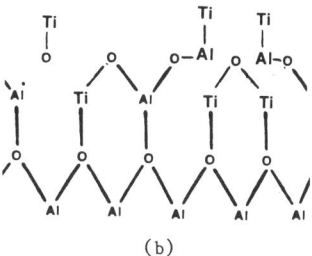

(a)   (b)

Figure 2. Models for the bonding of Ti to Al_2O_3. a) Ti deposited at 25 °C
b) Ti deposited at 1000 °C.

TEM Investigation

The first sample studied consisted of ~200 nm of Ti deposited onto an
Al_2O_3 substrate held at 800 °C. Figure 3 shows the microstructure which
consisted of ~100 nm diameter randomly oriented Ti grains. No reaction
layer could be seen at the Ti/Al_2O_3 interface. However, based on early XPS
experiments the reaction layer thickness under these conditions was on the
order of a few monolayers which is close to the limits of resolution of this
TEM. Another sample investigated had ~200 nm of Ti deposited at room
temperature and was annealed at 800 °C for 2 hours. The entire Ti film in
this sample had transformed to a new single phase product. This phase was
identified by composition and structure as ordered Ti_3Al with a composition
of 78 ± 5 Ti, 16 ± 3 Al and 6 ± 3 O wt pct (64 Ti, 24 Al and 12 O atomic
pct), Figure 4 [4,5]. The atomic ratio Al/O in the Ti_3Al layer should be
2/3, the same as in Al_2O_3. However, the measured atomic ratio Al/O is ~3/2
and this may imply that some O diffused out of the Ti_3Al during annealing in
the UHV chamber. The white area near the interface (arrow in Figure 4) is a
hole caused by was preferential thinning by the Ar^+ beam during sample
preparation. This was confirmed by noting that no diffraction pattern was
obtained from these areas when the electron beam was focused there.

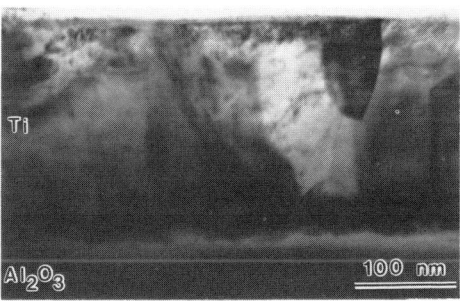

Figure 3. TEM micrograph of
Ti film deposited on an Al_2O_3
substrate held at 800 °C
without further annealing.
No reaction could be seen at
the interface.

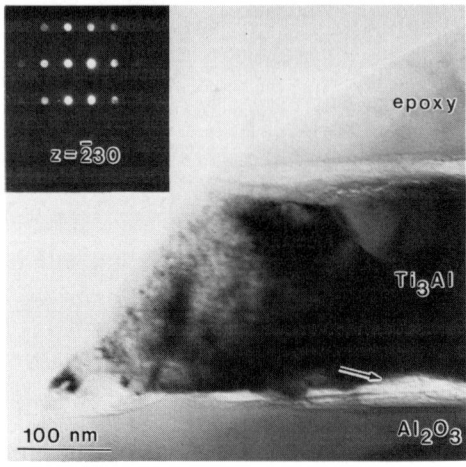

Figure 4. TEM micrograph of Ti film deposited on an Al_2O_3 substrate held at 25 °C and annealed for 2 hours at 800 °C. The entire Ti layer transformed to ordered Ti_3Al (78 ± 5 Ti, 16 ± 3 Al and 6 ± 3 O wt pct). The electron diffraction pattern was indexed to be from the ($\bar{2}$30) zone of ordered Ti_3Al.

CONCLUSIONS

1) For low coverages of Ti on Al_2O_3 at 25 °C no reaction was observed. At higher coverages, Ti-O bonds were formed, Ti + Al-O = Ti-O + Al.
2) At a deposition temperature of 1000 °C, Ti reduced Al_2O_3 with Ti-O and Ti-Al bonds forming, the reaction being Ti + Al-O = Ti-O + Ti-Al.
3) For 200 nm thick films of Ti deposited on Al_2O_3 at room temperature and annealed for 2 hours at 800 °C microcrystalline ordered Ti_3Al (78 ± 5 Ti, 16 ± 3 Al and 6 ± 3 O wt pct) was formed. Ti_3Al did not show any obvious orientation relationship with the Al_2O_3.

ACKNOWLEDGEMENTS

The authors would like to thank Dave Sokola and E.I. du Pont de Nemours and Co. for their assistance in this work.

REFERENCES

1) H. Lefakis, M. Liehr, G.W. Rubloff and P.S. Ho, in Interfaces and Phenomena, edited by R.H. Nemanich, P.S. Ho and S.S. Lau (Mater. Res. Soc. Proc. 54, Pittsburgh, PA 1986) pp. 133-138.
2) C.P. Lofton and W.E. Swartz, Jr, Thin Solid Films 52, 271-280 (1978).
3) Y.S. Chaug, N.J. Chou and H. Kim, J. Vac. Sci. Tech., 5A (4), 1288-1291 (1987).
4) E. Ence and H. Margolin, Trans. AIME, 221, 151-157 (1961).
5) M.J. Blackburn, Trans. AIME, 239, 1200-1208 (1967).

PART III

Interconnect Technology

LOW THERMAL EXPANSION POLYIMIDES AND THEIR APPLICATIONS

Shunichi NUMATA[*], Takao MIWA[*], Yutaka MISAWA[*], Daisuke MAKINO[**],

Junichi IMAIZUMI[***], and Noriyuki KINJO[*]

    *: Hitachi Research Laboratory, Hitachi, Ltd.
       4026, Kuji-cho, Hitachi-shi, 319-12, Japan
   **: Yamazaki Works, Hitachi Chemical Co., Ltd.
      4-13-1, Higashi-cho, Hitachi-shi, 317, Japan
 ***: Goshomiya Works, Hitachi Chemical Co., Ltd.
      1150, Goshomiya, Shimodate-shi, 308, Japan

ABSTRACT

Thermal expansion coefficients (TECs) for polyimides differ very much depending on their chemical structures. Polyimdes with a rod-like structure as their backbone chains have lower TEC values. This is attributed to restraining of thermal expansion by rod-like molecules within inter-molecular spaces, analogous to glass fibers in FRPs. The development of new polyimides which can closely match TECs of inorganic materials, such as metal or Si, can eliminate problems produced by thermal stress including warping, cracking or delamination.

A new multilevel interconnection system using multilayered dielectrics consisting of low thermal expansion polyimide and inorganic materials has been proposed as one future technology for submicron VLSIs. Consequently, adhesiveless, high quality flexible printed circuit boards have been developed using a polyimide with the same TEC as copper foil. Their most significant property is a high dimensional stability after heat treatments, such as in a soldering process. Furthermore, they have very high adhesion strength at elevated temperatures.

INTRODUCTION

It was believed for a long time that thermal expansion coefficients (TECs) of ordinary polymers were very high compared with those of inorganic materials, such as metals, ceramics, and glasses. Peculiar polyimides with low TECs, however, have been discovered in contradiction to the widely held belief (1-4). These materials are expected to be very useful for many kinds of elecronics applications because of their low thermal stress, as well as their excellent thermal stabilities, mechanical properties, etc.

In the present paper, chemical structure, a mechanism for lowering TEC, characteristics of the low thermal expansion polyimides, and applications to interlayer dielectrics of VLSIs and adhesiveless flexible printed circuit boards are reported.

EXPERIMENTAL

Samples

Polyamic acid varnishes were prepared by reacting different aromatic diamines with a stoichiometric amount of aromatic dianhydride in N-methyl-2-pyrrolidone at room temperature(5). Then they were coated onto glass plates and dried at 100°C for 1 h. The resulting films were fixed on iron frames and cured at 200°C for 1h, and at 400°C for another hour in nitrogen. This gave polyimide films.

Table 1   Thermal expansion coefficients of aromatic polyimides.

(unit : $\times 10^{-5} K^{-1}$)

Diamine \ Anhydride	(1)	(2)	(3)
(A)	—	2.10	0.26
(B)	3.20	2.94	4.00
(C)	—	3.95	3.19
(D)	0.04	2.59	0.58
(E)	3.48	3.95	4.00
(F)	1.61	—	—
(G)	0.59	2.17	0.54
(H)	0.20	1.54	0.56
(I)	1.37	4.91	4.64
(J)	0.56	1.83	0.59
(K)	—	—	1.72
(L)	1.58	1.60	1.13
(M)	2.16	4.28	4.56
(N)	4.15	5.24	4.61
(O)	4.57	4.50	4.18
(P)	5.76	5.36	4.85
(Q)	—	2.61	1.00
(R)	5.33	5.43	5.32
(S)	5.01	5.39	5.69
(T)	4.57	5.47	5.61
(U)	5.14	—	4.90

Measurements

The TECs were measured under the heating rate of 5°C/min in air, using a thermomechanical analyzer (TMA 1500, Shinku-Riko Co., Ltd.) between 25 to 400°C. Thermal decomposition properties of about 50 μm thick films were measured under isothermal conditions, using a thermo gravimetric analyzer (TGD-3000RHN). The decomposition temperature and activation energy were estimated by Arrhenius plots. Viscoelastic properties were measured using a Rheopexy analyzer (RPX-706, Iwamoto Seiki Co.) at a frequency of 10 Hz and 25°C. Density of the polyimide film was determined by using a liquid mixture of xylene and carbon tetrachloride at 30°C, which can suspend the film, in a tuning-fork type density meter (DMA 02C, Anton Paar). Wide angle X-ray diffractograms for polyimide films were measured using Geiger-Flex Rad (Rigaku Denki Co., Ltd.).

RESULTS AND DISCUSSION

Chemical Structure Characteristics

The TECs of various aromatic polyimides are summarized in Table 1. The values enclosed by a square indicate that they are lower than 2×10^{-5} K^{-1}. Most of the coefficients for polyimides are as high as those of ordinary organic polymers, and they are higher than those of inorganic materials, such as metals, glasses and ceramics. However, there are some aromatic polyimides having TECs lower than 2×10^{-5} K^{-1}. The lowest value is 4×10^{-7} K^{-1}, which equals that of quartz glass. It has already been noted that special polymer fibers which were highly stretched become ordered or crystallized and have low TECs (6-9). However, polymer films without stretching or extension or coatings which have low TECs are unknown.

Table 2 shows conformations of diamines and bisimide skeletons of low thermal expansion polyimides and some other polyimides(10-12). All of the

Table 2   Conformation of diamine and tetracarboxylic acid components.

diamine skeletons of the former are composed of only benzene or pyridine rings fused at the para-position. Furthermore, they include no bent structures such as a benzene ring fused at the meta-positions, or ether, methylene, thioether, or ketone linkages, etc. Similarly, the pyromellite bisimide skeleton is perfectly straight, and biphenyl bisimide seems to be almost linear. Therefore, conformations of these polyimides composed of linear diamines and bisimides would become necessarily almost linear. That is, the low thermal expansion polyimides should have rod-like structures. By contrast, there are no high thermal expansion polyimides without the above bent structures. These results suggest that the low values should be closely related to the linearity of the molecular structure.

Mechanism of Lowering Thermal Expansion Coefficient

Some possible mechanisms to explain the low TECs were proposed and investigated.
(1) Crystallizability: It is well known that TECs of polymers become smaller on crystallization caused by stretching, as mentioned above. Then, crystallizability of polyimides was studied by wide angle X-ray diffractograms. Measured results showed that low thermal expansion poly-imides tend to crystallize. However, as shown in Figure 1, some amorphous polyimides have low values, and some crystalline polyimides have high TECs. Therefore, crystallizability of polyimides does not seem to be the main factor in lowering TEC.

(2) Small free volume: Low thermal expansion polyimides with rod-like structures should have dense molecular packing, because of their small steric hindrance, thus, their free volumes appear as small. Since macro-scopic thermal expansion of polymers is closely related to an expansion of the free volume, it is reasonable that low TECs result because of this small

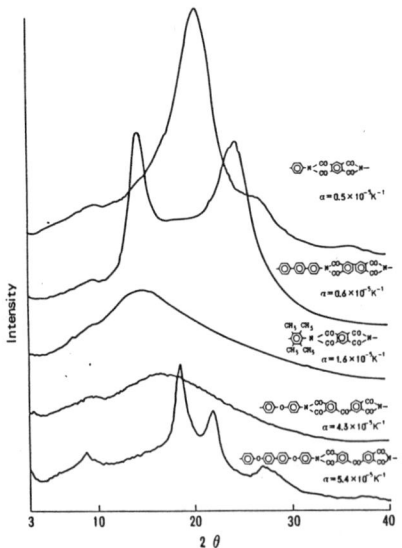

Figure 1 X-ray diffractograms for various polyimide films.

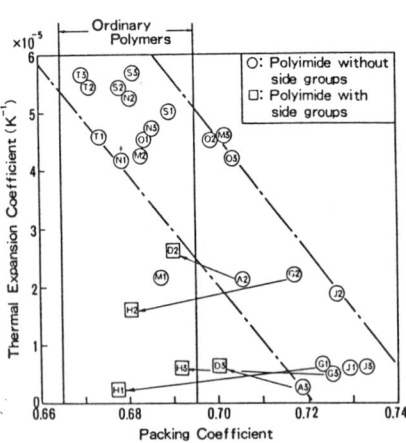

Figure 2 Relationship between thermal expansion coefficient and packing coefficient for various polyimides.

free volume. In order to clarify this point, examination of the relationship between thermal expansion and molecular packing coefficients for various aromatic polyimides was done.

In order to compare molecular packing, a "packing coefficient" parameter, which could be determined from density, has been proposed by Slonimskii et al.(13). Their investigation showed that most packing coefficients of ordinary polymers range from 0.665 to 0.695.

As expected, our results in Figure 2 show that most low thermal expansion polyimides have very high molecular packing coefficients, in contrast to the low values for the high thermal expansion polyimides. It seems that the larger the packing coefficients become, the lower the TECs become. However, in the case of rod-like polyimides with side groups on their molecular chains, even though their packing coefficients become smaller due to a hindrance effect from the introduced side groups, the TECs keep their small values. These results suggest that there is no correlation between packing coefficients and TECs (14).

(3) Restraint of thermal expansion by rod-like molecules: It is well known that glass fibers in FRP restrain expansion of the matrix resin, because it can expand very little in the direction of the fiber. Just like FRP, the rod-like polymer molecules restrict expansion within inter-molecular spaces at the molecular level. If the low TEC value is generated by this mechanism, these polyimide films should have very high moduli, because of the reinforcement effect by the rod-like polymer molecules. Figure 3 shows the relationship between TEC and modulus for various polyimide films. The low thermal expansion polyimide films have very high moduli and there seems to be a correlation between them. Hence, the low TECs are generated, because rod-like polymer molecules restrain the expansion within intermolecular spaces, at the molecular level (15).

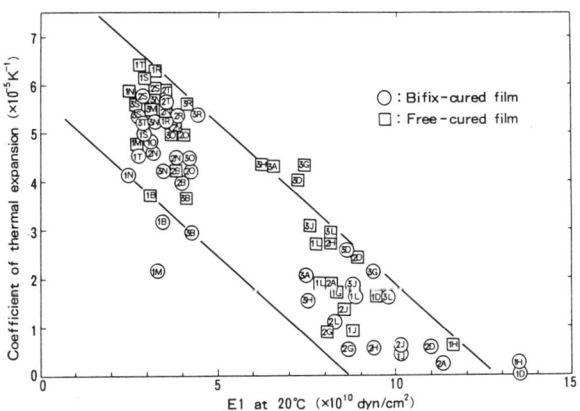

Figure 3   Relationship between thermal expansion coefficient and elastic modulus.

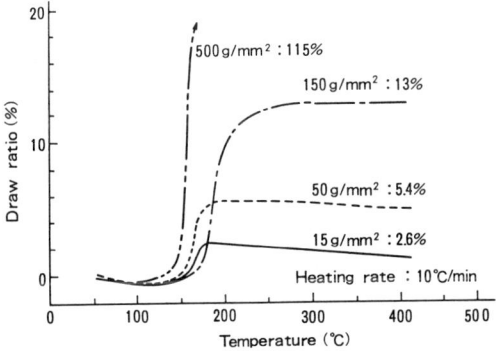

Figure 4   Drawing behavior of PIQ-L100 during imidization.

(4) Thermal shrinkage behavior of rod-like molecules: Uniaxilly stretched polyimide films were prepared by heating precursor polyamic acid films under stress during the imidization reaction for the low thermal expansion polyimide, PIQ-L100. PIQ-L100 is a modified copoly-imides based on one in the table 1. Figure 4 shows the dimensional changes during the imidization reaction with various stresses. Stretching of the films takes place from 150 to 200 °C.

Figure 5 shows the relationship between the degree of stretching, TEC of stretched film and stress during imidization reaction. When the degree of stretching is more than 13%, the TECs are negative. In particular, TEC of 113% stretched film is -1×10^{-5} K^{-1}. These results indicate that the rod-like polymer molecules have seemingly negative

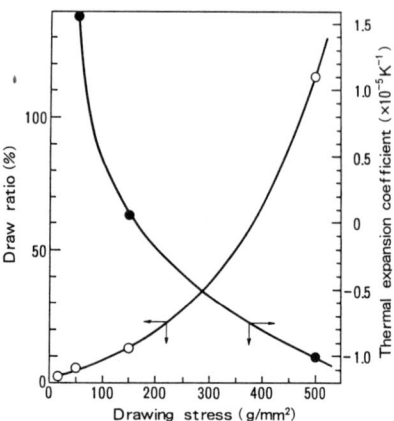

Figure 5 Drawing stress vs. drawing ratio and thermal expansion coefficient for PIQ-L100.

values in the molecular direction due to the thermal fluctuation of the linear polymer molecules. Thus, this characteristic of rod-like molecules is one important factor for lowering TEC.

Property Changes during Imidization Reaction

As mentioned above, it was seen that the low TECs were closely related to the rod-like structures of the polymer molecules. However, even if this consideration was true, the TECs of their precursor films (polyamic acids) should be as high as those of ordinary polymers, and the Young's moduli should be as small as those of ordinary polymers, because their confor-mations of polymer molecules are not rod-like structures, due to their amide bonds, as shown in Figure 6.

To elucidate the mechanism, property changes of the films during the imidization reaction were investi-gated(16). Figure 7 shows the relationship between percent conversion of imidization and heating temperature for the low thermal expansion poly-imides, PIQ-L100 and PIQ-L120, and the conven-tional high thermal ex-pansion polyimide, PIQ. They are manufactured by Hitachi Chemical Co., Ltd. The imidization reaction hardly proceeds below 100 °C, and it begins to occur exten sively from 150 °C. At

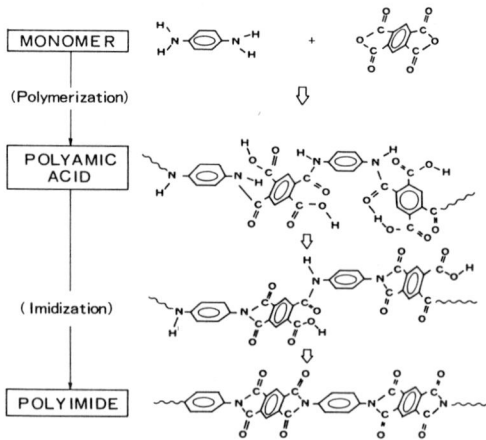

Figure 6 Conformational change during formation of polyimide.

250 °C the conversion of the imidization is about 90 %. But to complete the reaction, heating at higher temperatures, above the glass transition temperatures of their resulting polyimide, is required(17).

Figure 8 shows the relationship between the reaction temperature and mechanical properties, tensile strength and Young's moduli. As expected, the polyamic acid films of the low thermal expansion polyimides have low strength and moduli. However, with progress in the imidization reaction, values of strength and modulus for PIQ-L100 and L120 become higher, and their property changes continue to 250 or 300 °C, at which there is almost complete imidization. This result suggests that the rod-like structure is formed during the imidization reaction. By contrast, in the case of PIQ, its properties change very little.

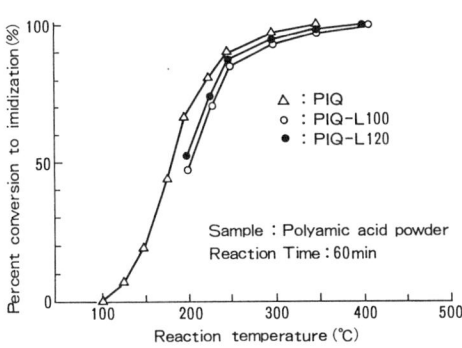

Figure 7   Imidization temperature vs. percent conversion.

Figure 9 shows changes in X-ray diffractograms for these polyimide films during the imidization reaction. According to these results, PIQ is an amorphous polymer from the polyamic acid state to after imidization. By contrast, in the case of PIQ-L100, it has an amorphous structure in the polyamic acid state, but it begins to crystallize after the imidization reaction is about 85 % completed. This result also suggests that the rod-like structure is formed during the imidization reaction.

Figure 10 shows the expansion behavior for various partially imidized films for PIQ, PIQ-L100 and PIQ-L120. The slope at the glass state (below Tg) for PIQ, i.e. TEC, changes hardly at all in spite of the progress of imidization. However, for the low thermal expansion polyimides, lowering of TECs is observed clearly. The changes in TECs of the polyimide films are summarized in Figure 11. The TECs of low thermal expansion polyimide films become lower during the imidization reaction, even though that of PIQ does not change at all. From these results we can confirm that the low TECs appear because of their rod-like polymer molecules.

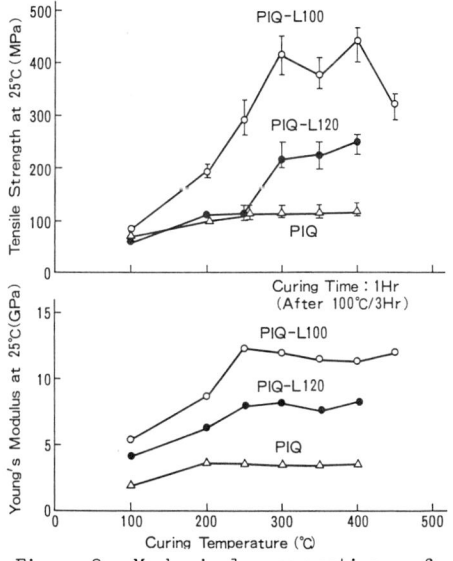

Figure 8   Mechanical properties of polyimide films cured at various temperatures.

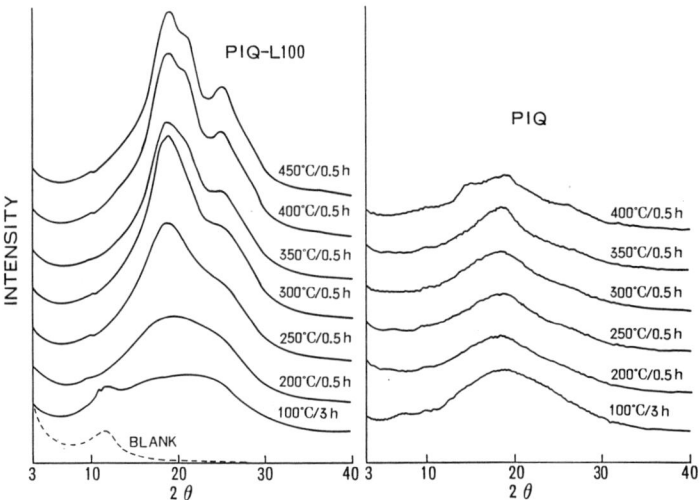

Figure 9   Changes in X-ray diffractogram for polyimide films
during imidization.

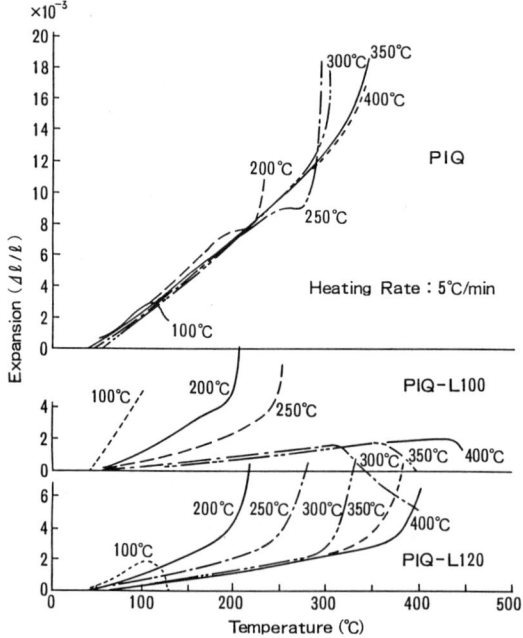

Figure 10   Thermal expansion of polyimide films
cured under various conditions.

Properties of Low Thermal Expansion Polyimide

Various properties of low thermal expansion polyimide PIQ-L100 are shown in Table 3. Its characteristics are as follows:
(1) High density.
(2) Small TEC, like metals and ceramics.
(3) High tensile strength and high Young's modulus which are about three times those of the conventional polyimide PIQ.
(4) High thermal decomposition temperature.
(5) Low absorbed moisture content and low water vapor permeability.

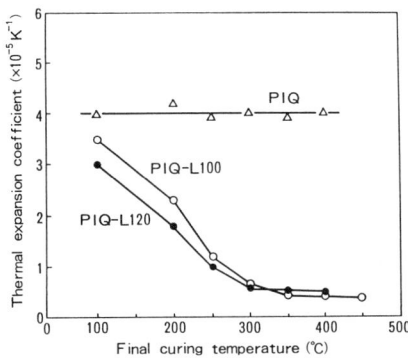

Figure 11 Thermal expansion coefficients of polyimide films cured under various conditions.

Table 3 Properties comparison for PIQ-L100 and PIQ

		PIQ-L100	PIQ
Density	(g/cm^3)	1.47	1.38
Coefficient of thermal expansion	(x10^{-5} K^{-1})	0.3	4.5
Tensile strength	(MPa)	380	130
Tensile elongation	(%)	22	20
Young's modulus	(GPa)	11	3.3
Decomposition temperature in air	(°C)	510	440
Activation energy of thermal decomposition in air	(kcal/mol)	50	35
Absorbed moisture content	(%)	1.3	2.3
Permeability constant	(x10^{12} g cm/cm^2 s cmHg)	0.5	20

APPLICATIONS

As mentioned above, the low thermal expansion polyimide PIQ-L100 has no problems with thermal stress, in addition to its other excellent properties. Therefore, it should have applications in a wide variety of fields. For example, it is expected to be very useful as interlayer dielectrics of VLSIs, barriers against alpha particles for memory devices, substrates for flexible printed circuit boards and magnetic disks, etc. A new multilevel interconnection system for fabricating submicron VLSIs by future technology(18,19) and adhesiveless high quality flexible printed circuits boards (FPC) (20) are described below.

Interlayer Dielectrics of VLSIs

LSIs using polyimides as interlayer dielectrics have been manufactured since 1973(21). The most significant advantage of

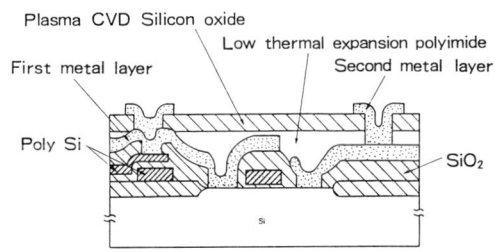

Figure 12 Schematic representaion of new multilevel interconnection system.

the polyimide interlayer system is its planar structure, compared with inorganics such as SiO₂. However, it has a problem in reliability against moisture penetration, which is inevitable with organic materials.

But, when the low thermal expansion polyimide is used, moisture penetration can be prevented with an inorganic layer fabricated on the polyimide layer, as shown in Figure 12. By contrast, when the inorganic layer is fabricated on the conventional polyimide having a high TEC, many cracks are produced, because of thermal stress due to difference in TECs. The results for reliability using resin molded test sample devices show that no failures are observed even after a 2500 h PCT (Pressure Cooker Test, at 120°C and 2 atm of water vapor) and after 2000 temperature cycles (-50 to 150°C).

Therefore, this new multilevel interconnection system has both high planarizability and very high reliability against moisture and heat cycle changes. This system is expected to be a future technology for submicron VLSIs.

Flexible Printed Circuit Boards (FPC)

Ordinary FPCs are fabricated by lamination of copper foil and polyimide film by adhesive and patterning the copper layer by photo lithography. However, they have problems in thermal stability, because they need thermally unstable adhesives with low Tgs to reduce thermal stress. Therefore, a new thermal stable FPC has been desired for some time.

FPCs are one good application for the low thermal expansion polyimides. Because their TECs can be controlled easily to match that of copper foil, problems in thermal stress can be solved, and then high thermally stable adhesiveless FPCs can be obtained.

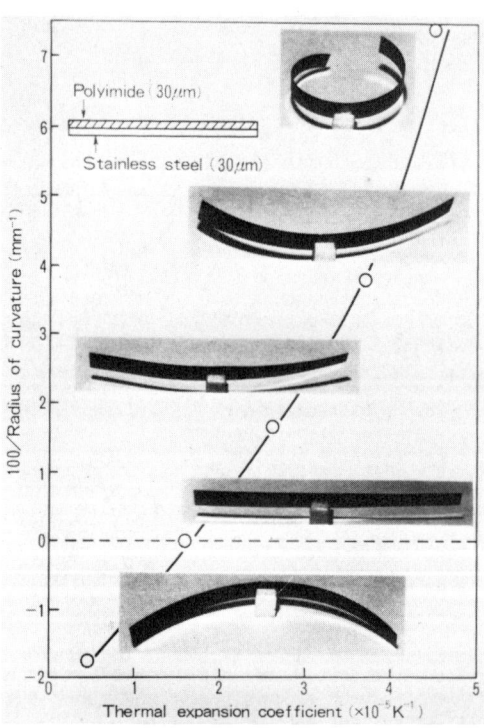

Figure 13 Relation between thermal expansion coefficient of polyimide film and radius of curvature of stainless steel.

Figure 13 shows the relation between the reciprocal radius of curvature of the copolyimide coated stainless steel foil and TEC of the copolyimide. The TEC of the stainless steel is 1.8 x 10⁻⁵ K⁻¹. In the case of high TEC copolyimides coated stainless steel, seriously curling is produced. In the case of low TEC polyimide, the stainless steel foil is curved in the other direction. When the TEC of copolyimide is like that of the substrate, a flat shape is obtained. Then this shape is kept at both elevated and very low temperatures.

Table 4 Properties comparison for new and conventional FPCs

Items		New FPC	Conventional FPC
Dimentional stability			
Thermal	(%)	0.15	0.4
Humid	(1/RH%)	1.0×10^{-5}	2.3×10^{-5}
Glasstransition Temperature	(℃)	350	390
Soldering stability		350℃/>1 min	260℃/1 min
Flexural Fatigue	(cycles)	120	70
Peel strength at 200℃	(kg/cm)	1.7	0.1
Tensile strength	(MPa)	350	200
Elongation	(%)	35	60
Modulus	(GPa)	6.0	3.5
Content of absorbed moisture	(%)	1.4	2.0
Surface resistivity	(Ω)	10^{14}	10^{14}
Dielectric constant	(−)	3.3	3.5
Dissipation factor	(−)	0.5	1.0

The TEC of copper foil is very close to that of stainless steel. Therefore, when the same copolyimide and the copper foil are combined into one body, a flat shape is obtained, too. Figure 14 shows a FPC fabricated using this approach. Its properties are summarized in Table 4. It has excellent dimensional stabilities at high temperature and in high humidity. There are no problems from soldering at about 300 °C. This enables connections between FPC and rigid low TEC glass or ceramic printed circuit boards, which are used in large scale computers, liquid crystal TVs, etc. Also, it enables direct bonding of silicon based semiconductor devices on this FPC. Another advantage of this FPC is its very high peel strength between copper foil and the polyimide film even at elevated temperatures, as shown in Figure 14.

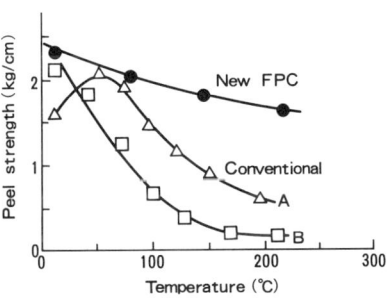

Figure 14 Peel strength between copper foil and polyimide film for various FPCs.

REFERENCES

1) S.Numata, S.Oohara, J.Imaizumi, and N.Kinjo, Polym. J., 17, 981 (1985)
2) S.Numata, S.Oohara, K.Fujisaki, J.Imaizumi, and N.Kinjo, J.Appl. Polym. Sci., 31, 101 (1986)
3) S.Numata, D.Makino, and N.Kinjo, Polym. Eng. Sci., in press.
4) S.Numata, K. Fujisaki, and N. Kinjo, Polymer, in press.
5) C.E.Sroog, J.Polym.Sci.:Macromole.Revs., 11, 161 (1976)
6) N.Inagaki and K.Nakagawa, Eng. Mat., Japan, 32 (4), 92 (1984)
7) Y.Takeuchi, F.Yamamoto, and Y.Shuto, Macromolecules, 1986, 19, p.2059
8) S.Rojstaczer, D.Cohn, and G.Marom, J.Mater.Sci.Lett., 3, 1028 (1984)
9) G.A.Orchard, G.R.Davise, and I.M.Ward, Polymer, 25, 1203 (1984)

10) R.A.Dien-Hart, and W.W.Wright, Makromol.Chem., 143, 189 (1971)

11) S.T.Wellinghoff, H.Ishida, J.L.Koenig, and E.Baer, Macromolecules, 1980, 13, 834

12) M.I.Bessonov, N.P.Kuznetsov, N.A.Adrova and F.S.Florinskii, Polym.Sci. U.S.S.R., 16, 2425 (1974)

13) G.L.Slonimskii, A.A.Askadskii, and A.I.Kitaigorodskii, Vysokomol. soyed.,A12, 494 (1970)

14) S.Numata and N.Kinjo, Proc.2nd SPSJ IPC, p.75, Tokyo, Aug., 1986
   S.Numata, K.Fujisaki, and N.Kinjo, Polymer, in press.

15) S.Numata, S.Oohara, J.Imaizumi, and N.Kinjo, Polymer Preprints, Japan, 34, No.4, p.888 (1985)

16) S.Numata, K.Fujisaki, and N.Kinjo, Polymer Preprints, Japan, 36, No.3, p.734 (1987)

17) S.Numata, K.Fujisaki, and N.Kinjo, Polyimides; Synthesis, Charaterization, and Applications, vol.1, pp.259-271, K.L.Mittal, Editor, Plenum Press, New York, 1984

18) Y.Misawa, N.Monma, S.Numata, and N.Kinjo, Semiconductor World, Japan, 5, No.11, 40 (1986)

19) Y.Misawa, N.Monma, S.Numata, and N.Kinjo, IEEE Trans. Electron Devices, ED-34, 621 (1987)

20) N.Kinjo and S.Numata, Nikkei Micro Devices, Japan, 1987, No.4, p.133

21) K.Sato, S.Harada, A.Saiki, T.Kimura, T.Okubo, and K.Mukai, IEEE Trans.Parts, Hybrid and Packaging, PHP-9, 176, Sep. 1973

ADVANCED MATERIALS FOR PRINTED CIRCUIT BOARDS

DAVID WEI WANG*
*IBM Corporation, Systems Technology Division, 1701 North Street,
Endicott, New York 13760

INTRODUCTION

The printed circuit board is an integral part of the electronic packaging hierarchy. Its use began more than 40 years ago, and the demand for printed circuit boards has increased in parallel with the growth of the electronics industry.[1] According to a recent forecast, the worldwide production of printed circuit boards will reach to over 19 billion U.S. dollars' worth by 1990.[2] With continuing demands for more interconnections, the multilayer circuit board industry is experiencing its fastest growth rate. Boards with more than 20 inner planes of circuitry are being manufactured with high reliability.

Based on dollar values, more than 90% of the circuit boards produced are in the rigid board category, where starting materials are based on thermosetting prepregs produced by a solution impregnation method. This article is a review of materials currently used in rigid composites.

PRINTED CIRCUIT BOARD LAMINATES

Except for a small volume of injection molded boards and flexible circuits, basic materials of a printed circuit board are copper foil clad laminates (Table I). From resin and glass cloth raw materials to finished boards, the manufacture of printed circuit composites is a rather complicated process. Its success depends on an effective integration of a wide range of materials, equipment, and engineering processes.[3,4] Traditional circuit board laminates are made from epoxy glass prepregs. However, with packaging trends leading to increased circuit densities, shorter propagation delays, elevated operating temperatures, and higher reliability, many advanced materials and processes are being developed to satisfy these requirements (Table II). New developments in these areas will have a great overall impact on interconnection technology.

RESIN SYSTEMS

FR-4 Resins

For more than 25 years, a variety of dicyandiamide (dicy) cured brominated epoxies, commonly known as FR-4 resins, has been in use as the matrix resin for circuit board laminates. This two-part resin system satisfies the need for a variety of products, ranging from simple single-sided boards to multilayer composites for mainframe computers. The resin, a mixture of bisphenol A type epoxies, has a bromine content of about 15-20% by weight to impart flame retardancy. Some formulations also include higher functionality resins to raise the glass transition temperatures (Tg)[5.6]. Typically resins are supplied in ketone solutions such as methyl ethyl ketone (MEK) or acetone. Because of the its poor solubility, dicy, the curing agent, has to be dissolved in another more polar co-solvent. Glycol mono-ethers and dimethylformamide (DMF) are often the choice. A small amount of amine catalyst, such as imidazoles, benzyldimethylamine,

Mat. Res. Soc. Symp. Proc. Vol. 108. ©1988 Materials Research Society

Table I.   Printed Circuit Laminate Materials

RESIN	REINFORCEMENT
Phenolic	Kraft Paper
Epoxy	Glass Fabric
Polyester	Quartz (Fused Silica) Fabric
Polyimide	Kevlar Fabric, Paper a
Bismaleimide-Triazine Resins	PTFE Fabric
Cyanate (Triazine) Resins	Copper/Invar/Copper, Kovar
Cross-linked Hydrocarbon Polymers	Alloy 42
Oxazolidone-Based Thermosets	Stainless Steel
Polytetrafluoroethylene (PTFE)	Graphite Fiber
Thermoplastics	Silicon Carbide/Aluminum

Table II.   Trends in Printed Circuit Board Packaging

REQUIREMENTS	TRENDS
Ease of Resin Handling and Treating	One Package/Solution Safe Low Boiling Point Solvent System Less Prepreg Defects (i.e. Voids, Recrystallization) Injection Molding Continuous Lamination
Better Prepreg Flow Control	Wider Processing Window Less Residual Stresses Lower Void Content
Higher Density	Fine Line and Space Surface Mount Multilayers More Interplane Connections Thinner Dielectric Spacing Smaller Holes, Higher Aspect Ratio
More Power	Heavier Copper Planes Better Thermal Management
Impedance Control	Thin, Uniform Dielectric Layers
Low Er Materials	Faster Signal Speed, Rise Time
Lower CTE	Surface Mount Direct Chip Attach Better PTH Reliability
Better Moisture Resistance	Improved Dimensional Stability and Insulation Resistance Lower Delamination Rate
Improved registration	Improved dimensional stability Improved impregnation methods
Higher Heat Resistance	Component Solder, Solder Rework Fine Line Repair
Less Smear	Simpler Hole Clean Operations

tetramethylenebutanediamine, etc. also is added to facilitate the curing reactions.

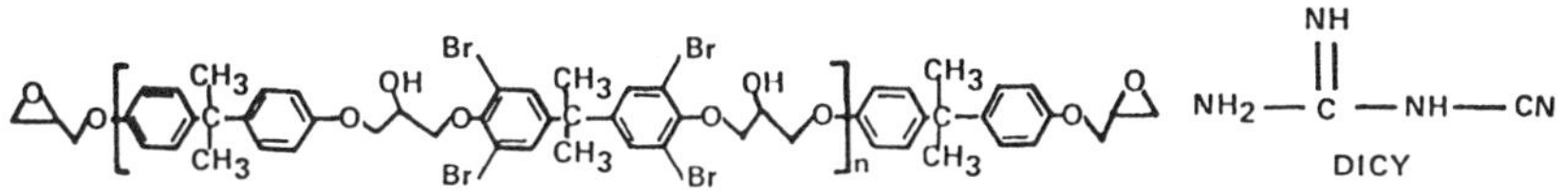

Brominated Epoxy Resin
Figure 1

After mixing, the FR-4 varnish requires several hours of aging before it can be used for solution impregnation. During the treatment process, resin is advanced to B-stage and prepreg's pick-up, flow and gel time are adjusted. A typical FR-4/E-glass prepreg has a Tg above 50°C, allowing room temperature storage for at least three months. The gel times are less than 5 minutes at above 150°C, and the pick-up values range from 40-75%. This is quite different from prepregs used in the advanced composite industry where properties such as low pick-up, long gel time, tack, and drap are usually emphasized.

Both hydraulic and autoclave presses are being used for circuit board lamination. The FR-4 prepregs reach full cure within 60-90 minutes when heated at 150°C-185°C. The physical properties of a typical FR-4 laminate are summarized in Table III.

Currently FR-4 is the resin most widely used in the printed circuit industry. However, there are several disadvantages related to its properties and processes:

1. It's a two-part resin system. The varnish requires aging, which doesn't have a mix-and-run capability.
2. Resin has a Tg well below the soldering temperature, most commonly around 250°C, used in assembly processes.
3. Dicy has to be dissolved in polar solvents, which generally causes environmental and health concerns.
4. Dicy has poor solubility in epoxies. Recrystallization often occurs during impregnation resulting in a heterogeneous B-staged system.
5. Resin generates a considerable amount of smears during drilling. Extensive hole cleaning has to be carried out in order to achieve reliable PTH interconnections.
6. Laminate moisture uptake affects dimensional stability, solder resistance, and insulation resistance.

Many of these disadvantages could be attributed to a lack of understanding of the dicy curing mechanism, which makes the control of reaction difficult and subsequently affects the properties of cured resin.[7-11] In addition to the relatively low Tg, some of the drawbacks of FR-4 are its limiting use in high-density multilayer boards, especially in the military and high-end computer areas.

High Glass Transition Temperature Resins

In a multilayer structure, good dimensional stability and correct registration are the primary concerns in achieving reliable interconnections. To minimize dimensional changes, a board material ideally should have a glass transition temperature higher than the maximum temperature encountered in the process. Therefore, a series of high Tg resins have been developed to meet the packaging needs. In general higher Tgs offer the advantage of better heat and solder resistances, and less total expansion during soldering. Materials such as polyimides also appear to produce less smear during drilling. One major disavantage is that using curing temperatures higher than in the $180^\circ C$ range may require special tooling and processes.

1. Polyimides

Because of their excellent high temperature resistance and good dielectric properties, many applications have been found for polyimides in electronic packaging. Two types of polyimides are widely used and can be differentiated by their chemical structure and cure mechanism.

Condensation Type Polyimides

Condensation type polyimides are often high molecular weight linear chain polymers. They produce volatile byproducts during the final stage of imidization reaction. The properties of a variety of polyimides have been studied extensively. In general, wholly aromatic polyimides are not soluble and not melt processable. Better processing capabilities can be achieved only by modifying the polymer chain structure. Condensation polyimides with a stiffer backbone also exhibit interesting low thermal expansion properties.[12] A polyimide derived from pyromellitic anhydride and 4,4'-bis(aminophenyl)-oxide(oxydianiline), commonly known as PMDA-ODA polyimide, is widely used in the electronics industry. It is available in film or polyamic acid (a precursor) solution form. The applications are primarily in first level packaging and flexible circuitry areas.

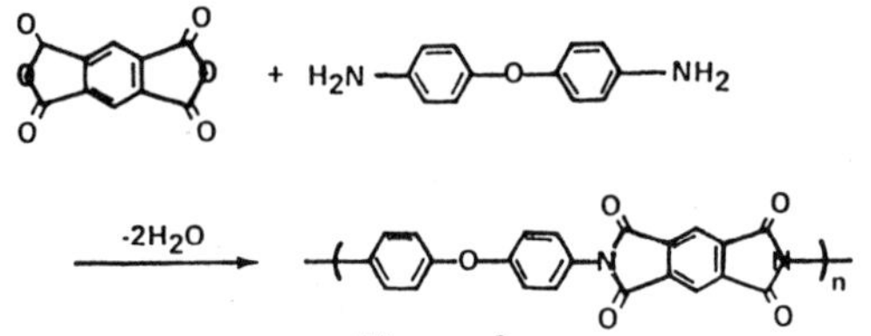

Figure 2

Addition Type Polyimides

Addition type polyimides have preimidized chemical structures that don't produce volatiles during cure.[13] Because they exhibit excellent thermal resistance and can be processed using the same equipment as conventional thermosets, this class of resins is finding increasing applications in aerospace and electronics industries. Acetylene-terminated polyimides are mostly used in advanced composites, and in chip and first level electronic packaging.[14] For printed circuit board applications, several commercial resins are commonly based on bismaleimide structure II wherein two maleimide groups are connected by a linking group R.

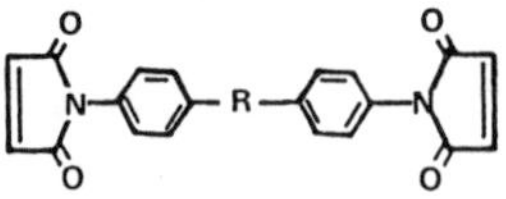

Figure 3

Maleimide undergoes homopolymerization through a radical or anionic mechanism to form brittle infusible materials.[15,16] It copolymerizes with allyl compounds without the need of free radical initiators.[17] Being a reactive dienophile, bismaleimides also react with divinyl benzene, bis-cyclopenta-dieneone, etc., to form Diels-Alder adducts.[18]

More commonly, through Micheal addition to the maleimide double bond, a diamine curing agent -- for example, methylene dianiline (MDA) -- is used to form cross-linked polymers.[19] Epoxies also can be blended with this type of resin to further modify its properties.[20,21] Kerimid® 601 is the polyimide thermosetting resin currently most in use.[22,23] The proposed cure mechanisms of an amine-cured bismaleimide resin are shown in Figure 4.

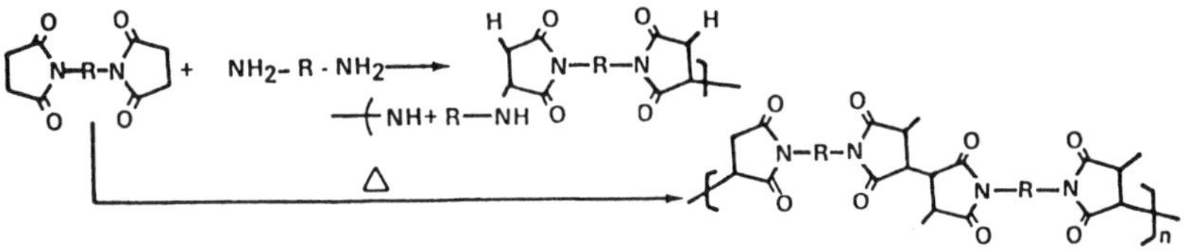

Figure 4

Since bismaleimides exhibit low solubilities in common organic solvents, high-boiling, polar solvents such as N-methyl pyrrolidone(NMP), dimethyl acetamide (DMAC), and DMF are frequently the choice in solution impregnation work. A relatively new resin system, Compimide™ 183 doesn't use MDA as the curing agent and appears to have better solubility characteristics.[24] Hitachi Chemical's polyimide resin reportedly can be processed in an ethylene glycol monomethyl ether/MEK mixture, but currently is only marketed as prepreg in the U.S.[25] The properties of three polyimide laminates are listed in Table IV. Laminate properties of circuit boards are strongly influenced by several factors, such as resin content, style of fabric, construction of test samples, test methods, etc. Values shown in this article are based on published results. Specific sample information shall be referred to the corresponding references.

In general, the process parameters of polyimide boards are quite different from those of conventional epoxies. Post cure is often required to obtain desirable properties. Since the polyimide group can be hydrolyzed by base, these laminates have to be processed in non-alkaline systems. With high Tgs and excellent thermal resistance, polyimide laminates exhibit much improved delamination resistance and reworkability. While the moisture absorption is higher than that of FR-4, the process is readily reversible. Better copper peel strength values are frequently achieved by the use of special grades of copper foils. Kerimid 601 also is known to exhibit low resin smear during drilling. This might be attributed to resin's high Tg and relatively brittle mechanical properties.

Currently, all polyimide resins and prepregs are supplied by foreign sources. Since the cost of polyimide resin is much higher than that of FR-4 resin, polyimide board applications are mostly limited to military and mainframe computer areas.

2. Cyanate (Triazine) Based Resins

The development of cyanate technology began in the 1960s. Monomers were first successfully synthesized from aromatic phenols.[27] Among several different routes for preparing cyanates, reaction of phenols with cyanogen halides is the most important process.[28] A cyanate resin

---

® Kerimid is a registered trademark of Rhone-Poulenc S.A.
™Compimide is a trademark of Boots Co. PLC.

Table III. Physical Properties of FR-4/E-Glass Laminates

Tg(C)	$110^{\circ}-130^{\circ}$
Dielectric Constant Er (1MHz)	4.0-5.5
Dissipation Factor (1MHz)	0.02-0.03
Surface Resistivity (Ω)	$10^{14}-10^{15}$
Copper Peel Strength (lb/in)	10-12
Water Absorption (24hr/RT)	0.3%
CTE (Below Tg ppm $^{\circ}$C) X,Y	16-20
CTE (Below Tg) Z	50-70

Table IV. Physical Properties of Polyimide Laminates

	Kerimid[23] 601	Hitachi[25] MCL-I-67	COMPIMIDE 183[26]
Cure Condition ($^{\circ}$C/Hrs)	180/2+P.C.	170/1.5	180/3+P.C.
Tg ($^{\circ}$C)	240-270	210-220	270
Er (1MHz)	3.9-4.5	4.5-5.0	4.6-4.7
Dissipation Factor (1MHz)	0.005-0.006	0.01-0.015	0.01
Surface Resistivity (Ω)	10^{14}	$10^{14}-10^{15}$	
Cu Peel Strength (lb/in)	6-8	8.4-9.0	6.7-7.8
Water Absorption (24hrs/RT)	0.3%	0.15-0.4%	1.0%
CTE (ppm/$^{\circ}$C) X,Y	13-15		15
CTE (ppm/$^{\circ}$C) Z	40-60	50	40

Table V. Physical Properties of BT and Cyanate Laminates

	BT GHPL-810[30]	CYANATE/EPOXY(4/6)[31]
Cure Condition ($^{\circ}$C/Hrs)	177/2	177/1
Tg ($^{\circ}$C)	200-230	183
Er(1 MHz)	4.2	4.4
Dissipation Factor(1 MHz)	0.005-0.01	0.011
Surface Resistivity (Ω)	$10^{12}-10^{14}$	
Cu Peel Strength (lb/in)	8-9	9.8
Water Abasorption(24 hr/RT)	0.1-0.4%	0.1%
Flammability	VO	VO

most commonly in use is based on dicyanates derived from bisphenol A, which is also a raw material for the epoxy resins.

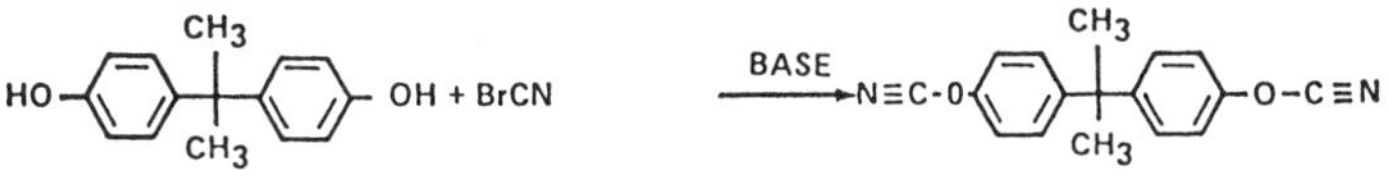

Figure 5

The first cyanate resin, Traizine A, was introduced to the circuit board industry in the mid 1970s.[29] By blending with methylene dianiline bismaleimide, Mitsubishi Gas Chemical Company later developed a BT (Bismaleimide-Triazine) resin system.[30] Pure, unblended materials weren't available until 1985 when Interez Inc. marketed a series of resin systems for both circuit board and advanced composite applications.[31]

Aliphatic cyanates readily isomerize to the corresponding isocynates and then subsequently can trimerize to form isocyanurates.[32] Catalyzed by organo-metallic salts, aromatic cyanates directly trimerize to form a stable cyanurate, s-triazine, ring structure.[33] Although in both cases, based on calculations, the isocyanurate form is thermodynamically more stable, aryl cyanates and aryl cynurates don't seem to isomerize.[34] Cyanate resins exhibit good solubility in organic solvents. They can be mixed with a wide variety of themosets such as epoxies, bismaleimides or acrylates to form compatible blends. Mixing with thermoplastics usually results in interpenetrating network type structures (see Figure 6).[35]

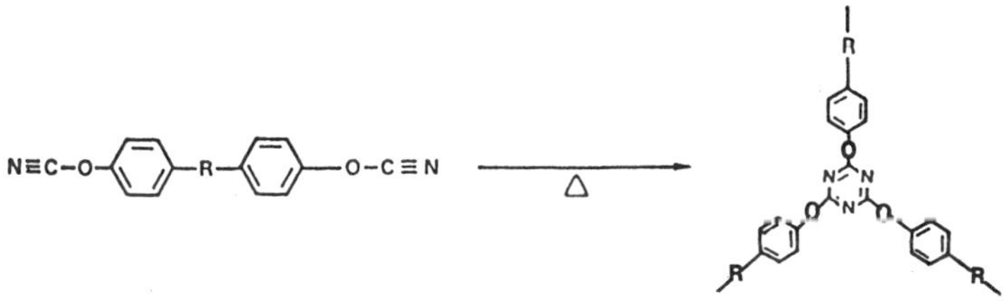

Figure 6

After being fully cross-linked, both cyanates and BT resins exhibit Tgs of at least 250°C. Usually they are blended with brominated epoxies up to 60% by weight to modify the Tg and achieve flammability rating. When an epoxy blend is used, biscyanates, in additon to trimerizing, copolymerize with the epoxy group to form oxazoline type structures. In the case of BT systems, a free radical initiator, such as cumyl peroxide, has to be added to promote the bismaleimide polymerization. In addition to the catalyst level, curing of the cyanate/epoxy system is influenced by the ratio of cyanate, epoxy, and hydroxyl groups in the resin mixture. Many complicated heterocyclic ring structures have been proposed as the products of BT-epoxy resin reactions, but it's difficult to verify them by spectroscopic methods. BT resins with high bismaleimide content isn't completely soluble in MEK. However, adding epoxies to the mixture seems to help in forming a stable MEK varnish.[36]

Table V summarizes the properties of cyanate-based laminates.

Blending with a high level of epoxies allows BT and cyanate resins to reach full cure at 177°C. Influenced by the curing chemistry, their flow characteristics are quite different from that of FR-4 resin.

Laminates are known to exhibit good resistance to conductive anodic filament (CAF) growth phenomenon.[37] They exhibit good moisture and delamination resistance and can be processed according to conventional methods. Copper peel strength is adequate, although a special grade of copper foil also is available. Drilling parameters have to be adjusted when processing thick BT laminates.

3. High Glass Transition Temperature Epoxies

Recently a series of non-dicy cured epoxies have been introduced by resin manufacturers. Quatrex™ 5010 and RSM-1151 are marketed by Dow Chemical Co. and Shell Chemical Co., respectively.[38,21] Both resins are one-package systems with Tgs in the 170-190°C range. Compared with FR-4 laminates, they exhibit higher thermostability. Moisture and solder resistances also are improved. Without needing high-boiling solvents and post cure, these resins appear to show some processing advantages over conventional polyimide resins.

REINFORCEMENT MATERIALS

Various kinds of reinforcement materials, ranging from kraft paper to graphite fibers, have been used for building printed circuit boards. They serve the purpose of providing mechanical strength to support organic dielectric materials and of controlling dimensional stability. Using each material's unique mechanical and thermal properties, one can tailor circuit board composites to reduce the CTE mismatch between different levels of interconnections. The dielectric constant and CTE of several common reinforcement materials are shown in Table VI.

Glass Fabrics

The majority of circuit boards are made from E-glass reinforced epoxies. Other grades of glass, such as S-glass and D-glass are only produced in much smaller volumes. They differ in chemical composition and physical properties (Table 7). A wide selection of glass fabrics with different thickness, weight, and yarn constructions are available. Light weave fabric made of fine yarn can produce prepregs with a final pressed thickness as thin as 0.001-0.002 inch. Table VIII lists the characteristics of some typical glass fabric styles.

Coupling Agents

Coupling agents are difunctional compounds in which each end interacts with either the resin or the surface of reinforcement materials. Organosilanes are the most commonly used coupler for glass fabrics.[40] They usually are applied as monolayers onto the glass surface and serve as an adhesion promoter between resin matrices and glass. By improving interfacial adhesion, couplers also have an effect on reducing the moisture absoption of laminates. To achieve good bonding strength, coupling agents are often developed for each special resin system. Examples of several frequently used coupling agents are shown in Table IX.

LOW DIELECTRIC CONSTANT LAMINATES

To meet the ever-increasing demand of data-processing speed, low

™Quatrex is a trademark of the Dow Chemical Company.

Table VI.   Properties of Reinforcement Materials

	Er(1 MHz)	CTE (ppm/C)
Glass	3.6-5.8	3.1-5.0
Kevlar Fabric	3.8	x,-2.0; z,60
Quartz (Fused Silica)	3.8	0.54
Cu/INVAR/Cu	Conductive	2-6
ALLOY 42	Conductive	4.0-5.2
Stainless Steel-400 Type	Conductive	10-11
Graphite Fiber	Conductive	x,-0.5- -0.9 y,17

Table VII. Properties of Fiberglass [39]

	E	S	D	Fused QUARTZ
Specific Gravity	2.54	2.49	2.16	2.2
TENSILE PROPERTIES				
Strength PSI X 10^3	500	665	350	500
Modulus PSI X 10^6	10.5	12.4	7.5	10
Elongation, %	4.8	5.4	4.7	
THERMAL PROPERTIES				
CTE PPM/^{o}C	5.0	2.9	3.1	0.5
Softening Point, ^{o}C	841	860	771	1670
ELECTRICAL PROPERTIES				
Er 1 MHz	5.8	4.53	3.56	3.8
Dissipation Factor 1 MHz	0.001	0.002	0.0005	0.0001

Table VIII. Glass Fabric Characteristics

STYLE	YARN COUNT	THICKNESS(IN)	WEIGHT (OZ/SQ YR)	WEAVE
108	60 X 47	.0022	1.40	PLAIN
1165	60 X 52	.0040	3.53	PLAIN
7628	44 X 32	.0067	6.00	PLAIN

Table IX.   Glass Fabric Finishes

TRADE NAME/DESIGNATION	STRUCTURE	RESIN
VOLAN®A	Methacrylic Chromic Chloride Complex	Polyesters, Epoxies
Z6040[1]	$CH_2-CH-CH_2-O-(CH_2)_3-Si(OCH_3)_3$ (with epoxide O)	Epoxies
Z6032[2]	$CH_2=CHC_6H_4CH_2NHCH_2CH_2NH(CH_2)_3Si(OCH_3)_3 \cdot HCl$	Epoxies
I-329[3]		Polyimides
CS-450[4]		Cyanates, BT

®VOLAN A is a Registered Trademark of E. I. DuPont de Nemours and Co.
1) Supplied by Dow Corning Corp.
2) Supplied by Burlington Glass Fabrics Co.
3) Supplied by Clark-Schwebel Fiber Glass Corp.

Table X.   Comparison of Low Dielectric Laminates [a]

MATERIAL	DIELECTRIC CONSTANT	DISSIPATION FACTOR
PTFE (10GHz) [42]	2.1	0.0004
PTFE/Microfiber (10GHz)[42]	2.2	0.0008
RO2800 (10GHz)[42] [b]	2.8	0.002
RO2500 (10GHz) [b]	2.5	0.0025
PTFE/Kevlar [43]	2.2-2.5	
PTFE/Glass	2.3-2.8	0.0008-0.002
Polyimide/Fused Quartz[43]	3.4	0.005
Polyimide/Kevlar[43]	3.6	0.008
BT-Polyimide/Fused Quartz[44]	3.0	0.0037
BT CCL-H870/Glass[44]	3.8-3.9	0.0017-0.0023
Rexolite 2200 [c]	2.6	0.001
Gore-Ply™ FR-4 Laminate[45]	2.6-2.8	

a. Measured at 1 MHz unless otherwise indicated.
b. Manufactured by Rogers Corp.
c. Manufactured by Norplex/Oak Corp.
d. ™ Gore-Ply is a trademark of W.L. Gore Associates, Inc.

dielectric constant materials are needed to decrease signal propagation delay and to reduce rise time. The relationship of the propagation delay of a transmission line with the effective dielectric constant of a circuit board is shown as follows:

$$t=3.33 \sqrt{Er} \text{ (picosecond/mm)}$$

From a structure-property relationship point of view, on the molecular level, a low dielectric material ideally should have a symmetrical structure resulting in minimum dipoles. It also should contain no ionic or charged species. Equally desirable are physical properties such as good dimensional stability, low moisture absorption, and low dependence of electrical properties on temperature and frequency. Among various kinds synthetic materials, fluoropolymers are the ones most commonly in use for building high-speed, low-loss boards requiring a dielectric constant less than 3.[41] In addition to glass, other types of reinforcements also are extensively used in this area. They include: Kevlar®, quartz (fused silica), ceramic, polytetrafluoroethylene (PTFE), etc. Properties of some examples of low dielectric constant laminates are summarized in Table X.

Although PTFE polymer exhibits excellent electrical properties, the material is thermoplastic, which requires a melt-processing temperature above 300°C. Low modulus, high CTE and poor dimensional stability also make its use difficult in a multilayer strucuture. Special drilling and surface activation processes also are needed.[46]

In the thermoset materials area, low dielectric constant resins are normally based on cross-linkable polyhydrocarbons such as polybutadienes and various styrenes coplymers.[47-50] Three cyanate based resins, BT2425G, BT2060BG, and Dow XU71787, also are marketed as low dielectric constant polymers.[51] A relatively new material, Gore-Ply prepregs are based on thermosetting resins reinforced with fabrics of expanded PTFE.[45] These offer some advantages in achieving a dielectric constant of 2.5-2.9 without making extensive changes in the conventional process.

LOW EXPANSION BOARD TECHNOLOGY

In the high-performance board area, the trend is to use surface mount technology to achieve high-density interconnection. Traditional E-glass reinforced boards exhibit CTEs ranging from 12 to 20 ppm/C. This value, although much higher than that of alumina ceramic's 6-7 ppm/C, is considered acceptable for attaching pin-grid type chip carriers. For surface mount components, especially leadless ceramic chip carriers and chip-on-board applications, one will need low-expansion boards to reduce the CTE mismatch.

Being the higher modulus member in circuit board composites, reinforcement materials play a key role in controlling the overall CTE. Fibers such as quartz (fused silica), Kevlar, and graphite are commonly used in low CTE board construction. Metal restraining cores include copper-Invar®-copper, Kovar, Alloy 42, stainless steel, etc. Conductive metals and graphite fiber-reinforced cores are used as power and ground planes.[52-55] Good thermoconductivity also makes them useful as heat sinks. Examples of low CTE laminates are shown in Table XI.

Quartz (fused silica) fiber, with excellent electrical properties and a low CTE, is an ideal electronic packaging material. However, it's expensive and supplied by only a single foreign source. Because of its hardness, the material also is difficult to drill. Another silicate based material, mica, also has been used for low CTE controls.[57-60]

®Kevlar is a registered trademark of E.I.DuPont de Nemours and Co.
®Invar is a registered trademark of the Carpenter Technology Corporation, Reading, PA.

Kevlar, a high modulus synthetic aramid fiber, has been extensively used in various kinds of composite applications. Products based on non-woven Kevlar paper also are available. Having the lowest CTE values, compared with other commercially available fibers, make it ideal for low expansion control.[61,62] However, the processing of Kevlar boards has to overcome serveral difficulties such as high moisture absorption of the fiber, rolled fiber in the PTH after drilling, high z-axis expansion, and poor adhesion between fiber and matrix resins.

Invar is an alloy of iron (64%) and nickel (36%), which exhibit a CTE of 0.5 ppm/C in the x-y plane. The CTE of copper-clad material varies from 2-6 ppm/C, depending on copper thickness. They are used as restraining ground and power planes. Good thermoconductivity also makes them a good heat sink. Special drilling and etching methods are needed to process boards containing copper-Invar-copper planes.[63-65]

Table XI. Comparison of Low CTE Laminates

MATERIAL	CTE (ppm/^{o}C)	
	X,Y	Z
Epoxy/Fused Quartz [43]	5-7	
Polyimide/Fused Quartz [42]	6-8	34
BT/Polyimide/Fused Quartz [44]	6	27
Epoxy/Kevlar [56]	9-10	89
Epoxy/Kevlar [43]	6-7	
Epoxy/Polyimide/Kevlar [56]	2.7-3.0	76
Polyimide/Kevlar [42]	3-7	83
Epoxy/Glass-Cu/Invar/Cu [56]	10-11	65
Epoxy/Graphite Fiber Laminate [52]	3-5	
Alumina Ceramic	6-7.5	
Silicon Chip	3	
Copper	17	

INJECTION MOLDED BOARDS

High molecular weight thermoplastic polymers are used for fabricating injection molded boards. They require high melt processing temperatures and special circuitization processes different from that of conventional copper-clad laminates. One of the biggest advantages of injection molded boards is structure and functional versatility, which allows the integration of various features such as connectors, clips, recesses, etc. into one package. The ability of molding stucture with via holes also eliminates the need for drilling and punching.

Engineering thermoplastics, mostly compounded with inorganic fillers, have heat distortion temperatures often higher than 200°C, thus resulting in good heat and solder resistance. In general, they also exhibit better electrical properties; for example, lower dielectric constant, dissipation factors, and higher resistivities, than FR-4 laminates. Resins frequently used for molded board applications include polyetherimide, polyethersulfone, polyarylsulfone, and polysulfone. Other materials such as polyphenylenesulfide, polyetheretherketone, polyamideimide, aromatic polyesters, polyethyleneterephthalate, acrylonitrile-butadiene-styrene copolymer, etc. also are been studied. The physical properties of molded materials are closely dependent on tooling and processing conditions. In the case of crystalline polymers, their properties are further influenced by the crystallinity and morphology achieved during processing. Materials selection is determined by specific end use requirements and available processing technology.

Although current technology allows the production of three-dimensional circuitry, complex multilayer structures are not available. For two-dimensional boards, the size is also generally limited to less than one square foot. In addition to achieving good dimensional tolerance, the key area in molded boards is circuitization technology. This includes new techniques in photoprocessing, surface activation, and plating areas. With the relatively high start-up tooling cost, injection molded boards are most suitable for high-volume productions.

SUMMARY

To meet requirements for higher circuitry integration and better processability, advanced materials such as high Tg resins, low dielectric constant polymers, and low CTE materials have been developed for use in high-performance boards. Injection molding technology is further expanding the use of thermoplastics in second level packaging. Although implementation of these materials often are influenced by overall packaging strategy and cost factors, development of new materials and processes will continue in order to meet future demands for packaging density and system performance. As a general trend, one can also expect an increase in the use of new specialty polymers in electronic packaging.

REFERENCES

1. M. E. Pole-Baker, Printed Circuit Fabr. 7(12), 26(1984); ibid., 8(1), 103(1985)
2. "PCB Information Service, Quarterly Report," BPA (Technology and Management) Ltd., Dorking, U.K., November 1986
3. D. P. Seraphim, D. E. Barr, W. T. Chen, and G. S. Schmitt, Advanced Thermoset Composites, edited by J. M. Margolis, Van-Norstrand Reinhold, New York (1986), p.110
4. D. P. Seraphim, L. C. Lee, B. K. Appelt, and L. L. Marsh, Electronic Packaging Materials Science, edited by E. A. Giess, K. N. Tu, and D. R. Uhlmann (Mater. Res. Soc. Symp. Proc. 40, 1985) pp.21-47
5. L. N. Chellis, U.S. Pat. 3,523,037(1970)
6. D. R. McGowan, Printed Circuit Fabr. 7(8), 53-59(1984)
7. T. F. Saunders, M. F. Levy, and J. F. Serino, J. Polym. Sci., A-1, 5,1609(1967)
8. S. A. Zahir, Adv. Org. Coat. Sci. Ser.4, 83-102(1982)
9. R. J. Jackson, A. M. Pigneri, and E. C. Galgoci, Electronic Materials and Processes, edited by N. H. Kordsmerier, C. A. Haper, and S. M. Lee (Intl. SAMPE Electron. Conf. Ser., 1, 1987) pp.204-212
10. B. R. LaLiberte, R. E. Sacher, and J. Bornstein, AMMRC-TR-81-30(1981)
11. G. L. Hagnauer and D. A. Dunn, J. Appl. Polym. Sci.26,1837-1846(1981)
12. S. Numata, K. Fujisaki, D. Makino, and N. Kinjo, Recent Advances in Polyimides Science and Technology, edited by W. D. Weber and M. R. Gupta, Mid-Hudson Section Soc. of Plat. Eng. Inc.(1987) pp.164-173
13. D. A. Scola, Intl. SAMPE Symp. Exhib. [Proc] 31,1844-55(1986)
14. P. M. Hergenroter, J. Macrmol. Sci. Rev. Macromol. Chem. C19(1) 1-34(1980); A. L. Landis and A. B. Naselow, Polyimides, Synthesis, Characterization, and Applications, edited by K. L. Mittal, Plenum Press, N.Y. vol.1(1984), p.39
15. E. J. Goldberg, U.S. Pat. 2,958,672(1960); M. Gurffax, and J. L. Locatelli, ibid. 4,111,919(1978)

16. G. F. D'Alelio, U.S. Pat. 3,890,272(1975)
17. S. A. Zahir and A. Renser, U.S. Pat. 4,100,140(1978) 18. F. W. Harris and J. K. Stille, Macromol. 1, 463(1968)
19. J. V. Crivello, J. Polym. Sci. Polym Chem. Ed., 1185-1200(1973)
20. W. F. Graham, U. S. Pat. 4,288,359(1981)
21. A. M. Pignerl, E. C. Galgoci, R. J. Jackson, and G. C. Young, Proc. of Tech. Program, Natl. Electron. Packag. and Prod. Conf. West '87,Anaheim, Calif. 1987, pp.29-45, ibid., in Electronic Materials and Processes edited by N. H. Kordsmerier, C. A. Haper, and S. M. Lee (Internatl. SAMPE Electron. Conf. Ser., 1, 1987) pp.657-667
22. M. Bargain, A. Combet, and P. Grosjean, U.S.Pat. 3,562,223(1971); M. Bargain and P. Gorsjean, ibid.,3,526,764(1971)
23. P. Bednar, IPC Technical Paper, TP-048(1975); J. W. Lamp and D. L. Boaz, AFML-TR-77-210, vol. 1 and 2 (1977)
24. H. D. Stenzenberger, U.S.Pats 4,211,800(1980), 4,303,799(1981)
25 A. Takahashi, S. Yokosawa, M. Wajima, and K. Tsukanishi, IPC Technical Paper, WCIII-14, (1984)
26. H. D. Stenzenberger, M. Herzog, W. Romer, S. Pierce, M. S. Canning, IPC Technical Paper, TP-602(1986); ibid. R. Sheiblich, and K. Fear, Natl. SAMPE Symp. Exhib. [Proc] 30, 1568-84(1985)
27. R. Storch and H. Gerber, Angew. Chem. 72. 1000(1060)
28. D. Martin and M. Bauer, Org. Synth. 61, 35-8(1983)
29. K. K. Weirauch, IPC Technical Paper, TP-066(1975)
30. M. Gaku, IPC Technical Paper, TP-283(1979); BT Resin-Glass Fabrics Multilayer Materials, Mitsubishi Gas Chemical Company, Inc.(1984)
31. F. A. Hudock and S.J.Ising, IPC Technical Paper, TP-618(1986)
32. K. A. Jensen and A.Holm, in Chemistry of Cyanates and Their Thio Derivatives, edited by S. Patai, vol. 1, J. Wiley, New York(1977) pp.569-618
33. D. A. Shimp, Intl. SAMPE Symp. Exhib. [Proc.] 32,1063-1072 (1987); M. Bauer, J. Bauer, and G. Kuhn, Acta Polym. 37, 715-719(1987)
34. V. A. Sergecv, V. K. Shitikov, and V. A. Pankratov, Russ. Chem. Rev. 48(1), 79-93(1979)
35. D. C. Prevorsek and D. C. Chung, U.S.Pat. 4,157,360(1979); D. H. Wertz and D. C. Prevorsek, Polym. Eng. and Sci., 25, 804(1985)
36. L. R. Daley and F. R. Christie, U.S. Pat. 4,456,712(1984)
37. J. P. Mitchell and T. L. Welsher, IPC Technical Paper, WC2-2A5(1981)
38. M. Brody, Printed Circuit Fabr. 8(12), 22(1985)
39 J. G. Mohr, in Handbook of Fillers and Reinforcements for Plastics, edited by H. S. Katz and J. V. Milewski, Van Norstrand, N.Y.(1978) pp.467-503
40 E. P. Plueddemann, Silane Coupling Agents, Plenum Press, New York (1982)
41. R. Keeler, Electron. Packag. & Prod., 26(1), 141(1986); J. Balde and G. Messner, Circuit World, 14(1), 11(1987)
42. D. L. Arthur, IPC Technical Paper, TP-585(1986); ibid. C. S. Jackson "Fluoropolymer Composite PWB for High Speed Digital Applications," presented at Intl. Electron. Packag. Soc. Symp, Orlando, Fl. October 1985.
43. C. A. Haper and W. W. Staley, Electron. Packag. & Prod. 25(11), 52(1985)
44. M. Gaku, N. Ikeguchi, and S. Ayano, Electron. Packag. & Prod. 25(12), 30(1985)
45. D. D. Johnson, U.S. Pat. 4,680,220(1987); ibid. Electron. Packag. & Prod. 27(2), 80(1987)
46. J. Johnson, Printed Circuit Fabr. 9(6), 56(1986)
47. A. J. Beuhler and J. A. Wrezel, U.S.Pat. 4,604,438(1986)
48. A. J. Beuhler, J. A. Wrezel, J. L. Maas, and R. C. Sundahl Jr., Electronic Packaging Materials Science, edited by K. A. Jackson, R. C. Pohanka, D. R. Uhlmann, and D. R. Ulrich (Mater. Res. Soc. Symp. Proc. 72, 1986) pp.223-229
49. N. Sawatari, I. Natanabe, H. Okuyama, and K. Murakawa, IEEE Trans. on Electrical Insulation. EI-18, No.2, 131(1983)

50. M. Saito and Oodaira, Europ. Pat. 0194655(1986) 860917 51. M. Gaku, H. Kimbrara, H. Muramoto, S. Higashida, and F. Sato, WO 8606085(1986); E. P. Loo and D. J. Murry, Europ. Pat. Appl. EP 014,754,882(1985)

52. W. M. Jensen, U.S. Pat. 4,318,954(1982); A. W. Noblett, et. al. IPC Technical Paper, TP-648(1987)

53. J. D. Leibowitz, U.S.Pats 4591659(1986); 4,689,110(1987)

54. J. D. Leibowtiz, W.Winter, and J.Kolkin, Electron. Packag. Prod. $\underline{25}$(6), 86(1985)

55. C. Thaw, R. Minet, J. Zemany, and C. Zweben, SAMPE J. 23(6),40–43(1987)

56. L. Hayes, Printed Circuit Fabr.,$\underline{7}$(12), 49(1984)

57. W. G. Reimann and L. E. Gates,IPC Technical Paper, TP-443 (1982)

58. B. Mahler, IPC Technical Paper, TP-441 (1982)

59. S. N. Hoda, Printed Circuit Fabr.,$\underline{9}$(11),109–112(1986)

60. D. D. Gasparaitis, M. W. Lauroesch, Proc. of Tech. Conf., Sixth Annual Intl. Electron. Packag. Conf. San Diego, CA. 1986, p.27–39

61. C. T. Brooks, Printed Circuit Fabr. $\underline{5}$(2), 32(1982)

62. R. A. Mogle and D. J. Sober, IPC Technical Paper, TP-475 (1983)

63. S. H. Lindblom, IPC Technical Paper, TP-617(1986); F. Gray et. al., $\underline{ibid}$, TP-547(1985)

64. J. R. Hanson and J. L. Hauser, Electron. Packag. Prod., $\underline{26}$(11), 48(1986)

65. F. Gray, L. Cartwright, and S. Lindblom, Printed Circuit Fabr., $\underline{9}$(2), 48(1986)

66. E.F.Timpane, IPC Technical Paper, TP-529(1985)

67. K. Quinn, J. Travis, and J. Ganjie, 19th Annl. Connectors and Interconnection Tech. Symp. Proc. 1986, pp.35–45; $\underline{ibid}$., in Electronic Materials and Processes edited by N. H. Kordsmerier C. A. Haper, and S. M. Lee (Intl. SAMPE Electron. Conf. Ser., 1, 1987) pp.668–680

68. J. Travis and J. Ganjei, Printed. Circuit. Fabr. $\underline{9}$(7), 89(1986)

69. W. Jacobi and M Krisch, $\underline{ibid}$. 50(1986)

70. J. Ganjie and D. C. Frisch, Circuit Manufacturing, $\underline{26}$(6), 39(1986)

MATERIALS FOR HIGH SPEED CIRCUIT BOARDS

R. W. SEIBOLD, R. T. LAMOUREUX AND S. H. GOODMAN*
Hughes Aircraft Company, P. O. Box 902, El Segundo, CA 90245

INTRODUCTION

During the past five years, the computer you bought with 128 Kbytes of memory has been upgraded by the availability of 40 Mbyte storage and modems which allow instant access to databases around the world. The continuing need to perform more complex missions for modern defense systems in more hazardous environments will call for substantially higher data rates than are available today.

This paper addresses the field of electronic packaging commonly referred to as level 2 packaging, i.e., the interconnections between the integrated circuit package and other elements of the subsystem package. Specifically, we will focus on materials for printed circuitry and related constructions which must transmit high speed electrical signals. As new materials are developed, package designs will be restricted less by materials innovations and more by fundamental physical limits. Systems analyses [1] of these limits are helping to define the directions materials investigations must follow.

We will address three major areas of concern, viz., thermal dissipation, dielectric properties, and controlled thermal expansion. Our approach will be to highlight fundamental design requirements, briefly summarize the current state-of-the-art, and finally, propose areas of research and development we believe are necessary for continued progress in each of these three areas.

THERMAL DISSIPATION

All electronics have an unwanted by-product of their operation - heat. This heat must be removed to keep devices within their operating temperature limits. Increased operating frequency and smaller package sizes demand increased heat dissipation.

Among the standard electrically insulating materials, beryllia has the highest thermal conductivity, k = 200 W/mK, comparing well with several metals, but it is extremely difficult to work with due to its toxicity. Alumina is nontoxic but has an order of magnitude lower conductivity. Silicon carbide has an intermediate conductivity, k = 90 W/mK, but behaves as a semiconductor under some conditions. These materials have been used as integrated circuit packages, inserts in packages, and fillers for polymers to increase their conductivity.

Heat transfer is governed by three intrinsic material properties: thermal conductivity, specific heat, and density. One can model all thermal problems by examining these intrinsic properties and one extrinsic property, the interface transfer coefficient. Specific heat and density are frequency independent, thermal conductivity is greatly influenced by

---

* Presenter

Mat. Res. Soc. Symp. Proc. Vol. 108. ©1988 Materials Research Society

purity, and the heat transfer coefficient is a strong function of the morphology of the interfaces.

Because electrons are the primary carriers of thermal and electrical energy in metals, the ratio of electrical to thermal conduction is approximately constant. Insulators, on the other hand, conduct heat through lattice vibrations, or phonons, because the electrons are not free to move in the lattice. When the total impurity count in the material is reduced below a few parts per million, the lattice motions can work in a cooperative manner to move heat very effectively. A good example is extremely pure diamond, with a thermal conductivity five times that of copper at room temperature. Other materials exhibit similar results when the lattice is not littered with impurity atoms.

The heat pipe, invented in 1942, exhibits thermal conductivity hundreds to thousands of times higher than normal materials by making use of the heat of vaporization and mass transport of low-boiling liquids. Recent research at Hughes Aircraft Company has shown that heat pipes can be used for efficient thermal management of high power printed wiring boards (PWBs) [2]. Research at Thermo Electron Corporation on the use of plastic materials for wicking in heat pipes allows inexpensive direct cooling of PWBs and introduces the possibility of using low-cost extended flexible channels to carry heat away from circuits to more remote cooling areas [3]. The major disadvantage of present heat pipes is their relatively large size, which precludes their use within the interior of PWBs, where much of the heat is generated.

High emissivity surface fins for greater surface radiation cooling, and immersion in dielectric insulating liquids, have been used for heat removal. Fluorocarbon liquids, in static or flowing modes, can be used in direct contact with functional circuits because they are nonflammable and have a high dielectric strength. These liquids have dielectric constants greater than one but sufficiently lower than a typical PWB material, which results in decreased degradation of signals going through the board.

Ultra-pure ceramics and metals represent a potentially fruitful area for new research. Aluminum nitride (AlN) has an intrinsically high thermal conductivity if it is fabricated with a low oxygen content. The theoretical conductivity has been reported as 320 W/mK (Figure 1) [4]. Commercial grades of AlN have values of 50 to 95 W/mK. A translucent grade has

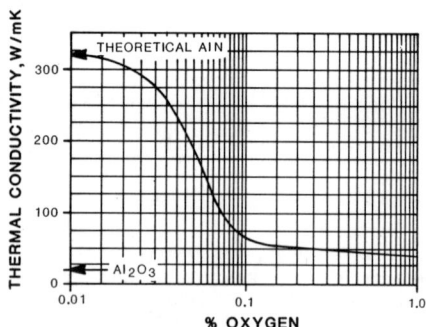

Figure 1.   Thermal conductivity of aluminum nitride as a function of oxide content [4].

recently come onto the market that has a thermal conductivity of 140 W/mK at room temperature and 130 W/mK at 100°C, eight times that of alumina (Al_2O_3) and one-half to two-thirds that of beryllia (BeO) [5]. Development of a grade of AlN, "Super SHAPAL," which has higher conductivity than 99.5 percent BeO has also been reported (Table I) [5]. This high conductivity is

Table I. Thermal Conductivity of Improved AlN [5]

Material	Thermal Conductivity (W/mK)		
	25°C	100°C	200°C
Super Shapal AlN	260	220	180
99.5 percent BeO	260	180	140

attained by severely controlling oxygen impurities both in the powder and during ceramic processing. Aluminum nitride with thermal conductivity >200 W/mK would be of great interest for high power microcircuit applications and electronic packaging (Figure 2). The lower than expected conductivity of available AlN grades is attributed to impurities of Al_2O_3, Y_2O_3, MgO, etc., at the grain boundaries, which act as thermal resistors in series. Research is needed to develop a process which would yield AlN with <2000 ppm of oxygen at the boundaries.

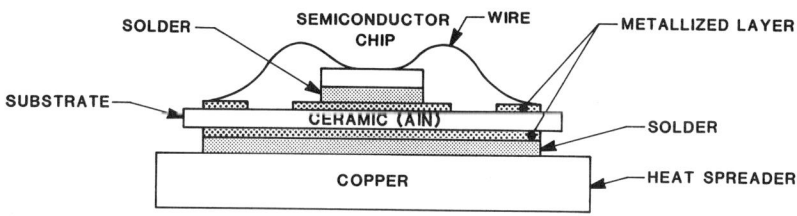

COURTESY TOSHIBA CORP.

Figure 2. Aluminum nitride as an electronic packaging substrate.

Efforts are underway in industrial laboratories to engineer sheet metal clad with electrical insulators to achieve optimum thermal and dielectric properties. At the Metallurgical Materials Division of Texas Instruments, progress has been made on developing package substrate materials that include a metal base coated with glass-ceramic, porcelain, or thermally conductive polymer coatings. Research at Allegheny Ludlum Corp. has resulted in a pure-alumina-clad stainless steel that offers superior strength and heat dissipation while retaining the processibility of an alumina surface [6]. Research is needed to develop other improved combinations of insulator-clad conductors.

One hypothetical material combination with high thermal but negligible electrical conductivity would be an organic plastic or polymeric matrix filled with very small glass or ceramic heat pipes as a filler material. The concept of a micro heat pipe was first developed at the Los Alamos National Laboratories. Research there is directed toward development of heat pipes in which the capillary channel is only slightly larger than the molecular diameter of the working fluid [7]. This is a promising avenue for continued research.

DIELECTRIC PROPERTIES

To appreciate the effects of materials on propagation of electromagnetic waves, or signals, in a PWB we must look at the speed with which the signal travels in the material. In free space it travels at the speed of light, $c = 3 \times 10^8$ m/sec. The relationship between wavelength (λ_0) and frequency (f) is

$$\lambda_0 = c/f. \tag{1}$$

In solids the propagation speed is equal to $v = c/\sqrt{\varepsilon}$, where ε is the dielectric constant of the material. Inverting this equation we find that the time for the signal to travel one foot is 1.016 nsec. In an epoxy/glass PWB the time is doubled because of its higher dielectric constant. Thus, if we want a low propagation delay on a circuit board, the parts must be very close together. Figure 3 illustrates signal delay for board materials.

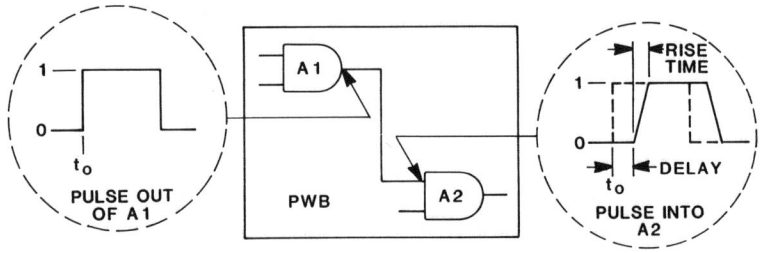

Figure 3. Propagation delay simplified.

In digital circuits, the rise and fall time of a pulse as well as the switching speed are important. Increasing the propagation delay has the effect of stretching pulses and reducing their slopes. This effect can cause some high frequency systems to fail because the rise time is not fast enough after the signal has traveled across the board (Figure 4).

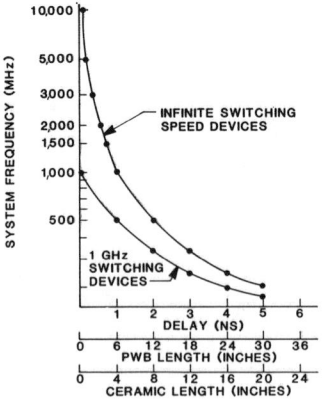

Figure 4. Propagation delay for PWB and alumina [8].

The electrical performance of electrical materials is controlled by three intrinsic variables; specific conductivity, σ, magnetic permeability, μ, and dielectric constant, ε. Three other terms frequently used, viz., stray capacitance, characteristic impedance, and signal attenuation, also play a role in limiting the upper frequency which can drive signals through a material. They serve to distort signal shapes, slow signals down and reduce the signal strength, but they are controlled more by circuit design than by material properties.

Conductors are characterized at high frequencies by skin depth, or the phenomenon of the signal traveling along the surface of the conductor rather than in its bulk. Skin depth is inversely proportional to the square root of the product of the conductivity, the permeability and the frequency. At 1.0 GHz, the signal travels in the top 3 microns of the conductor; the remaining 33 microns of typical conductor depth are essentially unused. Since path length is very important and the signal travels entirely on the surface, the surface and edges of the conductors must be kept smooth and scratch-free. The specific conductivity of the conductor should be as large as possible to reduce current loss as the signal is driven through it. Also, the magnetic permeability should be minimized to reduce the magnetic field around the conductor, thus reducing inductive crosstalk.

Both the specific conductivity and magnetic permeability of PWB materials are negligible; thus the dielectric constant becomes the major driver. Except at certain wavelengths, very close to absorption bands for the material, ε will be greater than one. To improve the high frequency properties of PWB materials, we need to reduce the dielectric constant from the present values of near 4.0 to less than 3.0.

The dielectric constant is a function of polarizability, molecular at the lowest frequencies, electronic at the highest. Lower dielectric constants will thus be found among nonpolar materials with tightly bound electrons.

Fluorocarbon-containing composites represent new advances in electronic packaging materials with substantially lower ε. One such development is epoxy resin reinforced with polytetrafluoroethylene (PTFE) fibers. Gore-Clad™ and Gore-Ply™, new materials developed by W. L. Gore and Associates, Inc., use expanded PTFE fibers to reinforce laminated composites and prepregs. The expanded PTFE fibers have a high degree of crystal orientation and a porosity of approximately 10%. These properties, reported by Gore [9], are shown in Table II.

Table II. Typical Expanded PTFE Fiber Properties.

Property	Value
Monofilament diameter	1 - 2 mils
Tensile strength	0.5 - 0.7 GPa (75 - 100 ksi)
Dielectric constant	1.8 - 2.1
Dissipation factor	0.0001

The fibers are also chemically inert, can withstand high temperatures, and are hydrophobic. Their cold flow is small due to the high tensile strength relative to that of unexpanded PTFE (approximately 35 MPa).

An alternative approach for exploiting fluorocarbon resins in electronic packaging is through the use of modified fluorocarbon matrix
resins, combined with conventional fiberglass and other reinforcements.
The Rogers Corporation recently developed RO-2800™, a fluorocarbon-based
rigid PWB material, which exhibits low expansion, compliancy sufficient
to relieve stresses on plated-through holes, and low dielectric constant.
Typical properties of RO-2800™ laminates reported by Rogers [10] are shown
in Table III.

Table III. Typical RO-2800™ Laminate Properties

Property			Value
Dielectric constant	10 GHz	23°C	2.8
Dissipation factor	10 GHz	23°C	0.002
Tensile strength			9 MPa (1.3 ksi)

Characteristics of resins suitable for fabrication of high-temperature PWBs with low dielectric constant include (1) a highly aliphatic
(non-polar) structure, (2) minimal presence of specific structures and
reactive end groups that increase polarity, (3) amenability to complete
cure, (4) easy processing, and (5) good mechanical properties, thermal
stability, and adhesion. Desirable structural segments in the polymers
that can contribute to low ε include aliphatics (straight chain, cyclo-
aliphatics, isopropylidene), fluoroaliphatics (except polyvinylidene
fluoride), aromatics, siloxanes, and olefinic groups. Undesirable structural moieties include hydroxyl, carboxyl, amide, urethane, and vinylidene
fluoride groups. Undesirable connecting and cross-linking groups which
can detract from low ε if present in sufficient amounts include esters,
maleimide end groups, and epoxies [11, 12].

The standard imides used for PWBs are bismaleimides having ε of
approximately 3.4. Recent data obtained at Hughes Aircraft Company on a
new fluorinated polyimide, Thermid FA7010, manufactured by the National
Starch and Chemical Corporation under license from Hughes, indicate ε
as low as 2.7.

Benzocyclobutenes, recently developed at the Dow Chemical Company,
constitute the basis of a new and versatile approach to the synthesis of
high performance polymers for electronics applications [13]. The basic
technology involves a family of thermally polymerizable monomers which
contain one or more benzocyclobutene groups per molecule. Depending on
their molecular weight and degree of functionality, these monomers can be
polymerized to yield either thermosetting or thermoplastic materials.
The derived polymers typically exhibit exceptional thermal stability
(>450°C in air) and the retention of useful mechanical properties at elevated (250°C) temperature.

The benzocyclobutenes (or BCB's) are endcaps which can be chemically
combined with various center (or "X") groups:

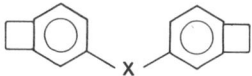

The polymerization of bis-benzocyclobutenes is a useful process because of the wide structural variety that can be built into the monomers and polymers. The properties of the group linking the two benzocyclobutene units are those which dictate the properties of the final polymer. The linking group can be polyamides, polyimides, polyesters, polysiloxanes and so on. By merely "end capping" the appropriate difunctional molecule, dielectric constants of <3.0 should be achievable.

The 1,2-polybutadiene resins have been shown to have excellent promise for yielding PWBs with low dielectric constant, but require formulation with other types of resins to improve processibility and thermal capability [14].

Commercial polybutadiene can exist as two different polymers (Figure 5). 1,2-PBD can cyclize when it is cured by free radicals, whereas 1,4-

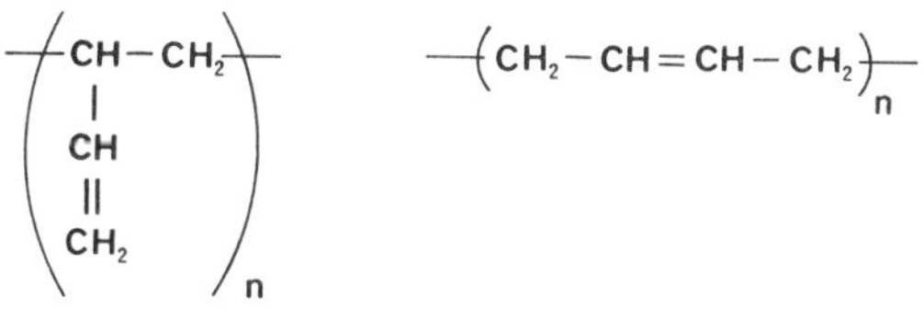

$$\left(\!CH\!-\!CH_2\!\right)_n \quad \quad \left(\!CH_2\!-\!CH\!=\!CH\!-\!CH_2\!\right)_n$$

1,2-PBD   1,4-PBD

Figure 5. Polybutadiene isomers

PBD will not. Such cyclization leads to products with a higher glass transition temperature (Tg).

At the Westinghouse Research and Development Center, an alloy or polyblend has been developed that consists of equal weights of two prepolymers. One prepolymer, designated MAPBD, was a maleated high-vinyl-content PBD (they used Ricon 154 1,2-PBD from Colorado Chemical Specialties, Inc.) The other prepolymer, designated CH, was an anhydride-extended cycloaliphatic epoxy resin with excess epoxy. During final cure, the two prepolymers were joined by a maleic anhydride-epoxy reaction, while peroxide-induced cure of the vinyl groups in the PBD raised the Tg to a reported 192°C. This polyblend was found to have ε of 2.7, but required improvements in shelf life [15]. Further research with these resins is warranted.

Work is presently underway at the University of Massachusetts to achieve blends of dissimilar thermosetting and thermoplastic polymers by controlling solubility parameters. One achievement has been successful blending of polybenzimidazole (PBI) and polyetherimide (PEI) [16]. Such work could be extended to develop polymers with properties tailored for PWB use.

Some work is being done to use electro-optical materials which utilize electrical inputs to modulate light. Fiber optics are used to transmit the resulting signals without the propagation delay of PWBs. Such devices are coupled with fibers either embedded in the board or mounted on its surface.

The use of superconductive materials in level 2 interconnections will have immediate payoffs in sensor technology for advanced space and airborne systems. Superconductive flex cable or striplines can be used to interface with focal plane array dewars. In these systems, thermal isolation of the cold stages from the "hot" processing electronics is difficult because of heat conduction through the connecting wires. A superconductive interface would provide the necessary thermal insulation, while maintaining good electrical conduction. This, in turn, would allow for design and manufacture of lighter and less expensive cryogenic systems. Other applications of interest are shown in Table IV.

Table IV. Potential Applications of Superconductive Materials

Device	Application	Advantage
Microwave guides	Radar arrays	Low loss, high directionality
Chirp filters	Radar	Low loss
CCD arrays	IR imaging	Tunable, fast response time
Optical/superconductor switch	IR detection, processing	High speed, low power
Maser/laser resonating cavities	Resonators	Low loss
Direct replacement of metal with superconductor	Circuits	Low noise, low power
Magnetic field sensors	Submarine detection	High sensitivity

CONTROLLED THERMAL EXPANSION

The most difficult mechanical problem facing the level 2 packaging engineer is combining materials with widely differing thermal expansion coefficients into a single reliable structure (see Figure 6). To do this, either the parts have to be bonded together with an adhesive strong enough to withstand the applied stresses or the structure has to be built with sufficient flexibility to allow for differential movement of the parts.

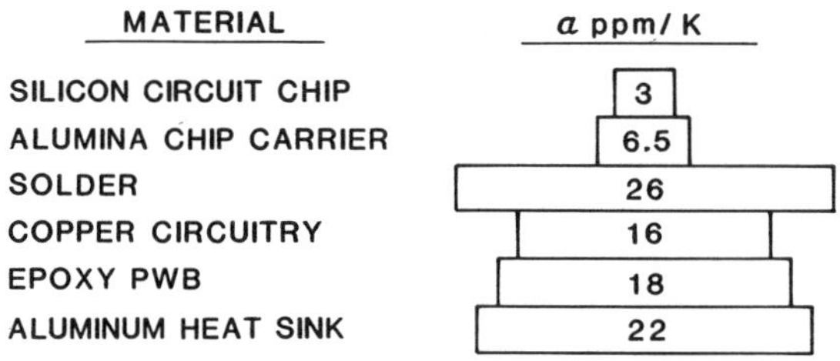

MATERIAL	a ppm/ K
SILICON CIRCUIT CHIP	3
ALUMINA CHIP CARRIER	6.5
SOLDER	26
COPPER CIRCUITRY	16
EPOXY PWB	18
ALUMINUM HEAT SINK	22

Figure 6. Thermal expansion pyramid.

For example, the CTE mismatch between a ceramic chip carrier (c) soldered to a PWB (b) causes a shear stress (τ) on the solder(s) over the temperature range ($T_2 - T_1$) according to the model (see Figure 7):

$$\tau = \frac{E_s}{4(1+ \nu_s)h} \cdot \frac{L_c}{L_s} \cdot \left| \alpha_c (T_{c2} - T_{c1}) - \alpha_b (T_{b2} - T_{b1}) \right| \quad (2)$$

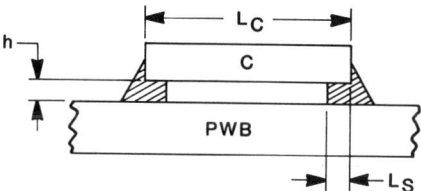

Figure 7. The LCC solder crack problem.

where α is the thermal expansion, ν is Poisson's ratio, and E is the elastic modulus. We can see that if we reduce the difference between the expansion coefficients or reduce the temperature swings, we can reduce the stress and extend the life of the solder joint. Even if the expansions of the two materials were the same, there would still be a stress on the solder joints due to power in the device heating it faster than the PWB. To reduce this effect, good thermal conductors that are also good electrical insulators are bonded between the board and the chip carrier to reduce the temperature difference. This has the added effect of restraining the board in the region of the chip carrier, putting less stress on the solder joints.

The above equation shows that the elastic moduli of the chip carrier and PWB are also important intrinsic properties of the materials which can be independently measured and used to calculate effects on new designs.

Current packaging has been tailored to silicon ($\alpha = 3$ ppm/K). Since we are reaching the speed limits of silicon devices, and are designing with materials such as GaAs ($\alpha = 5.7$ ppm/K), packaging materials will have to be examined again to confirm their utility. This complexity of design considerations has produced a dual need for (1) improved design tools, and (2) improved materials with tailorable CTEs. A computer program developed at the Boeing Electronics Company has been used to predict the CTE of various PWB designs. The results illustrated the dependence of solder joint life on PWB CTE and verified the use of a computer model to design surface mount PWBs [17].

The most common solution today for decreasing the CTE of PWBs is to utilize low CTE reinforcements, such as Kevlar and quartz fibers, in the laminates. This approach can reduce the CTE to approximately 5 ppm/K.

As discussed earlier, fluorocarbon composites can reduce stress on solder joints as a result of their low modulus of elasticity (<5% of epoxy/glass), which permits the PWB to deform before the solder joints fail.

Silicon carbide reinforced aluminum can exhibit CTEs one-third to one-half that of bulk aluminum. The SiC discontinuous reinforcements consist of small particles or whiskers. The CTE can be tailored by controlling the SiC content. SiC/Al has been developed as a successful replacement for Kovar microwave system packaging components at the General Electric Space Systems Division [18].

Polycrystalline silicon has also been used as a substrate material for mounting of high speed devices. The layers of conductor are provided by alternately depositing conductors and silica layers and defining the circuits photographically and chemically. This scheme eliminates thermal expansion mismatch with silicon devices, but it works only with silicon.

Low-CTE polyimide resins, recently developed at the Hitachi Chemical Company, Ltd., offer promise for developing improved low CTE packaging materials. High quality flexible printed circuits have been developed using polyimides with CTE equivalent to copper foil. Their most significant property is high dimensional stability after heat treatments, such as soldering. Furthermore, they have very high adhesive strength at elevated temperatures, even after heat aging [19].

Polymer-ceramic composites, based on sol-gel processing, offer promise for combining desirable features of polymers and ceramics [20, 21]. No efforts in this area, specifically directed toward electronic packaging, are known. This appears to represent a fruitful direction for research.

CONCLUSIONS

The rapid advancement of high speed integrated circuit electronics is driving the need for new materials for the interconnection of these devices to higher level systems. In order to keep pace with the forthcoming technological breakthroughs, materials research must continue to stress the evolution of new materials and improved properties of existing materials.

The heat generated by high speed circuitry must be removed through innovative methods of thermal dissipation. New materials such as aluminum nitride, ceramic coated metals, and microcapillary heat pipes are fruitful areas for investigation.

New, easy-to-process polymers having lower dielectric constants will provide a base of printed circuitry capable of handling high frequency data rate transmission. The potential for superconducting films, ribbons and substrate coatings has already stirred designers' imaginations as they create new, even smaller electronic device configurations. These devices will be able to operate faster and use less power.

Prevention of thermomechanically induced stress failure will only occur through the creative integration of new diverse materials. Until solder can become (if ever) a truly mechanical (as opposed to electrical) element in a package, we must continue development of materials with compatible expansion coefficients.

ACKNOWLEDGEMENT

The authors wish to acknowledge the assistance of Dr. N. H. Harris, Hughes Aircraft Company, who provided information on advanced ceramic materials for electronics.

REFERENCES

1. W. E. Pence and J. P. Krusius, IEEE Transactions on Components, Hybrids, and Manufacturing Technology CHMT-10 (2), 176-183 (1987).

2. A. Basiulis, H. Tanzer, and S. McCabe, Proceedings of the 6th Annual International Electronics Packaging Conference, San Diego, CA, November 17-19, 1986.

3. R. L. Matsu and N. J. E. Johnson, 1st International SAMPE Electronics Conference, Santa Clara, CA, June 23-25, 1987.

4. N. Iwase, The International J. for Hybrid Microelectronics, 7, (4) (1984).

5. N. Kuramoto, H. Taniguchi, and I. Aso, IEEE Transactions on Components, Hybrids, and Manufacturing Technology, CHMT-9 (4), 386-390 (1986).

6. Anon., Advanced Materials & Processes 131 (1), (1987).

7. G. P. Peterson, Aerospace America, 25 (10), 20-22 (1987).

8. W. S. Fujitsubo, presented at 1985 Hughes Aircraft Co. Printed Wiring Board Symposium, El Segundo, CA 1985 (unpublished),

9. W. L. Gore & Associates, presentation on Gore-Clad, February 1986.

10. D. J. Arthur and C. S. Jackson, Proceedings of the 5th Annual International Electronics Packaging Conference, Orlando, FL, October 21-23, 1985.

11. J. M. Butler, R. P. Chartoff, and B. J. Kinzig, 15th National SAMPE Technical Conference, October 4-6, 1983.

12. J. A. Brydson, Plastic Materials, 4th ed. (Butterworth Scientific, London, 1982).

13. R. A. Kirchhoff, C. E. Baker, J. A. Gilpin, S. F. Hahn, and A. K. Schrock, 18th International SAMPE Technical Conference, October 7-9, 1986.

14. W. P. Barie, Jr., J. J. Conroy, and J. E. Dereich, presented at Adhesives and Sealants Council Fall Seminar, Dearborn, MI, October 27-30, 1985.

15. M. P. Zussman, U.S. Patent No. 4,601,944 (22 July 1986).

16. F. Karasz, U. of Massachusetts (private communication).

17. R. L. Williams and A. N. Noblett, 1st International SAMPE Electronics Conference, Santa Clara, CA, June 23-25, 1987.

18. Anon., MMCIAC Current Highlights, 6 (3), December 1986.

19. T. Miwa, N. Kinjo, Y. Misawa, presented at Materials Research Society Fall Meeting, Boston, MA, Nov. 30 - Dec. 5, 1987.

20. H. K. Schmidt, Polymer Preprints, 28 (1), 102 (1987).

21. H. H. Huang, R. H. Glaser, and G. L. Wilkes, Polymer Preprints 28 (1), 434 (1987).

RESIN SELECTION CRITERIA FOR MOLDED ELECTRONIC PACKAGES

S. B. RIMSA
Amoco Performance Products, Inc., P.O. Box 409, Bound Brook, NJ 08805

ABSTRACT

The selection of the proper resin system for a molded PWB or electronic package must take into account considerably more parameters than simply the cost of and the ability of the polymer to fill the tooling. Molding of an electronic package is only the initial step in a multi stage sequence to produce a functional PWB or interconnect device.

It is of extreme importance that a manufacturer of molded electronic packages understand not only the fabrication properties of a resin, but also those mechanical, thermal, and chemical requirements of the circuit application and assembly operations.

A specific circuitry application process will usually require a specific set of substrate characteristics. These performance characteristics may relate to platability, thermal stability, modulus, compressive strength, or even adhesive bondability for a given procedure. The substrate requirements for several popular circuit application methods is reviewed.

Assembly presents perhaps the most demanding area of resin performance in electronic package or PWB fabrication. The thermal stresses encountered in wave soldering and vapor phase reflow will quickly differentiate among current engineering resin offerings. The importance of controlled moisture absorption and high temperature modulus retention is explored in addition to end use environmental factors which can be of special concern.

INTRODUCTION

Numerous papers have been presented highlighting the properties of specific processes and substrates for producing molded PWB's and related devices.

It is the objective of this presentation to examine specifically the thought process required to successfully utilize engineering thermoplastics in molded electronic package applications. The term "electronic package" is now rather generic in context referring to a wide variety of electronic componentry including printed wiring boards, chip carriers, connectors and combinations thereof. All of these devices are tied together via a common denominator as they basically provide an insulating support of electrically conductive pathways between circuit elements and external package sources.

CHOOSING A MOLDED PACKAGE

The decision to utilize a molded electronic package versus a more conventional assembly provides a departure from the confines imposed by traditional materials. The design engineers frame of reference is based largely upon his experience with planar FR-4 printed wiring boards in combination with discrete connectors and componentry. In considering new applications which could employ molded package technology, the design engineer is motivated by three driving forces.

First and foremost is the potential to reduce cost; this is accomplished by integrating electrical functionality into a molded housing to serve as a wiring substrate and enclosure simultaneously. Part consolidation where a connector shell and PWB may be molded as one assembly is another cost reduction example. Size reduction is the second driving force. Higher density packaging requirements may be well suited for the incorporation of a 3D molded package where high interconnect density per unit area is critical, i.e. "mold to fit" PWB's and the like. The requirement for upgraded electrical and thermal performance is a third driving factor in utilizing molded packages. The electrical property advantages offered by high temperature thermoplastics have been presented in the literature. In general high speed, lower loss assemblies will be realized due to the intrinsically lower Dielectric Constants and Dissipation Factors versus more conventional substrates such as FR-4.

From a property balance standpoint, the amorphous thermoplastic resins such as polysulfone (UDEL), polyarylsulfone (RADEL), polyethersulfone (Victrex) and polyether imide (Ultem) are preferred. These materials have been shown to exhibit the best overall performance profile for molded package fabrication where moldability, dimensional stability, platability and thermal stability are all important criteria.

RESIN SELECTION CRITERIA

The selection of the appropriate thermoplastic for use in a molded electronic package is predominantly application dependent. Every application has its own unique set of fabrication and end use criteria which must be carefully reviewed. The criteria of special concern include moldability, circuitizing, assembly, environmental effects, and economic considerations.

Moldability

The ability of a high temperature amorphous resin to fill an intricate part geometry is dependent upon a multitude of factors. The rheology of the resin at recommended processing temperatures, mold design and gating are all of critical importance. In general, amorphous resins are more viscous to process and flow less readily than the crystalline resins commonly used in the electronics industry (polyethylene terephthalate, Nylon 6, 66).

Many of the resins utilized for molded packages contain fillers and reinforcements added to increase mechanical strength, enhance platability or reduce resin cost. In general, filled and reinforced systems will be somewhat more difficult to process than unfilled counterparts of equal molecular weight. Some reinforcements, in particular glass fibers, tend to orient during molding and may induce part warpage. Platelet-like fillers have demonstrated superior performance to glass reinforced resins in warp resistance but may tend to exhibit a higher degree of brittleness.

It is important for the designer to consult the resin supplier early in the molded package design process. Recommendations for cavity thickness, gate placement, and runner system design are dependent upon part size and geometry and will be different for a 3 Dimensional versus a planar package. Mold shrinkage data must also be obtained prior to cavity fabrication and will normally be lowest for highly reinforced amorphous resins. Existing molds utilized with crystalline resins will usually not be suitable for use with amorphous packaging resins without modification.

Circuitization

A second criteria of resin selection is circuitization. The chosen method of conductor application to a molded package relates directly to the characteristics of the substrate. Semi- and fully additive processes are normally employed to circuitize molded PWB's and related constructions. Semi-additive and certain fully additive processes require the substrate to respond readily to adhesion promotion via swelling with an organic solvent followed by etching in an oxidizing acid. The latter sequence produces a micro-roughened topography upon which the subsequently deposited electroless metal can bond. The amorphous resins used in electronic packages vary in oxidative stability and their response to organic solvents, necessitating careful selection if swell and etch processing is utilized. Practical experience has demonstrated that polyarylsulfones appear to respond most readily to etching in oxidizing media followed by polyethersulfone, polysulfone, and polyetherimides. Response to the etching media can be varied significantly via the incorporation of a finely divided mineral filler at the expense of ductility loss, reduced mold flow, and opacity.

Non-traditional additive techniques demand yet a different property balance from the substrate. Polymer thick film systems are sensitive to the surface chemistry/uniformity presented by a molded substrate. Circuit transfer and thermal compression techniques demand specific levels of thermal stability and creep resistance to be present in the molded package.

Sputtering and ion plating appear selective to the base resin and the filler system employed. In addition, the 2 shot molding concept is being explored in conjunction with catalytic resins as a novel approach to 3D circuitization providing a platable molded conductor. Two shot molding, in particular, requires a special balance between the rheological, thermal, and mechanical characteristics of the resin systems employed. It is once again advisable to contact both the polymer supplier and circuitizer to make sure the candidate resin and process are compatible.

Assembly

The assembly of an electronic package places special demands upon the substrate. Assembly operations, in particular soldering, must be considered early in the resin selection process. A high continuous use temperature or Tg while important is not an absolute indication of substrate performance in wave or vapor phase soldering. Wave and vapor phase soldering place very different stresses upon a thermoplastic. Wave soldering presents a hot side/cold side exposure requiring only one face of the substrate to be in relatively brief contact with molten solder in the 500-515°F range. Vapor phase soldering in most instances is a much more aggressive exposure for amorphous resins. Total immersion in perflourinated vapors at 419°F results in instantaneous substrate heating analogous to "deep frying" the molded package. Equilibrium moisture content, moisture vapor transmission rate, and resin high temperature modulus retention become of extreme importance. The incorporation of fillers or reinforcements are commonly utilized to enhance modulus and reduce total contained moisture content in many resin systems.

Amorphous resins will absorb a small amount of moisture from the environment to usually necessitate predrying prior to vapor phase exposure and, in some instances, wave soldering. Drying criteria will be somewhat resin dependent, usually requiring a bake in the 250 - 300°F range prior to soldering.

In addition to thermal performance requirements, the mechanical properties of the resin system selected will be of significance in assembly. Most of the molded-in features and functionality provided by a molded electronic package require a balance between mechanical strength and ductility. Highly filled and/or reinforced systems in general, while providing high modulus/mechanical strength, usually lack ductility. Designs incorporating snap fits for positive component location or subassembly will normally require neat unmodified polymers or resins with only minor amounts (<10% by wt) of filler/reinforcement incorporation. Compounded amorphous resin systems with a tensile elongation to break of 3% or less will usually not provide a satisfactory mechanical property balance for molded package assembly.

Environmental Considerations

The performance of a molded electronic package in the end use environment is dependent upon how well the design engineer understands both the resin selected and the environment. Molded package performance will be affected by temperature, humidity, impact and environmental factors.

Most amorphous thermoplastics will be attacked by chlorinated and aromatic hydrocarbons, ketones and esters. Filled and/or reinforced resins will usually exhibit a greater degree of apparent chemical resistance than their unfilled counterparts. Radiation exposure most commonly in the form of ultraviolet light has a deleterious effect on some resin systems. When long term exposure of a molded package to weathering is a requirement, UV stabilized products should be utilized to enhance performance. Since the molded package in many instances may provide both support and protection for electrical circuitry, enclosure integrity and appearance is often of prime importance. Impact resistance in general will necessitate the utilization of unfilled or only lightly reinforced resin systems. Impact resistance information in the form of drop dart or tensile impact data is available from the resin suppliers to aid the molded package designer.

Temperature and humidity effects will also influence the performance and longevity of a molded electronic package. Long term exposure to elevated temperatures is a strong point of high temperature amorphous resins. The polymers utilized for molded package construction typically have UL continuous use temperature ratings in the 160-220°C range. Thermal excursion to temperatures above the heat deflection temperature (HDT) of the polymer should be avoided or distortion may be experienced. Moisture will also be reabsorbed by a molded package in the end use environment. A rapid thermal stress placed on a molded device that has equilibrated in a high humidity environment can lead to substrate deformation in certain instances. It is advisable to check the environmental equilibrium moisture content of a molded package prior to rapid thermal excursion (above the C.U.T.) if such service might be anticipated. A moisture level of 0.4% by weight or less will usually provide satisfactory performance but must be verified. End use testing of a molded package is highly recommended to reveal any unexpected environmental interactions.

Economic Considerations

To this point, we have discussed five different criteria to be examined in the resin selection process. The last we will discuss -- that of economics -- weighs heavily on all of the foregoing and again is application dependent. An example is helpful:

If a design engineer desires to integrate a connector and planar PWB in a single molded package, he must examine the previously mentioned criteria and define the costs associated with each. These costs are unique to each resin "system". I use the word "system" in that each resin requires different processing whether it be circuitizing methods, mold tooling, or assembly.

Thus, as a cost analysis is done, the uniqueness that each thermoplastic resin possesses, brings with it certain downstream costs. The options created by this matrix are many, thus eliminating any generalizations on cost. In an extreme case, the size of the part alone is enough to rule out the use of high cost, poor flow resins.

As anyone who has attempted to cost a molded package concept knows very well, the per pound cost of a substrate resin alone is not meaningful. It is in fact, the beginning of an interactive process which examines the cost associated with the "system" required.

CONCLUSIONS

Molded electronic packages provide the design engineer new flexibility for combining electrical functionality into package design. It has been the intent of this presentation to explore the driving forces for the implementation of molded packages and present an overview of the criteria for proper resin selection. High temperature amorphous packaging resins offer excellent thermal and electrical performance but must be selected and utilized with care. No one resin system will be suitable for all molded geometries, circuitizing methods, assembly techniques or end use environments. The design engineer should actively interface with the resin supplier to develop a mutual understanding of resin performance criteria and their interactions. Successful implementation of a molded electronic package should include end use testing to ensure satisfactory performance is realized and to identify any unpredicted environmental effects.

SURFACE MOUNT TECHNOLOGY (SMT) SUBSTRATE
MATERIAL REQUIREMENTS - A BRIEF REVIEW

A. MAHAMMAD IBRAHIM
Martin Marietta Laboratories,
1450 South Rolling Road, Baltimore, Maryland 21227

ABSTRACT

The state of the art for the printed circuit (pc) industry is the Surface Mount Technology (SMT) using leadless ceramic chip carriers (LCCCs). The
SMT technique is used to design, fabricate, and assemble affordable high-
speed, high-density electronic modules with reduced size and weight. However,
to take full advantage of the SMT, new high-performance printed circuit board
(PCB) substrate materials must be developed, especially if the goal is to
satisfy the high reliability required in military applications. The most
critical requirement is matching the in-plane coefficient of thermal expansion
(CTE) with the SMT-PCB's and the chip carriers (LCCCs) to reduce the solder-
joint stresses during thermal cycling. Solder-joint reliability can be improved significantly by tailoring the in-plane CTE to approximately 6-7 ppm/°C
(of the alumina chip carrier) using Kevlar and/or quartz fabric-reinforced
polymer composites, instead of conventional glass composites. Less attention
has been focussed on other important requirements, e.g., out-of-plane (Z-axis)
expansion, glass transition temperature (T_g), dielectric constant, and resin
microcracking, which also play important roles in the overall performance of
the substrate materials. For example, high Z-expansion puts additional strain
on the plated through holes (PTH), thereby affecting PTH reliability. A low
T_g significantly increases the amount of thermal stress imposed during thermal
cycling on solder joints and PTH, and leads to failures. This paper contains
a brief review of the requirements of a SMT-PCB substrate material, including
such critical parameters as: in-plane CTE, out-of-plane CTE, T_g, dielectric
constant, fiber-to-resin ratio, and resin microcracking, and their effects on
solder joint reliability, PTH reliability, dimensional stability, and electrical performance.

INTRODUCTION

The state of the art for the printed circuit (pc) industry is surface
mount technology (SMT) using leadless ceramic chip carriers (LCCCs). The
development of LCCCs has provided military electronics with an opportunity for
achieving higher packaging densities required for very high speed integrated
circuit (VHSIC) applications. The SMT with the LCCCs is used to design, fabricate, and assemble affordable high-speed, high-density electronic modules
with reduced size and weight, and improved electrical performance. In surface

mount devices (SMD), the LCCCs are soldered directly to the fabric composite substrate materials. Thus, to take full advantage of the SMT, new high-performance printed circuit board (PCB) substrate materials must be developed, especially if the goal is to satisfy the high reliability required in military applications. Consequently, SMT is driving the development of PCB substrate materials through improvements in thermal and electrical properties. This paper provides a brief review of the critical properties requirements of a SMT-PCB substrate material and their effects on performance variables such as solder joint reliability, plated through hole (PTH) reliability, dimensional stability, and electrical performance.

SMT-PCB SUBSTRATE REQUIREMENTS

Selection of an "ideal" PCB substrate material for VHSIC applications depends on several design parameters. Current designers for such applications are somewhat limited in their choices because there are few candidate substrate materials that possess the right combination of thermal, mechanical, and electrical properties required for high-reliability applications. The critical properties requirements of PCB substrate materials are:

o Low in-plane coefficient of thermal expansion (CTE)
   (ideally close to that of alumina, $\sim 6 - 7$ ppm/°C)

o Low out-of-plane CTE
   ($<$ 40 ppm/°C; ideally close to that of copper, 17 ppm/°C)

o High glass transition temperature, T_g
   ($>$ 180°C; ideally higher than the solder reflow temperature, 288°C)

o Low dielectric constant, ε_r
   (2-4, according to The Institute for Interconnecting and Packaging of Electronic circuits (IPC) Society)

o Optimum fiber-to-resin ratio

o None or minimal resin microcracking

The importance of the critical materials requirements listed above and their effects on the PCB performance are briefly outlined in the following sections.

In-Plane CTE (X- and Y-direction CTE)

Anyone who is familiar with SMT is well familiar with the CTE mismatch that exists between the SMT-PCB's and LCCCs as depicted in Fig. 1. The LCCCs are fabricated from 94% Al_2O_3 ceramic material with a CTE ranging from 5.9 to to 7.4 ppm/°C [1]. The in-plane (X-Y) CTE's of the conventional E-glass fabric-reinforced epoxy composite substrate materials range from 12 to 16 ppm/°C. This in-plane CTE mismatch results in shear stress on the solder joints during thermal and power cycling. After a few cycles, if the shear stress at the solder joints is sufficiently high, e.g, due to extreme temperature cycling from -55 to +125°C, the solder joints will fatigue and eventually crack resulting in electrical failures. To increase the reliability of the solder joint, the in-plane CTE of the substrate material should be tailored closely to that of the chip carrier (~ 6-7 ppm/°C), especially if the goal is to achieve highly reliable assemblies capable of operating in a wide temperature range in military environments.

The thermal expansion coefficients of fiber-reinforced composites depend on the properties of the matrix resin and reinforcing fiber, as well as the fiber-to-resin ratio. Like elastic moduli, the thermal expansion coefficients are very anisotropic in nature for oriented fiber-reinforced composites.

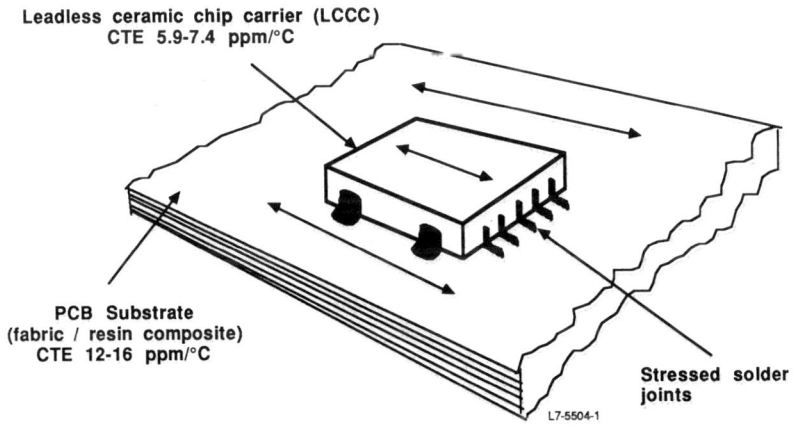

Fig 1. Schematic diagram showing the SMT/CTE mismatch problem.

For example, the linear thermal expansion coefficients of a unidirectional composite in the direction parallel to the reinforcing fiber ($\alpha_{\text{longitudinal}}$, α_L) may be predicted using Schapery's equation [2]. Schapery[2] derived this simple and accurate equation after reviewing the equations developed earlier by Halpin et al [3,4].

$$\alpha_L = \frac{\alpha_m E_m V_m + \alpha_f E_f V_f}{E_m V_m + E_f V_f} \tag{1}$$

where α, E, and V are the thermal expansion coefficient, elastic modulus, and volume fraction of matrix (m) and fiber (f), respectively. (In this equation it is assumed that the Poisson's ratios of matrix and fiber are close.) In general, the thermal expansion coefficient in the fiber (or longitudinal) direction, α_L, is small due to the mechanical restraints imposed by the fibers, which typically have very small thermal expansion coefficients, as shown in Table I.

Table I

CTE and Modulus of Reinforcing Fibers
Typically Used for PCBs

Fiber	Tensile Modulus (MSI)	CTE (ppm/°C)	
		Longitudinal Direction	Radial Direction
PBT*	55.0	-12.0 to -7	+ve (>50)
Kevlar-49	18.5	-5.0	60
Kevlar-29	12.0	-4.0	60
Graphite(T-300)	33.5	-1.0	16.7
Quartz	10.0	0.54	0.54
Nextel	22.0	2.6	2.6
S-Glass	12.5	2.6	2.6
E-Glass	10.5	5.5	5.5

------

* PBT = Polybenzthiazole (Self-reinforcing rigid-rod polymer)

It is well established in the literature that the in-plane CTE of the fabric reinforced composites is a fabric-dominated property. From Eq. (1), it is clear that low-fabric CTE and high-fabric modulus are required to achieve a lower in-plane CTE for the composite. As Table I shows, polybenzthiazole (PBT), Kevlar, and graphite fibers have anisotropic thermal expansion. However, according to Eq. (1), their negative CTE's and high modulus in the longitudinal (fiber) direction are extremely desirable for obtaining a low or tailorable in-plane CTE, which will result in improved solder-joint reliability. Typical in-plane CTE's of current composite substrate materials (gathered from the literature) are provided in Table II.

Table II

Literature In-Plane CTE's of
Composite Substrate Materials

Substrate	In-Plane (X-Y) CTE (ppm/°C)
Epoxy/E-Glass	12-16
Polyimide/E-Glass	11-14
Epoxy/S-Glass	13
Epoxy/Nextel	7.4
Epoxy/Kevlar (120)	6-7
Polyimide/Kevlar (120)	3-7
Kevlar (108) Composites	8-11
Quartz Composites	6-12

Out-of-Plane CTE (Z-direction CTE)

The out-of-plane (Z) CTE of fiber reinforced composites is a matrix-dominated property that predominantly depends both on the CTE and on the glass transition temperature, T_g, of the resin matrix. Low-CTE and high-T_g matrix resins generally result in a lower out-of-plane CTE for the corresponding laminates.

In general, the out-of-plane CTE's (below T_g) range from 40-70 ppm/°C for the conventional fabric reinforced epoxy and/or polyimide composite substrate materials, from 30 to 60 ppm/°C for quartz composites, and from 60 to 120 ppm/°C for Kevlar composites. These CTE values are much greater than that of copper, which has a CTE of 17 ppm/°C. When thermally cycled from -55 to +125°C or exposed to a solder reflow temperature of 288°C (550 °F), the CTE difference between the substrate and copper plating creates thermal stresses that lead to cracks in several locations of the plated through-holes (PTH) [1]. The PTH failure due to cracking can lead to electrical opens, which would shut down the whole system. Thus, PTH reliability can be increased by using substrates with low-Z-CTE's close to that of copper.

Glass Transition Temperature, T_g

T_g is defined as the temperature at which the polymer changes from a brittle, glassy state to a soft, rubbery state. At, below, and above T_g, the polymers undergo significant changes in their thermal, thermomechanical, static, and dynamic mechanical properties. For example, above T_g, the modulus decreases while the CTE increases; both are drastic changes. The popularity of high-performance composite substrates is due to their high T_g and the unique properties associated with it.

Although the epoxy composite substrates currently used have many advantages, as well as a long history of use within the pc board industry, they have some disadvantages, primarily their relatively low Tg (< 125°C) [5]. The

T_g is a critical material parameter affecting the dimensional stability of the pc board laminate; a higher T_g (well above the military thermal cycling range, -55 to +125°C) provides better dimensional stability. These low T_g's (close to the upper thermal cycling temperature) affect several facets of the performance of the PCB's, primarily by increasing the amount of thermal stresses on the solder joints and PTH during thermal cycling, ultimately leading to failures [1,6].

The effect of T_g on the CTE's can be demonstrated from our previous work on QUATREX epoxy/E-glass composite substrates (Table III) [7,8]. The T_g of this QUATREX epoxy is ~ 180°C.

Table III

Effect of T_g on CTE's of QUATREX Epoxy/E-Glass Composite Substrates[7]

Temperature Range (°C)	Mean CTE[*] (ppm/°C)	
	X-Y CTE	Z-CTE
25 - 140 (Below T_g)	15.7	69.4
25-250 (Through T_g)	11.5	125.4

As Table III shows, the out-of-plane (Z) CTE for the QUARTEX epoxy composite for the temperature range from 25 to 250°C is nearly double that of the 25 to 140°C range. The nearly two-fold increase in the CTE of the laminate as it passes through the T_g tends to cause PTH failures during processing and reworking of PCB's. This clearly demonstrates the need for high performance materials with T_g's well above the required operating temperature ranges. Even though the in-plane (X-Y) CTE is a fabric-dominated property, our results indicate the influence of matrix T_g in slightly decreasing the CTE in the fabric direction as well.

Furthermore, in general, as the T_g's increase, the CTE's decrease, as shown in Table IV for five E-glass-reinforced composites [9].

In addition, the T_g strongly depends on the extent of cure [10-12], so that as the T_g for any one resin system increases, its CTE decreases [13]. By increasing the T_g of the material, both the PTH and solder-joint reliability could be greatly improved.

Table IV

Effect of T_g on CTE's of
E-Glass Reinforced Composite Substrates[9]

Substrate Material	T_g (°C)	CTE (ppm/°C)		
		X	Y	Z
Epoxy/E-Glass	120	15	15	180
Multifunctional Epoxy/E-Glass	140	14	14	160
BT-Epoxy/E-Glass	190	13	13	90
Kerimid 601/E-Glass	270	12	12	60
Modified polyimide/ E-Glass	240	12	12	50

Dielectric Constant, ε_r

The dielectric constant, ε_r, determines how fast or slow an electrical
signal will travel from one point to another on the pc board. The signal
propagation velocity is inversely proportional to the dielectric constant, as
shown in Eq. (2),

$$\nu = \frac{C}{\sqrt{\varepsilon_r}}$$ (2)

where C is the velocity of light.

The ε_r of a fabric reinforced composite depends: on the ε_r of both the
matrix resin and fabric; on the fabric-to-resin ratio; on the temperature and
frequency of the experiment; on the physical conditioning of the sample before
and during the experiment; and finally, on the instrument and test method
used. The dielectric constants of the E-glass reinforced epoxy composites can
range from 4.5 to 8.0 depending on the epoxy, curing agents, catalysts, and
fillers; the temperature and frequency dependency is closely associated with
the thermal transitions, chain stiffness and crosslink density [14]. Typical
dielectric constants of some of the current PCB substrate materials (gathered
from the literature) are provided in Table V. In general, the dielectric
constants decrease with frequency and increase with temperature for polymer
composites [15].

The dielectric constant also depends on the extent of cure, as shown in
Table VI from our previous work on QUATREX epoxy/E-glass composite substrates [7] (with about 45% by weight resin content). As this table shows,
the relative dielectric constant for a sample that was cured for 15 min is
almost 10% lower than that of a sample cured normally (1 1/2 h). Although the
DSC kinetic investigations show almost a 90% cure within 15 min, which
resulted in a T_g as high as 171°C, the changes in the dielectric constant are

Table V

Literature Dielectric Constants of Substrate Materials
at 1 MHz

Substrate	ε_r (~ range)
Epoxy/E-Glass	4.5 - 5.0
Polyimide/E-Glass	4.2 - 4.6
Kevlar Composites	3.4 - 3.8
Quartz Composites	3.0 - 3.6
Teflon (PTFE) E-Glass	2.2 - 2.6

Table VI

Dependency of Dielectric Constant on Extent
of Cure in QUATREX Epoxy/E-glass Composite Substrates [7]

Property	15-min Cure	1-1/2 hr Cure (Normal Cure)
% Cure[a]	90	Fully cured
T_g[b]	171	180
ε_r[c] @ 1 kHz	4.64	5.08
@ 10 MHz	4.13	4.57

----

[a] Using Differential Scanning Calorimetry (DSC)
[b] Using Thermomechanical Analyzer (TMA)
[c] Using Impedance Analyzer at R.T.

quite significant. Since signal propagation velocity in VHSIC devices is inversely proportional to the ε_r of the PCB, these tabular data suggest the importance of a "proper cure" for epoxy or other thermoset matrix composites in obtaining the proper ε_r for that particular substrate material.

Resin Microcracking

Resin microcracking or the formation of small cracks in the resin at the yarn crossovers is observed frequently in both the electronic and the aerospace industry. It is especially prevalent in Kevlar and graphite-reinforced composites. The inherent anisotropy in the CTE's of the Kevlar and graphite fibers, as shown earlier in Table I, is largely responsible for the microcracking observed when these composites are exposed to thermal cycling. For example, heating causes the Kevlar and graphite fibers to shrink in their fiber (longitudinal) direction while expanding quite significantly in their radial direction. This type of anisotropic expansion of the fiber (or fabric) in its composite results in stresses at or near the fiber/resin interface that lead to microcracks. Microcracking significantly increases as the frequency and the temperature range of subsequent thermal cycling increases. In some cases, extensive microcracking may degrade the performance of PCB's.

In contrast to Kevlar and graphite fibers, the isotropic thermal expansion of quartz and E-glass fibers (Table I) results in improved resistance to microcracking, as shown in Table VII from our previous work on both thermoset QUATREX epoxy and thermoplastic polyphenylene sulfide (PPS-Ryton) composites [7,16].

Table VII

Comparison of the Microcracking Resistance of
Thermoset and Thermoplastic Fabric Reinforced Composites [7,16]

Composite	Resin Content (wt.%)	Mean CTE (ppm/°C)		No. of Cycles to Microcrack
		X-Y CTE	Z-CTE	
QUATREX Epoxy/E-Glass	45.0	15.7	67.0	No microcracks up to 1000 cycles
Ryton PPS/Quartz	38.6	4.3	21.9	No microcracks up to 300 cycles
Ryton PPS/Kevlar	40.1	-1.2	44.2	No microcracks up to 100 cycles
Ryton PPS/Graphite	42.5	1.9	63.3	0 (microcracks as-laminated)

SUMMARY

This paper focussed on the critical requirements of a surface mount technology printed-circuit-board (SMT-PCB) composite substrate. Such critical properties as in-plane CTE, out-of-plane CTE, glass transition temperature, dielectric constant, fiber-to-resin ratio, and resin microcracking, and their effects on the overall performance of the assemblies in terms of solder-joint reliability, plated through reliability (PTH), dimensional stability, and electrical performance were also briefly reviewed.

ACKNOWLEDGEMENT

The author wishes to thank J.D. Venables for his valuable suggestions and discussions.

REFERENCES

1. G.F. Love, IPC Technical Review, pp. 12-19 (1981).

2. R.A. Shapery, J. Composite Mater. 2, 380 (1968).

3. J.C. Halpin and N.J. Pagano, AFML TR68-(395).

4. J.E. Ashton, J.C. Halpin, P.H. Petit, Primer on Composite Materials: Analysis (Technomic Publishing Co., Stanford, Conn., 1969), pp. 88-91.

5. Z.N. Sanjana, J. Valentich, and J.R. Marchetti, IPC-TP-488, 1983.

6. J.K. Lake and R.N. Wild, Natl. SAMPE Symp. Exhib. 28, 1406 (1983).

7. A.M. Ibrahim, K.A. Green, B.B. Djordjevic, and J.D. Venables, Internatl. SAMPE Symp. 32, 1238 (1987).

8. A.M. Ibrahim, presented at IPC Fall Meeting, San Diego, CA, 1986.

9. S. Lee, IPC-TP-503, 1984.

10. A.M. Ibrahim and J.C. Seferis, in Interrelations Between Processing, Structure and Properties of Polymeric Materials, edited by J.C. Seferis and P.S. Theocaris (Elsevier Science Publishers, Amsterdam, 1984), pp. 325-341.

11. A.M. Ibrahim and J.C. Seferis, Natl. SAMPE Symp. Exhib. 28, 581 (1983).

12. A.M. Ibrahim and J.C. Seferis, Polymer Composites 6, 47 (1985).

13. A.M. Ibrahim, presented at 9th International Thermal Expansion Symp., Pittsburgh, PA, 1986.

14. H. Lee and K. Neville, Handbook of Epoxy Resins (McGraw Hill, New York, 1967), pp. 14-36.

15. J.M. Butler, R.P. Chartoff, and B.J. Kinzig, Natl. SAMPE Tech. Conf. 15, 660 (1983).

16. A.M. Ibrahim, T.K. Shah, L.J. Matienzo, and J.D. Venables, Internatl. SAMPE Symp. Exhib. 31, 669 (1986).

SURFACE MIGRATION OF LOW MOLECULAR WEIGHT PRODUCTS INDUCED BY HYGROTHERMAL FATIGUE IN EPOXY COMPOSITES

A.PORTO**, A.LICCIARDELLO*, D.GRASSO*,
A. CAVALLARO**, G.FERLA** and O.PUGLISI*
*Dipartimento di Scienze Chimiche, viale A.Doria 6, Catania, Italy.
** S.T.Microelettronica, Stradale Primosole, Catania, Italy.

ABSTRACT

The hygrothermal fatigue induces a complex phenomenology in the epoxy composites, with degradation of the linear coefficients of expansion.
Epoxy composites, (novolac type, with SiO_2 as filler, 75÷85 % in weight) have been exposed to high temperature and high pressure moisture.
A role important is played by the decomposition products formed during the aging. The Molecular Weight Distribution of these species has been followed by GPC and shows that there is partial depolymerization of the matrix after 24 h treatment time. It has been found that there is surface migration of low Molecular Weight species, both additives and decomposition products.

INTRODUCTION

The wide use of epoxy composites for microelettronic molding applications has stimulated the research on environmental aging effects on these material [1]. The most studied environmental degradation is that induced by the epoxy moisture high temperature interaction [2] and accelerated life tests have been developed to determine the effects of real time environmental aging [3]. The most dangerous effects induced by the aging are the depression of the glass transition temperature (Tg), the mechanical and chemical degradation [4] of the epoxy matrix and the segregation of some brominated additives at resin metal interface [5].
There is a strong correlation among these phenomena but little is known about the nature, the mechanism of formation and the possibility of surface migration of the degradation products. In this paper we report a study of the effects induced on commercial composite epoxy resins by the exposure to high temperature, and high pressure moisture environment, with enphasis on the dilatometric properties and the chemical degradation.

EXPERIMENTAL

The samples here studied were epoxy novolac composites with high thermal conductivity (supplied by NITTO and TOSHIBA) or standard thermal conductivity (TOSHIBA and PLASKON). The filler of the composites was crystalline SiO_2 (≈85% in weight) for the high thermal conductivity samples and amorphus SiO_2 (≈75% in weight) for the low thermal conductivity ones. The results shown in the figure refer to the NITTO sample but the qualitative behaviour of the other samples is similar. For the all samples the curing was done according to the manufacturer's suggestions. For the 48 hours extraction with tetrahydrofuran (THF) the samples were cured and not molded, obtaining a rather porous material. For 30 minutes extraction with THF the samples were molded and cured obtaining disks 5 cm diameter, 0.3 cm thickness. The exposure to the moisture was performed within a pressure pot (120°C, 2.0 atm.). The glass transition temperature was measured by means of a METTLER TC 10A equipped with a TMA 40 cell. In this cell a probe rests on the surface of pellet of the epoxy composites. By measuring the vertical displacement of the probe, as the temperature is increased, one gets a measure of the linear coefficient of expansion, which increases abruptly at the glass transition temperature. Gel Permeation Chromatography (GPC) was done with a VARIAN 500 L.C. equipped with two Ultrastyragel (TM) columns filled with a bed of highly cross-linked polystyrene. This bed is porous and with a well defined pore size distribution . The small molecules enter a large fraction of the pore structure and are delayed; the large molecules enter a small fraction of the pore structure or are completely size-excluded, so that at the end of the columns, the U.V. detector will detect first the larger molecules and last the smaller ones. The elution time scale is calibrated by injecting know M.W. polystyrene standards, but for chemical species other than polystyrene this scale is only indicative (see Figs.3 and 4).

RESULTS AND DISCUSSION

Fig. 1 shows the effect of the high temperature moisture-resin interaction on the thermal expansion of a commercial epoxy composite (see EXPERIMENTAL). As can be see the dilatometric curve is heavily affected by the treatment and the effect is detectable even with the one-hour exposure sample. This complex dilatometric behaviour is accompained by a corresponding change of the thermal expansion coefficient, α , shown in Fig.2. The figure shows that on going towards longer and longer exposure time there is the growth of an anomalus peak in the α -curve. Moreover, this anomalous behaviour is confined to the near Tg region, while for the temperature far from the Tg, the curves remain almost unchanged. The apparent Tg values, corresponding to the inflection point of the curves, undergo a notable depression ($\Delta T \sim 20°C$).

We have also found that the effects induced by longer exposure time, 48h and 72h, are almost superimposable to those of the 24h curves and therefore the corresponding curves are not shown in the above figures.

Before leaving Figs 1 and 2 there is to note the surprising behaviour of all the samples on rescanning. When the samples were rescanned their dilatometric curves improved in a rather surprising way and became superimposable to the "as-received" curves. The improvement occurred only for those sample that during the first scanning were heated at temperature higher than 200°C, while for sample heated up to T<180°C, the rescanning did not show any appreciable change. Therefore we must conclude that the dilatometric anomaly induced by hygrothermal treatment is reversible for heating at temperature higher than 200°C. This fact suggests that the improvement upon heating is not directly related to the desorption of water from the sample. On the contrary this anomaly might be in principle be related to the formation of decomposition products during the

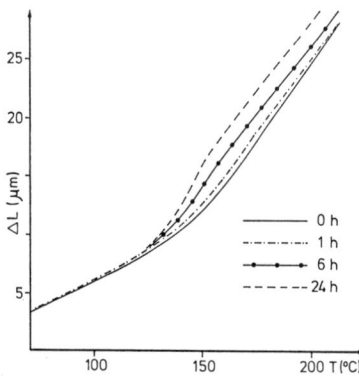

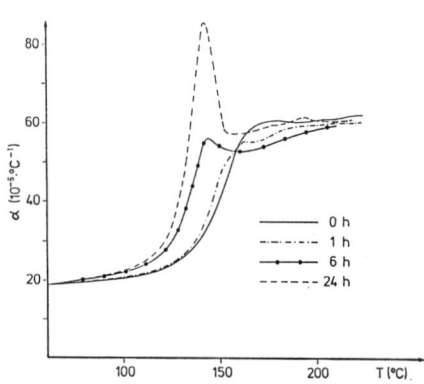

Fig.1 Thermal expansion vs temperature at different treatment time .

Fig.2 Influence of the treatment on the linear coefficient of expansion.

hygrothermal aging.It is well known that the presence of decomposition products in the bulk of the composite can cause considerable depression of the Tg [6] thus, it was interesting to see whether the hygrothermal aging induces chemical modification in the organic matrix of the composite. To do this we extracted the organic soluble material by soaking the epoxy composites in THF for 48 h. The solution so obtained were analyzed by GPC and the molecular weight distribution was obtained. Fig.3 shows the GPC traces of the soluble material extracted from the samples with different treatments. The amount of soluble material is rather different for the various cases and is indicated in Fig.3, normalized to the organic content of the "uncured" resins (100%). The uncured resin trace is shown in the bottom of the Figure and it can be considered as the reference for the M.W.D. of the pristine oligomers. When the resin is cured, reticulation occurs and the most part of the resin becomes insoluble in the solvent. Only a small fraction (1.3%) can be extracted and the corresponding GPC traces has a minor component at high M.W., while the most part of this soluble material is constituted by low M.W. species (see Fig.3). The inspection of the I.R spectrum of the mixture shows that typical peaks of this epoxy (904 cm^{-1}, 1602 cm^{-1}) are rather weak, while there are strong CH and CO bands. This can been understood considering that upon curing the resin becames highly insoluble and the most part of the extractable material is due to additives. In the upper part of Fig.3 is shown the GPC trace of the soluble material (4.3 %) extracted from the resin at 24h hygrothermal treatment. As can be seen there is a component at high M.W., very similar to the correspondig high mass M.W.D. of the uncured resin, shown in the bottom of the Figure. This suggests that the 24h treatment induces depolymerization of the organic matrix in agreement with the I.R. spectrum were the typical epoxy peaks are rather intense. Fig.4 shows the results of one experiment devised to show the presence of surface segregation phenomena in these systems. The epoxy composite is again soaked in THF but for only 30 minutes in order to extract only the soluble organic material lying in the outer regions of the composites. Indeed, the typical diffusionl length for THF in an epoxy matrix is $x=(2Dt)^{\frac{1}{2}}$ were t is the time (1800 s) and D is the diffusion coefficient. For similar cases [7] we extimate the D value ranging between 10^{-8} $cm^{2} \cdot s^{-1}$ and 10^{-11} $cm^{2} \cdot s^{-1}$, and the diffusion length turns out to range between $6x10^{-3}$ cm and $2x10^{-4}$ cm. Thus, the soluble organic material extracted from the epoxy composites gives information on the composition of the outer region of the sample. With reference to this point it was interesting to compare the GPC traces of these "surface"extractions

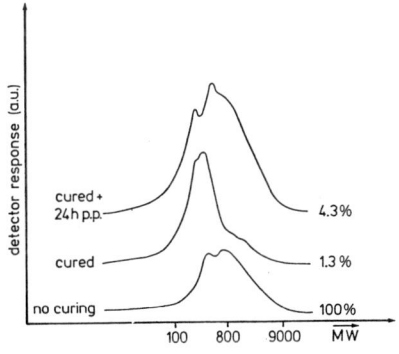
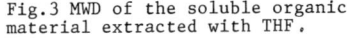
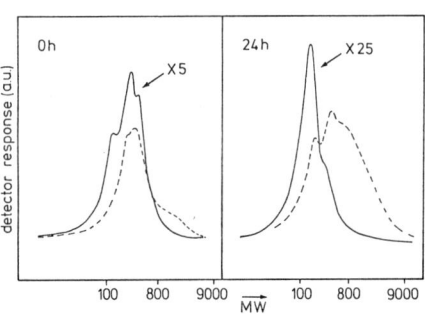

Fig.3 MWD of the soluble organic material extracted with THF.

Fig.4 MWD of the "Bulk"(--) and "Surface" (—) fractions. See text.

with those coming from the whole composite (Fig.3). This comparison has been reported in Fig.4 and gives two important informations:i) the amount of the soluble material extracted from the near-surface region is much more than that expected on the basis of the extimated penetration of the solvent; ii) the M.W.D. for the "surface" extraction is rich in low M.W. compounds, while the high M.W. component present in the "bulk" extracted is almost absent.

The I.R.analysis of the surface fraction of the as received cured sample (Øh) indicates that the mixture is mainly costituted of additives. On the contrary for the 24h sample, the I.R. shows peaks indicative of presence of both additives and epoxy degradation products.

We believe that this behaviour can be interpreted assuming that surface segregation of low molecular weight products occurs in the epoxy composites. During the 24h exposure at high temperature and moisture, the segregation enriches the near surface regions of the composite with low M.W. compounds that can be considered fast diffusers in the cross-linked epoxy matrix. These low M.W. are both additives and decomposition products of the epoxy.

In conclusion the above discussion shows that the hygrothermal fatigue induces a complex chemical and physical behaviour of epoxy composites. This phenomenology seems to be dominated by the phenomena of partial depolymerization (see Fig.3) and surface segregation of low M.W. products, (both decomposition products and additives). Much additional work has to be done in order to understand whether the results of the accelerated test here shown can be extrapolated to real time aging. In any case the phenomenology here described has to be taken in account for high temperature operating devices.

ACKNOWLEDGEMENTS

Financial support from "Progetto Finalizzato Materiali e Dispositivi per la Elettronica a Stato Solido" (CNR, Roma), is gratefully acknowledged.

REFERENCES

1. T.S. Ellis, F.E. Karasz, Polymer, 25, 644 (1984).
2. P. Moy, F.E. Karasz, Polym. Eng. & Sci., 20, 315 (1980).
3. E. Demuts, P. Shyprykevich, Composites, 15, 25 (1984).
4. P. Peyser, W.D. Bascom, J. Mater. Res., 16, 75 (1981).
5. A. Torrisi and S. Pignataro, Surf. Interf. Anal., 9, 441 (1986).
6. F.N. Kelley and F. Bueche, J. Polym. Sci., 1, 549 (1961).
7. Polymer Handbook, edited by J. Brandrup and E.H. Immergut (J. Wiley, N.Y. 1975).

PART IV

Polymer Interfaces and Adhesion

HIGH PERFORMANCE SILICONE GEL AS IC DEVICE CHIP PROTECTION

C. P. WONG
AT&T Engineering Research Center, Princeton, New Jersey 08540

ABSTRACT

Recent advances in IC device encapsulants and polymeric materials have made high reliability VLSI plastic packaging a reality. High performance silicone gel possesses excellent electrical and physical properties for IC protection. With their intrinsic low modulus and soft gel-like nature, silicone gels have become very effective encapsulants for the delicate larger chip size and wire-bonded VLSI chips. Recent studies indicate that adequate IC chip surface protection with high performance silicone gels in plastic packaging could possibly replace conventional ceramic hermetic packaging. This paper will review some potential IC encapsulants. Special focus will be placed on the high performance silicone gel, its chemistry, and its application as a VLSI device encapsulant.

INTRODUCTION

The rapid advances in integrated circuit (IC) technology has had a profound technological and economic impact on the electronic industry. The exponential growth of the number of components per chip, the exponential decrease of device dimensions [1] [2] and the steady increase in IC chip size have imposed stringent requirements not only on the IC physical design and fabrication, but also on the IC encapsulants. The increase of integration in Very Large Scale Integration (VLSI) technology has resulted in a large increase in chip size. Effectiveness in protecting these increasingly large IC devices, high performance encapsulants, such as silicones (elastomers and gels), polyimides, epoxies, silicone-polyimides, and polyxylylene (Parylene®) are being studied.[3] High performance silicone gels possess excellent electrical, chemical, and physical properties for this type of IC protection. With their intrinsic low modulus and soft gel-like nature, silicone gels have become very effective encapsulants for the delicate larger chip size and wire-bonded VLSI chips. Recent studies indicate that adequate IC chip surface encapsulation with high performance silicone gels in plastic packaging could possibly replace conventional ceramic hermetic packaging. [4-6] This paper will review some potential encapsulants. Special focus will be placed on the high performance silicone gel, its chemistry and its application as VLSI device encapsulant.

Device Encapsulation Background

The purpose of encapsulation is to protect the electronic IC devices and prolong their performance reliability. Moisture, mobile ions, (such as: sodium, potassium, chloride, fluorides), UV-VIS, alpha particle radiation, and hostile environmental conditions, are some of the possible modes of degradation or interaction which could negatively affect device performance or lifetime. Silicon dioxide, silicon nitride and silicon-oxy-nitride, which are commonly used as passivation layers, are excellent encapsulants. These passivation layers are known to have excellent moisture and mobile ion barriers of the devices. As for the sodium ion barrier, silicon dioxide is inferior to silicon nitride. However, the use of phosphorous-doped (a few weight percent) silicon dioxide has greatly improved its mobile ion barrier property. A thin layer (in 1-2μm thickness) of one of these dielectric materials is deposited uniformly on the finished device, except on the bond pad areas for bonding. Besides, due to the "edge effect" of passivation of IC device after wire bonding interconnection, we need the protection from the bond pad area. Unfortunately, the passivation layers (such as inorganic oxides and nitrides) are not 100% pinhole or crack-free. To ensure the device reliability, a conformal coating of organic encapsulant is usually used for device encapsulation. Epoxies, polyimides, polyxylylene (Parylene®), silicone-polyimides, and silicones (elastomers and gels) are usually used for this application. Typical properties of these potential encapsulants and their general advantages and disadvantages are listed in Tables I and II.

TABLE I

PHYSICAL PROPERTIES OF SOME POTENTIAL ENCAPSULANTS

Encapsulants	TCE* (ppm/°C)	Modulus (psi)
Epoxy	10 - 80	$1 - 5 \times 10^6$
Polyimide	3 - 80	1×10^6
Parylene	35 - 40	0.4×10^6
Silicone-Polyimide	5 - 100	0.4×10^6
Silicone Gel	200 - 1000	0 - 400

*TCE = Thermal Coefficient of Expansion

General Chemistry of Silicones (Elastomers and Gels)

The basis of commercial production of the silicones is that chlorosilanes are readily hydrolyzed to give disilanols which are unstable and condense to form siloxane oligomers and polymers. Depending on the reaction conditions, a mixture of linear polymers and cyclic oligomers is produced. The cyclic components can be ring opened to linear polymers and it is these linear polymers that are of commercial importance (See Figure 1). The linear polymers are typically liquids of low viscosity and, as such, are not suited for use as encapsulants. These must be cross-linked (or vulcanized) in order to increase the molecular weight to a sufficient level where the properties are useful. Two methods of cross-linking are used: those which can be classified as condensation cures and those which are addition cure systems. For electronic applications, only the room temperature vulcanized silicone which uses alkoxide condensational cure, and platinum catalyzed additional heat cure vinyl and hydride silicone systems are suitable for device encapsulation. RTV silicone elastomer, an organosiloxane, cures by a moisture initiated, catalyst assisted process. It is one of the most effective encapsulants used for temperature cycling and moisture protection of IC devices (See Figure 2).

TABLE II

POTENTIAL IC ENCAPSULANTS

Encapsulants	Application	Advantages	Disadvantages
(1) Epoxies			
	• Normal Dispensing or Molding	• Good Solvent Resistance • Excellent Mechanical Strength	• Non-Repairable • High Stress • Marginal Electrical Performance
(2) Polyimides			
	• Normal Dispensing (spin-coat)	• Good Solvent Resistance • Thermally Stable ($\sim 500°C$)	• High Temp Cure • Non-Repairable • High Stress
(3) Polyxylylene (Parylene®) (® Trademark from Union Carbide)	• Thermal Deposition (reactor)	• Good Solvent Resistance • Conformal Coating	• Thin Film Only • Non-Repairable
(4) Silicone - Polyimides			
	• Normal Dispensing	• Less Stress vs. Polyimide • Better Solvent Resistance vs. Silicone	• Higher TCE vs. polyimides • Thin Film Only
(5) Silicones (RTV, Gel)	• Normal Dispensing	• Good Temp. Cycling • Good Electrical • Very Low Modulus	• Weak Solvent Resistance • Low Mechanical Strength

Since World War II, silicones (organosiloxane polymers) have been used in a variety of applications where properties of high thermal stability, hydrophobicity, and low dielectric constant are necessary, eg., as encapsulants or conformal coatings for electronic components. In 1969, it was demonstrated that room temperature vulcanized (RTV) silicones exhibited excellent performance as moisture protection barriers for IC devices and a number of different RTV silicones have since been adapted for use in the electronics industry [7-16].

Ionic materials, whether from the device surface, encapsulation materials or the environment, affect the electrical reliability of encapsulated IC devices. For this reason, the silicones are subjected to intense purification. The concentration of Na^+, K^+, and Cl^- mobile ions is less than a few ppm, and alpha particle emission is less than 0.001 alpha/cm^2 hr. Thus it offers excellent alpha particle shielding for eliminating soft error in dynamic random access memory devices, such as; 64K, 256K, and megabit chips.[17] The drawbacks of RTV silicone as an IC encapsulant are both its poor solvent resistance and weak mechanical properties.[18] Highly fluorinated alkyl substituted siloxanes have shown an improvement in solvent resistance. However, a recently developed silicone

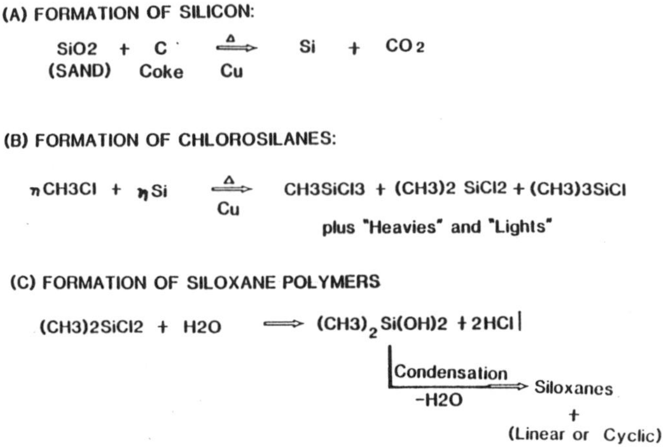

(A) FORMATION OF SILICON:

$$SiO_2 + C \xrightarrow[\text{Cu}]{\Delta} Si + CO_2$$
(SAND) Coke

(B) FORMATION OF CHLOROSILANES:

$$n\,CH_3Cl + n\,Si \xrightarrow[\text{Cu}]{\Delta} CH_3SiCl_3 + (CH_3)_2\,SiCl_2 + (CH_3)_3SiCl$$

plus "Heavies" and "Lights"

(C) FORMATION OF SILOXANE POLYMERS

$$(CH_3)2SiCl_2 + H_2O \Longrightarrow (CH_3)_2\,Si(OH)2 + 2HCl$$

$$\xrightarrow[{-H_2O}]{\text{Condensation}} \text{Siloxanes}$$
+
(Linear or Cyclic)

Figure 1. Preparation of Organosilicones.

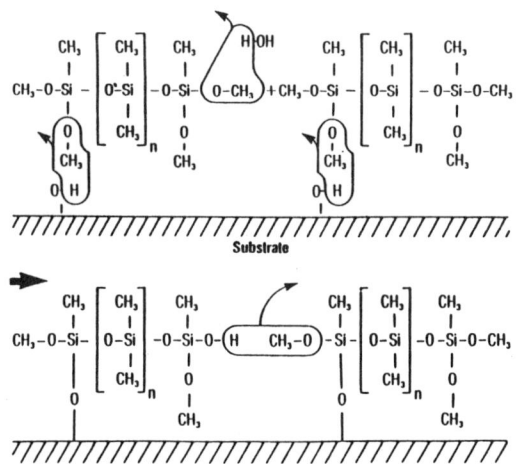

Figure 2. RTV Silicone Cure Reaction Mechanism.

material with high cross-linking and high filler loading system seems to significantly improve the solvent resistance of the silicone encapsulant.[19-20]

Heat curable silicone (either elastomer or gel) has become an attractive device encapsulant. Its curing time is much faster than the RTV-type silicones. Besides, heat curable silicone gels tend to have slightly better thermal properties than conventional RTV silicone. With its excellent jelly-like (very low modulus) intrinsic softness, silicone gel is a very attractive encapsulant in wire bonded large chip size IC devices. The two part heat curable system which consists of the vinyl and hydride reactive functional groups, and the platinum catalyst additional cure system provides a fast cure system without any byproduct (See Figure 3 for cure mechanism). This solventless type of heat curable silicone gel will have increased usage in electronic applications.

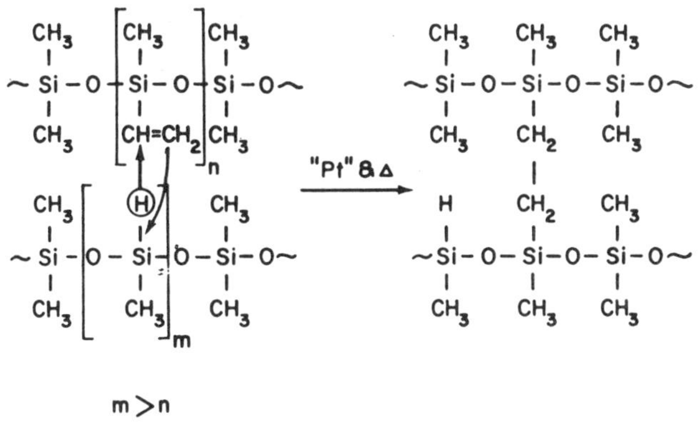

Figure 3. Heat Curable Silicone Cure Reaction Mechanism.

High Performance Silicone - A Potential for Replacement of Ceramic Hermetic IC Packages

Recent advances in silicone materials have made silicone gel a very attractive device encapsulant. Silicones, especially gels, possess excellent electrical properties and ultrapure chemical purity as device encapsulants. Due to their intrinsic low modulus and soft gel-like nature, silicone gels have become very effective encapsulants for the delicate large chip size and wire-bonded VLSI chip.[4-6] AT&T, Hitachi, Burroughs, GM Delco, to name a few, have been using silicone gels for sometime with excellent results. These silicone gels have excellent environment acceleration test results on silicon-on-silicon CMOS module devices (See Table III).[5]

By addition of an excess of inert silicone oil (methyl-terminated polydimethyl siloxane), Hitachi scientists [4-6] have improved all the environment acceleration test results (See above Table III). The excess of the inert silicone oil passivated the IC device and provided the protection in all these environment testings. At AT&T, we have studied silicones as device encapsulants for some time.[7-13] We have found that certain additives, such as crown ethers and cryptates, could effectively immobilize the mobile ions and improve the encapsulant electrical performance.[21-22]

We have also developed a very sensitive micro-dielectric measurement to define the degree of cure of the silicone gel. This microdielectrometry which utilized the Micromet miniature IC sensor and a wide range of frequencies to monitor the polymer cure by loss factor (E") measurement is a very attractive technique to quantify the degree of silicone gel cure.

TABLE III

ENVIRONMENTAL ACCELERATION TEST RESULTS
OF SI ON SI CMOS MODULE

TEST ITEMS	CONDITIONS	RESULTS (No. of Failures)
High Temperature Operation	125°C 1000hr.	0/32
High Temperature Storage	150°C 1000hr.	0/22
Low Temperature Storage	-35°C 1000hr.	0/22
High Humidity Storage	65°C/95% R.H.	0/22
(High Humidity Bias) cycling	85°C/85% R.H. 1000hr.	0/22
Temperature Cycling	-35°C <==> 150°C 1000 cycles	0/22
Thermal Shock	-35°C <==> 125°C 100 cycles	0/22

Ref: T. Yamada, et. al., IEPS, 557 (1985)

In addition, we have also developed a unique procedure to reduce the cured silicone gel hardening which is the major cause of the wire-bond breakage of the encapsulated delicate wire-bond devices. Results of these techniques will be discussed.

Experimental

A. Trapping of Mobile Ions (Such as Na^+, K^+, HCl, Cl^-)

The low level of contaminant ions (such as, sodium, potassium) in ppm level which are present in the silicone was determined by atomic absorption. Stoichiometric amounts of crown ethers or cryptates were first dissolved in a small amount of xylene, then mixed in with the silicone. The material was mixed overnight to allow the crown ether to chelate the mobile ions within the silicone resin (See Figure 4) prior to coating on the triple track test device for accelerating Temperature Humidity Bias (THB) Testing. Results are shown in Figure 5.

B. *Curing of Silicone Gel*

A thin layer (approximately ⁻5-10 mils thick) of silicone gel was coated on the Micromet mini-dielectric sensor. The silicone gel was cured at 150°C. The loss factor (E") at 0.05, 1,100, 1000, 10,000 HZ frequencies were monitored with curing time. The result is shown in Figure 6.

C. *Poisoning of Cured Silicone Gel to Prevent Gel Hardening*

The silicone gel of components A and B, at a prescribed mixing ratio, was filled into a polyurethane pouch of ⁻0.35 inch thickness and cured to the prescribed schedule. Then, the cured gel within the pouch was exposed to an ammonia gas or ammonia solution at a closed container for overnight. Results are shown in Figure 7.

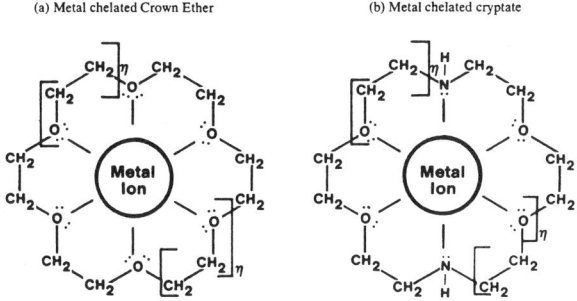

Figure 4. Metal Chelated Crown-Ether and Cryptate.

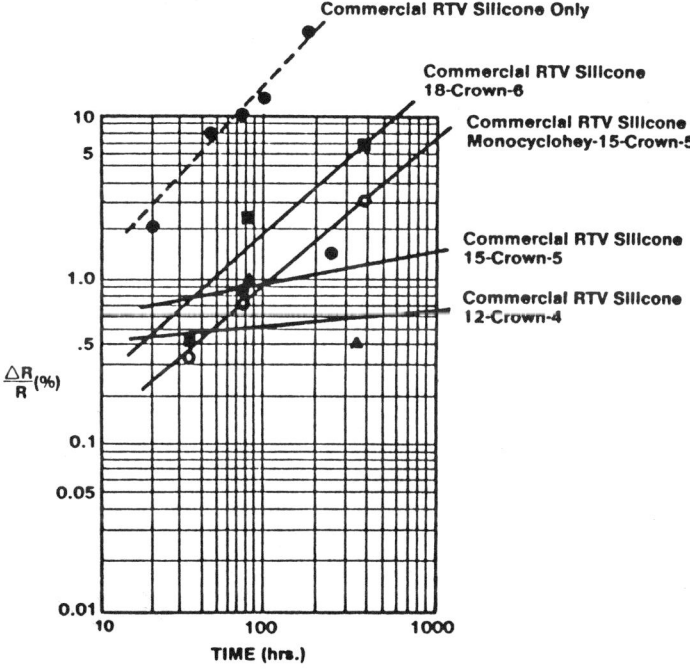

Figure 5. THB Testing Results of Crown-Ether Added Silicone

RESULTS & DISCUSSIONS

The results of the THB testing on Triple Track Resistor (TTR) testing showed that the inclusion of crown ethers and cryptates into a silicone formulation dramatically enhanced the electrical reliability of our test vehicles (Figure 5). In the triple track resistor experiment, we grounded the two outer tracks and biased the center track. Then we measured the resistance change between the centers of the conductor lines. This process measures the degree of "electro-oxidation." Leakage currents due to impurities can cause the resistor to anodize. The change of the resistance with respect to the original resistance will increase with time. This is mainly due to the oxidation process. The less the resistance changes with testing time, the better the encapsulant material will be. This data adds further evidence that sodium and potassium ions contribute to the failure of devices in as much as the crown ethers with the smaller cavities outperform those with larger cavities. The 12-crown-4, with a cavity diameter of 1.8*angstrom* may be more suitable for complexing Na^+ with an ionic diameter of 0.95*angstrom*.[21] The 15-crown-5, with a cavity of 2.7*angstrom*, may be most effective for K^+ at 1.35*angstrom* (See Figure 4). In our experiment, the sodium and potassium ions both seem to have been trapped within the crown ether quite securely even under the most severe testing conditions. Formation of 'ion-pairs' between the metal chelated cation-crown ether and halogens counterion is quite feasible. The pairing of metal chelated crown ether with halogen ions [22-23] (i.e. Cl^-) would be beneficial in the trapping of chloride contaminant materials. Since most halogens are potential harmful contaminants in an encapsulant material, the formation of 'dendrites' - which increases the leakage current between conductor path is greatly enhanced by the presence of halogens especially under high electrical potential bias, temperature and humidity. Such an argument has been confirmed, and has been well documented.[24]

Cryptates have also been known to coordinate hydrochloric acid in solution chemistry.[25] Hydrochloric acid and chloride ion have been associated with the metal migration in the IC devices. The addition of these cryptate compounds to the silicone encapsulant may thus also have potential as HCl and Cl^- scavengers. This finding may also be used to prevent silver, gold and copper ions migrations in electronic devices. The thermal and hydrolytic stability of crown ethers and cryptates must be taken into consideration, however, when chosen as additives. Also, since these compounds may be potentially hazardous to our health, caution must be taken in using these types of compounds. Ways to eliminate the leaching of cryptates from the encapsulant were proposed by incorporating the cryptate into the backbone or grafting into the substituent side chain of the siloxane polymer. This has been shown to be feasible.[26] The use of crown ethers and cryptates to eliminate contaminant ions may have some potential application in trapping contaminant ions.

In the microdielectric measurement experiment, the loss factors of the silicone gel increase rapidly at the beginning of the cure process (See Figure 6). This is mainly due to the increase in oven ramping temperature (from room temperature to 170°C) which generates the thermal activation of ionic conductivity, as we can see from the cure temperature profile. However, the loss factors (E") with all frequencies (0.05, 1, 100 HZ) decrease rapidly after the temperature stabilizes at 170°C. We can attribute this decrease in E" to the tightening of the silicone gel network and reduction of the ionic conductivity. (Please note the loss factor is in log scale). After ~80 minutes heating at 170°C, there is no change in E" of the material, which is a good indication of complete curing of the silicone gel. This microdielectric loss factor measurement provides a fast

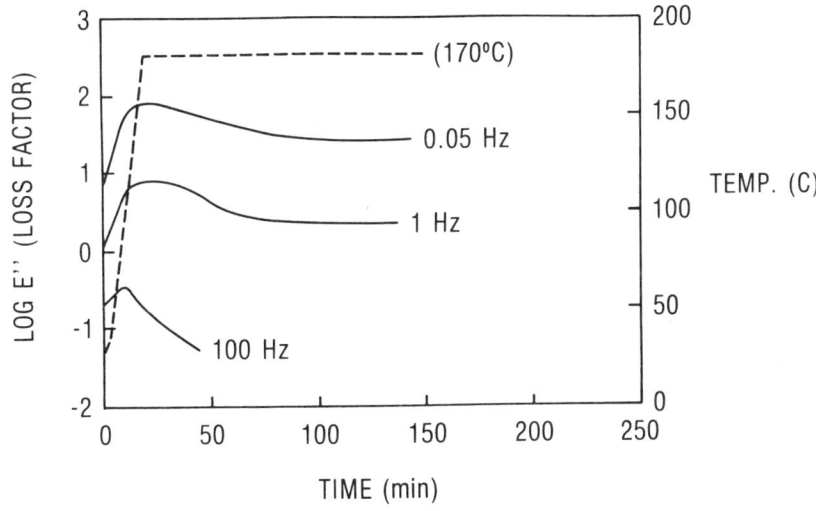

Figure 6. Microdielectric Measurement of Silicone Gel Cure Reaction

and real time measurement of degree of cure of the silicone gel.

Most of the silicone gels chemistry are based on reaction of silicon vinyl (Part A) and silicon hydride (Part B) addition cure mechanism. The addition of the hydride from the silicon hydride to the vinyl and the formation of carbon-silicon crosslinking is the key to the reaction mechanism. A trace amount (a few ppm) of platinum catalyst is needed to facilitate the crosslinking (See Figure 3). However, most of the silicone gel formulations have an excess of silicon hydride remaining after the crosslinking of vinyl and hydride reactive functional groups (by FT-IR measurment of the cured gel). With the remaining reactive platinum catalyst in the cured silicone gel, this excess silicon hydride remaining from the cured silicone gel will further react with moisture in the ambient to generate silicon hydroxyl groups and hydrogen gas. These silicon hydroxyl groups will further react with the excess silicon hydrides and/or themselves to increase the crosslinking density which results in a higher crosslinking and hardening gel (See Figure 8). As the silicone gel hardens, the modulus of the gel increases. This increase of modulus causes the wire-bond breakage of the delicate devices. When silicone gels are subjected to elevated temperature aging, the gel hardening rate will also increase.

To prevent this gel hardening, we have developed a simple process to poison the reactive platinum catalyst. A heteroatom gas (such as, ammonia) is used to diffuse through the cured silicone gel. This ammonia gas will react with the reactive platinum and poison the catalyst. Since there is only a few ppm platinum is used in the gel system and the diffusion rate of ammonia gas through the silicone gel is relatively fast, only a brief exposure to ammonia is required. This catalyst deactivation process is quite simple and straightforward. [We have used a polyurethane pouch to contain the silicone gel and measured the defraction force of the silicone gel-filled pouch to determine its hardening

184

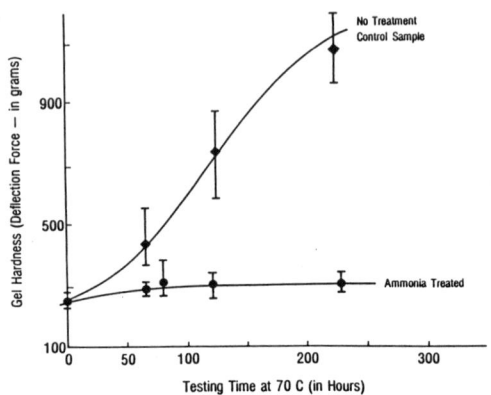

Figure 7. Result of the Ammonia Treated Silicone Gel

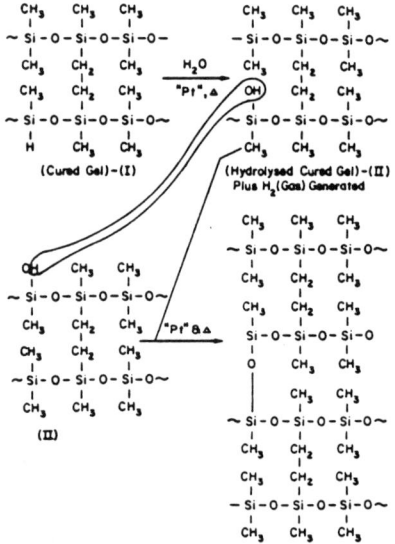

Figure 8. Silicone Gel Hardening Reaction Mechanism

effect]. Results of the deactivated silicone gel are shown in Figure 7.

Both ammonia treated and untreated pouches were placed in a 70°C oven for accelerating testing. There is a distinct difference between the treated and untreated samples. The ammonia treated silicone gel materials shows a slight initial increase in hardness which could be attributed to the initial gel hydrolysis or hardening prior to ammonia treatment. However, thereafter, it remains at constant hardness. At the same time, exponentially at the 70°C temperature of the study. The material stress (S) is a function of modulus (E) times the difference in thermal co-efficient expansion (TCE) of the materials, and the difference in temperature (dT) of T_1 and T_2. The expression of this type of equation is as follows:

$$S = \kappa \int_{T_1}^{T_2} dT \, (\Delta TCE) E$$

The increase in modulus due to hardening could generate enough stress to cause wire-bond breakage and device failure. Experiments in coating the test chip with ammonia-treated silicone gel show over 1000 cycles at -40°C⟷150°C without any wire-bond breakage on multiwire-bonded mirror test chip. However, the untreated control samples show high wire-bond failure. In addition, the thin coating of the silicone gel is an excellent sodium or chloride mobile ions barrier. (This result will be subjected to future discussion). In addition, the THB performance of the soft gel is excellent at 85°C/85 % R.H. and 10Vdc bias of the accelerated condition - which is shown in Figure 9.

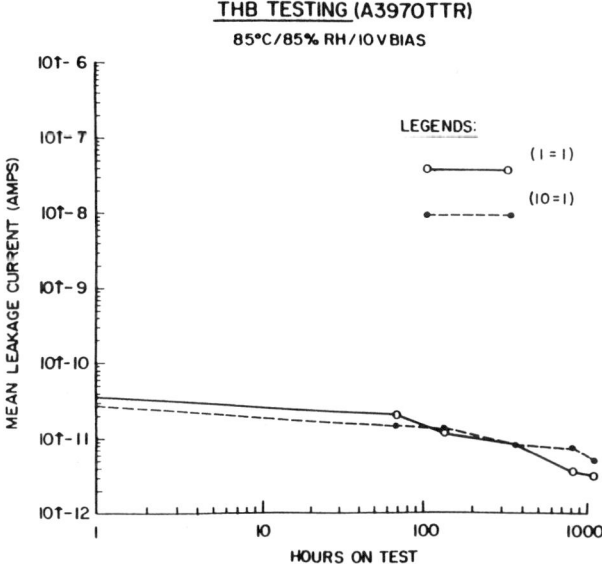

Figure 9. THB Testing Results of Silicone Gel.

CONCLUSIONS

We strongly believe that the high performance silicone gel possess an excellent IC chip protection and have a potential to replace the ceramic hermetic package of IC.

REFERENCES

1. G. Moore, "VLSI, What Does the Future Hold?", Electronics Australia, 42, 14 (1980).

2. S. M. Sze, Ed., "VLSI Technology," McGraw-Hill, New York, New York (1983).

3. C. P. Wong, "Integrated Circuit Encapsulants," A Chapter in "Polymers in Electronics," Second Edition of The Encyclopedia of Polymer Science and Engineering, Vol. 5, 638 (1986), John Wiley and Sons Publisher, New York, New York.

4. K. Otsuka, Y. Shirai, K. Okutani, "A New Silicone Gel Sealing Mechanism for High Reliability Encapsulation," IEEE Transactions on Components, Hybrids, and Manufacturing Technology, Vol. CHMT-7, No. 3, P. 249 (1984).

5. T. Yamada, et. al., International Electronic Packaging Society, IEPS, 557 (1985).

6. C. P. Wong, "Can IC Surface Protection Replace Hermetic Ceramic VLSI Packaging?", p. 45, Program and Extended Abstracts, 5th VLSI Packaging Workshop, Co-spoonsored by IEEE-CHMT Society and National Bureau of Standards, Nov. 17-18, 1986, Paris, France.

7. M. L. White, "Encapsulation of Integrated Circuits," Proc. IEEE, Vol. 57, p. 1610 (1969).

8. R. G. Mancke, "A Moisture Protection Screening Test for Hybrid Circuit Encapsulants," IEEE Trans. Components, Hybrids and Manufacturing Technology, Vol. 4, No. 4, 492 (1981).

9. D. Jaffe, N. Soos, "Encapsulation of Large Beam Leaded Devices," Proc. Electronic Components Conf., 213 (1978).

10. C. P. Wong, "High Performance RTV Silicones as IC Encapsulants," The International Journal for Hybrids and Microelectronics, Vol. 4 (2), 315 (1981).

11. C. P. Wong, D. E. Maurer, "Improved RTV Silicone for IC Encapsulants," National Bureau of Standards, Special Publication 400-72, Semiconductor Moisture Measurement Technology, 275 (1982).

12. C. P. Wong, "Improved Room-Temperature Vulconized Silicone Elastomers as Integrated Circuit Encapsulants," "Polymer Materials for Electronics Applications", American Chemical Society Symposium Series, No. 184, 171 (1982).

13. C. P. Wong, D. M. Rose, "Modified RTV Silicone as Device Packagings," 33rd Electronic Components Conference Proceedings, 505 (1983).

14. C. P. Wong, "Integrated Circuit Devices Encapsulants," An Intensive Short Course in "Polymers in Electronics," University Extension, University of California at Berkeley, 1-30, August (1983).

15. C. P. Wong, "Thermograviometric Analysis of Silicone Elastomers and IC

Device Encapsulants," Chapter 23 in "Polymers in Electronics," American Chemical Sociiety Symposium Series, No. 242, 285 (1984) and references therein.

16. C. P. Wong, "Effects of RTV Silicone Cure in Device Packagings," Polymer Science and Engineering Precedings, American Chemical Society, Vol. 55, 803 (1986).

17. J. E. Riley, "Ultrahigh Sensitive Uranium Analyses Using Fission Track Counting: Further Analysis of Semiconductor Packaging Materials," J. Radioanalytical Chemistry, Vol. 72, 89 (1982) and references therein.

18. W. Noll, "Chemistry and Technology of Silicones," Academic Press, New York (1968).

19. C. P. Wong, United States Patent No. 4,564,562 (Jan. 14, 1986).

20. C. P. Wong, United States Patent 4,592,959 (June 3, 1986).

21. C. J. Pederson, *J. Amer. Chem. Soc., 89,* 7017 (1967).

22. C. P. Wong, United States Patent 4,271,425 (June 2, 1981).

23. C. P. Wong, United States Patent 4,278,784 (July 14, 1981).

24. K. W. Michael, R. G. Antonen, *Proceedings of the Soc. of Hybrids and Microelectronics Conf.,* p. 253, Anaheim, CA (1978).

25. J. M. Lehn, B. Dietrich, J. P. Savuage, *Tetrahedron Letters,* 2885 (1969).

26. C. P. Wong, United States Patent 4,396,796 (Aug. 2, 1983).

CHEMISTRY OF ADHESION AT THE POLYIMIDE-METAL INTERFACE

M. Grunze[*], W.N. Unertl, S. Gnanarajan and J. French, Laboratory for Surface
Science and Technology and Department of Physics, University of Maine, Orono,
ME 04469.
[*]Present address: Lehrstuhl fuer Angewandte Physikalische Chemie, Institut
fuer Physikalische Chemie, Universitaet Heidelberg, Im Neuenheimer Feld 253,
6900 Heidelberg, West Germany.

ABSTRACT

This article describes recent studies of the chemistry of adhesion between thin (d $\geq$ 11 Å) polyimide films and silver and copper substrates, and
the structural changes in the polymer when polyamic acid is imidized to polyimide. The thin polyamic acid films were formed by vapor phase deposition of
1,2,4,5-benzenetetracarboxylic anhydride (PMDA) and 4,4-oxydianiline (ODA)
under high vacuum conditions and subsequent imidization by heating in vacuum.
Both ODA and PMDA are at least partially dissociated upon adsorption onto
clean copper and silver and with increasing film thicknesses react to form
the polyimide precursor, polyamic acid. Heating to T $\geq$ 425 K leads to polymerization to form polyimide films which are thermally stable to about 700 K.
Polyimide films with mean thicknesses as small as 1.1 nm have been fabricated
in this way and their bonding to the substrate as determined by x-ray photoemission studies is summarized. Infrared reflection absorption data gives
further evidence that the polyimide bonds to the substrate via fragmented
PMDA. Changes in the surface topography and molecular structure of the films
during imidization are demonstrated by scanning tunneling micrographs and infrared reflection absorption data.

INTRODUCTION

Polyimides (PI) are high temperature polymers that have a unique combination of thermal stability, low dielectric constant, chemical inertness and
easy processibility into coatings or films. The most commonly used polyimides are those formed by the reaction of 4.4'-diaminodiphenyl ether (oxydianiline (ODA)) and 1,2,4,5 benzenetetracarboxylic anhydride (pyromellitic di
anhydride (PMDA)). In microelectronic device applications [1-3], these are
used in both packaging [1,4] and as insulating interlevel dielectrics [1,5].
The successful adhesion between PI and metals is essential in these applications and the physical and chemical factors which contribute to the adhesion are of fundamental interest. In the absence of extrafacial inhomo-
geneity (eg. stress free films) the strength of the adhesive couple is dependent directly on the physics and chemistry at the polymer/metal interface
[6]. This has prompted a number of investigations to probe the microscopic
origins of the adhesive bonding. Utilizing surface science techniques such
as X-ray photoemission spectroscopy (XPS) and near edge x-ray absorption fine
structure (NEXAFS), the electronic core and valence structure of PMDA-ODA
polymer/metal interfaces have been studied [7-12]. Other techniques such as
transmission electron microscopy [13], Rutherford backscattering spectroscopy
[14] and electron energy loss spectroscopy [15,16] have also been utilized in
the study of polymer/metal interactions.
Studies of thin metal films deposited on the surface of a much thicker
(usually bulk) polyimide phase have provided the main source of chemical information [7-12,17]. For example, room temperature deposition of chromium
leads initially to bonding to the PI substrate, possibly via the carbonyl
groups, and subsequently with increasing chromium coverage to formation of a
carbide like carbon species [18]. Similarly, other electropositive metals
such as aluminum, [11] titanium [12] and nickel [7] also appear to react
through this carbonyl entity. Copper [7,9,11] and silver [7], however, show

only a weak interaction with the oxygen in the ether part of the chain.

The second method for producing a metal/polyimide interface is spin coating or evaporating the polymer precursor (PAA) onto a supported metal film, prior to curing and the formation of polyimide. In the spin coating process the polyamic acid is dissolved in a polar solvent, e.g. N-methylpyrrolidone (NMP), whereas in the evaporation process both polymer constituents PMDA and ODA are deposited onto the substrate [19-21]. Bulk polyimide/metal interfaces formed in this way have been shown to produce a marked increase in, for instance, the peel strength of a PI/copper oxide interface compared to conventional metal deposition [17]. The precursor/metal interfacial reaction is apparently much stronger compared to that of the metal/PI interfacial reaction where the polymer is fully cured prior to metal deposition.

The presence of the polar solvent must also be considered in a comparison of spin coated and vapor deposited films. As demonstrated recently by Kim, et al. [17] and Kowalczyk, et al. [22] for polyimide/ copper oxide interfaces, the presence of the solvent NMP leads to formation of cuprous oxide particles in the polymer film, whereas interfaces prepared by the solventless method, i.e. vapor deposition of the organic constituents or copper evaporation onto fully cured polyimide, showed no such precipitates.

The way the interface is prepared needs to be considered if any valid comparisons are to be made between the adhesion of different metals/polyimide interfaces. The fundamental variations between each type of interface will essentially reflect the electronic and therefore chemical properties of the bulk metals as compared to those of metal atoms, clusters or very thin metal films. However, the formation of the polymer (from the dissolved precursor or from the vapor codeposited constituents) will also play a significant role in determining the specific bonding.

This brief report is organized as follows. First, we summarize our previous results on the interaction of the pure polyimide constituents PMDA and ODA with clean silver and copper surfaces. Codeposition of PMDA and ODA leading to polyamic acid formation is discussed in relation to the adsorption of the pure constituents. As will be shown, imidization of polyamic acid films involves structural changes in the polymer film resulting in an average orientation of the polyimide chains parallel to the substrate. We finally show the first data on the topography of thin vapor-deposited polyimide films as determined from scanning tunneling micrographs.

EXPERIMENTAL

Our experimental setup and the procedures to produce ultra-thin polyimide films by vapor deposition has been described in detail previously [21,23]. In brief, the constituents PMDA and ODA were deposited from heated quartz tubes in a UHV chamber onto the substrate held at room temperature, followed by heating in vacuum to imidize the film. The various stages of deposition and reaction were followed by x-ray photoelectron spectroscopy using a hemispherical electron energy analyzer (Leybold Heraeus EA11) operated with a resolution of 0.92 eV as measured on the Ag 3d 3/2 emission. The infrared data were recorded with a home-built vacuum IR Reflection Absorption Spectrometer with a grating monochromator and with a commercial FTIR Spectrometer (Mattson Cygnus 100). The IR data have been corrected for the transmission of the instrument so that relative intensities can be compared over the whole spectral range. The STM results at various deposition and curing stages were obtained with a Nanoscope (Digital Instruments NanoScope I) operated in air.

RESULTS AND DISCUSSION

Both PMDA and ODA undergo fragmentation on clean polycrystalline silver and copper surfaces and on a Cu(111) substrate [20,23,24]. An important aspect for evaluating the possible bonding situation of polyamic acid and subsequently polyimide with the metal substrates is the degree of dissociation, i.e. whether the fragments still retain functional groups which can react with ODA or PMDA to effectively interlink the polymer film with the substrate. The degree of fragmentation on the surface was determined from the XPS data by the C 1s, O 1s and N 1s binding energies and an evaluation of the stoichiometry of the surface phase, as discussed in detail in references [21,23]. A model for the bonding configuration of the molecular fragments consistent with the core level binding energies and the stoichiometry of the surface layer can be derived, but needs to be tested by direct structural probes. In the following we summarize our previous XPS data and show IR reflection absorption data for ultra-thin films of PMDA, ODA, polyamic acid and polyimide on polycrystalline copper which support our previous findings that polyimide adhesion occurs via a chemically bonded fragmented layer of PMDA and possibly ODA.

Adsorption of PMDA and ODA on Silver and Copper Surfaces

The XPS data taken for PMDA adsorbed on polycrystalline silver exhibited distinctive changes as a function of film thickness, i.e. monolayer to multilayer adsorption. The spectra taken for film thicknesses in excess of d ~ 11 Å clearly showed the expected binding energies and intensities for molecularly condensed PMDA. For monolayer coverages, however, the interpretation of the C 1s and O 1s band revealed a deficit of one CO moiety of PMDA, indicating the reaction of one anhydride functionality with the surface leading to the release of one CO molecule into the gas phase. Angular resolved XPS measurements further suggested a tilted bonding geometry of the PMDA fragment with respect to the substrate. The resulting model of PMDA adsorption on silver at room temperature is shown in Fig. 1. A tilted bonding geometry of the PMDA fragment has also been observed on a Ag(110) surface in a NEXAFS study [25].

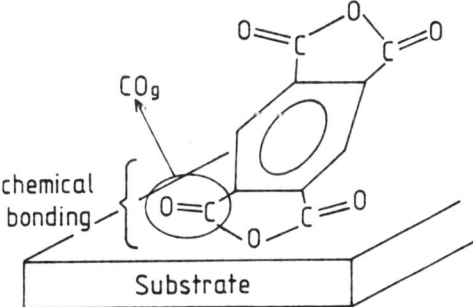

Figure 1.   Proposed bonding model for PMDA on a polycrystalline silver surface.

The bonding for PMDA on silver is not fully understood but probably involves interaction of the oxygen atoms of one fragmented anhydride group in a monodentate or bidentate configuration and possibly of a carbon atom in the phenyl ring with the substrate. The bonding to the substrate is relatively strong, since heating the layer does not lead to desorption but to further decomposition [23]. This bonding configuration is such that the undisturbed anhydride group of the PMDA fragment can react with ODA and thus link a polyimide chain to the substrate. Indirect evidence for such a bonding configuration in the polyimide/metal interface has been obtained in our studies of ultra-thin (11 Å < d < 35 Å) polyimide films on polycrystalline silver and copper surfaces. The O 1s data showed spectral features which are consistent with the interfacial presence of fragmented PMDA and ODA. Further support

for such a model was derived from the observation that pure PMDA and ODA layers undergo decomposition upon heating to temperatures necessary to imidize
the polyamic acid, whereas in the case of thin polyamic acid films no evidence was found for a dissociative reaction in the interface. This result
indicates that the PMDA (and ODA) fragments are thermally stabilized by the
polyimide overlayer, which in turn means that a chemical bond must exist between interface and polymer film.

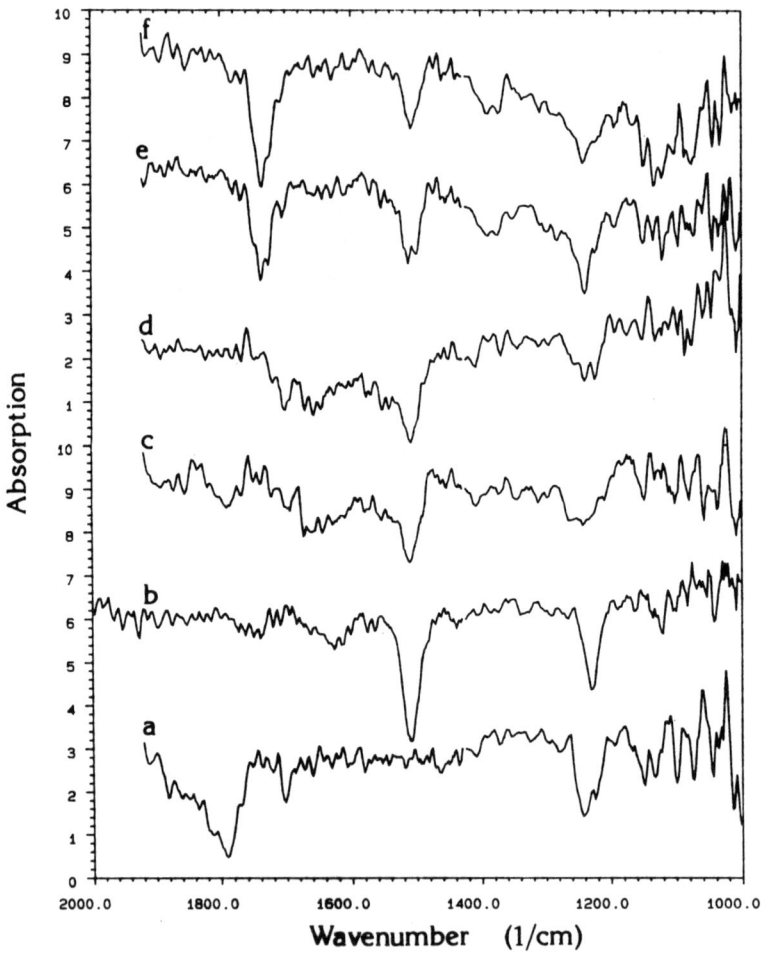

Figure 2.   Infrared Reflection Absorption data with a polycrystalline copper
substrate.   a) PMDA, T = 300 K; b) ODA, T = 300 K; c) Codeposited
ultra-thin film of PMDA and ODA, T = 300 K; d) Film heated to 375
K for 15 min.; e) Film heated to 440 K for 15 min.; f) Film
heated to 473 K for 15 min.

Whereas the stoichiometry of the PMDA surface complex on silver was
well defined, our XPS experiments on copper indicated additional decomposi-

tion involving the loss of two carbonyl groups [24]. IR reflection absorption data for a thin film (1-2 monolayer) of PMDA on polycrystalline copper are shown in Fig. 2a. The spectra displayed were taken after dosing the substrate at 220 K and subsequently heating to 300 K. A comparison of the IR reflection absorption spectrum to that of bulk PMDA [26] or dissolved PMDA [27] reveals that the phenyl ring modes (expected at 1368 cm^{-1} and 1460 cm^{-1}) are missing or very weak, whereas the symmetric and asymmetric carbonyl stretching modes at 1790 cm^{-1} and 1860 cm^{-1} (although in a different intensity ratio and not resolved as in the bulk compound) and the ν_{C-O-C_1} anhydride stretching mode (or B_{2u} ring mode of the phenyl ring) at (1240 cm^{-1}) are present (see Table I). In addition, a small absorption band is observed around 1703 cm^{-1}, which does not exist in the PMDA bulk spectra.

Table I

Expt (cm^{-1})	Literature [27] (cm^{-1})	Mode Assignments [30,31]
		PMDA
1085	1085	
1152	1152	
1240	1240	$\nu_{(COC)}$ stretch or ring mode (B_{2u})
----	1368	Ring mode
----	1460	Ring mode (B_{2u}) (ν_{17})
1703	----	Tentatively assigned to a $_{OCO}$ stretch in a formative type surface bond
1790	1770	Carbonyl $\nu_{(C=O)}$ stretch (sym)
1860	1858	Carbonyl $\nu_{(C=O)}$ stretch (asym)
		ODA
1110	1115	
1225	1225	$\nu_{(COC)}$ asym stretch or ring mode (B_{2u})
1275	1282	
1325	1320	
1390	1380	
1510	1500	Ring mode (B_{1u}) (ν_{13})
1625	1620	NH_2 scissor mode or ring mode (Ag)(ν_2)

In molecular PMDA, the phenyl ring and the anhydride groups are all in the plane of the molecule. Thus, if the molecule would physisorb in a configuration parallel to the surface, the in-plane phenyl ring modes as well as the carbonyl and anhydride stretching modes should not be detectable due to the dipole/image dipole cancellation of the effective dipole moment on the metal surface. For a configuration where the PMDA molecular plane is perpendicular or tilted with respect to the surface plane, the above mentioned modes should be active. However, our results show that only the carbonyl and anhydride modes are active, whereas the phenyl ring modes are not detected. This, consistent with the XPS results, reveals that the PMDA molecule must undergo fragmentation in the adsorbate phase, and the absence of the phenyl ring modes might even indicate that concurrent with the loss of carbonyl groups the phenyl ring opens due to a strong interaction with the metal substrate.

The present data are not sufficient to derive a bonding model, in particular since it is not obvious that the adsorbate phase is homogeneous. However, we want to point out some characteristic features. The decrease in the asymmetric carbonyl stretching frequency, when compared to the undissociated condensed molecules at 220 K [26] reveals that to some extent the anhydride functionalities, which include two carbonyl groups, are effected upon heating from 220 to 300 K. Secondly, we find a band at 1703 cm^{-1} which is at a frequency where the carbonyl stretching frequency of formiate species have been observed [28], indicating the possibility of a similar bonding situation for the fragmented PMDA molecule. Most importantly, however, the fragmented PMDA layer still contains anhydride functionalities as indicated by the 1240 cm^{-1} C-O-C stretching frequency. These anhydride functionalities therefore are accessible to a reaction with oxidianiline. Heating the ultra-thin PMDA film to 373 K results in the loss of essentially all molecular vibrations [26]. This is consistent with XPS data, which show sublimation and further fragmentation of PMDA above room temperature.

Oxidianiline was found to decompose partly on both silver and copper surfaces at room temperature in our XPS studies. Contrary to PMDA on silver, where a defined surface species was isolated, ODA is adsorbed in a mixed phase possibly consisting of molecular ODA, dissociated oxygen and oxidianiline fragments [23,24].

The IR reflection absorption spectra of a polycrystalline copper surface with a 1-2 monolayer ODA film (Fig. 2b.) shows, albeit with different intensities than in the dissolved molecular compound [27], all molecular vibrations. Compared to the molecular spectra, the NH$_2$ scissor at 1625 cm^{-1} and NH$_2$ stretching modes (not shown here) are very broad indicating possibly hydrogen bridge bonding to the surface oxygen (originating from partial dissociation of ODA) or within the adsorbate phase. The presence of all molecular vibrations furthermore indicates a structural and/or chemical non-homogeneous phase, as was also inferred from the XPS data.

Codeposition of PMDA and ODA: Polyamic Acid Formation and Imidization of the film

Codeposition of ODA and PMDA onto the metallic substrates held at room temperature leads to polyamic acid formation [20,21,23,24]. The film can contain excess PMDA and ODA molecules, depending on the flux of the two constituents, but the excess molecules will evaporate when the film is heated. In the imidization reaction, water is formed and released into vacuum. The XPS results showed that imidization of the film begins at T ~ 400 K as evidenced by a decrease in the hydroxyl O 1s band. However, in all our XPS experiments we were unable to produce a pure polyimide film. Spectroscopic evidence suggested that the films formed by vapor deposition always contain some unreacted or incompletely reacted oxidianiline molecules or fragments [21,23].

The XPS experiments on the formation of ultra-thin polyimide films (11 Å < d < 35 Å) on silver and copper further revealed that the interface between the polyimide and the metal contains PMDA and/or ODA fragments which are stabilized against thermal decomposition by the presence of the polyimide film. If pure ODA or PMDA monolayer films are heated in vacuum above T ~ 470°C, dissociation into carbonous species was observed.

The same conclusion, i.e. the stabilization of the interfacial layer of fragmented ODA and PMDA is derived from our IR experiments. The vibrational mode assignments are given in Table II. Fig. 2c shows the IR reflection spectrum of an ultra-thin codeposited layer of PMDA and ODA. Both the bands of molecular ODA and PMDA fragments are present, in addition to a more pronounced broad NH$_2$ scissor mode band around 1660 cm^{-1} and, not shown here, bands due to hydroxyl group formation around 3600 cm^{-1}. Also, the band at 1703 cm^{-1}, which is identified with fragmented PMDA on the surface, is visible. Heating the film results in more complete polyamic acid formation. The spectra taken at T = 373 K (Fig. 2d) of the same film are dominated by the ODA ring mode, a broad NH$_2$ scissor mode region, and a persistence of the mode originating from fragmented PMDA at 1703 cm^{-1}. Note that the carbonyl stretching frequencies are absent within the signal to noise level, indicating that in the thin polyamic acid film the dynamic dipole moments associated

with the carbonyl stretches must have a strong component parallel to the surface plane. At T > 425 K, imidization sets in and leads to dramatic changes
in the spectra. The bands at 1117 cm^{-1} and 1383 cm^{-1} are due to the imide
(OC)$_2$N C mode and imide C-N stretch, respectively. The strong increase in
the carbonyl stretching frequency indicates a structural rearrangement in the
film as compared to the polyamic acid film. Note that the PMDA fragment vibration at 1703 cm^{-1} is persistent as a shoulder on the strong carbonyl band.

TABLE II

	25°C	PAA 100°C	150°C	PI 200°C	Mode Assignments [30]
(1)	1085	1085	1082	1082	
(2)	1104	----	1100	----	
(3)	----	----	----	1125	1117 (OC)$_2$N C mode (imide)
(4)	1152	1152	1150	1150	
(5)	1245	1245	1245	1245	1252 Ether linkage $\nu_{(COC)}$ stretch (asym).
(6)	----	----	1385	1385	1383 $\nu_{(C-N)}$ stretch (imide)
(7)	1510	1510	1510	1508	1505 Ring Mode (ODA)
	1545	1545	1540	----	
	1625	----	----	----	
(8)	1700	1700	1705	----	
(9)	----	----	1738	1738	1725 Carbonyl $\nu_{(C=O)}$ stretch (sym)
	----	----	1770	1770	1779 Carbonyl $\nu_{(C=O)}$ stretch (asym)
(10)	1790	----	----	----	Carbonyl mode in polyamic acid
(11)	1860	----	----	----	Carbonyl mode in polyamic acid

The polyimide spectra (2e, f) show all the features of a fully cured
polyimide film [19] and no indications for a substantial amount of isoimide
formation. The same observation, i.e. that vapor deposited polyamic acid
converts nearly quantitatively to polyimide, has been reported originally for
thick films [19].
Our IR data thus reveal that spectral features associated with PMDA
fragments in the interface persist when a thin polyamic acid film is
imidized, and secondly, during imidization structural changes occur in the
polymer film.
The average orientation of the polymer chain can be inferred indirectly
by changing the polarization of the light with respect to the surface plane.
In films where the thickness does not exceed the screening length of the
metal electrons only those modes are detected in an IR reflection absorption
experiment which have a component of their dynamical dipole moment perpendicular to the surface. In thicker films other modes will also be detected, yet
their relative intensity ratio will still reflect the average orientation of
their dynamical dipoles.
In Fig. 3 we show the IR reflection absorption spectra of a thick polyimide film taken with p-polarized light at an incidence angle of 85° with respect to the surface normal (3a) and a FTIR Reflection absorption spectrum
with unpolarized light at an incidence angle of 45°. The relative polarization of the light can be changed by refraction effects at the polyimide/vac-
uum interface. This would change the direction of the p-component of the
electric field vector in the polymer film to a larger angle (49°) with respect to the surface normal. This, however, does not alter the general model
of the polymer orientation discussed below [26].
The FTIR spectrum is partially obscured by residual CO$_2$ gas phase bands
between 1800 and 1500 cm^{-1}. Both spectra show the same bands, but the intensity ratio between the carbonyl stretching band (1725 cm^{-1}), the ODA ν_{13} ring
modes (1505 cm^{-1}) and the imide bands (1383 cm^{-1} and 1117 cm^{-1}) and the ν_{COC}

ODA mode is distinctly different. In spectrum 3a, only those vibrations with a dipole moment perpendicular to the surface will be detected, whereas in spectrum b, all dipole orientations should be active, since the incident light was not polarized.

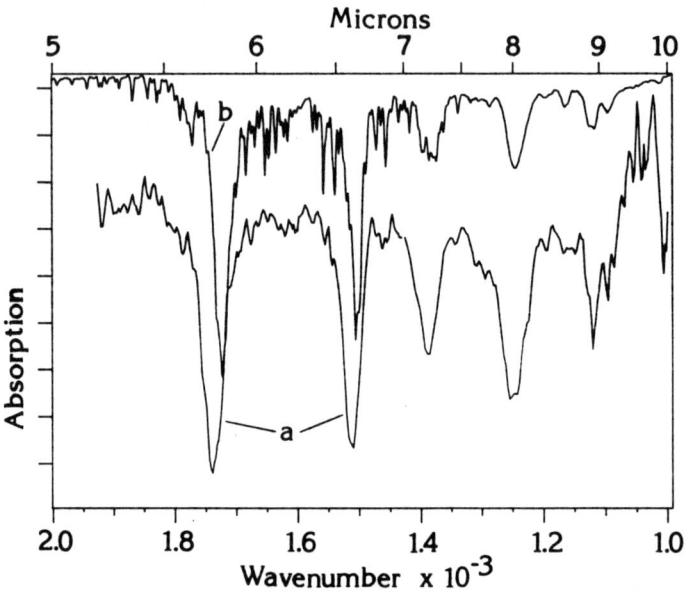

Figure 3.   Infrared Reflection Absorption data for a polyimide film about 100 Å thick. a) P-polarized light, incidence angle 85°. b) Unpolarized light, incidence angle 45°.

The carbonyl stretching frequency at 1725 cm^{-1} and the out of plane aromatic ODA ν_{13} mode at 1505 cm^{-1} both have a dynamical dipole moment perpendicular to the chain axis of the polymer. As schematically shown in Fig. 4, the polyimide chain is not straight, but due to the bond angle in the ODA molecules exhibits a "zig-zag" orientation. Russell [29] reported x-ray diffraction data on spun-on polyimide films, from which he concluded that the polyimide chains are oriented parallel to the surface. However, from his data, which give the projection of the polyimide repeat unit onto the surface plane, it cannot be determined whether the "zig-zag" structure is in the plane or perpendicular to the plane of the surface. Since the vibrational modes which have a dynamical dipole along the chain axis are intense in the p-polarized spectra, we have to conclude that the direction of the chains must have a component perpendicular to the surface, which is the case for a zig-zag orientation perpendicular to the substrate as indicated in Fig. 4. When both polarizations of the light contribute to the spectrum (3b), the relative intensity ratio between the out of plane modes (with respect to the polymer chain) and the axial modes changes. It appears that the out of plane modes (ν_{CO}, ν_{13}) are more intense due to the additional s component of the light. The dynamical dipole moment of the ν_{CO} mode and the ν_{13} phenyl modes are close to perpendicular to each other due to the staggered configuration of the ODA and PMDA constituents of the chain. Yet the intensity ratio between these two bands does not change as a function of polarization, which means that both the PMDA and ODA moieties must be inclined by approximately the same angle with respect to the surface normal. Such a configuration would result in a polarization independent absorption ratio of these two bands and an equal enhancement of absorption when the light is unpolarized. Since the polymer chains will have random azimuthal orientation in the configuration shown in Fig. 4, the intensity ratios between the axial modes will be largely independent of both polarization and incidence angle as observed experimentally.

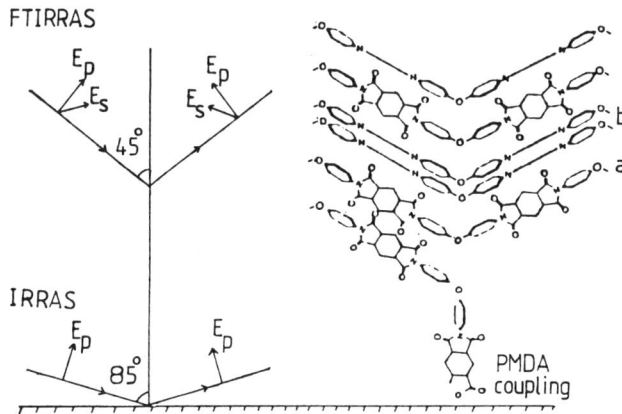

Figure 4.   Orientation of the polyimide chains as determined from the infrared reflection absorption data. The polyimide chains are oriented parallel to the surface, the "zig-zag" of the chains is on average in the plane of the surface normal. The orientation of the PMDA and ODA units is in between the two orientations labelled **a** and **b**. Chemical bonding to the substrate is indicated by a PMDA fragment adsorbed on a silver surface.

Our IR data thus supports the orientational model shown in Figure 4 where the rotational orientation of ODA and PMDA units of the chains with respect to the polymer axis must be between the two structures indicated as a and b.

In Figure 4 we also show the postulated interlink between the polyimide film and a PMDA fragment on a silver surface. This model is consistent with our previous XPS and NEXAFS data for PMDA adsorbed on silver and with the infrared absorption data described above.

Topographic Changes in Thin Polyamic Acid Films During Imidization

During imidization of the polyamic acid the average thickness of the film decreases by up to 60% [20,23,24], due to loss of material and structural changes in the film. XPS results on the ultra thin polyimide (d < 40 Å) films on silver, copper and gold further revealed that the film must become discontinuous.

To study these topographical changes we have used a Digital Instruments Nanoscope I to obtain scanning tunneling microscope (STM) images of polyimide films at various stages during curing. Figure 5 shows two examples. In this case the substrate is a gold film evaporated onto mica. The codeposited PMDA-ODA film was heated in vacuum below the imidization temperature (390 K for 15 minutes) (Fig. 5a) and then was cured at 425 K in vacuum for 15 minutes (Fig. 5b). After each heating, the film was removed for examination by STM and FTIR. The uncured film (Fig. 5a) was about 100 Å thick based upon attenuation of the gold XPS signals. The area shown in the STM images is about 200 Å by 180 Å and the topographic relief in 5b is about 5 Å. The uncured film shows no topographic features, the apparent structure in the pho-

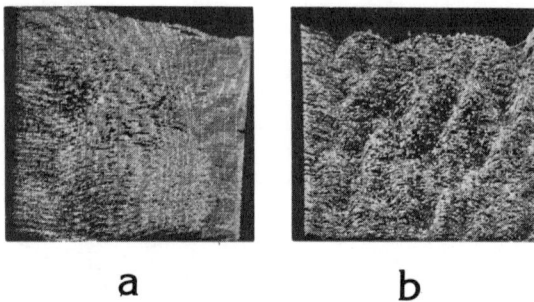

a   b

Figure 5.   Scanning tunneling micrographs of **a**: a thin polyamic acid film
             ~100 Å thick on a gold substrate and **b**: for the same film par-
             tially imidized at 425 K for 15 minutes. The regions displayed
             are 200 Å x 180 Å. The ridges are about 35 Å apart and the re-
             lief in surface topography is about 5 Å. The tunneling current
             was about 2 nA.

tograph is due to Moire' interference on the oscilloscope screen of the dis
play. In the partially cured film, however, topographical features are detected. Parallel ridges spaced about 35 Å are clearly visible in several regions but the detailed local structure varies indicating that the film structure is not coherent across the region of the image. Thus, the STM image in
Fig. 5b supports the structural model shown in Fig. 4. A comparison of the
STM images taken before and after imidization thus demonstrate that the
structural changes in the polymer film also induce topographic changes.

SUMMARY

Our XPS and IR studies show that the two polyimide constituents undergo
partial fragmentation on silver and copper surfaces. In the case of PMDA the
interaction with the substrate is dominated by the anhydride functionalities,
in the case of ODA a less specific interaction seems to occur. The interfacial PMDA fragment retains its ability to interact with oxidianiline and thus
is able to chemically interlink the polyimide with the substrate. Whether
ODA fragments can also act as a chemical interlink is at present not clear.
Imidization of the vapor phase deposited polyamic acid results in a
structural change of the polymer chains with respect to the surface, which in
turn causes topographical changes in the film surface. In the cured polyimide film, the polymer chains are in average oriented parallel to the surface, with their "zig-zag" structure having a strong component in the direction of the surface normal.

ACKNOWLEDGEMENTS

This work was supported by the Office of Naval Research. We gratefully
acknowledge the help of T. Strunskus and J. Baxter in preparing the films for
the STM studies and G. Strasser for help in the design and construction of
the IRRAS apparatus.

REFERENCES

1.   "Polymer Materials for Electronic Applications," ACS Symp. Ser., <u>184</u>
      (1982).
2.   "Polyimide - Synthesis, Characterization and Application," Vol. 1, Ed.
      K.L. Mittal, Plenum (1984).

3. A.M. Wilson, Thin Solid Films, 83 (1981) 145.
4. R.A. Larsen, IBM J. Res. Develop., 24 (1980) 268.
5. S. Mastroianni, Solid State Tech., May, (1984) 155.
6. K.L. Mittal, in "Microscopic Aspects of Adhesion and Lubrication," Tribology Series Vol. 7, Ed. J.M. Georges, Pub. Elsevier, Amsterdam (1982).
7. N.J. Chou and C.H. Tang, J. Vac. Sci. Technol. A, 2 (1984) 751.
8. J.L. Jordan, P.N. Sanda, J.F. Morar, C.A. Kovac, F.J. Himpsel and R.A. Polak, J. Vac. Sci. Technol., A, 4 (1986) 1046.
9. P.N. Sanda, J.W. Bartha, J.G. Clabes, J.L. Jordan, C. Feger, B.D. Silverman and P.S. Ho, J. Vac. Sci. Technol., A, 4 (1986) 1035.
10. J.L. Jordan, C.A. Kovac, J.F. Morar and R.A. Pollak, Phys. Rev. B (in press) (1987).
11. P.S. Ho, P.O. Hahn, J.W. Bartha, G.W. Rubloff, F.K. LeGoues and B.D. Silverman, J. Vac. Sci. Technol. A, 3 (1985) 739.
12. F.S. Ohuchi and S.C. Freilich, J. Vac. Sci. Technol. A, 4 (1986) 1039.
13. N.J. Chou, D.W. Dong, J. Kim and A.C. Liu, J. Electrochem. Soc. 131 (1984) 2335.
14. C. Chauvin, E. Sacher, A. Yelon, R. Groleau and S. Gujrathi, preprint (1987).
15. N.J. DiNardo, J.E. Demuth and T.C. Clarke, Chem. Phys. Letts. 121 (1985) 239.
16. J.J. Pireaux, C. Gregoire, P.A. Thiry, R. Caudano and T.C. Clarke, J. Vac. Sci. Tech. A, 5 (1987) 598.
17. Y.-H. Kim, G.F. Walker, J. Kim and J. Park, International Conference on Metallurgical Coatings (San Diego) p. 23-27 (1987).
18. J.L. Jordan, C.A. Kovac, J.F. Morar and R.A. Pollak, Phys. Rev. B, July 15 (1987).
19. J.R. Salem, F.O. Sequeda, J. Duran, W.Y. Lee and R.M. Yang, J. Vac. Sci. Technol. A, 4 (1986) 369.
20. M. Grunze and R.N. Lamb, Chem. Phys. Letts., 133 (1987) 283.
21. R.N. Lamb, J. Baxter, M. Grunze, C.W. Kong and W.N. Unertl, Langmuir (in press) (1987).
22. S.P. Kowalczyk, Y.H. Kim, G.F. Walker, J. Kim, preprint.
23. R.N. Lamb and M. Grunze, J. Chem. Phys., submitted.
24. M. Grunze, J.P. Baxter, C.W. Kong, R.N. Lamb, W.N. Unertl and C.R. Brundle, to be published in Proceedings of the American Vacuum Society Topical Conference on Deposition and Growth: Limits for Microelectronics, Anaheim, CA, November 1987.
25. R.N. Lamb, J. Stöhr, M. Grunze, in preparation.
26. S. Gnanarajan, J. French, M. Grunze, in preparation.
27. The Aldrich Library of FT-IR Spectra, Charles T. Pouchert, Ed., Aldrich Chemical Company, Inc., 1985.
28. E.M. Stuve, R.J. Madix and B.A. Sexton, Surf. Sci., 119 (1982) 279.
29. T.P. Russell, Journal of Polymer Science, Polymer Physics edition, Vol. 22, (1984) 1105.
30. H. Ishida, S.T. Wellininghoff, E. Bauer, and L.L. Koenig, Macromolecules 13 (1980) 826.
31. M.C. Tobin, "Laser and Raman Spectroscopy," Wiley Interscience, New York, 1971.

CHARACTERISTICS OF POLYETHYLENE THIN FILMS
DEPOSITED BY IONIZED CLUSTER BEAM

HIROAKI USUI, KOUJI NUMATA, HITOSHI DOHMOTO, ISAO YAMADA AND TOSHINORI TAKAGI
Ion Beam Engineering Experimental Laboratory, Kyoto University,
Sakyo, Kyoto 606, Japan

ABSTRACT

Polyethylene thin films were deposited by the ionized cluster beam (ICB) method. The dielectric, resistivity, and breakdown field measurements showed that the ICB polyethylene films have excellent properties as an electrical insulator. The characteristics of Au/polyethylene/Si MIS diodes and MISFETs indicated that the ICB method can control the film-substrate interface property. The SIMS and ESCA analyses showed that the ICB films have pure and stable chemical structure.

INTRODUCTION

Recent development of new organic compounds has opened up a new application areas for organic materials in functional electronic devices. For such applications, new organic film formation processes are desired to enable flexible control of film quality. We have shown that the ionized cluster beam (ICB) method can be effectively utilized for organic film deposition. The materials deposited by ICB so far include anthracene [1], polyethylene [2], and Cu-phthalocyanine [3,4].

Our previous experiments have demonstrated that low molecular weight polyethylene (viscosity averaged molecular weight $\bar{M}_v$=900) deposited by ICB has the following features [5]: 1) The deposited films have nearly the same average molecular weight as the source material, though the molecular weight dispersion is smaller. 2) The film IR absorption is identical to that of the pure polyethylene. 3) The film crystallinity is improved by the ionization and acceleration of polyethylene. The electron diffraction of a film deposited on Si(111) showed a single crystalline spot pattern, c-axis being normal to the substrate surface. 4) The film-substrate adhesion is improved by the use of accelerated ions.

This paper provides additional information on the characteristics of the ICB polyethylene films. Electrical properties of the films were investigated by measuring dielectric constant, resistivity, and breakdown field. The film-substrate interface property was studied by measuring the characteristics of MIS diodes and MISFETs fabricated on Si substrates. SIMS and ESCA analyses were also used to clarify the chemical property of the ICB polyethylene films.

DEPOSITION OF POLYETHYLENE BY ICB METHOD

The ICB method is an ion assisted film formation technique developed by Takagi et al. in 1972 [6]. Detailed description of the method has been reported elsewhere [7]. In short, the source material is vaporized in a crucible and ejected into vacuum through a nozzle under conditions under which the vapor forms a continuum supersonic nozzle beam. Due to the adiabatic cooling in the ejection process, the vapor molecules form clusters by homogeneous nucleation. The clusters are partially ionized by electrons and accelerated toward the substrate. The formation process of organic clusters has been studied for anthracene. [8] Typical organic cluster size is estimated to be about 10 molecules.

In the work reported here, the films were prepared as the following method. The source material was low molecular weight polyethylene (Mitsui Petrochemical Industries, Hi-wax 400P, $\bar{M}_v$ = 4000 and density 0.97 g/cm^3).

The crucible temperature T_c was raised slowly, degassing the source material, up to a deposition temperature of 330°C. The electron current for ionization, I_e, was either 10 or 20 mA, and the electron acceleration voltage for ionization, V_e, was fixed at 50 V. The ion acceleration voltage, V_a, was varied between 0 and 2000 V. The substrates, either Corning 7059 glass or Si wafer, were ultrasonically cleaned in acetone and trichloroethylene. Si substrates were lightly etched by hydrofluoric acid after this process. Prior to the deposition, the substrates were heated to 120°C for one hour in vacuum, and then cooled down to room temperature. A typical deposition rate was about 8 nm/min, and the ion current density was 4 to 8 nA/cm^2, depending on I_e and V_a. Films of 130 to 200 nm thickness were deposited in a background pressure on the order of 5×10^{-6} Torr.

MOLECULAR WEIGHT AND CRYSTALLOGRAPHIC PROPERTIES

The molecular weight distributions of the source material and the film were examined by GPC. The source material had a number average molecular weight $\overline{M}_n$ of 2300 and a ratio of weight averaged molecular weight $\overline{M}_w$ to $\overline{M}_n$ of 2.3. The deposited film exhibited a lower molecular weight ($\overline{M}_n$ = 540) and a smaller dispersion ($\overline{M}_w/\overline{M}_n$ = 1.1). [9] Compared to the low molecular weight polyethylene [5], the higher molecular weight materials tend to undergo cracking in the vaporization process. The feature of nearly monodispersed molecular weight would be advantageous to improve the crystallinity.

The film crystallinity was improved by the ionization and acceleration of the polyethylene as reported before. [5] The film crystallinity was also influenced by the substrate surface condition. Figure 1 shows electron diffraction patterns of the films deposited on NaCl (100) substrates using a neutral beam (a) and an ionized beam at I_e = 10 mA and V_a = 500 V (b). There was no special substrate treatment after cleavage in the air. The film deposited using a neutral beam showed a ring pattern indicating randomly oriented polycrystals. The film deposited by an ionized beam indicated a spot pattern, characteristic of single crystalline orientation. On the other hand, the films had higher crystal orientation when the substrates had been preheated at 120°C for 1 hour in vacuum. The film deposited by neutral beam showed a fiber structure, the c-axis being oriented parallel to the substrate surface in two directions. These results suggest a surface cleaning effect by the ions due to the sputtering of weakly adsorbed impurities. [10]

ELECTRICAL PROPERTIES

Polyethylene is known as an excellent electrical insulator. The insulating characteristics are correlated to such properties as film packing density, molecular structure, chemical impurities and film morphology. The dielectric constants of ICB polyethylene films were measured by preparing Au/polyethylene/Au sandwich cells of 1 mm diameter on the glass substrates. The Au electrodes were deposited by conventional evaporation in another chamber. The measurement was made with a Hewlett Packard 4260A universal bridge. The film deposited at I_e = 10 mA and V_a = 200 V had a dielectric constant of 2.3, which is equal to the value of bulk polyethylene. No

(a)   (b)

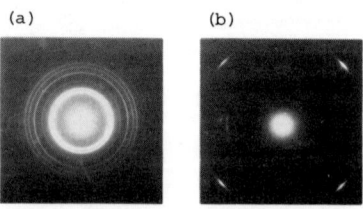

Fig. 1 Electron diffraction patterns of polyethylene films deposited on NaCl substrates without ionization (a), and with ionized beam at I_e= 10mA, V_a=500V (b).

frequency dependence was observed in the range from 100 Hz to 20 kHz. This means that the film is physically dense and includes no polar impurities. The dielectric loss, tan δ, at frequencies higher than 1 kHz was less than 1×10^{-3}, below the detection limit of the bridge. The dielectric loss increased at lower frequencies. Similar results were obtained for samples prepared under other conditions. The film deposited at $I_e = 10$ mA and $V_a = 1$ kV, condition coinciding with the optimum for crystallinity, showed the smallest tan δ. [9] The dielectric constant exhibited no change in the temperatures below 100°C. However, the tan δ was reduced by heating the film to around 100°C.

The current-voltage (I-V) characteristics and the break down field strength were measured for Au/polyethylene/p-Si(100) MIS cells. The cells were 0.3 mm diameter and the polyethylene layers were 130 nm thick. The cells prepared at $I_e = 10$ mA and $V_a = 1$ kV or $I_e = 20$ mA and $V_a = 500$ V showed ohmic conduction at electric fields lower than 7×10^6 V/cm, and Schottky emission conduction at higher fields. The film resistivity in the ohmic region was from 10^{15} to 10^{16} Ωcm. The I-V characteristics of the films deposited at other conditions did not fit to the ideal ohmic or Schottky emission curves and showed lower resistivities. The breakdown field strength of the ICB films was from 8×10^6 to 1×10^7 V/cm, which was comparable to that of bulk polyethylene. These results show that the ICB method can produce uniform and pinhole free polyethylene films even at a thickness of 130 nm.

INTERFACE CHARACTERISTICS

The film-substrate interface property is of particular importance for electronic devices. The polyethylene-Si interface was studied by measuring high frequency (1 MHz) capacitance-voltage (C-V) characteristics of Au/polyethylene/p-Si (100) MIS diodes. Figure 2 shows the C-V curves of the cells prepared without ionization of polyethylene (a), and with ionized polyethylene at the optimum condition (b). It is seen that the C-V characteristic was improved, and a larger capacitance change and a smaller flatband shift were obtained in the sample (b). The interface state density of the sample (b) estimated by Terman's method was 1.3×10^{11} cm^{-2}eV^{-1} at the middle of the bandgap.

For a further investigation into the interface property, polyethylene gate insulator MISFETs were fabricated on n-Si(111) substrates. As illustrated in Fig. 3, the MISFETs have a simple Schottky source-drain structure. The Au electrodes were deposited by conventional evaporation in another chamber. The gate length and width were 50 μm and 1 mm, respectively. Figure 4 shows typical characteristics of the MISFETs. The sample (a) was fabricated with neutral polyethylene beam, and (b) with an ionized beam at $I_e = 10$ mA and $V_a = 1$ kV. The leakage current through the source and drain Schottky junctions has been subtracted. The use of the ionized beam improves considerably the MISFET characteristics. The source-drain current of sample (b) showed the typical saturation characteristic, indicating the formation of an inversion layer. For gate voltages higher than 30 V, the square root of the saturation current increased proportionally to the gate voltage. These improvements of the interface properties may arise from the surface cleaning effect by the accelerated ions.

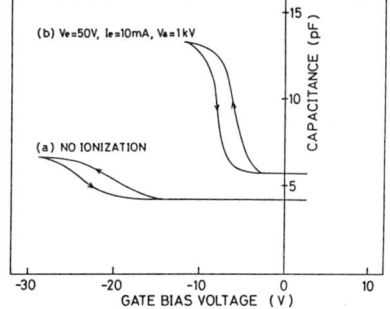

Fig. 2 C-V characteristics of Au/Polyethylene/Si MIS diodes fabricated without ionization (a), and with ionized beam at I_e=10mA and V_a=1kV (b).

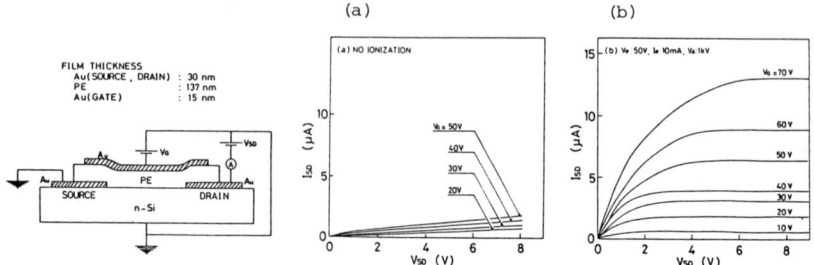

Fig. 3 Structure of polyethylene
gate insulator MISFET.

Fig. 4 Typical characteristics of the MISFET
fabricated without ionization (a), and with
ionized beam at I_e=10mA and V_a=1kV (b).

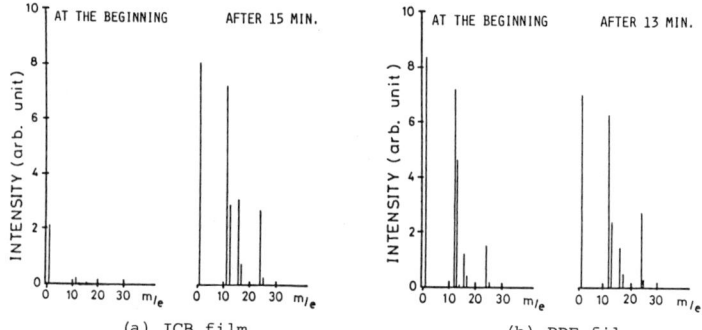

(a) ICB film

(b) PPE film

Fig. 5 SIMS spectra of ICB polyethylene deposited at I_e = 10 mA and
V_a = 1 kV (a) and PPE film (b) at different Ar ion etching time.

SIMS AND ESCA ANALYSIS

The electrical properties of the polyethylene films are related to their
crystallinity and chemical purity. For example, plasma polymerized films
sometimes suffer from large dielectric loss because of impurities and
residual radicals in the films. [11] In order to further understand the ICB
films, they were examined by using SIMS and ESCA analyses. For comparison,
plasma polymerized ethylene films (PPE) were also analyzed. PPE films were
formed by a magnetically enhanced parallel plate glow discharge. The chamber
was evacuated to 1×10^{-6} Torr, and ethylene monomer gas was introduced at a
rate of 10 sccm to a pressure of 5×10^{-2} Torr. A low frequency discharge (20
kHz, 100 mA) caused film deposition at a rate of 5 nm/min on substrates
placed between the parallel electrodes.

The SIMS analysis was made with an ANELVA Model SIMS-300N using Ar ion
bombardment. Figure 5 shows the secondary negative ion spectra obtained at
the beginning and after 15 or 13 minutes sputtering by a 3 keV - 0.6 μA Ar
beam. The main peaks were H^-, C^-, CH^-, O^-, OH^-, C_2^-, and C_2H^-. The
intensity of secondary C^- ions are plotted in Fig. 6 versus the sputtering

time. The yield of all the above mentioned ions was different for ICB and PPE films. The ICB polyethylene did not give a secondary ion yield as high as that of PPE until it was irradiated by the Ar beam for over 10 minutes. On the other hand, the PPE film showed a high negative ion yield from the beginning of the irradiation. The yield of PPE decreased slightly with irradiation time. This difference was more pronounced at a lower probe ion energy. By 200 eV Ar sputtering, the ICB polyethylene emitted few secondary negative ions during the first thirty minutes, while the PPE film yielded considerable secondary negative ions from the beginning.

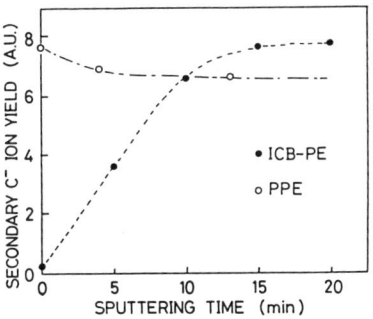

Fig. 6 Secondary C⁻ yield of the polyethylene films versus Ar ion etching time.

The ESCA analysis was made with a Shimazu ESCA-750. The films were sputter etched by 500 eV - 10 μA Ar ion beam during the analysis. Figure 7 shows the ESCA spectra of the ICB film and of the PPE film at different Ar etching times. It is seen that the ICB film has a much lower oxygen content than the PPE film. There was a tendency that the ICB film deposited at higher ion acceleration voltage contained a smaller amount of oxygen. The O_{1s}/C_{1s} peak area ratio of as deposited ICB films were 0.014 for the film deposited by neutral beam, 0.009 for the one deposited at V_a = 200 V, and 0.008 for the one at V_a = 1 kV. The PPE film had the O_{1s}/C_{1s} ratio of 0.17. For comparison, the O_{1s}/C_{1s} ratio of commercially obtained polyethylene films was 0.14 for a low density polyethylene (Mitsui Petrochemical Industries, Mirason S-927-5, $\bar{M}_v$ = 9x10⁴, degree of crystallinity 50%) and 0.15 for a high density polyethylene (Mitsui Petrochemical Industries, Hi zex 7000F, $\bar{M}_v$ = 2x10⁵, degree of crystallinity 70%).

As the Ar etching proceeds, the oxygen contents of the ICB films increased slightly at the beginning and subsequently decreased. The oxygen signal of the ICB film is nearly at the detection limit, and is therefore not necessarily coming from the film. The oxygen contents of the PPE and commercially available films substantially decreased with Ar sputtering. Another point of interest is the full width at half maximum (FWHM) of the C_{1s} peak. The C_{1s} peak of the as deposited ICB film shows a symmetrical feature

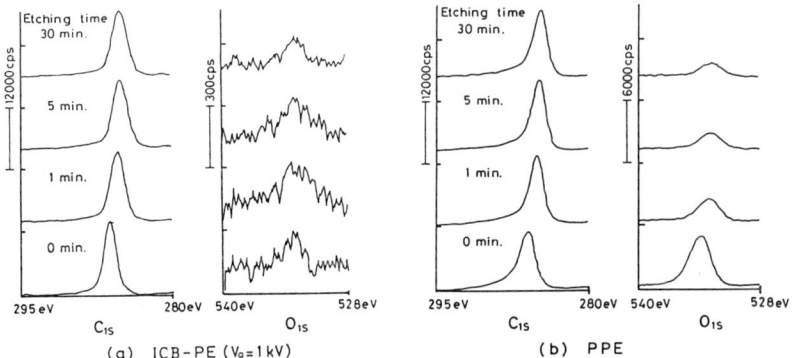

(a) ICB-PE (V_a=1kV)   (b) PPE

Fig. 7 C_{1s} and O_{1s} ESCA spectra of ICB polyethylene deposited at I_e = 10 mA and V_a = 1 kV (a) and PPE film (b) and PPE films at different ion etching times.

and has a FWHM of 1.55 eV, indicating the high purity of the chemical structure. The FWHM increases to 1.80 eV after 30 min Ar sputtering. This suggests that the chemical structure of the ICB film may be modified by the Ar beam. On the other hand, the C_{1s} peak of the PPE film is not symmetrical. The FWHM of the as deposited PPE film is 1.95 eV, which decreases to 1.80 eV after the 30 min Ar sputtering. This could be because the impurity bonds in the PPE are removed by the ion bombardment. Commercially available polyethylene films gave the similar results as the PPE film.

According to the preliminary results of SIMS and ESCA analyses, a possible explanation for the difference between the ICB film and the PPE film is as follows. The PPE film contains a larger amount of chemical impurities or residual radicals that cause surface adsorption of impurities. Such impurities may induce instability in chemical structure and lead to a high secondary ion yield from the beginning of the ion irradiation. The secondary ion yield and FWHM of the ESCA peak decrease as the surface impurities are removed by the sputtering. On the other hand, the ICB film has higher purity and crystallinity, leading to a stable chemical structure. The ion irradiation, however, damages the chemical bonds and crystallinity, and may also cause oxidation. The film surface becomes unstable like that of the PPE, giving higher secondary ion yield and larger ESCA signal FWHM. This kind of radiation damage is also occurring for the PPE film concurrently with the impurity desorption. Eventually, these processes reach an equilibrium, giving a similar results for any films. The chemically pure nature of the ICB polyethylene may reflect the plasma-free and low energy ion processing feature of this method.

CONCLUSION

The polyethylene films deposited by the ICB method have monodispersed molecular weight distribution. The crystal orientation is improved by applying the ion acceleration voltage. The electrical properties of the ICB film are characterized by low dielectric loss, high resistivity, and high breakdown field strength. The ion acceleration is effective to improve the film-substrate interface characteristics. The SIMS and ESCA analyses indicate that the ICB polyethylene films have high purity and chemical stability. This property of the ICB film is in consistency with the excellent electrical characteristics of the ICB film.

REFERENCES

1) H. Usui, M. Naemura, I. Yamada, and T. Takagi, Proc. 6th Symp. Ion Sources and Ion-Assisted Technol. (Inst. Elect. Engs. Japan, Tokyo, 1982) p. 331.

2) H. Usui, I. Yamada, and T. Takagi, Proc. Int. Ion Eng. Cong. -ISIAT'83 & IPAT'83- (Inst. Elect. Engs. Japan, Tokyo, 1982) p. 1247.

3) H. Usui, M. Naemura, H. Nakanishi, I. Yamada, and T. Takagi, Proc. 8th Symp. Ion Sources and Ion-Assisted Technol. (Res. Group Ion Eng., Kyoto Univ. 1984) p. 271.

4) F. Nakanishi, H. Takata, H. Usui, I. Yamada, and T. Takagi, Proc. 8th Symp. Ion Sources and Ion-Assisted Technol. (Res. Group Ion Eng., Kyoto Univ., 1985) p. 449.

5) H. Usui, I. Yamada, and T. Takagi, J. Vac. Soc. Technol. A4, 52 (1986).

6) T. Takagi, I. Yamada, M. Kunori, and S. Kobiyama, Proc. 2nd Int. Conf. Ion Sources, 1972, Vienna (Österreichische Studiengesellshaft für Atomenergie, Vienna, 1972) p. 790.

7) T. Takagi, Mat. Res. Soc. Symp. Proc. 27, 501 (1984).

8) H. Usui, I. Yamada, and T. Takagi, Proc. Int. Workshop Ionized Cluster Beam Technique (Res. Group Ion Eng., Kyoto Univ., 1986) p. 63.

9) H. Dohmoto, F. Nakanishi, H. Usui, I. Yamada, and T. Takagi, ibid., p. 171.

10) T. Takagi, J. Vac. Sci. Technol. A2, 382 (1984).

11) P.D. Buzzer, D.S. Soung, A.T. Bell, J. Appl. Poly. Sci. 27, 3965 (1982).

ETCHING AND SURFACE MODIFICATION OF POLYMERS IN CF_4/O_2 PLASMA DISCHARGES

P. M. Scott*[$], S. V. Babu*, R. E. Partch**, and L. J. Matienzo[+]

*   Department of Chemical Engg, Clarkson Univ., Potsdam, NY 13676
**   Department of Chemistry, Clarkson Univ., Potsdam, NY 13676
[+]   IBM Corporation, Systems Technology Division, Endicott, NY 13760

ABSTRACT

Some results of simultaneous measurement of polyimide etch rates, temperature, and O- and F- atom emission intensities in CF_4/O_2 plasma discharges are described. Several process variables have been investigated both during steady and unsteady plasma reactor operation. Atomic compositions on the polymer surface have been determined as a function of time of exposure to the plasma using ESCA, and the C 1S spectra analyzed to identify the various C-O and C-F bonds.

INTRODUCTION

Etching of a variety of polymers in plasma discharges plays an essential role in current VLSI device processing and electronic package manufacturing [1]. The compatibility of such "dry" processes, as opposed to processing with liquid solvents and etchants, with sub-micron design rules and a high degree of process automation has led to the widespread use of plasma processing. To mention a few examples, pattern transfer from an imaging resist film to the underlying planarizing layer and thus to the substrate is best carried out in a low pressure plasma [2]. Drill smear in the through-holes of multilayer printed circuit boards can be stripped by exposure to CF_4/O_2 plasma discharges [3]. Consequently, there is a considerable interest in elucidating the underlying chemistry of the polymer etching process. A recent review [4] cogently summarizes the present status of the field: Battey [5] studied the stripping of poly(isoprene)-based photoresists at 13.56 MHz in a barrel reactor and demonstrated the role of oxygen atoms in the etching process. Recent ESR measurements of the O^3P signal from microwave discharges [6] confirmed Battey's observations. Addition of fluorocarbon species like CF_4 at low concentrations to oxygen plasmas significantly enhances polymer etch rate [7]. A maximum in the etch rate has been reported [7,8] when typically 10-25% of the feed gas contains CF_4, the rest being O_2. The substantial increase in the etch rate is caused by the ability of F-atoms to abstract hydrogen atoms [10,11] from the polymer as well as the enhancement of the oxygen atom concentration in the gas phase [10,12]. As the fluorine concentration is further increased in the gas phase, surface etching is inhibited through competition with O-atoms for surface sites and reduced O and O_2 concentration. Higher CF_4 concentrations lead to the formation of fluorine rich polymer on the substrate as determined by surface analysis [8,9]. It has been recently suggested that the ratio of O-atom to F-atom concentration is most significant in the etching of several polymers [8].

The application of spectroscopic and surface analytical techniques, as described by these results, has provided only a general and qualitative understanding of the interaction between the gas phase reactive species in the plasma discharge and the primary and secondary radical sites on the polymer surface. However, a quantitative

---

$ Current address: IBM Corporation, Owego, NY 13827

Mat. Res. Soc. Symp. Proc. Vol. 108. ©1988 Materials Research Society

description of the detailed kinetics is still unavailable. For example, specifics of the competition between fluorination of the surface and etching, as well as their dependence on the structural aspects of the polymer are lacking. Similarly, the exact role of temperature and reactor residence time on the kinetics needs to be elucidated to achieve optimal reactor performance.

In this paper, simultaneous measurement, under steady and unsteady reactor conditions, of substrate temperature and several gas phase and surface species is described. The experiments were performed in CF_4/O_2 plasma discharges at 13.56 MHz using polyimide. The resulting data are used to elucidate the relative importance of etching and surface fluorination.

EXPERIMENTAL

A Plasma-Therm PD-2484 plasma etch/reactive ion etch parallel plate plasma reactor powered by the 13.56 MHz radio frequency generator was used. The power output of the generator was operator controlled with a maximum output of 3000 watts. The aluminum plasma chamber had a total flow volume of 48.9 liters and 22 inch aluminum electrodes which were spaced 1 inch apart. Temperature of the chamber walls and the lid was controlled by a two channel Bay Voltex Tempryte-HT 6820 heat exchanger. Feed gas flow to the chamber was metered by Brookes 5850 mass flow controllers. The gas entered the chamber through a gas ring around the base of the bottom electrode and flowed radially inward toward the pump port. The vacuum system consisted of an Alcatel 2063 CPI mechanical pump assisted by a Leybold Hereaus Turbovac 450 turbomolecular pump. A base pressure of 10^{-5} torr could be reached in about 10 minutes. A CPI Cryogenics cryopump further lower the pressure to 10^{-6} torr.

A PT Analytical LES-300/84 laser interferometry system mounted on top of the chamber gave instantaneous updates of the etch rates. A PARC EG&G OMA II model 1460 diode array optical emission spectrometer was used to monitor the emission from various excited species in the plasma through a sapphire window. The spectrometer could be used to take full scans of the spectrum (200 to 1000 nm wavelength), or to monitor up to 8 different spectral bands, measuring and storing their intensities on disk. A Luxtron 1000Å fluoroptic thermometer was used to monitor the temperature of the samples as they were being etched. The probe was coated with etch resistant, dark colored teflon and was affixed with the tip exposed on the etched surface.

ESCA data were obtained on a Perkin-Elmer Physical Electronics (PHI) model 560 instrument with a double pass cylindrical mirror analyzer using Mg K x-rays as the source. A flood gun set at 0.3 mA emission was used for charge neutralization of insulating surfaces. Binding energies were referenced to the hydrocarbon peak at 284.6 eV. Atomic compositions were calculated from peak areas using PHI's sensitivity factors. A Gaussian-Lorentzian fitting routine with inelastic background correction was used to fit ESCA data.

Silicon coupons were spin-coated with DuPont 5878 polyimide. The samples were placed on the bottom electrode and monitored for etch rates and temperatures. Care was exercised in affixing the temperature probe to the coupon to avoid spurious results. Optical emission data were simultaneously gathered. The spectral lines monitored, once every second, were the atomic oxygen line at 844.5 nm (a triplet which was not resolved under the experimental conditions), the fluorine emission at 704 nm, and, for purposes of actinometry, the Ar 750.4 nm line. The ratios of the peak areas O 845/Ar 750.4 and F 704/Ar 750.4 were automatically calculated, plotted, and stored on disk. Coupons withdrawn from the reactor were immediately analyzed by ESCA after treatment under various steady and unsteady reactor operation conditions.

STEADY STATE RESULTS

A rather extensive set of data has been collected [13] and will be fully reported elsewhere [14]. Due to limitations on space, only a few of the typical results will be presented and discussed here. In our study, maximum etch rates are obtained around 2-5% CF_4 in contrast with earlier reports as discussed elsewhere [13,14]. Figure 1 shows a temperature history for a sample treated with a 10% CF_4/90% O_2 plasma mixture. The maximum temperature reached during etching is a function of the feed gas composition, as shown in Figure 2. The large confidence intervals are a result of somewhat erratic results, perhaps because the probe was single-point calibrated at 50°C. In any case, the highly exothermic nature of the etching process is apparent. Steady state optical emission and atomic surface composition measurements yielded data similar to those of earlier investigations [8]. It is now possible to evaluate the kinetic mechanism proposed recently [11]. The associated expression for the etch rates is given by

$$\text{Etch Rate} = \frac{k_1\,[R*][O]+k_2[R*][O_2]+k_3[ROCF_x]}{1+k_4[CF_4]} \quad (1)$$

where the k_i are some effective rate constants as described in [11], and [] denotes the concentration of the species enclosed in the brackets. Furthermore, R* is a polymer radical generated by the abstraction of a hydrogen atom from the polymer RH and $ROCF_x$ is the species generated by the reaction between the alkoxy radical RO* and the CF_x radical (x = 1, 2, or 3).

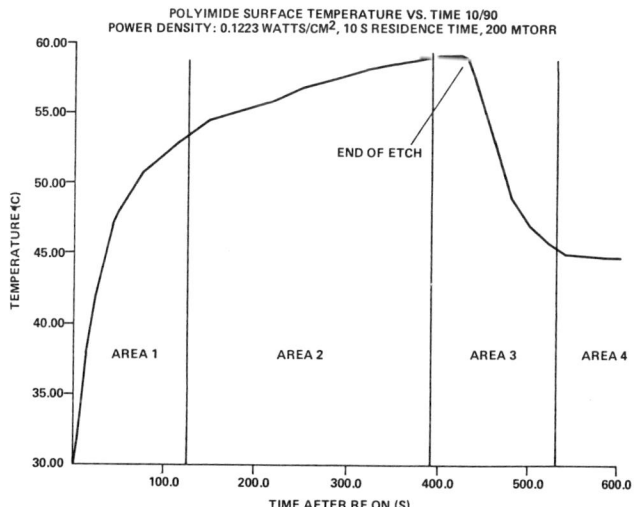

POLYIMIDE SURFACE TEMPERATURE VS. TIME 10/90
POWER DENSITY: 0.1223 WATTS/CM2, 10 S RESIDENCE TIME, 200 MTORR

Figure 1

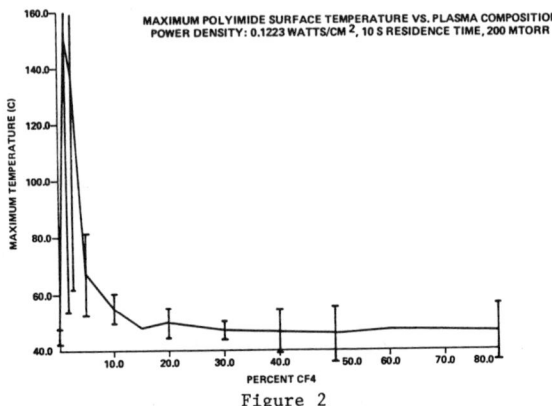

Figure 2

An approximate version of (1) has been shown earlier [15] to represent the general shape of the scaled etch rate vs feed gas composition curve quite well. In that earlier paper, the necessary concentrations and rate constants were obtained numerically by solving the Boltzmann equation for the electron energy distribution function and then solving the nonlinear coupled steady state balance equations for the species of interest.

Here, the gas phase concentrations are available from actinometry and [R*] and [ROCF$_x$] can be approximated from the ESCA measurements. The necessary details are described in [13] and are discussed at length elsewhere [14]. Of course, the rate constants in (1) are unknown a priori. They have been treated as parameters to be fixed by fitting the measured etch rate. One resulting curve is shown in Figure 3. The fit is quite satisfactory except for the magnitude of the peak etch rate. Unfortunately, however, the resulting set of rate constants is not unique and several sets of widely differing rate constants generate curves similar in shape and similar least squares deviations.

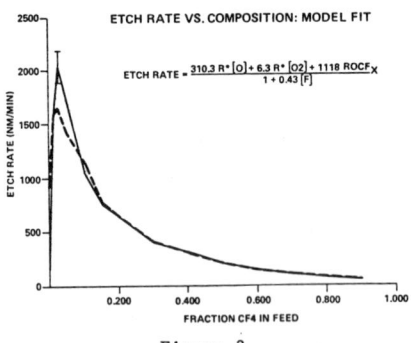

Figure 3

TIME-DEPENDENT MEASUREMENTS

Again only a limited set of data will be reported here. Equilibration of the oxygen and fluorine atomic emission intensities in a 90% CF$_4$/10% O$_2$ plasma is shown in Figure 4. The equilibrium process for both species spans several seconds. This is in sharp contrast to the almost instantaneous attainment of the final emission level by oxygen atoms in a pure O_2 plasma [13,14]. Furthermore, the atomic surface compositions reach equilibrium in about 10 seconds in pure O_2 plasma, in about 50 seconds in 10% CF_4/90% O_2 plasma, and in about 2-5 seconds in a 90% CF_4/10% O_2 discharge. As an illustration, an analogous result is shown in Figure 5. Here the fractional peak areas of curve resolved C 1s spectra are plotted as a function of exposure time in a 90% CF_4 plasma. A more complete description is provided elsewhere [13,14].

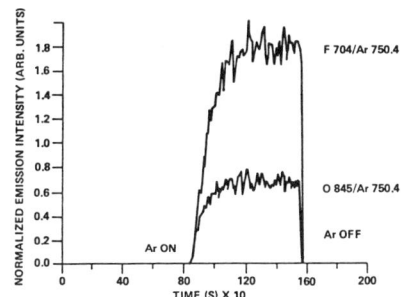

Figure 4

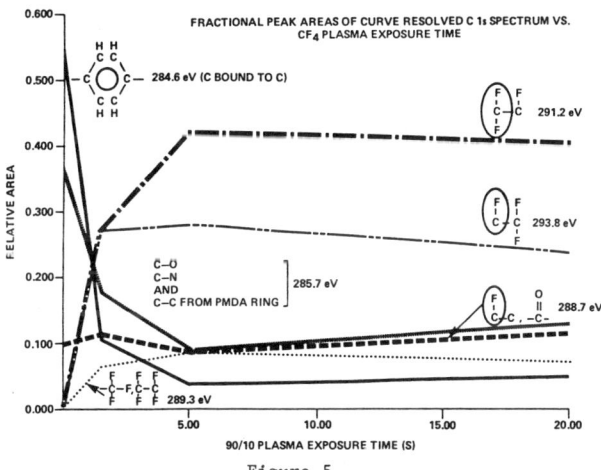

Figure 5

Polyimide is, of course, highly aromatic in structure, and has a large number of Π-electrons. The electronegativity of F-atoms ensures that they will bond preferentially with these Π-electrons to be followed, only later, by hydrogen atom abstraction. This sequence has been analyzed kinetically by integrating the appropriate species balance equations for fluorine rich plasmas to understand the time dependent evolution of the various types of CF_x species on the surface [13,14].

Acknowledgments

The authors thank J. Hoffarth, IBM Corporation, Endicott, N.Y., for sponsoring this work.

References

1. L. F. Thompson, C. G. Willson, and M. J. Bowden, (Editors), "Introduction to Microlithography", (American Chemical Society, Washington, D.C., 1983).

2. See for example, W. Pilz, T. Sponholz, S. Pongratz and H. Mader, Microelectronic Engg. 3, 467 (1985).

3. D. Cantwell, Printed Circuit Fabrication , Dec. 1983 and Jan. 1984.

4. S. J. Moss, Polym. Degrad. Stab., 17, 205 (1987).

5. J. F. Battey, IEEE Trans. Electron Dev., ED-24, 140 (1977), J. Electrochem. Soc., 124, 437 (1977).

6. J. M. Cook, J. J. Hannon and B. W. Benson, in Sixth Int. Symposium Plasma Chem. p. 616 (Montreal, 1983).

7. See for example, G. Turban and M. Rapeaux, J. Electrochem. Soc., 130, 2231 (1983).

8. F. D. Egitto, V. Vukanovic, F. Emmi, and R. S. Horwath, J. Vac. Sci. Technol. B3, 893 (1985).

9. J. Dedinas, M. M. Feldman, M. G. Mason, and L. J. Gerenser, in Proceedings of the First Int. Conf. Plasma Chem. Tech., p. 119, H. Boenig (Editor), (Technomic Press, Lancaster, Penn. 1983); T. Yagi, A. E. Pavlath, and A. G. Pittman, J. App. Poly. Sci., 27, 4019 (1982); and M. Anand, R. Cohen, and R. F. Baddour, Polymer, 22, 361 (1981).

10. See for example, D. L. Flamm and V. M. Donnelly, Plasma Chem. Plasma Process 1, 317 (1981) and references cited there.

11. S. V. Babu, L. A. Tiemann, and R. E. Partch, in 7th Int. Symp. Plasma Chem. p. 1025 (Eindhoven, 1985).

12. R. d'Agostino, F. Cramarossa, S. de Benedictis and G. Ferraro, J. App. Phys., 52, 1259 (1983).

13. P. M. Scott, M. S. Thesis (Clarkson University, 1987).

14. P. M. Scott, S. V. Babu, R. E. Partch and L. J. Matienzo, (to be published).

15. V. Srinivasan, M. S. Sivasubramanian, and S. V. Babu, in 7th Int. Sym. Plasma Chem., Proceedings, p. 1405 (Eindhoven, 1985).

THE CRACKING AND DECOHESION OF THIN FILMS

A. G. EVANS, M. D. DRORY AND M. S. HU
Materials Department, College of Engineering, University of California,
Santa Barbara, California 93106

ABSTRACT

The cracking and decohesion of thin films can be characterized by critical values of a non-dimensional parameter governed by the residual stress, the film thickness and a fracture resistance. This article describes the status of understanding concerning the magnitude of this number for various types of adherent film on either brittle or ductile substrates. Important effects of elastic properties, substrate thickness and yield strength are described.

INTRODUCTION

Thin films of metals, ceramics and polymers are typically subject to appreciable residual stress.[1,2] Such stress can interact with small flaws to cause film decohesion. The decohesion process is governed by a critical non-dimensional parameter,[3-6]

$$\Omega_c = K_c / \sigma_0 \sqrt{h} \tag{1}$$

where σ_0 is the stress in the film, h is the film thickness and K_c is the fracture resistance. The parameter Ω is referred to as the decohesion number. The critical magnitude of the decohesion number, Ω_c, and the appropriate K_c (interface, substrate, film) depend on the sign of the residual stress, the ductility of the film and substrate and their relative elastic moduli. The known modes of decohesion are summarized in Table I. This multiplicity of possible events demands quantitative solutions that provide the values of Ω_c and K_c pertinent to each. Then, definitive decisions regarding decohesion can be reached, based on the physical properties of the film, substrate and interface.

For conditions of residual tension in the film, the behavior of adherent brittle films on brittle substrates is exemplified by results obtained for Cr films on glass.[6] Decohesion in this case involves film cracking, followed by cracking of the substrate parallel to the interface. The appropriate K_c is then the substrate value and Ω_c is governed both by the Σ ratio of the film modulus, E^f, to the substrate modulus, E^s, and the relative crack depth beneath the interface, λ (Fig. 1). For this case, film

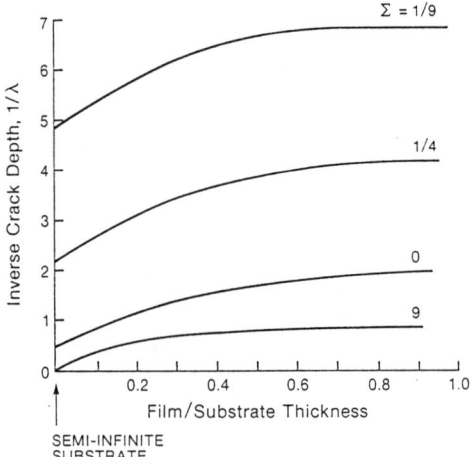

Fig. 1. The variation in crack depth beneath the interface according to the criterion, $y = 0$, as functions of the substrate thickness and the elastic modulus ratio

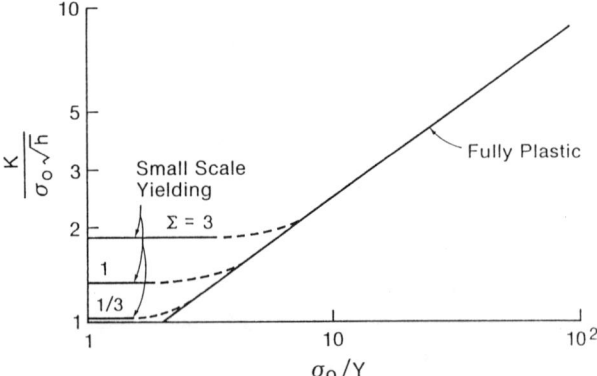

Fig. 2. Trends in the film cracking number (for adherent films) with substrate yield strength and elastic properties

cracking occurs with sufficient facility that film decohesion nearly <u>always</u> obtains when $\Omega = \Omega_c$. Preliminary studies have also been conducted for brittle substrate systems with adherent thin films having relatively high toughness: notably, polymers on glass[9] and alumina on glass.[7] In these cases, decohesion initiates at specimen <u>edges</u>, but still extends through the substrate <u>parallel</u> to the <u>interface</u>. Consequently, for these cases, K_c again refers to the substrate and Ω_c is again given by Eqn. (2). However, now, the decohesion probability Φ is quite low at $\Omega = \Omega_c$, and furthermore, Φ is sensitive to flaws introduced at the edge by machining, etc. It remains to establish the trends in decohesion probability with the severity of edge flaws.

Some aspects of film fracture have been studied in systems with adherent <u>brittle</u> films on <u>ductile</u> substrates: such as Cr films on Al.[10] In such systems, a multiplicity of parallel cracks propagate across the film at a critical stress. This process can again be described by the non-dimensional parameter Ω_c where K_c now refers to the <u>film</u>. The magnitude of the film cracking number, Ω_c, depends on the response of the interface and the substrate. For adherent films, a range of values governed by substrate yield strength and the elastic properties has been determined (Fig. 2).

Studies of films subject to residual <u>compression</u> have also been conducted.[3,11] These studies have analyzed the buckling and post-buckling behavior of films above interface flaws. Experiments have been used to compare predicted and measured trends in decohesion, initiating from circular flaws at the interface. These comparisons have established a self-consistent view of the decohesion problem for the case wherein both the film and substrate are brittle (i.e., ZnO on Si). For this case,

$$\Omega_c \approx 1.46 \qquad\qquad (2)$$

where the K_c contribution to Ω_c refers to the <u>film</u>.

Some of the underlying fundamentals of the fracture process concern trends in the stress intensity factor $|K|$ and the phase angle of loading, ψ, which represent the mixity of shear and tensile displacements on the crack surface.[*] Film decohesion studies have established that <u>film</u> and <u>substrate</u> cracking always proceed along crack trajectories wherein $\psi = 0$.[6,7] Consequently, K_c in Eqn. (1) is the Mode I (opening) value, K_{IC}. However, when cracking occurs along the <u>interface</u>, the crack typically grows with $\psi \approx \pi/4$.[12] Then, the fracture

---

[*]The stress intensity factor can be used interchangeably with the strain energy release rate, $\mathcal{G}$, using $|K|^2 = E\mathcal{G}$.

condition must be characterized by a locus of critical stress intensity factor $|K_c|$ with ψ.

TABLE I

MODES OF THIN FILM DECOHESION

RESIDUAL STRESS	FILM	SUBSTRATE	INTERFACE BONDING	DECOHESION MECHANISM(S)
Tensile	Brittle	Ductile	Good	Film Cracking: No Decohesion
			Poor	Film Cracking → Interface Decohesion
	Ductile	Brittle	Good	Edge Decohesion In Substrate
			Poor	Edge Decohesion At Interface
	Ductile	Ductile	Poor	Edge Decohesion At Interface
	Brittle	Brittle	Good	Film/Substrate Splitting → Substrate Decohesion
			Poor	Edge Decohesion At Interface (Higher Film Toughness) Film Cracking → Interface Decohesion
Compressive	Brittle	Ductile	Good	Buckle Propagation In Film
			Poor	Buckle Propagation At Interface
	Ductile/ Brittle	Brittle	Good	Substrate Splitting
			Poor	Buckle Propagation At Interface
	Ductile	Ductile	Good	No Decohesion
			Poor	Buckle Propagation At Interface

REFERENCES

1. R. W. Hoffman, Physics of Thin Films 3, (1966) 211.

2. M. Janda, Thin Films 142, (1986) 37.

3. A. G. Evans and J. W. Hutchinson, Intl. J. Solids and Structures 20, (1986) 451.

4. R. M. Cannon, R. M. Fisher and A. G. Evans, MRS Proceedings 54, (1986) 799.

5. M. D. Thouless, A. G. Evans, M. F. Ashby and J. W. Hutchinson, Acta Met. 35, 1987) 1333.

6. M. Hu, M. D. Thouless and A. G. Evans, Acta Met., in press.

7. M. D. Drory, M. D. Thouless and A. G. Evans, Acta Met., in press.

8. J. W. Hutchinson, to be published.

9. P. Ho (private communication).

10. M. S. Hu and A. G. Evans, to be published.

11. C. Rossington, D. B. Marshall, B. T. Khuri-Yakub and A. G. Evans, J. Appl. Phys. 56, (1984) 2639.

12. P. G. Charalambides, J Lund, R. M. McMeeking and A. G. Evans, J. Appl. Mech., in press.

POLYMER ENCAPSULATED MICROELECTRONICS:
MECHANISMS OF PROTECTION AND FAILURE

J.E. ANDERSON*, V. MARKOVAC*, AND P.R. TROYK**
* Research Staff, Ford Motor Company, Dearborn, MI 48121
** Pritzker Institute, Illinois Institute of Technology,
   IIT Center, Chicago, IL 60616

ABSTRACT

    This research relates electrochemical failure of encapsulated
microelectronics to surface moisture and to surface impurities. It draws
heavily on "osmotic blistering", a phenomenon known to produce corrosive
failure of coated/painted metals [1].

    Leakage currents were measured on aluminum comb specimens as a function
of relative humidity (RH) and temperature. We studied both bare combs and
combs encapsulated with polysiloxane. 9.05 Vdc bias was used. We
performed "contamination-by-design" experiments by deliberately introducing
known amounts of NaCl, $CaCl_2$ and sucrose onto the comb surface. Results
were compared with corresponding data taken on well-cleaned specimens.

    Our principal findings are

    (1) Under dry conditions (RH<1%), small leakage currents are observed,
ranging between 1-10 pA, which are insensitive to surface contamination
levels. This implies that solid surface impurities per se do not promote
electrochemical IC failure.

    (2) In extremely moist environments (RH>99%), surface-contaminated
samples exhibit large leakage currents, ranging between 1 and 10 μA, that
are roughly proportional to surface loading.

    (3) Different surface chemical compounds produce leakage-current steps
at specific RH values, corresponding to solid-to-saturated solution
transitions. For example, $CaCl_2$ exhibits a transition at 21% RH; NaCl at
75% RH. At RH values above the transition, aqueous droplets, or vacuoles,
were observed at surface sites occupied by solid deposits. The RH location
of the transitions is largely unaffected by the presence or absence of
polymer encapsulant. Leakage current steps were typically four to six
orders of magnitude. The size of the step change varied between bare and
encapsulated samples, and with surface loadings.

    (4) Variable temperature studies, performed at constant external water
vapor, exhibited step decreases in leakage current at temperatures
corresponding to saturated solution to solid transitions.

    (5) Sucrose, a nonelectrolyte, exhibited a leakage current step
similar to those observed with $CaCl_2$ and NaCl. The size of the sucrose
step change was significantly less than that observed with the
electrolytes.

    (6) Electrochemical attack patterns varied among the different
chemical compounds. For example, as shown in Fig. 1, $CaCl_2$ exhibited
anodic attack on alternate metallization lines. NaCl produced attack on
all metallization lines.

    A full report of this work will appear elsewhere [2].

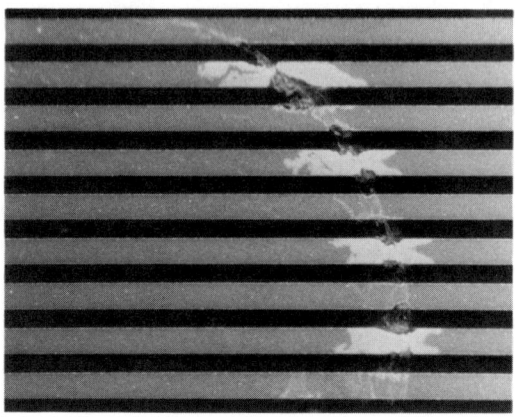

Figure 1. Photomicrograph of aluminum metallization, "contaminated-by-design" with $CaCl_2$. This specimen was not encapsulated. Attack on alternate lines illustrates selective anodic attack under external dc bias. 32X magnification.

"Osmotic Blistering," Vacuole Formation and Growth

"Osmotic blistering" is a well-documented phenomenon known to produce corrosive failure of coated/painted metals [1]. The stable thermodynamic form of solid compounds changes with RH. Certain compounds, such as $CaCl_2$, form one or more solid hydrates at different external RH values. All compounds exhibit a critical relative humidity, $(RH)_{crit}$, where the compound or its solid hydrate become unstable relative to a saturated aqueous solution. At relative humidities above $(RH)_{crit}$, solutions of specific composition represent the equilibrium thermodynamic state.

$(RH)_{crit}$ values may be calculated as follows. The water vapor pressure, $p(m)$, over an aqueous solution is given by

$$(RH) - p(m)/p(0) = exp(-\pi(m)V_w/RT) \qquad (1)$$

where $p(0)$ is the vapor pressure of pure water, V_w is the partial molar volume of water, and $\pi(m)$ is the osmotic pressure of a m molal solution [3]. $\pi(m)$, in turn, is given by

$$\pi(m) = (RT \ \nu\phi/1000)m \qquad (2)$$

where ν is the number of ions formed in electrolyte dissociation and ϕ is the osmotic coefficient [3]. $(RH)_{crit}$ values are obtained from Eqs. (1)-(2) by setting m equal to its saturated solution value. Table 1 lists representative values.

Table 1

(RH)$_{crit}$ Values and Saturated Solution Osmotic Pressures
(Calculated for 25° C)

Compound	(RH)$_{crit}$ (percent)	π (atm)	Compound	(RH)$_{crit}$ (percent)	π (atm)
Sucrose	85	222	$CaCl_2$	29	1670
KCl	84	232	KOH	19	2250
NaCl	76	373	NaOH	14	2710
$AlCl_3$	75	383	LiCl	11	3000

Vacuole formation; i.e., the solid to saturated solution transition, is favored by the enormous osmotic pressures of saturated solutions which range between 100 and 10000 atm. Typical values, calculated using Eq. (2), are given in Table 1. We present an analysis elsewhere [2] showing that an encapsulant modulus of similar magnitude would be required to impede vacuole formation and growth.

Experimentally, aqueous vacuoles, or "blisters," are observed to grow at humidities above (RH)$_{crit}$ at sites occupied by crystals (vide Fig. 2).

Figure 2. Photomicrograph showing vacuoles on an unencapsulated comb specimen. 32X magnification.

Leakage Current Measurements

Leakage currents were measured on Al comb specimens of the type shown in Figure 1. Metallization lines were 6600 μm long x 140 μm wide x 1 μm thick; they were separated by 114 μm. The combs were prepared on a 10μm thick SiO_2 substrate on microelectronic grade Si wafer. The combs were attached electrically to an Airpax Model ADI-140-0059 chip carrier with 25μm dia. gold wire. Leakage currents were measured with a Keithley Model 610A Electrometer. A transistor battery provided 9.05 Vdc bias. Further details of the experiment and sample handling are given in the full text [2].

Specimens were "contaminated" by evaporating known volumes (droplets) containing fixed concentrations of NaCl, CaCl$_2$, or sucrose. 7 x 10^{-8} mho-cm water was used to prepare the solutions. Evaporation produced reproducible, but nonuniform, surface loadings: deposits tend to precipitate around the perimeter of the original droplet.

Measurements were performed in a flow tube/environmental chamber. Humidity was controlled by blending dry N$_2$ and N$_2$ saturated in water. A EEG Model 660 Dewpoint Hydrometer monitored humidity in the gas leaving the chamber. A tube furnace, placed around the flow tube, allowed temperature variation while also acting as a Faraday shield. A dc supply powered the furnace.

Representative results, obtained at constant temperature and variable RH are shown in Fig. 3. Corresponding data, taken at fixed external water vapor concentration and variable temperature, are shown in Fig. 4.

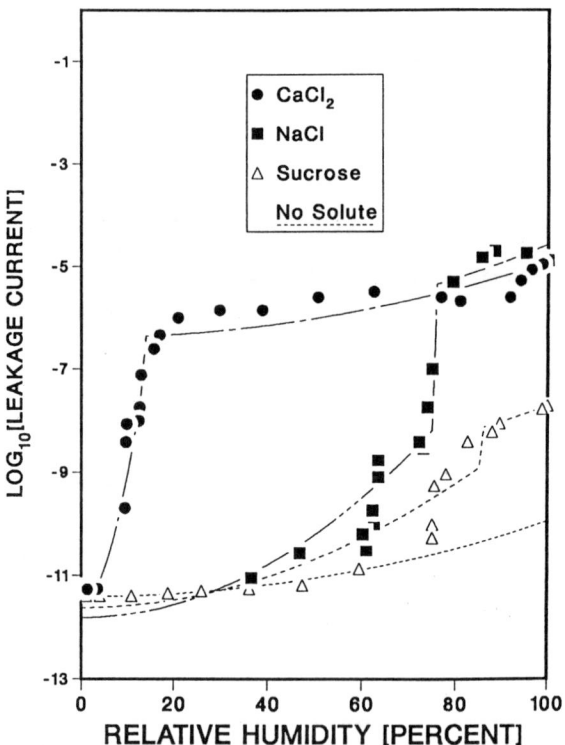

Figure 3. The variation of steady-state leakage current with external relative humidity for samples prepared by evaporating droplets of 0.01 M CaCl$_2$, 0.01 M NaCl, 0.01 M sucrose and a droplet of distilled water. The specimen was encapsulated with polysiloxane.

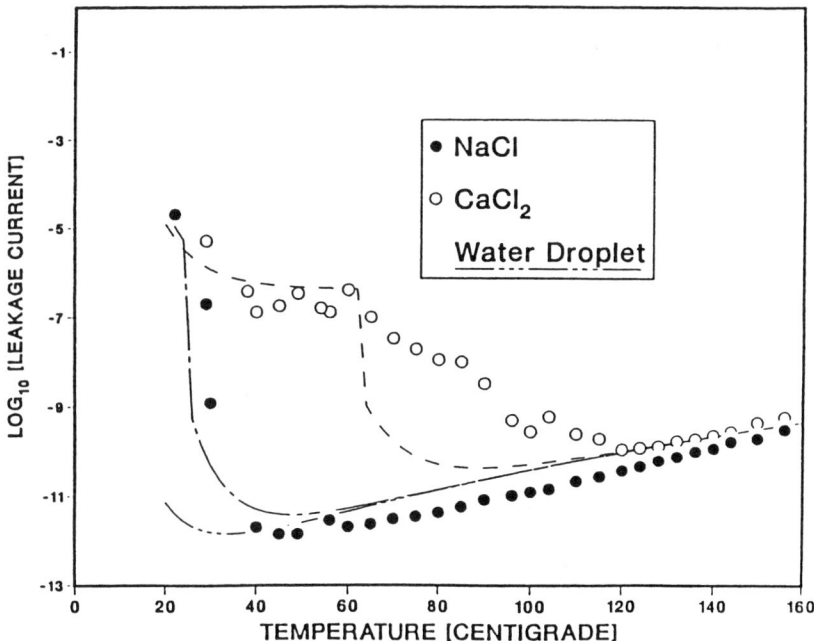

Figure 4. The variation of leakage current with temperature for encapsulated specimens prepared by evaporation of 0.01 M $CaCl_2$ and 0.01 M NaCl droplets. A constant external water vapor of 17.5 mm Hg (100% RH at 20°C) was maintained.

References

1. W. Funke, Prog. Organ. Coatings 9, 26 (1981).

2. J.E. Anderson, V. Markovac and P.R. Troyk, IEEE Trans. Components, Hybrids, and Manuf. Technol., to appear (1988).

3. R.A. Robinson and R.H. Stokes, Electrolytic Solutions (Academic Press, New York, 1959), pp. 24-30.

CHROMIUM-POLYIMIDE INTERFACE CHEMISTRY

M. J. Goldberg, J. G. Clabes, A. Viehbeck and C. A. Kovac
IBM Thomas J. Watson Research Center P.O. Box 218, Yorktown Heights, NY 10598

ABSTRACT

We have studied the reactions occurring for the deposition of chromium onto pyromellitic dianhydride - 4,4'-oxydianiline (PMDA-ODA) polyimide for a wide range of metal coverages, using X-ray photoelectron spectroscopy as well as solution studies involving model compounds. A multi-step, coverage-dependent, mechanism is proposed which includes an initial electron transfer from Cr to PMDA-ODA, followed by formation of Cr-O and and Cr-N polymer-bound intermediates. At higher metal coverage, Cr-oxides and Cr-nitrides are formed which may activate the polymer for subsequent formation of Cr-carbide. No evidence was obtained in support of a model involving initial formation of a π-arene complex between chromium atoms and PMDA-ODA. In the course of this work we have shown that similarities exist between the chemistry of metal atoms in condensed phases and the reactions occurring between metal atoms and surfaces.

INTRODUCTION

With increasing miniaturization in microelectronic packaging and the stringent processing and operating requirements placed upon metal-polymer interfaces, detailed studies of these systems are important. The modification of polymer surfaces by evaporated or sputtered electropositive metals has been shown to improve adhesion between metal and polymer [1]. Previous X-ray photoelectron (XPS), UV photoelectron (UPS), near-edge X-ray adsorption fine structure (NEXAFS), and electron energy loss (EELS) studies have examined the primary site of reaction when metal atoms are deposited onto the surface of various polymers [2-9]. For metals such as Ti [8] and Cr [7] there appears to be a general trend for initial charge transfer, complexation and/or bond scission involving the PMDA carbonyl groups. In contrast, recent theoretical studies, which have examined the energetics of possible initial bonding modes [10-12] between various metals and polymers, conclude that formation of π-arene complexes between Cr and PMDA-ODA, rather than direct interaction with the carbonyl groups, lead to the observed changes in the various XPS core level spectra. It is important to establish the nature of the re-action in the very early stages of metal deposition in order to better understand the events oc-curring at higher coverages. This is addressed by in-situ exposure of initial reaction products to reactive ligands, spectral comparison with electrochemically reduced PMDA-ODA films, and solution model studies to differentiate between possible reaction pathways and products formed. The variation in metal-polymer reactivity and product distribution with changing deposition conditions offers further insight regarding the initial interface chemistry. We report herein a de-tailed picture of the interaction of chromium metal with PMDA-ODA, illustrating a general trend for the interaction of electropositive metals with polyimides.

EXPERIMENTAL

Mat. Res. Soc. Symp. Proc. Vol. 108. ©1988 Materials Research Society

General Data. UV-Visible spectra were recorded on a Perkin-Elmer Lambda 3B spectrophotometer. Infrared spectra were taken using an IBM Instruments IR44 fourier transform spectrometer. Spectra were recorded over 128 or 256 scans at 2.0 cm^{-1} resolution.

Reagents and Materials. Pyromellitic dianhydride (PMDA) was twice recrystallized from methyl ethyl ketone prior to use: mp 288 - 290 °C (lit. 285 - 287 °C). N,N'-di-(n-butyl)pyromellitic diimide (PMDA-BA) was made by the condensation of pyromellitic dianhydride (PMDA) and n-butylamine in N,N'-dimethylacetamide (DMAc) and recrystallized twice from acetone prior to use. Tris(acetonitrile)chromiumtricarbonyl was prepared in the standard manner [13,14], by refluxing hexacarbonylchromium in acetonitrile overnight. Synthesis of (arene)tricarbonylchromium complexes followed the method of Pauson and Mahaffy, with slight modifications where noted [15].

XPS Experiments. XPS apparatus and gas infusion system used for in-situ carbon monoxide dosing studies have been described in detail elsewhere [7]. PMDA-ODA films were prepared by spinning commercial polyamic acid diluted to 1% in NMP on plain Si wafers. Continuous coatings of 10-20 nm thickness, as measured by mechanical profiling, were obtained which helped to avoid significant charging problems. For the electrochemical experiments PI films were spun onto conductive substrates and imidized by thermal or chemical dehydration. PMDA-ODA films were electrochemically reduced under potentiostatic conditions (-1.1 V and -1.7 V vs. saturated calomel electrode (SCE) reference) in a 0.1 M tetrabutylammonium fluoroborate/acetonitrile solution. Electrodes were made from platinum or stainless steel foil. A three-compartment cell was used with anode and cathode compartments separated by a fritted glass disk. An analogous procedure was used for the reduction of PMDA-BA (n-butyl amine) monomer solutions. All electrochemistry was done in a nitrogen atmosphere maintained at 25 °C. Reduced films were transferred to the XPS equipment in a nitrogen environment; reduced solutions were handled accordingly during absorption studies in an IBM IR-44 fourier spectrometer.

Reaction of PMDA-BA with $Cr°(CO)_6$. PMDA-BA (0.2 g, 0.6 mmol) and hexacarbonylchromium (0.26 g, 1.2 mmol) were dissolved in a mixture containing 5 mL of tetrahydrofuran (THF) in 55 mL of p-dioxane. The solution was degassed thoroughly prior to refluxing. After heating for 235 hours under nitrogen, both solvent and unreacted $Cr°(CO)_6$ were removed under vacuum. The resulting dark green, air sensitive, solid was manipulated in a nitrogen glove box: IR (THF, CaF_2) ν(C=O) 1621, 1590 cm^{-1}; IR (DMF, CaF_2) ν(C=O) 1591, 1569, 1534, 1333 cm^{-1}; UV-Vis (DMF) 714, 647, 413, 319, 265 nm. After exposure to oxygen: IR (THF, CaF_2) ν(C=O) 1773, 1725, 1394 cm^{-1}; UV-Vis (DMF) 317, 310, 266 nm.

Reaction of PMDA-BA with $Cr°(CH_3CN)_3(CO)_3$. PMDA-BA (0.5 g, 1.5 mmol) was added to a solution containing 0.3 g (1.1 mmol) of $Cr°(CH_3CN)_3(CO)_3$ in 75 mL of THF. After degassing, the solution was refluxed for 18 hours under nitrogen. Solvent removal (in vaccuo) again led to a dark green precipitate which gave a green solution when dissolved in DMF: UV-Vis (DMF) 712, 645, 412, 316, 265. After exposure to oxygen: UV-Vis (DMF) 317, 309, 266 nm.

RESULTS AND DISCUSSION

A PMDA-ODA repeat unit, along with the corresponding core level XPS assignments (in eV), as determined by others [16], is shown below.

Figures 1 and 2 show the changes occurring in the C1s, O1s and N1s core level XPS spectra with increasing chromium coverage, deposited at ambient temperature (25 °C) and 250 °C, respectively. The corresponding Cr2p XPS spectra are not shown because, even though spectral changes are observed, they are unspecific as to the oxidation state and coordination geometry of the metal at very low coverages. Over the course of the metal deposition coverage-dependent changes are observed. Earlier work by this group [2,7,9] and others [3-6] has substantiated the PMDA group as the site of initial reaction with Cr atoms. The changes which occur upon deposition at ambient temperature, shown in Figure 1, have been discussed in detail previously [2,7,9] The observed shifts in the C1s and O1s spectra and lack of changes in the N1s spectra, at low Cr coverage (0.03 nm), are the same as those observed for films reduced electrochemically or by in-situ cesium deposition [7], as discussed below.

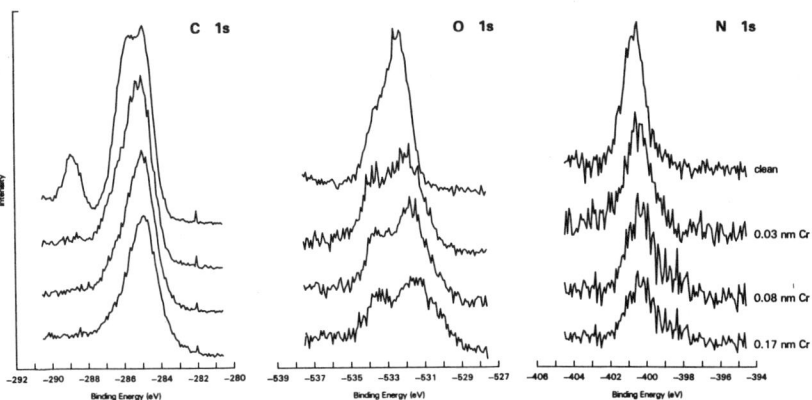

Figure 1. Changes in the C1s, O1s, and N1s core level XPS spectra of PMDA-ODA as a function of increasing Cr coverage, deposited at ambient temperature (25°C).

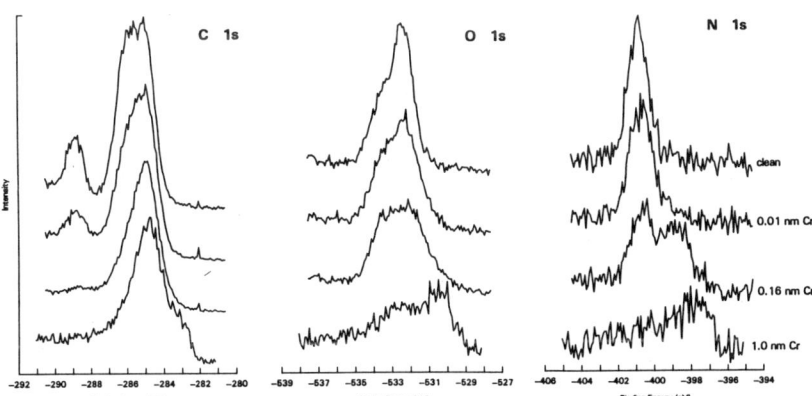

Figure 2. Changes in the C1s, O1s, and N1s core level XPS spectra of PMDA-ODA as a function of increasing Cr coverage, deposited at 250°C.

While the same low coverage (0.01 nm Cr) trends are also observed for deposition at 250 °C (Fig. 2), distinctly new features are seen with increasing coverage. The most significant change is observed in the N1s core level spectra, where at 0.16 nm Cr there is a peak of substantial area at 399 eV as well as a peak of much smaller intensity at 397.5 eV. This shift of 2 eV (from 401 to 399 eV) is greater than that expected for a "nearest neighbor" effect, induced by reaction of Cr with the PMDA carbonyl groups. This is, instead, more consistent with charge localization onto the nitrogen atom, either through direct metal interaction or imide ring opening. At higher metal coverage (1.0 nm) the O1s and N1s spectra are dominated by peaks at 530.3 and 397.5 eV, respectively. These peaks are ascribed to chromium oxides and nitrides by comparison with authentic samples and from the published work of others [17,18]. At the same coverage there is a new C1s peak of smaller intensity at 283 eV which is assigned to chromium carbide [18]. Others have recently suggested that this peak may be due to the formation of π-arene complexes of chromium with the ODA portion of the polymer [12], a conclusion which is inconsistent with results which are discussed in detail below. The appearance of this 399 eV peak in the N1s spectrum independent of similar changes to the C1s and O1s core levels indicates a strong temperature dependence for the reaction of Cr with PMDA-ODA. The question of whether the reaction of chromium with carbon, oxygen and nitrogen occurs along parallel or consecutive reaction pathways may be answered by more detailed studies of the temperature dependence of the reaction. As inferred by relative intensities, the formation of chromium-oxide and chromium-nitride appears to precede and is perhaps responsible for the subsequent formation of chromium-carbide.

Taken together, the trends observed in all core level spectra, in conjunction with data presented in a previous paper [7], indicate a stepwise process involving initial charge transfer from Cr to the PMDA group. This is followed by the formation of polymer-bound Cr-O (531.4 eV) and Cr-N (399 eV) intermediates and finally bond scission to form polymer-independent species such as chromium-oxide (530.3 eV), chromium-nitride (397.5 eV) and eventually chromium-carbide (283 eV). A step-wise mechanism has also been proposed by Freilich and Ohuchi to explain changes occurring in the XPS spectra of PMDA-ODA upon the deposition of Ti atoms [8].

Let us consider the very initial stages of the reaction in more detail. Two likely mechanisms which have been proposed are: (1) complex formation between a zero-valent Cr atom and π-orbitals of PMDA; and (2) electron transfer from Cr to PMDA-ODA, resulting in a reduced polyimide species. Separate investigations of these two possible mechanisms are given below.

π-Complex Formation

First, consider the case of π-complex formation between Cr° atoms and PMDA-ODA polyimide. Qualitatively, both kinetics and thermodynamics would favor formation of π-arene complexes with the diamine (ODA) portion of PMDA-ODA as opposed to the anhydride (PMDA) portion. Thermodynamically, arene rings substituted with amino or methoxy groups, like ODA, will form more stable complexes than those directly conjugated to carbonyl groups [19-20]. Stoichiometrically, there are two arene rings in the diamine portion of the molecule to only one on the PMDA portion, which would favor attack by chromium in a kinetically controlled reaction. This implies an unusually high selectivity for chromium to react with the PMDA group despite both kinetic and thermodynamic considerations. Nevertheless, since the PMDA group has been shown experimentally to be the initial site of reaction with chromium we will focus our discussion on this [2-9].

Formation of π-complexes could be identified by reaction with donor ligands such as carbon monoxide, which is known to react readily with and form stable complexes of the sort Cr°(CO)$_x$. In the case of chromium atoms reacting with PMDA to form stable arene π-complexes, the three most likely products to be formed would be an unsaturated PMDA-Cr°, a PMDA-Cr° complex whose coordination sphere has been filled by interaction with carbonyl π-bonds from surrounding groups ((PMDA)(R$_2$C=O)$_3$Cr°), or a bis(PMDA)Cr°. All of these would be expected to react with carbon monoxide. PMDA-Cr° is highly coordinatively unsaturated. Gas phase studies of Cr°(CO)$_6$ have shown that when one carbonyl ligand is photolyzed off to give

$Cr°(CO)_5$, the strong tendency to fill the empty coordination site is such that complexes involving inert compounds such as cyclohexane have been observed [21]. Therefore, one would expect a complex having the chromium atom six electrons (three ligands) unsaturated to react at a diffusion controlled rate when exposed to carbon monoxide. In the case of $(PMDA)(R_2C=O)_3Cr°$, the transient nature of π-bonded complexes of this sort would lead to carbonyl ligand displacement by carbon monoxide to form a more stable $(arene)Cr°(CO)_x$ complex.

The most stable arene complex that can be formed from the reaction of chromium with PMDA would involve two PMDA arene rings bonded to one chromium atom. This is, however, the most difficult to form given the known conformational proximity and inter-chain distance of PMDA groups in fully cured PMDA-ODA [22]. Therefore, if such complexes were initially formed upon chromium deposition we would expect carbon monoxide to displace one, if not both, of the co-ordinated arene rings. In all cases reaction with carbon monoxide could lead to stable $(arene)Cr°(CO)_3$ complexes or perhaps displace the chromium from the polymer surface completely in the form of $Cr°(CO)_6$, leading to chemical changes easily observable by XPS.

No changes are observed in the XPS spectra after deposition of chromium (0.01 nm) and subsequent exposure to 5 torrmin of CO. This result offers no positive evidence for the formation of π-arene complexes.

Electron transfer mechanism

Another possible reaction that can occur when chromium atoms are deposited onto polyimide is electron transfer from metal to polymer. Polyimides have been shown to be electroactive with reduction potentials of -0.72 and -1.38 volts (vs. SCE) to form the corresponding radical-anion and dianion, respectively [23,24]. Due to the highly delocalized nature of the PMDA LUMO [11], a one-electron reduction results in a shifting and splitting, in the FTIR spectrum, of all PMDA carbonyl groups from 1723 cm^{-1} to lower wavenumbers [24]. The corresponding oxidation potential for the preferred $Cr°$ to Cr^{+2} couple (-0.91 volts) is such that it is thermodynamically feasible for chromium to do two one-electron reductions, and perhaps form a structure like that shown below.

A model of this sort also explains the high efficiency of small amounts of chromium to quench those XPS peaks associated with the PMDA group. In this case one chromium atom could, directly and indirectly, affect eight carbonyl groups. Figure 3 shows the corresponding XPS spectra of a PMDA-ODA film before and after electrochemical reduction to the corresponding radical-anion [24]. One can clearly see the similarities with those core level spectra at low coverages of chromium. There is the characteristic depletion of the carbonyl carbon intensity at 289 eV in the C1s spectrum. The respective O1s component is also distinctly modified without recognizable change of the ether part. The N1s spectrum appears asymmetrically broadened to lower binding energy. The tetra(n-butyl)ammonium counterion appears as an additional emission in the lowest binding energy state of the C1s spectrum and a shoulder at higher binding energy in the N1s spectrum. We have also observed similar spectral changes for reduction of PMDA-ODA by in-situ deposition of cesium metal, described elsewhere [7].

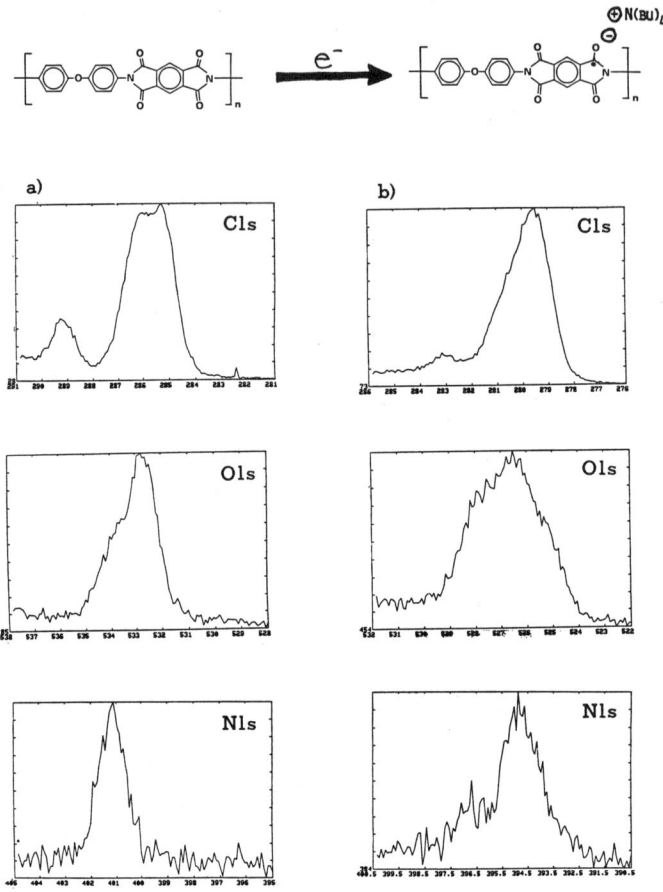

Figure 3. C1s, O1s and N1s core level XPS spectra of a PMDA-ODA film before (a) and after (b) electrochemical reduction.

Model compounds such as the diimide formed from the reaction of PMDA and n-butylamine (BA) have one and two electron reduction potentials very similar to that of cured PMDA-ODA, due in part to the lack of substantial delocalization through the imide nitrogen [23,24]. In an attempt to synthesize $(arene)Cr°(CO)_3$ complexes similar to those discussed above for both XPS and NEXAFS studies, PMDA-BA was reacted with $Cr°(CO)_6$ and $Cr°(CH_3CN)_3(CO)_3$ in a variety of solvents under a variety of conditions. Although these reactions were done under conditions which should favor formation of an $(arene)Cr°(CO)_3$ complex, the only products obtained were those derived from chromium reduction of PMDA-BA (see Experimental), as identified by their characteristic UV-Vis spectra [7,23], which are also identical to that of an electrochemically reduced PMDA-ODA film [23,24]. The comparability of all these spectra leads to two important conclusions.

First, the fact that the chromium carbonyl complexes reacted with PMDA-BA have oxidation potentials lower (-0.45 for $Cr^\circ(CO)_3(CH_3CN)_3$ and 1.05 for $Cr^\circ(CO)_6$) [25] than that for chromium metal (-0.91 volts) and still result in diimide reduction strongly supports the claim of an electron transfer from chromium to the PMDA portion of PMDA-ODA as the initial reaction product upon metal deposition. This is also consistent with the carbon monoxide experiment described above since oxidized chromium is somewhat inert towards reaction with such ligands. Similar electron transfer initiated reactions have been observed by Freilich and Ohuchi for the solution reaction of a Ti° suspension with a model compound similar to PMDA-BA [8]. Secondly, the similarity between the electrochemical and spectroscopic behavior of the PMDA-BA monomer and that for PMDA-ODA films implies that there is no significant difference between charge delocalization over the PMDA backbone in solution, where counter-ions are solvated and free to migrate from one site to another, and that in a solid film where the anion-cation ion pair is essentially "locked" in a particular configuration. Thus, for reactions of PMDA-ODA involving charge donation into the delocalized LUMO, the inherent chemistry is not drastically changed upon going from solution to a surface, and one can then draw upon related solution and matrix isolation work in the literature to explain possible reactions occurring at the polyimide surface.

Examining the literature concerning metal atom reactions of Cr° atoms with carbonyl containing compounds, one notes a trend similar to what we discussed above, concerning the site and mechanism of these reactions. Elschenbroich and co-workers [26] found that when a frozen matrix containing chromium atoms and benzophenone is allowed to warm from liquid nitrogen to room temperature, a number of products are isolated, the majority of which have undergone reaction with chromium at the carbonyl group. Gladysz and co-workers [27] have observed efficient removal of oxygen atoms by chromium atoms from "activated" carbonyl compounds such as amine oxides and sulfoxides. This observation correlates well with what we have found for deposition of chromium atoms onto polyimide films which have been pre-activated by cesium metal; that being an enhanced cleavage of the $C=O$ bond and formation of products derived thereof [28].

CONCLUSION

A number of important conclusions can be drawn from the work described herein. In addressing the initial objective, a coverage-dependent step-wise mechanism has been documented for the deposition of chromium onto PMDA-ODA polyimide. Strong support has been given for an initial electron transfer from metal to polymer, via in-situ and model studies. This accounts for the observed high selectivity toward reaction with the electroactive PMDA group as opposed to the ODA moiety. Further reaction with deposited metal results in formation of Cr-O and Cr-N polymer-bound intermediates. With continuing metal deposition bond scission occurs, leading to Cr-oxides, Cr-nitride and Cr-carbide. In general, there is a trend for initial electron transfer in the reaction of these and other electropositive metals with electroactive polyimides. Perhaps most important is that the chemistry observed here for the deposition of metal atoms onto a polymer surface under a variety of conditions is inherently the same as that seen by others for systems (i.e. matrix isolation studies and metal atom reactions) which are shown to be governed by the same basic principles.

REFERENCES

1. J.M. Burkstrand, J. Appl. Phys. 52, 4795 (1981); Phys. Rev. B 20 (12), 4853 (1979); J. Vac. Sci. Technol. 20 (3), 440 (1982).

2. J.L. Jordan, C.A. Kovac, J.F. Morar, R.A. Pollak, Phys. Rev. B. 36 (3), 1369 (1987).

3. N.J. Chou, D.W. Dong, J. Kim, A.C. Liu, J. Electrochem. Soc. 131, 2335 (1984).

4. N.J. Chou, C.H. Tang, J. Vac. Sci. Technol. A2, 751 (1984).

5. P.O. Hahn, G.W. Rubloff, J.W. Bartha, F. LeGoues, R. Tromp, P.S. Ho, Mat. Res. Soc. Symp. Proc. 40, 247 (1985).

6. N.J. Dinardo, J.E. Demuth, T.C. Clarke, Chem. Phys. Lett. 121, 239 (1985).

7. J.G. Clabes, M.J. Goldberg, A. Viehbeck, C.A. Kovac, J. Vac. Sci. Technol. (in press).

8. F.S. Ohuchi and S.C. Freilich, J. Vac. Sci. Technol. A4 (3), 1039 (1986).

9. J.L. Jordan-Sweet, C.A. Kovac, M.J. Goldberg, J.F. Morar, J. Chem. Phys. (in press).

10. P.N. Sanda, J.W. Bartha, B.D. Silverman, P.S. Ho, A.R. Rossi, Mat. Res. Soc. Symp. Proc. 40, 238 (1985).

11. A R. Rossi, P.N. Sanda, B.D. Silverman, P.S. Ho, Organometallics 6, 580 (1987).

12. R.C. White, R. Haight, B.D. Silverman, P.S. Ho, Appl. Phys. Lett. (in press).

13. D.P. Tate, J.M. Augl, W.R. Knipple, Inorg. Chem. 1, 433 (1962).

14. B L. Ross, J.G. Grasselli, W.M. Ritchey, H.D. Kaesz, Inorg. Chem. 2, 1023 (1963).

15. P.L. Pauson and C.A.L. Mahaffy, Inorg. Synth. 19, 154 (1979).

16. L.P. Buchwalter and A. Baise in Polyimides Vol. 1, edited by K.L. Mittal (Plenum, New York, 1983), p. 357.

17. Handbook of X-Ray Photoelectron Spectroscopy, edited by G.E. Muilenberg (Perkin-Elmer Corp., Minnesota, 1979), p. 72.

18. J.G. Clabes, unpublished results.

19. N.N. Travkin, V.P. Rumyantseva, B.G. Gribov, J. Gen. Chem. USSR 41, 1929 (1971).

20. N.N. Travkin, B.G. Gribov, V.P. Rumyansteva, B.I. Kozyrkin, B.A. Salamantin, J. Gen. Chem. USSR 40, 2669 (1970).

21. R. Bonneau and J.M. Kelly, J. Amer. Chem. Soc. 102, 1220 (1980).

22. T.P. Russell, J. Poly. Sci. (Poly. Phys. Ed.) 22, 1105 (1984).

23. S. Mazur, P.S. Lugg, C. Yarnitzky, J. Electrochem. Soc. 134, 346 (1987).

24. A. Viehbeck and M.J. Goldberg (unpublished results).

25. J.W. Hershberger, R.J. Klinger, J.K. Kochi, J. Amer. Chem. Soc. 104, 3034 (1982).

26. C. Elschenbroich, J. Heck, F. Stohler, Organometallics 1, 1399 (1982).

27. S. Togashi, J.G. Fulcher, B.R. Cho, M. Hasegawa, J.A. Gladysz, J. Org. Chem. 45, 3044 (1980).

28. M. J. Goldberg and J. G. Clabes (unpublished results).

XPS STUDIES OF CHROMIUM ATOMS INTERACTING WITH A PMDA-ODA POLYIMIDE SURFACE

R. Haight, B.D. Silverman, R.C. White,
P.S. Ho and A.R. Rossi

IBM Research Division
T.J. Watson Research Center
Yorktown Heights, NY 10598

ABSTRACT

XPS data obtained during the initial stages of vapor deposition of chromium atoms onto a PMDA-ODA polyimide substrate is presented. Utilizing the results of molecular orbital calculations it is emphasized that the data is consistent with the formation of a charge transfer complex between the deposited metal atoms and the PMDA unit of the polymer.

INTRODUCTION

One distinguishing feature of the technologically important polyimides is their aromatic backbone. This holds consequences, not only for the detailed microscopic structure of these materials, but also for the interpretation of data obtained from the many techniques utilized to probe surface sensitive properties during interface formation. Perhaps the most widely studied polyimide to date has been the PMDA-ODA polyimide (1-6). The present paper will therefore focus on the results of experiments performed on this polymer.

Figure 1a. shows the repeat unit of the PMDA-ODA polymer. It can be considered to be composed of two parts, each exhibiting a structural similarity to the monomeric precursors utilized in synthesis of the polymer, e.g, pyromellitic-dianhydride and oxydianiline - hence the acronym, PMDA-ODA.

Cursory examination of this structure leads one to infer that the only relatively free degrees of freedom expected are associated with the dihedral angles defining the orientation of the planes of the aromatic units of the polymer with respect to each other. However, even these degrees of freedom are restricted. For example, since the phenyl group of the ODA unit is adjacent to the PMDA unit, it is prevented from being in the same plane as PMDA due to steric repulsion between hydrogen atoms of ODA and the carbonyl oxygen atoms of PMDA. An X-ray scattering study (7) of a related molecular fragment does indeed show these two planar units to be rotated by approximately 60 degrees with respect to each other. Furthermore, it is well known that due to the steric crowding between the two phenyl groups composing oxydianiline, the planes of these groups are rotated by approximately 45 degrees with respect to each other. Wide angle X-ray studies (7) have indicated the presence of microscopic packing that is consistent with a type of local smectic ordering. A series of self diffusion studies (8) suggests that one might expect the fully imidized material to be quite immobile, perhaps exhibiting little if any self diffusion. In the present paper we will review XPS data obtained during the initial stages of chromium deposition onto a fully imidized PMDA-ODA polyimide surface. Our interpretation of these data has de-

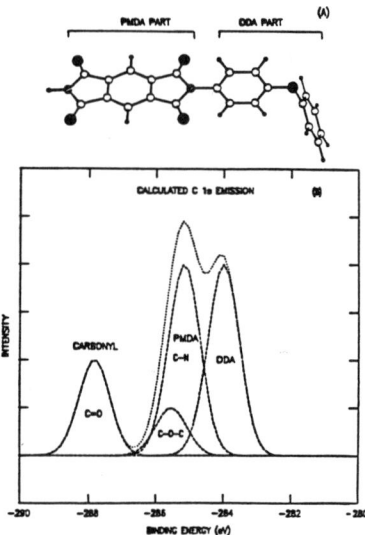

Figure 1. a) Repeat unit of the PMDA-ODA polymer. Carbon, oxygen and nitrogen atoms are given by open, solid and hatched circles respectively. Hydrogen atoms are the small open circles. b) Calculated carbon 1s XPS spectrum of the PMDA-ODA polyimide surface. Assignments are shown as discussed in the text.

pended upon inferences drawn from the structural studies. From these studies one surmises that the "solid state" aspect of the problem holds important consequences for any model building involving the interaction between the polyimide and a deposited metal atom. During the initial stages of metallic deposition, one expects the situation to be not unlike that found in the matrix isolation studies of organometallics involving transition metals (9). For such cases, the formation of the organometallic structures are determined to a significant degree by interactions between the constituents of the matrix. One therefore expects to find a variety of structures, e.g, half-sandwich (10) as well as distorted structures that bear little resemblance to the fully saturated highly symmetric zero valent organometallic chromium structures encountered in the gas-phase.

In what follows it will be shown that the results obtained during the initial stages of deposition, i.e., in the relatively dilute metal atom regime, are fully consistent with that expected from the complexing of a single metal atom with the PMDA unit of the PMDA-ODA repeat unit. In many respects the results obtained from molecular orbital calculations performed on this simple unsaturated complex are similar to the calculated results we have obtained on a wide variety of metal atom-PMDA ligand complexes.

The consistency of the data with the results of molecular orbital calculations obtained from examining the complexing of a single metal atom with a PMDA unit depends upon four distinct features of the XPS data.

For this low metal atom coverage regime, we have observed:

1. Essentially no or little change in the position or intensity of the low binding energy side of the major peak of the carbon 1s spectrum while other features of the carbon 1s spectrum change significantly.

2. A decrease and broadening of the carbonyl oxygen 1s levels with essentially no or little change in the ether oxygen peak.

3. Not only a decrease and broadening of the carbonyl carbon 1s peak but a concomitant shift of intensity out of the high binding energy side of the major peak to lower binding energy.

4. Essentially no or little change in the position of the nitrogen 1s peak.

These features have been listed in this particular order for a specific reason. Features 1. and 2. have enabled us to unambiguously determine that initially, chromium atoms interact with the PMDA unit of the polymer and not with the ODA unit. Features 3. and 4. yield information concerning the details of this interaction i.e., between the chromium atoms and the PMDA unit. Historically, examination of changes in the XPS spectrum, as metals have been deposited onto a polyimide surface, have been primarily concerned with the decrease and broadening of the carbon 1s carbonyl peak, i.e., the first of the observations listed of feature 3.

EXPERIMENTAL METHODS

Care was taken to obtain fully imidized spun-on films since one expects polyamic acid to interact oxidatively with chromium. Solutions of the polyamic acid were spun on silicon wafers that had been coated with aluminum. Films of thicknesses of 100-300 Angstroms were obtained. Such thin films reduce charging effects on the energy positions of the spectral peaks. The films were subsequently heated at 85 °C to drive off the solvent. They were then annealed in UHV for 30 minutes at 350 °C to complete the imidization process. This procedure resulted in highly reproducible X-ray and ultraviolet photoemission spectra. Photoemission measurements were performed in ultrahigh vacuum using monochromatized AlKα and nonmonochromatized MgKα radiation. The core level spectra were obtained at room temperature and for a grazing exit angle geometry which enhanced surface sensitivity. Chromium evaporation was performed in-situ at room temperature.

THEORETICAL METHODS

Molecular orbital calculations were performed on the one-to-one complex, i.e., one chromium atom complexed with one PMDA unit, previously described (11) as well as on other complexes. Some of the complexes investigated involving PMDA dimers were chosen with geometries consistent with local crystalline smectic order. The GAMESS set of molecular orbital programs (12) was utilized with a split basis for the chromium atoms. Calculations on the one-to-one complex were checked with more extensive calculations utilizing the 3-21G basis (13) for all of the atomic constituents. For the one-to-one complex, the only minima found when moving the chromium atom in a plane parallel to the PMDA plane were above and in the center of the six-membered ring as well as in the center of the five membered ring. The relative high coordination and molecular orbital matching of the phases makes this simple to understand (11). For example, location of the Cr atom above and in the center of the six-membered ring allows the dxy orbital to overlap significantly with the antibonding π levels of the carbon atoms of PMDA. Such an arrangement results in a low energy and hence stable configuration. A number of different complexes were investigated that involved the chromium atom or atoms interacting directly with one or more of the carbonyl oxygen atoms. All such complexes investigated exhibited significantly

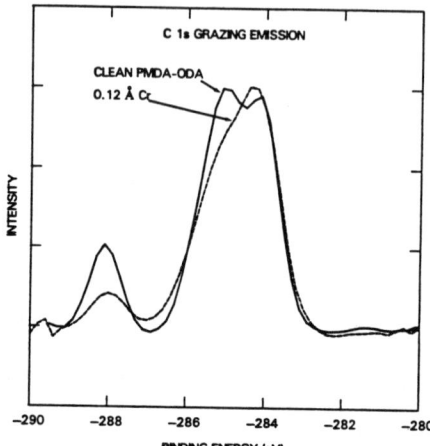

Figure 2. Carbon 1s XPS spectra of the PMDA–ODA polyimide surface (solid curve) and interface with chromium (dashed curve) after one initial deposition. Note that 0.12 Å coverage corresponds to roughly 1 Cr atom for three repeat units.

higher total energies than found for complexes involving chromium atoms that were highly coordinated with the aromatic ring structures.

Since the calculations were *ab-initio* the calculated core level positions enable one to estimate core level ionization energies in the Koopman's approximation. We have also performed Δ SCF calculations on the PMDA unit and found that even though relaxation, i.e., electron screening of the core hole shifts each of the core levels significantly, i.e., by about 10 eV., all the core hole shifts are approximately equal. The Δ SCF calculations, therefore yield relative XPS ionization energies that are similar in value to those obtained in the Koopman's approximation.

As stated, numerous calculations were performed for all reasonable half-sandwich geometries as well as for dimer structures involving two PMDA units. We have found that the calculated XPS shifts are qualitatively the same for a broad class of structures, however the total energies are different. The most energetically stable structures found have always involved the highest possible coordination of the chromium or chromium atoms with the carbon atoms of the PMDA unit or units.

CHROMIUM ON THE PMDA-ODA POLYIMIDE SURFACE:XPS Interpretation

XPS data obtained for the clean polyimide surface contains one important clue concerning interaction sites involving the first chromium atoms that arrive at the polymer surface. Fig. 1b. shows the calculated carbon 1s core level spectrum of the clean material. The calculated states have been broadened to simulate the experimental resolution and assignments are shown. The main peak is composed of essentially a doublet with the low energy component arising solely from carbon atoms on the ODA unit. It should be mentioned that an early suggestion of Buchwalter

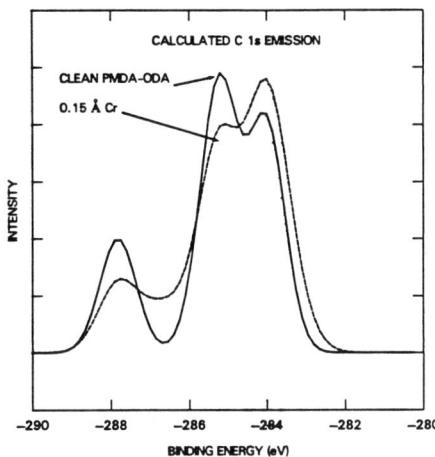

Figure 3. Calculated Carbon 1s XPS spectra within the Koopman's approximation for the PMDA-ODA polyimide surface and for the interface formed after one initial deposition of chromium.

and Baise (14) concerning the origin of the main peak together with subsequent molecular orbital calculations (15) were essential for such identification. The figure also shows the peak at 288 eV. arising from the carbonyl carbon atoms. Shakeup features have also been observed at higher binding energy (not shown) and have been discussed in detail elsewhere (16). Fig. 2 shows the affect on the carbon 1s spectrum after a deposition of 0.12 Angstroms of chromium. This corresponds to approximately one chromium atom per three polymer repeat units. One sees that there has been essentially no change in the low binding energy side of the major peak. Since this peak results from emission from carbon atoms composing the ODA part of the repeat unit, we conclude that the initial interaction of chromium with the polymer involves interaction sites with only the PMDA unit. Furthermore, one sees not only a reduction and broadening of the carbonyl carbon intensity upon this initial deposition, but a shifting of the high binding energy side of the major peak to lower binding energy. Since the major component of the high binding energy side of this peak arises from the carbon atoms that compose the central benzene ring of the PMDA unit, we conclude that the chromium atoms are interacting significantly with these atoms as well as with the carbonyl carbon atoms.

Figure 3 is a calculated XPS spectrum within the Koopman's approximation. The core levels have been broadened to simulate the instrumental resolution. We have superposed levels obtained from a half-sandwich complex with the chromium atom centered above the six-membered PMDA ring and from an unreacted polymeric repeat unit. One notes that the salient features of experiment are reproduced. First, one observes the decrease and broadening of the high binding energy carbonyl carbon atom peak. Concomitant with this is a shifting of intensity out of the high binding energy side of the major peak to lower intensity. These shifts arise due to charge transfer from the chromium atom to the lowest unoccupied molecular orbital of the unreacted PMDA unit.

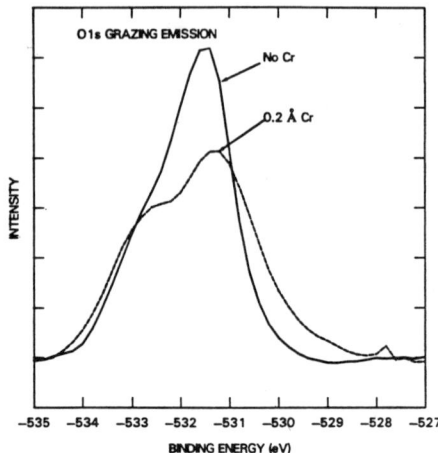

Figure 4. Oxygen 1s XPS spectra of the PMDA-ODA polyimide surface and interface with chromium after one initial deposition.

Actually, the energy match between the d-levels of the chromium atom and the LUMO of PMDA will lead to such shifting of the core levels of the organic ligand independent of where the chromium atom is placed. Such shifting and lowering of the total energy is however enhanced when the orbital overlap between the chromium d-levels and the PMDA LUMO is optimized. This is apparently the case for the chromium atom when it resides at the site of high symmetry above the central benzene ring of the half-sandwich (11). One expects the polymer structure in the vicinity of the surface/interface to present a number of different potential binding sites, some in the locale of significant disorder, others in regions of local crystalline order. The local geometry surrounding the chromium atom or chromium atoms will indeed hold strong implications with respect to the types of local structures formed. We expect, however, that the pattern of XPS shifts will be similar for the different structures reflecting similar distributions of charge transferred to the LUMO of the organic ligands complexed with chromium. This is what we have found in the numerous calculations that we have performed on different chromium-organic ligand complexes.

Fig. 4 shows the oxygen 1s XPS spectrum for the clean polymer and for one low chromium deposition. Prior to deposition, the spectrum is fit well by only two Gaussians. The more intense lower binding energy peak is attributed to the electron rich carbonyl oxygen atoms whereas the higher binding energy peak can be attributed to the ether oxygen atoms. Upon initial deposition, as expected and consistent with the carbon 1s shifts, the feature due the carbonyl oxygen atoms broadens and shifts to lower binding energy. The peak at higher binding energy arising from the ether oxygen atoms remains fixed. Calculated core level shifts of the carbonyl oxygen atoms are consistent with the experimental observations.

Finally, Fig. 5 shows the XPS spectrum arising from the nitrogen 1s core levels. Data is shown for the undeposited polymer surface as well as for a surface upon which 0.2 Angstroms of chromium has been deposited. As expected nitrogen displays only a single peak since there is only one unique chemically bonded consituent in the repeat unit of the polymer. Initial chromium

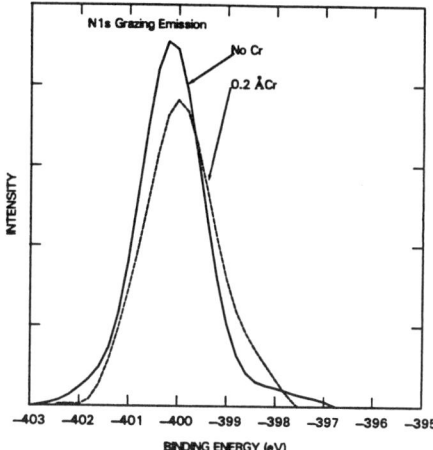

Figure 5. Nitrogen 1s XPS spectra of the PMDA-ODA polyimide surface and interface with chromium after one initial deposition.

coverage results in only a small shift of this peak to lower binding energy. This is consistent with the zero amplitude on the nitrogen atom of the PMDA LUMO. As a result of this, essentially no charge is transferred to the nitrogen atoms during complex formation. The only shifts of the nitrogen 1s core levels are due to charge transferred to adjacent atomic sites and as a consequence such shifts are small. This lack of core level shift upon initial complexation with chromium is, perhaps, one of the most telltale pieces of evidence that complex formation during the initial stages of chromium deposition has occurred with a polymeric backbone that has retained its structural integrity, i.e., has not broken apart or ring-opened. If indeed during the initial stages of chromium deposition the imide ring was opened and involved in an oxidation/reducton interaction with chromium, one would expect the nitrogen core levels to sense that change.

CONCLUSIONS

In the preceding, we have attempted to interpret a surface science problem utilizing the concepts of organometallic chemistry. The utilization of such concepts in connection with the interpretation of data obtained on systems involving deposition of organic ligands on metallic surfaces has been previously developed (17). One expects such connection between the two different disciplines to be even closer for surface science problems involving the deposition of metal atoms on an organic-polymeric substrate. Even though we do not have, at present, a complete picture of the structural details associated with the polymeric substrate, and consequently of the metal-polymer structures formed during deposition, we have shown the XPS data to be consistent with some very general features of the formation of metal-organic ligand charge transfer complexes.

REFERENCES

1. N.J. Chou and C.H. Tang, J. Vac. Sci. and Technol. A2, 751 (1984).

2. P.S. Ho, P.O. Hahn, J.W. Bartha, G.W. Rubloff, F.K. LeGoues, and B.D. Silverman, J. Vac. Sci. and Technol. A3, 739, (1985).

3. J.W. Bartha, P.O. Hahn, F.K. Legoues, and P.S. Ho, J. Vac. Sci. and Technol. A3, 1390, 1985.

4. P.N. Sanda, J.W. Bartha, J.G. Clabes, J.L. Jordan, C. Feger, B.D. Silverman and P.S. Ho, J. Vac. Sci. and Technol. A4, 1035, 1986.

5. J.L. Jordan, P.N. Sanda, J.F. Morar, C.A. Kovac, F.J. Himpsel, R.A. Pollack, J. Vac. Sci. and Technol. A4, 1046, 1986.

6. R.C. White, R. Haight, B.D. Silverman and P.S. Ho, Appl. Phys. Lett., 51, 481, 1987.n

7. N. Takahashi, D.Y. Yoon and W. Parrish, Macromolecules, 17, 2583, 1984.

8. E.J. Kramer, W. Volksen, and T.P. Russell, Mater. Res. Soc. Symp. Proc., 72, 195, 1986.

9. P.L. Timms, Proc. R. Soc. Lond, A396, 1, 1984.

10. M.P. Andrews, S.M. Mattar and G.A. Ozim, J. Phys. Chem., 90, 744, 1986.

11. A.R. Rossi, P.N. Sanda, B.D. Silverman and P.S. Ho, Organometallics, 6, 580, 1987.

12. M. Dupuis, D. Spangler and J.J. Wendelowski, NRCC Software Catalogue, Program QG01, Vol. 1, 1980.

13. K.D. Dobbs and W. J. Hehre, Journ. of Comp. Chem., 8, 861, 1987.

14. L.P. Buchwalter and A.I. Baise, in Polyimides: Synthesis, Characterization and Applications, K.L. Mittal, Ed., Plenum, New York, 1984, Vol. 1, p.537.

15. B.D. Silverman, P.N. Sanda, P.S. Ho and A.R. Rossi, J. Polym. Sci., Polym. Chem. Ed., 23, 2857, 1985.

16. R. Haight, R.C. White, B.D. Silverman and P.S. Ho, to be published.

17. M.R. Albert and J.T. Yates, in the "Surface Scientists Guide to Organometallic Chemistry", American Chenical Society, Washington, DC 1987.

THE PROMOTION OF METAL/POLYMER ADHESION BY
ION BEAM ENHANCED DEPOSITION

IH-HOUNG LOH AND J.K. HIRVONEN,Spire Corporation, Bedford, MA 01730
J.R. MARTIN, Spectrum Associates, Cambridge, MA
P. REVESZ AND C. BOYD, General Ionex, Newburyport, MA

ABSTRACT

The adhesion between metals and polymers is normally weak because they have drastically different electronic properties and interfacial bonding between metals and organic polymers is not readily affected by conventional coating techniques. The ion beam enhanced deposition (IBED) approach to improve adhesion focuses on the promotion of non-equilibrium "inter-diffusion" and the creation of nucleation sites at the interfaces by low energy ion bombardment before and during metal deposition onto the polymer substrate. *

The results of our work show a significant improvement of adhesion at the metal (Au)/polymer (Teflon) interface.

INTRODUCTION

Good adhesion can be formed at the interface between mutually reactive materials, or can be produced by molecular interdiffusion at such interfaces. However, weak adhesion is obtained between non-reactive, immiscible materials, such as metals and organic polymers. The problem is even more serious when the interfacial bonding is to withstand the effects of water diffusion through the polymer coating. Since water is a strong hydrogen bonding agent, it tends to destroy secondary bonds including Van der Waals forces at the metal-polymer interface.

It has been shown that dramatic thin-film adhesion enhancement can be produced by irradiating the film/substrate interface with ion beam mixing. There are a few experimental reports on metal-polymer adhesion promotion[1,2]. The basic mechanisms responsible for metal-polymer adhesion enhancement however remains a fertile field of study, both experimentally and theoretically.

The concept of ion beams improving adhesion focuses on:

(i) the radiation serves primarily to disrupt the electronic structure of chemical/metallic bonding of interface atoms, enabling new bonding configurations to develop an adhesion layer where atoms of both film and substrate species are stably bonded.

(ii) individual atoms from the surface coating film are struck by energetic beam ions and are driven into the substrate by elastic recoil to develop an intermixed interfacial layer.

IBED is an ion assisted coating process, as depicted in Figure 1, in which the metal is evaporated onto the substrate while being simultaneously bombarded with energetic (100 - 1000 eV) ions. The resulting coatings have a graded interface at the coating boundary and can provide much greater resistance to delamination than conventional coating processes. IBED is especially attractive for producing adhesive metal coatings onto polymer substrates because of its features given in Table I.

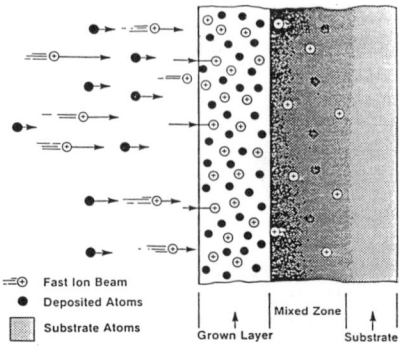

Figure 1. Schematic Representation of IBED System Geometry

Table I. Comparison of Ion Beam Mixing (IBM) and Ion Beam
Enhanced Deposition (IBED) Techniques

Ion Beam Mixing

- Maximum coating thickness limited by ranges of mixing ions
- High energy process (typically 100 - 200 KeV)
- Low currents (typically 1 - 20 $\mu A/cm^2$)
- Mixing ions capable of producing greater recoil energy
- Post deposition process
- Ions have isotopic purity

Ion Beam Enhanced Deposition

- No limition on coating thickness
- Low energy process (100 eV - 1000 eV)
- High currents possible (up to 1 mA/cm^2)
- Less heating than IBM at given beam current
- Process also contributes to improved film properties during growth (i.e., few pinholes, controlled stress)
- Non-mass analyzed beam

EXPERIMENTAL METHODS

The IBED facility, shown schematically in Figure 2, consists of a end station mated to a low energy, high current Kaufman ion source. The endstation is subdivided into two separate chambers: one is a cryopumped chamber which houses the sample holder and a lower chamber which contains an electron beam evaporator pumped by a four inch diffusion pump.

The evaporation of metal (copper, gold or silver) is set at a rate of nominally 1-3 nm/s as controlled by a quartz crystal deposition rate controller positioned in the same plane as the sample. Before and during the evaporation, 400 eV Ar ions bombarded the substrates at a flux of 25 micro- amps/cm^2.

During IBED operation, a thin wire is normally held in front of the sample shielding a portion of the substrate from the argon beam (horizontally) and another portion from the metal deposition (vertically). Thus, there are adjacent stripes of material with and without ion bombardment to provide built-in controls for adhesion tests.

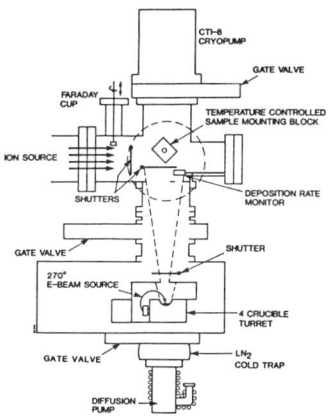

Figure 2. Schematic of IBED Apparatus

After the IBED process, adhesion of the films were tested with both the traditional "Scotch Tape Test" as well as a scratch method. The tape test is simply performed by pulling off a piece of Scotch tape, which was firmly pressed onto the film. The qualitative assessment of how much metal remained attached to the substrate served to indicate the metal/polymer adhesion. For the scratch tests, a modified microhardness tester was used. On each specimen, a series of scratches were made with different tips with varying Moh's hardness. The scratched film/substrate was then evaluated for any sign of adhesive failure by optical microscopy.

RESULTS AND DISCUSSIONS

The metal/polymer combinations investigated, the corresponding IBED parameters and adhesion tests applied and comparative results of IBED and non-IBED areas are summarized in Table II.

Interestingly, no adhesion improvements were observed on the IBED prepared Au/Plexiglass system. It was the only system that showed a difference between the IBED and non-IBED regions for the tape test, which was obviously too mild to differentiate between adhesion of other systems except for Au on Teflon. We suspected that the surface of plexiglass is degraded by low energy ions, and the low molecular weight material thus produced on the metal/plexiglass interface weakens the adhesion.

Table II. Various Metal/Polymer Combinations Prepared via IBED Process

Metal	Polymer	Ions	Current (μA/cm^2)	Thickness (nm)	Adhesion Tests	Comparison IBED	Non-IBED
Cu	Kapton	Ar	25	370	Tape	Adhered	Adhered
Ag	Kapton	Ar	25	360	Tape	Adhered	Adhered
Cu	Teflon	Ar	25	400	Tape	Adhered	**Failed**
Cu	Teflon	Ar	25	400	Scratch	Adhered	Failed
Ag	Teflon	Ar	25	400	Tape	Adhered	**Failed**
Ag	Teflon	Ar	25	400	Scratch	Adhered	Failed
Au	Teflon	Ar	25	400	Tape	Adhered	Failed
Au	Teflon	Ar	25	400	Scratch	Adhered	Failed
Au	Lexan	Ar	25	40	Tape	Adhered	Adhered
Au	Plexi-glass	Ar	25	40	Tape	Failed	Failed

With the Teflon substrate/Au film, the tape easily pulled the Au film from the unirradiated areas, whereas the Au film deposited with the IBED process still remained intact on the Teflon surface. Normally no metal can adhere well to the Teflon due to the low surface energy. Several other attempts, including presputtering of the Teflon before metal deposition [2] and ion beam mixing after deposition [1,3,4,5], have improved the adhesion of metals to Teflon. In our work, similiar adhesion promotion through the IBED process was observed on Au/Teflon, Cu/Teflon and Ag/Teflon pairs. The optical micrographs of these results are clearly revealed in Figure 3. The scratch method provides the desired comparative information as shown in Figure 4. The metal films without IBED preparation were disrupted from the scratch test, while the IBED prepared region showed a strong evidence of adhesion.

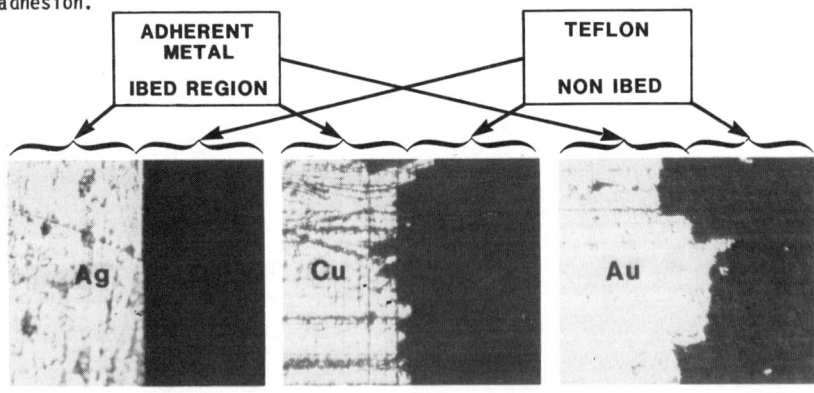

Figure 3. Optical Micrography of Specimens After Scotch Tape Test. Strong adhered metal remains on the left IBED region.

←IBED - Cu/TEFLON→|←Cu/TEFLON→

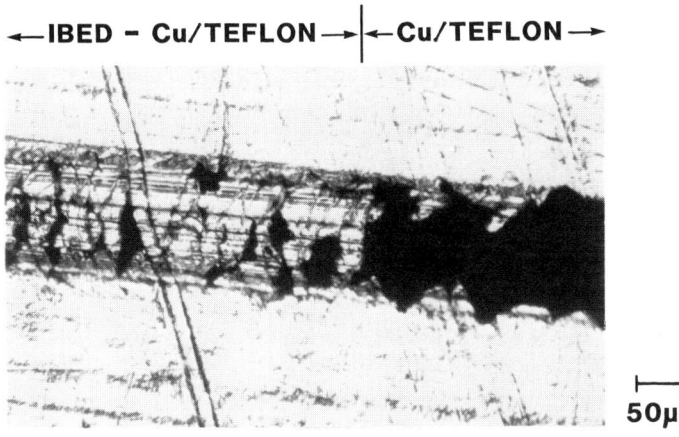

50µm

Figure 4. Optical Micrography of Scratches Made with 7.5 Moh's Tip.
Severe damage is observed on the non-IBED region (on the right
hand side)

An epoxy peel test was run on the IBED prepared 1 micron thick gold
coating on Teflon. The evidence of cohesive failure from the Teflon region
was observed from the ESCA analysis of the peeled interface, which
confirmed our previous statement that strong interfacial bonding between
gold and Teflon was formed by IBED process. The atomic mixing of the gold
and Teflon at the interface by IBED process was also evidenced by the
composition profile obtained by sputter ESCA spectroscopy shown in Figure
5. Rutherford backscattering was performed on two areas of the Teflon
following removal of the gold by epoxy. The surface beneath the IBED Au
layer can be seen to contain 6×10^{14} Au/cm^2 whereas the area under the
gold evaporated without the aid of the ion beam shows no discernible Au
peak (Figure 6). This is consistent with the IBED forming a mixed
Teflon/Au interlayer promoting adhesion.

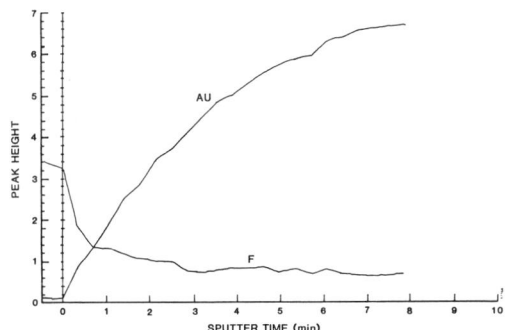

Figure 5. Compositional Profile Obtained by Sputter ESCA Spectroscopy of
the Au/Teflon Interface

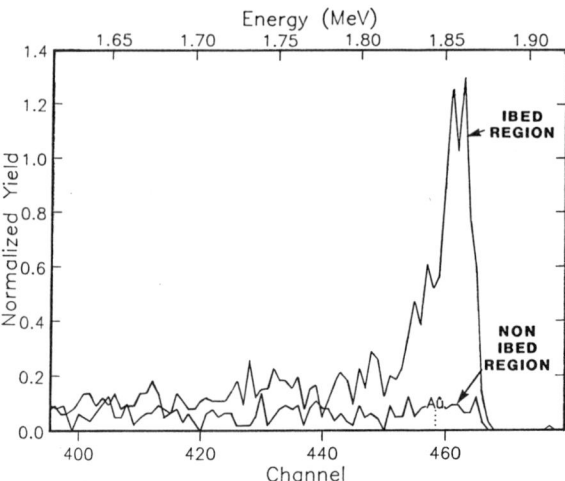

Figure 6. RBS Spectrum of Teflon Substrate After Epoxy Pull Adhesion Test.

FUTURE WORK

Further work has been directed to have an understanding of the ion beam enhancement effect on the chemistry, stoichiometry and bonding structure of the polymer/metal interface phase.

ACKNOWLEDGEMENTS

The authors would like to thank Dr. Theodore J. Reinhart for his suggestions. This work is partially supported by DOD SBIR Phase II (Contract No. F33615-87-C-5220).

REFERENCES

1. C.J. Sofield, C.J. Woods, C. Wild, J.C. Riviere, and L.S. Welch, MRS Symp. Proc.,25, 197 (1984)

2. C.A. Chang, K.C. Lin, J.E.E. Baglin, G. Coleman, and J. Park, MRS Symp. Proc., 93, 36 (1987).

3. J.E.E. Baglin, MRS Symp. Proc. 47, 3 (1985).

4. J.E.E. Baglin and G.J. Clark, Nucl. Instr. Met., B7/8, 881 (1985).

5. B.T. Werner, T. Vreeland, Jr., M.H. Mendenhall, Y. Qui, and T.A. Tombrello, Thin Solid Films, 104, 163, (1983).

HYDROGEN DIFFUSION BETWEEN PLASMA-DEPOSITED SILICON NITRIDE-POLYIMIDE POLYMER INTERFACES

Son Van Nguyen
International Business Machines Corporation
General Products Division
San Jose, California

Mike Kerbaugh
International Business Machines Corporation
General Technology Division
Essex Junction, Vermont

ABSTRACT

A Nuclear Reaction Analysis (NRA) for Hydrogen technique was used to analyze the hydrogen concentration near Plasma Enchanced Chemical Vapor Deposition (PECVD) silicon nitride-polyimide interfaces at various nitride-deposition and polyimide-polymer-curing temperatures. The CF_4 + O_2 (8% O_2) plasma-etch-rate variation of PECVD silicon nitride films deposited on polyimide appeared to correlate well with the variation of hydrogen-depth profiles in the nitride films. The NRA data indicate that hydrogen-depth-profile fluctuation in the nitride films is due to hydrogen diffusion between the nitride-polyimide interfaces during deposition. Annealing treatment of polyimide films in a hydrogen atmosphere prior to the nitride film deposition tends to enhance the hydrogen-depth-profile uniformity in the nitride films, and thus substantially reduces or eliminates variation in the nitride plasma-etch rate.

I. INTRODUCTION

Low-temperature plasma-deposited silicon nitride films have many applications in integrated circuit fabrication, such as multilayer resist lithography [1], side-wall image transfer [2], interlevel dielectric for multilevel metallization structures [3], and passivation layers [4-6]. The presence of hydrogen and its bonding in plasma-deposited silicon nitride films have dramatic effects on physical [7,8] and electrical [9,10] properties of the films.

Mat. Res. Soc. Symp. Proc. Vol. 108. ©1988 Materials Research Society

In many passivation and lithography applications in integrated circuit fabrication, polyimide polymer is normally used in conjunction with PECVD silicon nitride films because of its thermal stability. The PECVD film is usually deposited on cured polyimide polymer, then subsequently masked by a photolithographic technique and etched by dry (plasma) processing to form desirable patterns on the substrate. In a recent plasma-etching study of PECVD silicon nitride deposited on cured polyimide polymer, we observed substantial variation in the nitride film etch rate with depth thickness, especially near the nitride-polyimide interfaces, as compared to the more uniform etch rate of PECVD silicon nitride films on thermal oxide or silicon substrates processed under the same conditions (Figure 1). As a result, substantial over-etching of the nitride layer is required, and that leads to poor control of micron and submicron profiles in many lithographic and packaging applications.

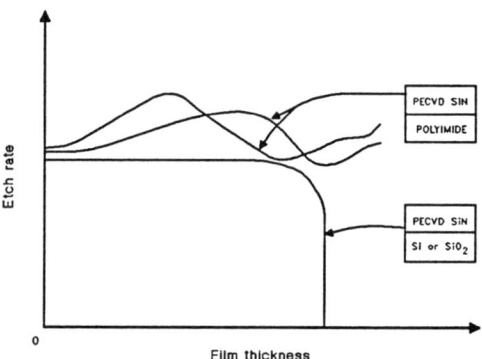

Note: Both PECVD SiN layers deposited under same conditions

Figure 1: Variation of etch rate of plasma silicon nitride
films with depth thickness (CF_4 + O_2 plasma)

In an effort to understand this etch-rate variation, a Nuclear Reaction Analysis (NRA) technique was used to analyze the hydrogen concentration of PECVD nitride deposited at various temperatures on silicon substrates or cured polyimide polymer films. The polyimide layers were also cured (annealed) in hydrogen or nitrogen atmospheres at various temperatures prior to the nitride deposition. The etch rate of PECVD nitride films in CF_4 + O_2 (8% O_2) plasma was also studied as a function of hydrogen concentration in the deposited films. Film compositions and bonding structures were studied as a function of deposition temperature using Auger and Fourier Transform Infrared (FTIR) measurement techniques.

II. EXPERIMENTAL PROCEDURE

Silicon nitride films were deposited on bare silicon p-type substrates or on cured polyimide polymer film (1-2 μm thick) spin-on silicon p-type substrates using a plasma-deposition system described elsewhere [11]. Before the nitride deposition, the polyimide films were cured in either hydrogen or nitrogen ambient for 30 minutes at temperatures equal to or slightly higher than the nitride-deposition temperatures.

Film thickness and refractive index were measured using a He-Ne laser ellipsometer (wavelength = 632.8 nm) together with Tally step techniques. FTIR, Auger, and NRA analysis techniques together with the $CF_4 + O_2$ plasma etching of the nitride films have also been described in earlier papers [7, 11, 12]. It should be noted that, during the hydrogen-depth profile study of PECVD silicon nitride deposited on a polyimide layer, only the hydrogen concentration in the nitride layer can be determined accurately. This is due to the unknown composition of cured polyimide and the degradation of the polyimide films under high-energy ^{15}N bombardment of NRA technique.

III. RESULTS AND DISCUSSION

A. Film Deposition

The deposited silicon nitride films show good film thickness and refractive index uniformity within each bath ($3\sigma = 11\%$). The deposition rate was reduced slightly from 10 to 9 nm/min as the substrate temperature varied in the 300-390°C step increases. As-deposited films in this temperature range have refractive indices ranging from 1.92 to 1.98. The film's refractive index increased monotonically with increased substrate temperature, which may indicate some densification in the film structure. Measurement of films deposited on bare silicon substrate and polyimide polymer surfaces under the same conditions indicated no significant difference in film thickness of the nitride layer.

B. FTIR and Auger Analyses

The representative IR spectrum of plasma silicon nitride film deposited in the 300-390° C range is shown in Figure 2. In general, the Si-H bond density (absorption intensity area unit/film thickness) decreased as the substrate temperature increased. (See Table I.) However, in the case of N-H bond density, the variation is somewhat more irregular with temperature, suggesting that Si-H bonds in silicon nitride films are more sensitive to the variation of substrate temperature. This result is also consistent with our previous study on hydrogen bonding and depth-profile variation of plasma silicon nitride deposited at the 25-300° C substrate temperature range [9, 10]. We observed no significant changes in total Si-N bond density and vibrational wavelength at this deposition temperature range.

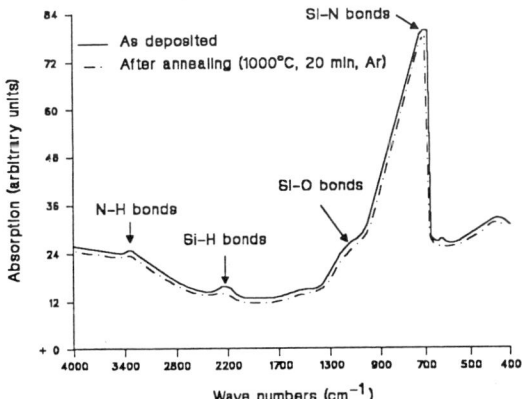

Figure 2: FTIR spectra of PECVD silicon nitride

Table I. Normalized hydrogen bond density in plasma-deposited silicon nitride at 300-390° C

Substrate temperature (°C)	N-H (3440-3200 cm^{-1})	Si-H (2400-2100 cm^{-1})
300	5.4	5.7
315	4.1	5.4
330	5.7	5.3
345	3.9	5.1
360	4.0	5.0
375	4.3	4.7
390	4.0	4.3

Arbitrary bond density = Absorption intensity area unit/nm film thickness

Note: Measurement does not include difference between

Si-H and N-H vibrational absorption coefficient (γ). γ Si-H = 1.4 γ N-H

Auger depth profiles of plasma silicon nitride (n = 1.92-1.98) deposited at the 300-390° C substrate temperature range showed only small variations in silicon and nitrogen concentrations from sample to sample. The average silicon concentration ranged from 45-50 while the nitrogen concentration range was 55-60 atomic percent. Overall, the Auger results showed only a small variation in film composition, and more surface oxygen contamination at higher deposition temperatures. In addition, we observed no significant difference in the bonding compositions between the silicon nitride films deposited on polyimide polymer and silicon substrate surfaces in the 300-390° C deposition temperature range.

C. Nuclear Reaction Analysis for Hydrogen and Plasma Etching

The analysis of hydrogen concentration showed an excellent linear correlation between total hydrogen concentration with substrate temperature 300-390° C. The hydrogen concentration in silicon nitride films decreased monotonically from 28 to about 21 atomic percent as substrate temperature increased from 300 to 390° C (Figure 3). A representative hydrogen depth profile of the nitride film on silicon substrate at 335° C shows that the hydrogen concentration is also quite uniform with only about 1 atomic percent variation (Figure 4). This result appeared to correlate well with the Si-H bond density measured by FTIR.

Etching results of plasma-deposited silicon nitride films in CF$_4$ + O$_2$ (8% O$_2$) plasma showed the etch rate also decreased monotonically with decreased hydrogen concentration in the film, i.e., increased substrate deposition temperature. The etch rate decreased from 51 nm/min for the film with 28 atomic percent hydrogen concentration to 43 nm/min for the film with 21 atomic percent hydrogen concentration. There is also marked linear correlation between etch rate and hydrogen concentration of the silicon nitride film, as seen in Figure 3. Since the silicon and nitrogen concentration and the Si-N bond density of plasma-deposited nitride films did not vary significantly in the 300-390° C substrate temperature range, the above data suggested that variation in plasma-etch rate of silicon nitride films, deposited in the 300-390° C range, with depth profile is a direct result of hydrogen-concentration variation in the nitride films. Therefore, the plasma-etch-rate variation of PECVD silicon nitride deposited on polyimide, as shown in Figure 1, may also be due to the variation of its hydrogen-depth-profile concentration.

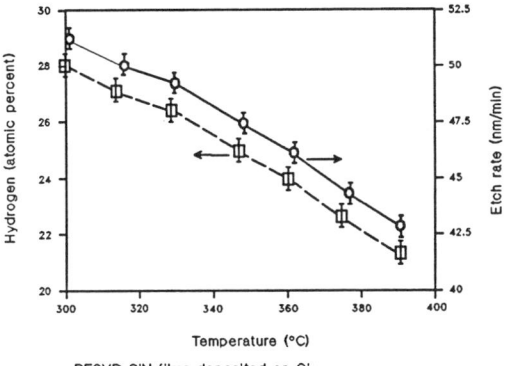

Figure 3: Variation of etch rate with [H] and
PECVD deposition temperature

The hydrogen-depth-profile study of PECVD silicon nitride deposited on polyimide polymer films, annealed in nitrogen or hydrogen ambient prior to nitride deposition, appeared to support the idea that etch-rate variation is due to nonuniformity in hydrogen-depth-profile concentration in the nitride layer. Figure 4 shows the hydrogen-depth profile of silicon nitride films deposited at 335° C on Si substrates and polyimide polymer films cured in nitrogen and hydrogen ambient at 335° C prior to nitride deposition. The hydrogen-depth profile of silicon nitride deposited on silicon substrates is more uniform as compared to those deposited on polyimide films. Furthermore, the hydrogen-depth profile of silicon nitride films deposited on spin-on polyimide polymer layers annealed (cured) in hydrogen ambient is more uniform compared to those annealed in nitrogen ambient. The hydrogen concentrations in the nitride films deposited on polyimide polymer layers are also 2 to 5 atomic percent higher than those deposited on silicon surfaces. Similar results were obtained for various polyimide-annealing and silicon-nitride-deposition temperatures. The annealing of polyimide films in hydrogen

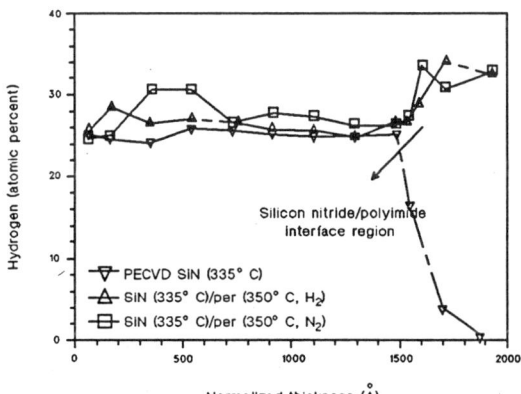

Figure 4: Hydrogen depth profiles of silicon nitride films deposited
under the same conditions on silicon substrate and polyimide
films cured in N_2 or H_2

ambient always resulted in more uniform silicon nitride hydrogen-depth profiles and slightly higher hydrogen concentration than films deposited on silicon substrates. Since the bonding and composition of the films deposited on polymer are similar to those on silicon substrates, as indicated by FTIR and Auger data, the extra hydrogen concentration observed in the nitride films deposited on polymer is probably due to the hydrogen diffusion from the polyimide to the nitride layer. This diffusion may happen during and/or after the nitride deposition.

The overall results clearly showed that the variation in etch rate of plasma-deposited silicon nitride on polyimide polymer is due to the nonuniformity of hydrogen-depth-profile concentration in the nitride. Experimental data indicated that this nonuniformity was caused by the hydrogen diffusion between polyimide and silicon nitride during and/or after the nitride deposition. If the polyimide films were cured in hydrogen ambient, instead of nitrogen, prior to the nitride deposition, the amount of hydrogen diffused from the polyimide layer to the nitride film was smaller, and the hydrogen depth profile of the nitride layer was more uniform. Accurate determination of hydrogen concentration in the polyimide is not possible due to the degradation of polyimide under high-energy irradiation of ^{15}N under nuclear reaction. Since polyimide polymer contains substantially higher hydrogen concentrations (25-40 atomic percent), compared to nitride films, we may speculate that the diffusion caused by the hydrogen concentration gradient between the nitride-polyimide layer and the hydrogen diffusion probably occurred during and/or after the nitride deposition. The curing of polyimide films in a hydrogen atmosphere may stabilize the polyimide polymer, and minimize or stabilize the diffusion rate between interfaces. At this time, the diffusion mechanism between the nitride-polyimide layer is still unknown.

IV. SUMMARY

Strong linear correlation was observed between the $CF_4 + O_2$ plasma etch rates and the average hydrogen concentration in PECVD silicon nitride layers deposited in the 300-390° C substrate temperature range. The plasma etch-rate variation with depth thickness in PECVD silicon nitride deposited on polyimide polymer films can also be correlated with the hydrogen-depth-profile concentration variation in the nitride layer. The hydrogen-concentration variation in the nitride layer is probably due to the hydrogen diffusion from the polyimide to the nitride layer. Curing of polyimide polymer films in hydrogen ambient, instead of nitrogen, appeared to stabilize the polyimide, minimized the hydrogen diffusion between the nitride-polyimide interfaces, and resulted in more uniformity of the hydrogen-depth profile in the silicon nitride layer; thus it also improved the etch-rate uniformity with depth thickness in the nitride layer.

V. ACKNOWLEDGMENTS

It is our pleasure to acknowledge the assistance of Dick Courchain in silicon nitride deposition, as well as Mark Lavoie and Steve Pierce of the Chemical Analysis laboratory for Auger and FTIR analyses. The significant assistant of Professor W. Lanford at State University of New York at Albany in nuclear reaction analysis of hydrogen study is also greatly appreciated.

REFERENCES

1. J. Underhill, S. V. Nguyen, M. Kerbaugh, and D. Sundling, Proc. of Society of Photo-Optical Instrumentation Engineers, 539, 83 (1985), and references therein.

2. H. Trumpp and J. Greschner, United States Patent 4502914.

3. H. Kotami, H. Harada, and K. Tsukamoto, Report 1A-R-5, 29th Meeting of Japan Society of Applied Physics, Tokyo, April 1982.

4. H. M. Maes and G. L. Heynes, Electrochemical Society Extended Abstracts, 83-1, 178, May 1983 Meeting, San Francisco.

5. K. Takasaki, K. Koyamo, and M. Tagaki, Electrochemical Society Extended Abstracts, 80-2, 767, October 1980 Meeting, Hollywood, Florida.

6. J. T. Yue, W. P. Funsten, and R. V. Taylor, Proc. of 23rd Reliability Physics Meeting, 126, IEEE, March 1985, Orlando.

7. W. Lanford and M. J. Rand, J. Applied Phys., 49, 2473 (1978).

8. S. V. Nguyen, P. Pan, and S. Burton, J. of Electrochemical Society, 131, 2348 (1984).

9. S. V. Nguyen, J. of Electronic Material, 16, 275 (1987).

10. S. V. Nguyen and S. Fridmann, J. of Electrochemical Society, 134, 2324 (1987).

11. S. V. Nguyen, W. Lanford, and A. L. Rieger, J. of Electrochemical Society, 133, 970 (1986).

12. W. A. Lanford, Solar Cell, 2, 351 (1980).

PART V

Silicon and Hybrid Circuit Interconnects

MATERIALS AND PROCESSES AT THE LEADING EDGE OF PHOTOLITHOGRAPHY

GARY N. TAYLOR
AT&T Bell Laboratories, 600 Mountain Avenue, Murray Hill, New Jersey 07974

Abstract

The evolution of optical lithography to higher resolution patterning with step-and-repeat exposure methods has resulted from the precise fabrication of higher numerical aperture lenses, more precisely engineered and controlled mechanical and electrical subsystems, and new lenses, systems components and light sources which enable efficient printing at shorter wavelengths. Optical resist technology is undergoing a parallel evolution which is sharpening this capability even further. This paper outlines resist materials and processes at the leading edge of photolithography which will enable printing near and perhaps at the theoretical resolution limits. Subtopics discussed include resists sensitive to 248.4 nm excimer laser radiation, planarization methods, multilevel resist materials, and surface-functionalized plasma-developed resists. Finally, some conjecture about the practical resolution limits of optical lithography is presented.

Introduction

The evolution of microelectronics to larger, higher-speed devices having smaller minimum features continues at a frantic pace despite the enormity of the problems in building such structures and in problems it has created in chip-to-chip communication and optimum exploitation of those new device properties. Many technological advances have contributed to this progress including multilevel metallization schemes, reactive-ion etching (RIE), laser and low temperature processing, etc. But, perhaps of greatest importance has been the steady evolution of optical lithography from the "many micron" status of the early seventies, to reaching the one micrometer barrier in the 80s and bringing the submicrometer regime into production reality today.

A primary contributer toward breaking this barrier has been the development of step-and-repeat exposure tools [1] which have recently been improved by the precise fabrication of higher numerical aperture (NA) lenses having wider fields [2], new system components, alignment methods and qualification procedures [3], more precisely engineered and controlled mechanical electrical and computer subsystems [2,3] and new light sources which enable efficient printing of smaller features with shorter wavelengths (λ) of light [4].

After many years of "status quo" behavior, the evolution of resist material has begun to play an increasingly important role in this surge to subμm dimensions and will play an even greater role in extracting all of the *practical* resolution at the limit of optical lithography around 0.2 μm. Two general areas of resist research have evolved: (1) solution-developed positive and negative resists sensitive to shorter wavelengths of light and intended primarily as single layer systems (2) multiple-layer, dry-processed resist materials having a planarizing layer in order to minimize topographic affects on focus. This review concentrates on the latter topic since it is only this approach which is envisioned as capturing all of the resolution inherent in very short wavelength projection exposure systems. The resolution limit for single layer, 1 μm thick solution-developed resists is in the 0.4−0.5 μm range or more than a factor of 2 greater than diffraction limited resolution with 193 nm light. First a discussion of deep-UV resist trends is presented. This is followed by a section on planarization theory and methods and a discussion of various multilevel resist schemes which show promise for use at the resolution limits. Finally, dry-developed multiple- and even single-layer resists are discussed which may operate beyond the conventional Rayleigh limits. The more complete version of this paper will have a very detailed section on solution-developed resists and will be published elsewhere.

Deep–UV Resists

The advantage of using shorter wavelengths of light derives from the relationship between minimum linewidth w in a line and space array, wavelength λ and lens numerical aperture NA such that

$$w = k\lambda/(NA) \tag{1}$$

where k is a material and process dependent constant with values ≥ 0.5, the theoretical value at best focus. Since resolution is linearly dependent on λ considerable effort is being expended on constructing exposure systems which emit high fluxes of short wavelength radiation. Optimum resolution is obtained with high NA lenses. But, higher NA lenses and shorter wavelengths lead to restrictions on field size and depth of focus, $\delta_z = \lambda/[2(NA)^2]$. Rewriting Equation 1 and substituting for NA we obtain

$$\delta_z = \frac{w^2}{2k^2\lambda} \tag{2}$$

Thus, for constant λ and process conditions (k), $\delta_z \propto w^2$. At shorter wavelengths this is softened somewhat by the λ^{-1} dependence. Also, δ_z is highly dependent on the resist and its processing. Some guidelines for w and δ_z dependence are summarized below for k=0.5 and NA=0.40. Diffraction-limited resolution is calculated as 0.25 μm at 193 nm with $\delta_z \approx \pm 0.3$ μm and is 0.32 μm at 248 nm where $\delta_z \approx \pm 0.5$ μm. These latter two values at 248 nm compare favorably with the limited experimental data available for nonoptimum resists [5].

From the time that Bruning [6,7] and King [8] first discussed the benefits of using shorter wavelengths, lithographers have been plagued by the lack of high brightness deep-UV sources. The Hg arc, the workhorse of UV and visible light lithography, has very weak polychromatic output in the deep-UV and can only be employed with scanning projection printers usually having 1:1 reflective imaging. Even in this instance highly sensitive resists must be used because of the low light levels. Excimer laser sources have provided new hope for deep-UV lithography since they provide both the high power and narrow spectral bandwidths needed in the deep-UV region. The properties of some excimer lasers in the vacuum and deep-UV regions are summarized in Table 1 [9].

The ArF and KrF lasers have the greatest promise and the latter has provided the basis for new step-and-repeat exposure tools [5,10] operating at 248.4 nm. Pol, et al. [5] were the first to describe such systems using all quartz optics and line-narrowed KrF laser light.

Table 1

Some Properties of Excimer Laser Sources For the Deep
and Vacuum UV Exposure Regions

Excimer Species	λ (nm)	Bandwidth (Å) Free	Bandwidth (Å) Narrowed	Average[a] Power (W)	Energy Per Pulse (mJ)	Max Rep Rate (Hz)	Compatible Glasses
KrF	248	5	0.01[55]	100	300	500	Quartz
KrCl	222	–	–	5	120	150	Quartz
ArF	193	5	0.10	40	180	400	Quartz, MgF_2, CaF_2
F_2	157	–	–	0.05	6	10	CaF_2

a. Free running

Initial experiments with a variety of resist materials highlighted one of the main problems with lithography at 248.4 nm. None of the commercially available resists or materials developed specifically for use in the deep-UV satisfied all the criteria necessary for optimum resolution and performance. A summary of these properties is given in Table 2.

Table 2

Resist Parameters for Lithography at 248.4 nm

Property	Value and Description
1. Sensitivity	≤ 100 mJ/cm^2
2. Contrast	$\gamma > 2$
3. Absorbance	$\leq 0.3/\mu$m
4. Etch Resistance	Equal to that of HPR-204
5. Resolution	≤ 0.5 μm lines and spaces
6. Wall Profile	$\geq 80°$

Orvek, et al. [11] evaluated 4 positive photoresists by contact printing at 248.4 nm. Their results are presented in Table 3. The best properties were found for Shipley MP2400-17 even though it was not as transparent as a Hitachi resist based on *p*-cresol novolac resins which have the lowest absorption of any novolac resin.

Table 3

Properties of Resists Exposed by Contact Printing
at 248.4 nm

Type	Thickness (μm)	Transmission @ 248 nm (%)	Sens. (mJ/cm^2)	Resol'n (μm)
Shipley 1400-17	0.66	3	–	–
AZ 4050	0.56	15	230	0.5
Hitachi 5000P	0.43	60	2000	0.4
Shipley 2400-17	0.43	42	230	0.4

Similar conclusions also were obtained by Bell Labs workers [12] who examined a variety of commercial resists and poly(methacrylate) systems. The latter had their photochemical properties and absorbance characteristics tailored for deep-UV use [13-16]. The resists were evaluated on multilayer substrates which eliminated exposure due to light reflected from the substrate. Those methacrylate copolymers which undergo chain scission upon photolysis had substantially poorer sensitivity when irradiated with low fluences of KrF laser radiation compared to those measured previously [13-15] using continuous radiation from Hg-arc sources (Table 4). The

Table 4

Comparison of Sensitivities for Positive Deep-UV Resists
with KrF and Hg-Arc Sources

Resist	Absorbance in an 0.5 µm Film	Sensitivity (mJ/cm^2) KrF Laser (248.4 nm)	Hg-Arc (240 nm)
1. Poly(methyl methacrylate-co-oximinobutanone methacrylate)	0.12	1000	130
2. Poly(methyl methacrylate-co-oximinobutanone methacrylate-co-methacrylonitrile)	0.10	350	60
3. Poly(methyl methacrylate-co-indenone)	0.40	450	60
4. AZ 2415	0.51	100	80-100
5. MP 2400-17	0.50	90	20-200
6. Poly(methyl methacrylate-co-methacrylic acid) + 20% o-nitrobenzyl cholate	0.36	120-180	265

other resists (examples 4-6 in Table 4) had approximately the same sensitivity for either radiation source and functioned by dissolution-inhibition processes in which a single photon causes a unimolecular reaction of the inhibitor to give base-soluble products. The other methacrylates (1-3 in Table 4) are less sensitive to KrF radiation by factors ranging from 6-8, whereas only a maximum difference of 2 is expected since the Hg-arc exposures were conducted on reflecting Si substrates. In each methacrylate case photolysis results in formation of free radicals. These are highly absorbing in the deep-UV region whereas the resists have only moderate to weak absorption in this region. The authors suggest [12] that two photon processes may be responsible for the reduced sensitivity of these three methacrylate resists.

Subµm resolution was obtained with all three methacrylates undergoing chain scission. However, the best results were obtained with AZ 2415 and MP 2400-17 resists from Shipley Co. These materials have a relatively high absorbance and are not bleached upon exposure. Nonetheless, 0.5 µm lines and spaces were resolved in 0.46 µm thick resist over the entire 10×10 mm field when exposed on a SiO$_2$/hard-baked HPR-204 trilayer resist structure. At best focus, 0.35 µm features were resolved, but had shallow 55° wall profiles. Near-vertical wall profiles have been obtained by Toshiba workers [10] using PMMA and PMGI positive resists which have extremely weak absorption at 248 nm and thus require very high exposure doses (5-30 J/cm^2 depending on the fluence).

Modeling studies using the SAMPLE program [17], A, B, C parameters for MP 2400-17 measured at 248.4 nm and dissolution rates measured as a function of dissolution inhibitor concentration have shown that the sloped wall profile and low contrast for these resists result mainly from the high absorption of the resist film and the absence of bleaching at 248 nm. The novolac resin itself is quite absorbing while both the 1,2-diazonaphthoquinone-5-sulfonate ester sensitizer and the indene carboxylic acid photo-product have intense absorption at 248 nm. Improved resins and sensitizers are needed and considerable research is now being directed toward solving these problems.

Single layer resist applications require high T$_g$ aromatic resins to provide adequate etch resistance during further processing steps. This places a lower limit on absorbance of about

$0.10 \mu m^{-1}$. The absorbance values for $1 \mu m$ films of various resin materials are summarized in Table 5. The novolac resins have rather high absorbance whereas the hydroxystyrene polymers have particularly low absorbance. However, unlike the novolacs, the latter are very soluble in dilute aqueous TMAH or alkali metal hydroxide solutions and show little discrimination

Table 5

Absorbance Values for $1 \mu m$ Films of Various Polymers

Polymer	Absorbance
o-Cresol Novolac	1.10
m-Cresol Novolac	0.60
p-Cresol Novolac	0.45
Poly(α-methyl-4-hydroxystyrene)	0.40
Poly(α-methyl-4-t-butylcarbonatostyrene)	0.22
Poly(p-hydroxystyrene)	0.19
Poly(styrene)	0.13
Poly(methyl methacrylate-co-methacrylic acid)	0.00

(contrast) when conventional dissolution inhibitors are added to them [18].

Workers in Japan [10] recently reported results for image reversal of AZ 5214 when exposed to 248 nm light. They were able to achieve $0.5 \mu m$ line and space resolution but at high doses (>200 mJ/cm^2 compared to <50 mJ/cm^2 at 365 nm). This presumably results from the very high absorbance of AZ 5214 at 248 nm ($2.0/\mu m$) and very little bleaching during exposure.

Sensitizer requirements for use in the conventional positive resist mode are that it should be hydrophobic, generate acid or small molecules upon exposure, be soluble in the host resin at concentrations of at least 20-30%, have a moderate extinction coefficient ($\varepsilon \approx 3000$ l/mole-cm) and photodegrade with high yield ($\Phi \geq 0.2$) to give weakly absorbing products ($\varepsilon < 500$ l/mole-cm). Willson and his coworkers [19,20,21] have expended considerable effort in searching for suitable deep-UV sensitizers. They first investigated photolabile diazoderivatives of Meldrum's acid. These materials had good absorption ($\varepsilon \approx 9000$ l/mole-cm) and reactivity ($\Phi \approx 0.3$), and decomposed upon photolysis to give nonabsorbing N_2, CO, CO_2 and ketone products. The exposed regions were more soluble in aqueous base than unexposed regions probably because of excess free volume created by the generation and loss of the volatile products. $1 \mu m$ patterns were resolved using Meldrum's diazo in Varcum 6000 p-cresol novolac resin. However, variable results were obtained because of sensitizer loss (sublimation) during prebaking. Other related materials proved to be nonvolatile and had appropriate absorption properties [20]. Diazopiperdinedione had somewhat improved properties and when exposed in a base-soluble methacrylate host resin gave positive tone patterns with 200 mJ/cm^2 exposures of 248 nm light and about $1 \mu m$ resolution. Nearly complete bleaching occurred for such materials and the somewhat disappointing resolution may be due to incomplete conversion of the ketene intermediate to the acid. Further studies led to the investigation of open chain phosphorus diazoketones [21] and optimum results were obtained with diazophenacyldimethylphosphonate.

Multilevel Resist Technology and Planarization

The previous sections have discussed a variety of aspects concerning resists used in single layer form. As long as resolution limits are not pushed single resist layers will be preferred because they can be used with relative ease. However, when resolution limits are pushed toward feature sizes less than $0.5 \mu m$, pattern line widths will vary from the dimensions desired because of poor image contrast and resist regions which are not in focus. The latter problem has contributions from the imaging apparatus (δ_z), substrate topography which can equal δ_z at very small pattern dimensions, wafer flatness and the thickness of the resist film which can constitute a considerate fraction of δ_z. In a recent paper Markle [22] estimates that at least 40% of the focus budget needed for $0.5\,\mu m$ resolution is consumed by resist thickness variations. This will become even greater as the exposing wavelength is pushed to even shorter dimensions. Thus, planarization is necessary in photolithography for printing at diffraction-limited sub $0.5\,\mu m$ dimensions in thick resist films. For analogous reasons, trilayer resist lithography [23] having organic and inorganic processing layers beneath a thin imaging layer was invented to surmount step coverage problems first encounted in x-ray and e-beam technologies. Similar advantages for photolithography also were found and the highly absorbing bottom organic layer was observed to reduce or eliminate problems caused by light reflected back through the imaging layer during exposure. Consequently, many experiments have been conducted with some form of multilayer lithography. The main advantage of multilayer resist schemes is higher resolution as evidenced by the sub 1000Å Si-fin array fabricated from PMMA patterns exposed on a Ge/polyimide multilayer structure by e-beam lithography [24] and high-aspect-ratio patterns made by the deep-UV stepper exposure of MP 2400-17 on the SiO_2/HPR-206 multilayer structure coated over topography [5]. These $0.5\,\mu m$ line and space patterns were made from $0.5\,\mu m$ features in the imaging layer having a considerably nonvertical profile. Such trilayer processing using CHF_3 reactive-ion etching (RIE) to transfer the pattern into SiO_2 and O_2 RIE for pattern transfer to hard-baked HPR-206 (HB206) can produce $0.25\,\mu m$ line and space patterns with little (≤ 0.05 μm) linewidth loss from imaging layers having wall profiles $\geq 60°$.

Development of highly etch resistant, absorbing planarizing layers is essential to utilizing all of the resolution inherent in the incident light image. Until recently, topographic planarization of all feature sizes ($< 500\,\mu m$) on a device structure was not considered to be a serious issue mainly because of the moderate resolution $1-2\,\mu m$ and large depth of focus of the exposure tools. Spin-coated planarizing layers such as HB206, PMMA and polyimide were recognized as being able to planarize small features ($< 10\,\mu m$), sometimes to a high degree ($> 70\%$). Results from such early studies are concisely summarized in a recent review article on planarization of substrate topography by Stillwagon [25]. More recently, a general awareness of the need for planarization over larger dimensions ($> 20\,\mu m$) has developed since most circuits contain substrate levels that have some large open regions ($> 10^4\,\mu m^2$) next to densely patterned regions. Problems in patterning small features in both dense and sparsely occupied adjacent zones have been observed especially at the contact window level where windows are opened at multiple heights on the topography.

Thermally induced flow of polymers has been employed in order to obtain higher degrees of planarization for larger dimension features. Workers at Futurrex Corp. [26] prepared a soft, low molecular weight polyester alkyd resin which gave 70% planarization for $100\,\mu m$ line and space features when heated at 200°C. Pampalone and coworkers [27] studied a low-melting o-cresol novolac resin spin coated over $100\,\mu m$ square islands separated by $100\,\mu m$ spaces. Spin coating gave conformal films over these features. Heating at 225°C gave a planarization value of 85% compared to 15% for a positive photoresist and 30% for a less thermally reactive m-cresol novolac resin. The latter two materials crosslinked readily upon heating which raised the film viscosity. Normally, viscosity decreases exponentially with temperature for high viscosity materials. Thus, crosslinking reduced the flow obtained at higher temperatures above T_g and gave lower planarization than expected.

Stillwagon and coworkers [28] have recently studied the basic phenomena involved in the planarization process. They considered the forces acting on a liquid film during the spin coating and leveling (flowing) processes and found that the film profiles over isolated features on the substrate that are narrower than about $50\,\mu m$ should be at least partially planar while the profiles over features wider than this should be conformal. Leveling of the profiles over features wider than $50\,\mu m$ can only occur after spinning ceases, and the time t required to level the profiles was estimated to be

$$t \propto \frac{kw^4 \eta}{\gamma h_o^3} \tag{3}$$

where w is the feature width, η is the film viscosity, γ is the surface tension of the film, h_o is the initial film thickness and k is a feature dependent constant. Equation 3 was tested by applying liquid aliphatic epoxy films onto topographic substrates by spin coating and curing (hardening) the films after various (leveling) times. The film profiles over topography were measured and the amount of leveling or planarization was determined. The leveling time scaled with η and h_o as predicted by Equation 3, but exhibited a $w^{\frac{1}{2}}$ dependence rather than the theoretical w^4 dependence. It was speculated that the disagreement between theory and experiment occurred because the measurements were made with features having different surrounding topographies.

More recently, Stillwagon and Larson [29] observed the leveling of isolated square holes and long trenches by non-volatile silicone fluids. Planarization was determined as a function of leveling time by measuring film thickness changes in the centers of the features. When the leveling of features having similar surrounding topography, e.g., isolated holes, was compared the leveling time was found to scale with w^4, γ and h_o^3 as shown in Equation 3.

Using 2 μm thick films of low viscosity, aliphatic epoxies 75 and 250μm wide features were planarized by about 90 and 70-80%, respectively [28]. Such techniques can be used to planarize each chip site to $\leq$20% of the maximum topographic height when features $\leq$300 μm in width are present. The proper choice of material and curing method can also provide other desirable properties such as high thermal stability and resistance to reactive-ion etching. Planarization materials and technology can now provide the lithographer with the flattest possible surface for lithography over high-aspect-ratio topographic chip sites. It can also aid in various multilevel metallization processes in which etchback planarization is a vital step [30].

Although trilayer technology has the obvious advantage of enhancing photoresist resolution, it has had no practical production success. Process engineers have vigorously resisted its implementation because of the large number of process steps and the high capital investment required. Both factors seriously affect device cost with the former being manifested in lower yields. These points are summarized in Table 6 which compares some performance parameters for single and multilevel resist processes. Note the severe cost and simplicity penalties associated with both the RIE and deep-UV trilayer processes. Their complexity has driven researchers to more simple, but still complex, bilayer techniques. Recent developments in this area are outlined below.

Organic–on–Organic Deep–UV Bilayer Systems

Some years ago B. Lin [31] described the first deep-UV bilayer process, termed portable conformable masking (PCM). In this technology a deep-UV absorbing and UV-sensitive positive resist is coated over a planarizing layer (such as PMMA), which is weakly absorbing in the deep UV, is degraded by deep-UV light, but is insensitive to the UV radiation used to expose the top layer. Exposure and development of the upper layer followed by flood exposure and development of the bottom layer gave high-aspect-ratio patterns. Hewlett-Packard workers [32,33] have utilized a Kodak 820 positive photoresist top layer over a PMMA or poly(methylisopropenyl ketone) bottom layer, containing a few percent of coumarin dye to absorb the incident G-line exposing light, in a production process. With such a combination

Table 6

COMPARISON OF PERFORMANCE PARAMETERS FOR SINGLE AND
MULTILEVEL RESIST PROCESSES

Process Type Parameter	Positive Single Layer	Image Reversal (NH_3)	Image Reversal (5214)	Deep UV Bilayer	Contrast Enhanced Bilayer	Organo-Metallic RIE Bilayer	Chalcogenide Bilayer	RIE Trilayer	Deep UV Trilayer
Sensitivity (mJ/cm^2)	10-100	100-300	100-300	20-100	200≤400	50-300	40-200	>20	>20
Resolution (µm)	0.75	0.60	0.50	0.50	0.50	0.50	0.25	≤0.50	0.50
Resist Thickness (µm)	1.5	1.5	1.5	2.5	2.0	2.0	2.0	2.0	2.5
Profiles Achieved[a]	OC UC	V,OH UC,OC	V,OH UC,OC	V,OH, UC,OC	V, UC	V, OH	V, OH	V, OH	V, OH
Levels Req Bias Control	1	1	1	2	1	2	2	3	3
Est. Defect Level[b]	1	2	2	3	2	4	5	5	5
Number of Dry Process Steps	0	0	0	0	0	1	2	3	2
Tool Cost[b]	1	3	2	3	2	4	5	5	4
Simplicity[b]	1	3	3	3	2	3	4	4	5

a.   OC = Overcut; UC = Undercut; V = Vertical; OH = Overhang.
b.   Graded in levels with 1 = best; 5 = worst.

they solved or minimized some of the problems encountered previously including the intermixing of the two layers, poor imaging layer profile and exposure by UV light reflected from the substrate.

Recently Lin, et al. [34,35] have reported further refinements to the PCM process. For optimum etch resistance, high temperature baking of the bilayer resist is needed before substrate etching. This is achieved by using a more thermally stable lower layer such as crosslinked P(MMA-co-MAA-co-MAAnh) [34] and a high T_g molding film coated over the imaging layer. For most positive photoresists low molecular weight PMMA is used as the molding layer after exposure and development of the imaging layer. The molding layer is applied and the composite structure is hardened by heating to 200°C. Flood exposure and development in a solvent that dissolves both the mold and exposed bottom layers gave higher-aspect-ratio patterns with improved linewidth control and top layer profiles.

An extension of the PCM technique is the use of poly(vinyl-p-t-butyl benzoate) sensitized with an onium salt [36] as the imaging layer above a poly(N-methyl glutariumide) PMGI [37] lower layer. With mid-UV imaging radiation and deep-UV flood exposure of the bottom layer, high-aspect-ratio, thermally stable (to 250°C) subµm bilayer structures were fabricated. Such techniques may be useful for generating sub 0.5 μm resolution patterns. However, the fabrication path will be difficult due to problems associated with imaging layer profiles and linewidth changes associated with bottom-layer processing.

Organometallic RIE Bilevel Processes

In 1979 Taylor and Wolf [38] conducted a basic study of the influence of polymer structure, elemental composition and bonding on resistance to polymer etching in an O_2 plasma. One of the main findings of this work was that organometallic polymers containing Si were not etched appreciably whereas organic polymers such as PMMA and polystyrene were etched at moderate rates (640 and 270Å/min, respectively). This finding opened entirely new approaches to resist development [39] (plasma processing) and showed that inorganic-containing polymers could be used as etch masks for underlying organic polymers. From this evolved what is now known as O_2 RIE bilayer technology which was first discussed by Hatzakis and coworkers [40,41] in 1981 and reported in 1982 [42]. The organometallic RIE bilayer process is similar to that for inorganic imaging layer resist analogues employing Ag-sensitized chalcogenide inorganic layers [43]. The thin imaging layer is exposed and developed by conventional techniques. Then the pattern is transferred to the bottom layer (usually hardened positive photoresist) by O_2 RIE techniques. The resolution of the final pattern is determined by the resolution of the imaging layer pattern and the selectivity between the two layers during O_2 RIE. Selectivity [rate (organic layer)/rate-(organometallic layer)] is dependent on the type and amount of inorganic species present in the imaging layer, the chemical nature of the planarizing layer (aromatic organic layers have moderate O_2 RIE rates), the device topography, the thickness of the resist layers and the exact plasma processing conditions. Taylor, et al. have modeled a typical low-pressure, moderate-bias O_2 RIE bilayer situation to assess the relationship between linewidth loss, imaging layer wall profile and etch selectivity [44]. For 0.5 μm resolution, wall profile and etch selectivity must exceed values of 70° and 30, respectively, in order to hold linewidth losses to <10% of the linewidth. More stringent tolerances are required (80° and 40) for 0.25 μm features. These are difficult parameters to achieve in new resist materials having low-to-moderate inorganic content. It is not surprising that the number of publications in this field is extensive [45].

An interesting glimpse of the influence of inorganic content on etch selectivity has been described for Si-containing polymers [44]. These have been favored for investigation because of the high O_2 etch resistance of SiO_2, the ready availability of a host of organosilicon polymers and the ease of preparing new Si-containing materials. Several trends are noted. First, a threshold of a few wt % Si is needed to obtain significant selectivity (>1.5). Next the authors note that between 5 and 10 wt % Si, an approximately linear dependence of ln(selectivity) on wt % Si is found and reaches a maximum selectivity of about 12. Between 10 and 18 wt % Si a transition region is encountered in which selectivity becomes more weakly dependent on Si content. Finally, above 20 wt % Si, ln(selectivity) again is linearly dependent on wt % Si but the dependence is less than twice that observed in the 5-10 wt % region. The only polymers that vary strongly from the correlation are the aliphatic polysilanes which have weak backbone Si-Si bonds and appear to degrade more readily than expected. Their aromatic analogues which are not cleaved readily by high energy and UV radiation behave as expected and have higher selectivity although their Si content is lower than those for the aliphatic polysilanes.

This behavior can be rationalized in the following manner. Once a threshold of a few wt % Si is exceeded an etch-resistant layer containing SiO_2 forms at the top of the resist [46,47]. This layer is discontinuous and becomes more resistant to etching as Si content increases. This surface coverage or coalescence of etch-resistant, SiO_2-rich regions dominates the etching process. Above ~10 wt % Si where high SiO_2 coverage occurs another process appears to become rate limiting. This process is probably the sputtering of SiO_2 and steady state conversion of underlying organosilicon species to SiO_2 which becomes more efficient with polymers having more Si [47,48]. The exception is the aliphatic polysilanes. Etching conditions and the types of inorganic species present are expected to alter the shapes of such plots. High-pressure, low-bias etching conditions will extend the high-slope region to lower Si percentages as will more etch-resistant inorganic species. For organosilicon polymers the breakeven point for pattern transfer of 0.5 μm features from the upper layer appears to occur at ~20 wt % Si (selectivity ~30) if maintaining linewidth (<10% loss) is critical.

A summary of materials approaches to this problem was given by Ohnishi, et al. [45] and Shaw, et al. [49] in articles about their most recent studies. From these papers and papers on the etching of organosilicon polymers by NEC workers [47,48] it is clear that Si polymers having ≥18 wt % Si are needed for high selectivity. The most recent examples of materials in this category are discussed below.

Polysiloxanes have been known as negative e-beam resists since the early 1970s [50]. They have extremely high silicon content which is surpassed only by their oxygen-free cousins, the polysilanes. Hatzakis, et al. [42] employed them as thin e-beam sensitive layers in the first O_2 RIE bilayer schemes. This IBM group has continually refined their materials and recently reported results for some UV and deep-UV sensitive materials having reactive groups attached to the siloxane chain [49,50]. Terpolymers of diphenylsiloxane, dimethylsiloxane and methylvinyl siloxane are negative, deep-UV resists and have high sensitivity (12 and 40 mJ/cm^2) with high contrast ($\gamma \geq 1.5$) when mixed with 10 wt % 2,2-dimethoxy-2-phenylacetophenone or dicumylperoxide photoinitiators, respectively. Isolated 0.5 μm lines were resolved by contact printing at 222 nm (KrCl excimer laser) while ≥0.75 μm line and spaces were resolved on the PE-500 printer operating in the deep-UV mode. When a 1,2-diazonaphthaquinone-5-sulfonamidopropyl group was substituted for the vinyl group in the siloxane terpolymer, UV-sensitive negative resists also were obtained. These had ~10 mJ/cm^2 sensitivity at 405 nm with γ=1.4. 0.6 μm line and space features have been resolved in 1.5 μm thick bilayer films of this material using a 2000Å thick upper Si layer. Etch selectivities were very high (40-50:1) because of the high Si content (~34 wt %).

One drawback is that these materials must be used in very thin (≤2000 Å) films in order to minimize swelling distortions upon solution development and distortions and cracking during O_2 RIE pattern transfer. This would normally create a defect problem due to pinholes. However, the low T_g values of these polysiloxanes allow pinhole removal by annealing at moderate temperatures before and after development. This "healing" process occurs via a creep mechanism which closes holes faster than other feature types having the same minimum dimension. It is still unclear whether creep itself causes significant changes in feature linewidths at sub 0.5 μm dimensions.

Such questions concerning the low T_g values of polysiloxanes have prompted other groups to search for higher T_g analogues. Workers at NTT [52] and Hitachi [53] have used the T-resin polysilsesquioxanes, which have a tree-like structure and either alkoxy or hydroxy end groups. Unlike the silicone gums, they are solids and have high T_g values. The NTT group [54] prepared chloromethylated poly(phenylsilsesquioxane) which by itself is a sensitive e-beam resist. Nucleophilic displacement of chlorine by a methacrylate group gave the ester which was efficiently crosslinked when exposed by UV light in the presence of *bis*-azide sensitizers. The resist containing 40 mole % methacrylate groups is sensitive to deep-UV, I-line and G-line radiation and is most sensitive in the deep UV region (20, 40 and 500 mJ/cm^2, respectively). With 5 wt % *bis*-azide sensitizer, it had an O_2 RIE selectivity of about 20 versus positive photoresist. 0.5 μm lines and spaces were fairly well resolved using 10X I-line lithography, while 0.3 μm lines and 0.5 μm spaces are delineated using deep-UV lithography. Linewidth loss during O_2 RIE was ≥0.1 μm, making this resist marginal for ≤0.5 μm features.

Hitachi researchers [53] have added polysilsesquioxanes to conventional positive photoresists to provide O_2 RIE resistance. In this instance both a glass resin and a cyclic siloxane were added to achieve high Si content, maintain compatibility with the positive photoresist and permit development in aqueous base. When 50 wt % of these silicon resins was added, O_2 RIE selectivity 20 versus OFPR-800 and 10 when 33 wt % was added. The latter had sensitivity equal to that of OFPR-800 and the Hitachi group was able to resolve 0.75 μm line and space features using exposures on a G-line reduction printer. However, linewidth losses were too high for use in the 0.5 μm regime.

Higher Si-content glass resins have even better O_2 RIE selectivity. Rosilio and coworkers [55] first investigated such materials containing methyl and vinyl groups as sensitive, but low contrast e-beam resists. These resins had an etching selectivity of about 60 versus HPR-204 resist at a Si content of 40.7 wt %. 0.25 μm line and space features were transferred to the lower organic layer with excellent fidelity. Brault, et al. [56] have obtained similar results with vinyldimethylsilyl end-capped glass resins having pendant methyl groups and ~42 wt % Si. Photolytically sensitive materials with such high Si content have not been disclosed. End capping with trimethylsilyl groups has also proved effective in stabilizing the T-resin materials to thermal crosslinking [57]. The Fujitsu group [57] has used such techniques to improve the stability of SNR-type polymers [52-54] for which optimum resolution of about 0.5 μm was achieved.

Another class of materials which has higher Si content than the siloxane resins is the polysilanes having a Si-Si backbone structure and aliphatic or aromatic side chain groups. The parent poly(dimethylsilane) is intractible; however, soluble species may be prepared as either nonsymmetric homopolymers [58] or as copolymers [59]. Both types have been used as mid- and deep-UV resists since the Si-Si bonding results in a strong $\sigma \rightarrow \sigma*$ transition whose λ_{max} and ε values depend on the degree of polymerization n and the substituents bonded to Si.

Hofer, et al. [60] were the first to study these materials as O_2 RIE bilevel resists. They were able to obtain 100 mJ/cm^2 sensitivity in the mid-UV region with 2000Å thick poly(cyclohexylmethylsilane). The long wavelength transition was bleached during exposure which produced lower molecular weight polymer by a chain scission process. O_2 RIE selectivity was about 30 versus positive photoresist and submicrometer patterns were obtained. Zeigler and coworkers [61] exposed polysilane copolymers with 248 nm excimer laser light and were able to ablatively pattern thicker films of these materials which functioned as good etch barriers for O_2 RIE pattern transfer to underlying organic layers.

More recently Taylor, et al. [44] studied thin films of these same copolymers as mid- and deep-UV resists. Best results were obtained for the 50:50 mole % copolymer of dimethylsilane and cyclohexylmethylsilane having 30.1 wt % Si. This resist had a sensitivity of 450 mJ/cm^2 and γ=2.3 in the mid-UV using isopropanol developer, but the images were distorted by swelling and formation of some insoluble material. When exposed in the deep-UV at 248 nm a better developer was needed to solubilize the residue in exposed regions; sensitivity was 260 mJ/cm^2 and γ≈1.7. Features as small as 0.8 μm were resolved, but residue was still evident in exposed regions. Smaller features were not completely resolved. The authors attributed this behavior to the very high unbleached absorption at 248 nm, crosslinking as well as chain scission and distortion due to swelling upon development. Even though the etch selectivity was as high as 38, pattern transfer to underlying organic layers couldn't be achieved without excessive linewidth changes which resulted from the shallow wall profiles of the fine-feature patterns.

Ohnishi, et al. [62] at NEC have adopted another approach in using Si-Si bonded structures in resists. They prepared monodisperse allyldimethylsilyl derivatives of poly(α-methylstyrene) and sensitized them to mid-UV excitation by adding 2,6-bis-(4'-azidobenzal)-4-methylcyclohexanone. The negative resists had optimum sensitivity of 12 mJ/cm^2 and high contrast (γ=2.2) resulting mainly from the narrow molecular weight distribution of the polymer. Subμm resolution was obtained. However, O_2 RIE selectivity was low (~10) when one Si per monomer repeat unit was used, but was much higher (>20) when a second Si was inserted in the sidechain structure. The authors believe that ~20 wt % Si is the threshold needed to minimize linewidth loss for patterns with dimensions of ~0.5 μm.

Hitachi researchers [63] have prepared somewhat analogous positive resists having the poly(p-disilanylenephenylene) structure. The Si content was 25 wt % and a high O_2 RIE selectivity of 30 was observed. The material degrades to silanols when exposed to UV light and has high sensitivity (<100 mJ/cm^2) and good resolution (<1 μm).

Finally, Gozdz and coworkers at Bell Communications Research [64] have reported a new deep-UV resist prepared from poly(1-trimethylsilylpropyne). Free radical bromination gave copolymers with various Br and Si contents. Irradiation with deep-UV light in air followed by heating in air gave positive tone patterns. Sensitivity was best for 0.1-0.2 Br atoms per repeat unit (25 mJ/cm^2 at 260 nm). Etch selectivity versus HPR-204 was at least 20 and 0.5 μm line and space patterns were resolved by contact printing. The mechanism of solubilization was not investigated but presumably involved oxidation of the unsaturated polymer chain to give a more polar polymer that is soluble in polar developers (butanol).

When all factors are considered in O_2 RIE bilayer technology, the use of Si appears to be limited to only those materials having both high contrast (γ>2) and high Si content. For 0.5 μm features and resists containing less than 30 wt% Si processing tricks such as high temperature O_2 RIE processing [48], that can provide somewhat higher selectivity, appear necessary. This also can be achieved by plasma anodization of certain high Si-content resists that are degraded via chain scission processes when exposed to e-beam or vacuum-UV radiation. Thus, Gozdz and coworkers [65-67] have observed that poly(ω-alkenyltrimethylsilylsulfones) such as the butenyl derivative are sensitive (2-3 μC/cm^2) high-resolution (≤0.5 μm) e-beam resists containing about 15 wt % Si. Under normal O_2 RIE conditions the polymer was removed at a high rate. But, if samples were passivated by high-bias (−700V) O_2 RIE or high pressure O_2 RIE for short times (<2 min), subsequent normal O_2 RIE had a much improved selectivity of about 20 versus hardened HPR-206 photoresist. This surface anodization did not occur for novolac, poly(methacrylate) and poly(styrene) resists containing equivalent amounts of Si bound to the polymer or Si mixed with it as glass resin polymer [68]. Plasma anodization did occur when somewhat volatile silicon monomers such as bis-acryloxybutyltetramethyldisiloxane were mixed with organic polymers. O_2 RIE of such mixtures resulted in an initial thickness loss and a gradual slowing of polymer erosion until etching ceased at about 40% total thickness loss. During etching it is believed that Si-containing monomer diffuses to the surface where it is converted to SiO_2 and functions as an increasingly effective etching mask. The anodization process with the poly(alkenylsilylsulfones) probably occurs in an analogous manner involving UV and vacuum-UV photolysis (initiated by light emitted in the plasma) or ion impact which may degrade the polysulfone to low-molecular-weight mobile Si species that rapidly diffuse to the polymer surface and react there to form SiO_2. Perhaps similar behavior can be designed into other resist systems if such a mechanism were really operative.

Other alternative methods for improving O_2 RIE selectivity are to prepare resists containing other inorganic atoms that may impart higher selectivity or to use completely inorganic resists. The latter prospect is a largely unexplored field except for the Ag-sensitized chalcogenide resists [43]. This class of materials is sensitive to UV and deep-UV light and has been widely investigated. Devices have been fabricated using Ag-sensitized Ge_xSe_{1-x} bilayer structures [69]. But, the process is not practical despite its edge-sharpening properties and high resolution because of the extreme care required in depositing the Ge_xSe_{1-x} layer and sensitizing it with silver. The process has never reached a state of high reliability and appears prone to defect formation.

Only one good example of sensitive, inorganic, etch-resistant resists has appeared to date. Kudo and coworkers at Hitachi [70] studied isopolytungsten acid i-WO_3 and heteropolytungsten acid h-WO_3 prepared by reacting W or WC with 15% H_2O_2 in water. The resulting WO_x films could be coated on underlying organic layers from aqueous solutions to give $0.01-0.1$ μm thick films. Deep-UV exposure gave negative tone patterns with doses of 120 mJ/cm^2 for 500Å thick films and ~900 mJ/cm^2 for 1000Å thick films. Dilute sulfuric acid was used to develop the unexposed regions and features as small as 0.4 μm were obtained by contact printing at high doses (3 J/cm^2). Insolubilization by light is believed to result from decomposition of peroxotungsten linkages and formation of polymeric W-O-W chains. O_2 RIE selectivity versus OFPR-800 was >50. Applicability to excimer laser printing at 248 and 193 nm seems remote since even thin WO_x films are too highly absorbing at these wavelengths. However, the extremely high etch selectivity indicates the potential benefits of such materials which are likely to provide fertile ground for future work.

Gas–Phase–Functionalized Plasma–Developed Resists

If one could combine the properties of planarizing materials with the etch resistance of inorganic or organometallic materials used in multilevel resist schemes, one would once again have a single resist layer with imaging capabilities at the resolution limits of optical lithography. Dry development is needed to achieve this goal and the fundamentals involved are outlined in a series of papers by Taylor and coworkers [71-73]. The basic concepts of dry development are as follows. Irradiation induces chemical changes in the exposed regions. Heating or some other process such as gas-solid reactions with gaseous inorganic species [74] affords regions with additional modifications in chemical bonding and the type of atomic species present. Such changes amplify those initially caused by the radiation and afford differences in the rates of removal of the exposed and unexposed regions in an oxygen plasma. If the removal rate R_u of the unexposed region is less than that for exposed regions (R_e) negative tone patterns are obtained; positive tone occurs for the converse situation when $R_e < R_u$. Clearly, maximum etching selectivity will result in higher resolution. Therefore, dry-developed resist schemes have focused on selectively incorporating etch resistant inorganic species in organic polymer hosts [39]. The most recent efforts have concentrated on using gas-solid reactions to introduce the inorganic species (usually Si) after photolysis. This is sometimes called gas-phase functionalization [74] a technique which developed from the combination of ion beam exposure of organic polymers with In^+ and Ga^+ ions and O_2 plasma development using the implanted inorganic species as a masking layer [75]. 1500Å resolution was obtained in 1 μm thick resist layers when $\geq 5 \times 10^{15}$ In^+ ions/cm^2 were implanted into organic films followed by development via O_2 RIE. Gas-phase functionalization, or the reaction of organic groups in the polymer with inorganic or organometallic gaseous species after exposure, was then devised in order to decrease the high doses needed in the ion implantation experiments. Either tone may be obtained and that which results is dependent on whether exposure removes groups which react with the inorganic-containing species (positive tone) or creates groups which react with such species (negative tone). Optimum effects are expected for those materials which are selectively functionalized on the *surface* or in the *near surface* region (topmost $1000-2000$ Å). In principle, this can be achieved by controlling the permeation and reactivity of the metal-containing reagent or by choosing radiation sources, their energies and film absorption properties to maximize near-surface exposure.

This approach will ultimately enhance resolution. In optical lithography use of short wavelengths (<220 nm) and materials with high ε values (>10^5 l/mole-cm) gives >90% absorption in the top 1000Å of organic films. This minimizes bulk resist contributions to exposure and depth-of-focus problems and eliminates all exposure from reflected light. Low (20 keV) ion exposure with heavy ions (Ga^+) gives 100% exposure within the topmost 200Å as would low energy electron exposure (<1 keV). Traditional e-beam (>10 keV) and x-ray (4-20Å) exposures are not as applicable because of the weak absorption of the high energy radiation by organic materials. Thus, optimum near-surface exposure will be achieved with either low energy charged-particle sources or high-to-moderate energy photon sources (100-3000Å).

The effect of permeation on functionalization is simply demonstrated with a bilayer analogue [76]. Negative photoresist containing cyclized-polyisoprene and *bis*-azide photosensitizer (Selectilux N60 from E. Merck and Co.) is first exposed with a broad-band UV source at doses >1000 mJ/cm^2. This exposure crosslinks the resist to a high degree. Then a thinner N60 negative photoresist layer is coated over the thicker crosslinked layer. Spinning solvent does not permeate into the lower crosslinked layer because of the high crosslink density. Exposure of the top layer with doses of <50 mJ/cm^2 followed by solution development affords rubbery negative-tone images with ~1.5μm resolution. Treatment of the patterned bilayer structure with gaseous $SiCl_4$ gives irreversible sorption of $SiCl_4$, but only in the upper patterned layer. In this case the $SiCl_4$ reacts with the N–H bonded products and also complexes with unreacted *bis*-azide. RBS and O_2 RIE experiments show that no $SiCl_4$ is sorbed into the lower layer. This difference is attributed to more rapid $SiCl_4$ diffusion into the upper layer that has a much lower T_g (<25°C). The lower, hardened layer has a T_g greater than 50°C. Permeation rate increases of a least 1000 fold have been observed when T_g is exceeded. O_2 RIE of the patterned and treated bilayer sandwich resist affords a protective SiO_2 layer on exposed regions and high aspect ratio patterns.

Similar radiation-induced changes in permeation rate would be expected to find use in gas-phase-functionalization schemes. Initial results by Follett and coworkers [77] with PMMA showed that exposed regions selectively sorbed more $(CH_3)_2SiCl_2$ reagent than unexposed regions because of microporosity generated in the film by the e-beam exposure. Other permeation-controlled materials will be detailed later in this section.

Some of the initial results with this technique were explored at Bell Labs [74] using 10 keV e-beam exposures and reaction of olefinic sites, destroyed or generated by the radiation, with B_2H_6 gas. Positive- and negative-tone images were obtained when various poly(chloroacrylates) were used. However, the selectivity during O_2 RIE development was poor (<2) and only thin final films (≤2000 Å) having 0.5 μm resolution were obtained.

These results lead Wolf and coworkers [78] to investigate negative photoresists containing *bis*-azide sensitizer. Upon photolysis the azide groups were converted to amines and aziridines. The authors surmised that these products would react with gaseous silyl halide reagents to lock silicon selectively into the irradiated areas and give negative tone when developed by O_2 RIE. However, just the opposite result, positive tone, was obtained when Selectilux N6O was irradiated with mid-UV radiation, treated with $SiCl_4$, organosilyl chlorides or $SnCl_4$ and developed by O_2 RIE. Etch selectivity reached a maximum of 5:1 and 0.6μm resolution features with near-vertical-profiles were obtained. Tone inversion resulted from incorporation of Si or Sn into the unirradiated regions as well as in exposed areas and was sensitive to ambient humidity. Rutherford backscattering spectroscopy (RBS), UV spectroscopy, IR spectroscopy and structure-property relationships were used to study the mechanism. The metal halide was observed to react with the amine groups as hypothesized but also complexed with unreacted *bis*-azide. The latter hydrolyzed rapidly to give MO_x, a good etch mask, whereas M–N groups in the exposed regions did not hydrolyze rapidly. Surprisingly, the latter species did not form MO_x during O_2 RIE while the MO_x in unexposed areas formed a good etching mask. Thus, positive tone was obtained.

The previous example failed to give negative-tone patterns for several reasons: (1) low selectivity during functionalization (2) low concentrations of inorganic atoms incorporated (3) improper binding for maximum etch selectivity. Other research groups have had more success in overcoming these problems using silicon organometallic functionalizing reagents and phenol reactive groups.

IBM workers [79,80] employed acid catalysts generated photolytically or direct photo fries reactions to produce phenolic or acidic hydroxyl groups throughout an organic film with low doses of deep-UV radiation (<50 mJ/cm^2). The resulting OH groups reacted readily with hexamethyldisilazane (HMDS) and other reactive silylating agent and the incorporated silicon reached concentrations of ~15 wt %. Consequently, selectivity was high (25) during O_2 RIE development and subµm patterns were obtained. Kodak researchers [81] studied the complicated thermal reactions involved in the deprotection and functionalization steps and found that substantial volume changes occurred. First, contraction was observed during deprotection followed by swelling during functionalization. Some HMDS was incorporated in unreacted form and continued to exude from the resist after processing had ended.

Coopmans, Roland and coworkers [82] have used similar functionalization reactions to achieve even more impressive results at conventional UV wavelengths. A recent summary of their work has appeared in Solid State Technology [83] and additional papers have addressed various aspects including the mechanisms responsible for selective functionalization [84] and RIE processing [85,86]. The basic steps in the dry-development process, called DESIRE, are as follows. A phenolic polymer such as a novolac resin or poly(p-hydroxystyrene) is mixed with a triester of 2,1-diazonaphthoquinone-5-sulfonic acid or is reacted with 2,1-diazonaphthoquinone-5-sulfonyl chloride to give a functionalized resin that is also a very highly absorbing resist. A dye also is added such that all the incident 436 nm light is absorbed in the film. After photolysis at moderate doses (100-200 mJ/cm^2) the film is reacted with HMDS or other silylating agents at >100°C.

Two processes occur simultaneously according to the detailed mechanistic work of the Philips group [87] which followed functionalization by monitoring the Si-O infrared absorption at 920 cm^{-1}. The first is reaction of the resin OH group with HMDS. Best results were obtained with poly(p-hydroxystyrene) resin. Thermal crosslinking of unexposed regions prevented diffusion of HMDS in these areas [85]. Reaction of indene carboxylic acid groups with HMDS did not occur and selectivity was very high compared to m-cresol novolac resin hosts. Best results with HMDS were obtained at 140°C using doses >125 mJ/cm^2 and silylation times of about 6 minutes. Silicon was incorporated in only the topmost 2500Å of the resist and resulted in a 10% increase in the thickness of the reacted region. Contrast was highly dependent on treatment time and temperature and was greatly improved when a C_2F_6 plasma descum was inserted in the processing sequence. This step also removed a thin Si residue formed on the surface of unexposed regions which retarded plasma removal and left "grassy" residue in etched areas. Etch-selectivity dependence on O_2 plasma etching conditions was studied in detail [86]. Maximum O_2-RIE etching selectivity was 15 at −200V bias in a hexode RIE system, was 17 in a planar reactor at −380V bias and was 30 in a magnetron-assisted reactor at low bias (~−50V), moderate power (1000 W) and high flow (150 sccm). Because of the high absorption, near-surface functionalization and high contrast, due mainly to high etch selectivity, excellent patterning was achieved. 0.75 µm and 0.6 µm lines and spaces were resolved on reflective substrates and over a 1 µm high step with aspect ratios exceeding 2. Features as small as 0.4 µm were printed using 365 nm light over topography. This is beyond the diffraction-limited resolution calculated for a resist that responds to an aerial image having a modulation transfer function (MTF) equal to 0.60. Since this resist has a contrast of at least 5.6, it can resolve features printed with aerial images equivalent to k values (Equation 1) equal to 0.45. Thus, this type of resist and process appear capable of working at lower image contrast than most previous resists. Improved materials may enable practical (reproducible) γ values of >10 to be achieved. This would afford k values as low as 0.3 and at 193 nm a limiting resolution of about 0.10µm would be expected. This is very close to the resolution achieved with combustion-developed, excimer-laser-exposed patterns (193 nm) in carbon films irradiated in an O_2 atmosphere [88,89]. In this instance 0.11 μm lines and spaces were obtained using a 36 fold reduction, reflective printing apparatus.

Latitude is of course unknown at such diffraction-limited resolution as is defect density and related issues. Undoubtedly, as linewidths continue to drop control will become more difficult. But, the use of shorter wavelengths of radiation, methods that achieve more planar surfaces for imaging, thin imaging layers and surface-sensitive, higher contrast systems will ease these burdens. The future is bright and 0.2 μm production resolution for optical lithography seems possible.

Acknowledgement

I thank Larry E. Stillwagon for proofreading assistance and for help in correlating the various parts of the manuscript.

References

1. D. Widmann and H. Binder, *IEEE Trans. Elect. Dev.*, Vol ED-22, 467 (1975); J. Bruning, *J. Vac. Sci. Technol.*, *17*, 1147 (1980); G. Ittner, *Proc. SPIE, 100*, 115 (1977).
2. K. Ushida, M. Kameyama and S. Anzai, *Proc. SPIE, 633*, 17 (1986).
3. S. Wittekok, H. F. D. Linders, F. J. van Hout and R. A. George, *ibid., 772*, 100 (1987).
4. U. Miller and H. L. Stoner, *Solid State Technol.*, *28*, 1, 127 (1985); V. Pol, *ibid., 30*, 1 (1987).
5. V. Pol, J. H. Bennewitz, G. C. Escher, M. Feldman, V. A. Firtion, T. E. Jewell, B. E. Wilcomb and J. T. Clemmens, *Proc. SPIE, 633*, 6 (1986).
6. J. H. Bruning, *J. Vac. Sci. Technol.*, *16*, 1925 (1979).
7. J. H. Bruning, *Semicon. Int'l, 4*, 4, 137 (1981).
8. M. C. King, *IEEE Trans. Electron Dev.*, ED-26, 711 (1979).
9. K. Jain, *Proc. SPIE, 774*, 115 (1987).
10. M. Nakase, T. Sato, M. Nonaka, I. Higashikawa and Y. Horiike, *ibid., 773*, 226 (1987).
11. K. J. Orvek, S. R. Palmer, C. M. Garza and G. E. Fuller, *ibid., 631*, 83 (1986).
12. T. M. Wolf, R. L. Hartless, A. Shugard and G. N. Taylor, *J. Vac. Sci. Technol.*, *B5*, 396 (1987).
13. R. L. Hartless and E. A. Chandross, *ibid., 19*, 1333 (1981).
14. C. W. Wilkins, Jr., E. Reichmanis and E. A. Chandross, *J. Electrochem. Soc.*, *127*, 2510, 2514 (1980).
15. E. Reichmanis and C. W. Wilkins, Jr., "Polym. Mat'ls for Electronic Applications", E. D. Feit and C. W. Wilkins, Jr., eds., ACS Symp. Ser. *184*, 30 (1983).
16. C. W. Wilkins, Jr., E. Reichmanis and E. A. Chandross, *J. Electrochem. Soc.*, *129*, 2552 (1982).
17. R. K. Watts, T. M. Wolf, L. E. Stillwagon and M. Y. Hellman, in "Polymers for High Technology", M. J. Bowden and R. S. Turner, eds., ACS Symp. Series *346*, 292-305 (1987).
18. M. J. Hanrahan and K. S. Hollis, *Proc. SPIE, 771*, 128 (1987).
19. B. D. Grant, N. J. Cleccek, R. J. Tweig and C. G. Willson, *IEEE Trans. Electron. Dev.*, ED-28, 1300 (1981).
20. N. J. Clecak, D. R. McKean, R. D. Miller, T. C. Tomkins and C. G. Willson, US Patent 4522911 (1985).
21. C. G. Willson, R. D. Miller, D. R. McKean, N. Clecak, and L. A. Pederson, *Proc. SPIE, 771*, 2 (1987).

22. D. A. Markle, *Proc. SPIE, 774*, 108 (1987).
23. J. M. Moran and D. Maydan, *J. Vac. Sci. Technol.*, *16*, 1620 (1979).
24. D. M. Tennant, L. D. Jackel, R. E. Howard, E. L. Hu, P. Grabbe and R. J. Capik, *ibid.*, *19*, 1304 (1981).
25. L. E. Stillwagon, *Solid State Technol.*, *30*, 6, 67 (1987).
26. Product information on PC1-1500 planarizing material furnished by Z. Sobzzak, Futurrex, Inc.
27. T. R. Pampalone, J. J. Dipazza and D. P. Kanen, *J. Electrochem. Soc.*, *133*, 2394 (1986).
28. L. E. Stillwagon, R. G. Larson and G. N. Taylor, *Proc. SPIE, 771*, 186 (1987); *J. Electrochem. Soc.*, *134*, 2030 (1987).
29. L. E. Stillwagon and R. G. Larson, *J. Appl. Phys.*, submitted for publication.
30. R. E. Gakly, S. J. Rhodes, E. Armstrong and A. Marsh, *Proc. IEEE V-MIC Conf.*, New Orleans, La., June 21,22 (1984), pp 23.
31. B. J. Lin, *Proc. SPIE, 174* (1979); *J. Electrochem. Soc.*, *127*, 202 (1980).
32. K. Bartlett, M. Chen, G. Hillis, R. Trutna and M. Watts, *Proc. SPIE, 394*, 49 (1983).
33. M. P. C. Watts, *ibid.*, *469*, 2 (1984).
34. F. S. Lai, B. J. Lin and Y. Vladimirsky, *J. Vac. Sci. Technol.*, *B4*, 426 (1986).
35. B. J. Lin, *Proc. SPIE, 771*, 180 (1987).
36. H. Ito, C. G. Willson and J. F. J. Frechét, *ibid.*, *771*, 24 (1987).
37. M. P. DeGrandepre, D. A. Vidusek and M. W. Legenza, *ibid.*, *539*, 103 (1985).
38. G. N. Taylor and T. M. Wolf, *Polym. Eng. Sci.*, *20*, 1087 (1980).
39. For a review of plasma-developed resists see G. N. Taylor and T. M. Wolf, in VLSI Electronics, N. G. Einspruch, ed., Vol 6, Academic Press, New York, 1983, pp 217-253.
40. M. Hatzakis, Paper presented at the 16 Int'l Symp. on Electron, Ion and Photon Beam Science and Technology, Dallas, Texas, May-June 1981.
41. M. Hatzakis, J. Paraszczek and J. Shaw, Proc. Int'l Conf. on Microlithography, ME 81, A Oosenbrug. ed., Lausanne, Switzerland, Sept. 28-30, 1981, pp 386-396.
42. J. M. Shaw, M. Hatzakis, J. Paraszczek, J. Lintkus and E. Babich, "Proc. Reg. Tech. Conf., Soc. Plast. Eng., Mid-Hudson Sect., Ellenville, NY, Nov. 8-10, 1982, pp 285-295.
43. For a compendium of recent papers on Ag-sensitized chalcogenide resists see "Proc. Symp. on Inorganic Resist Systems", D. A. Doane and A. Heller, eds., Vol 82-9, The Electrochemical Soc., Inc., Pennington, NJ, 1982.
44. G. N. Taylor, J. M. Zeigler and M. Y. Hellman, *Proc. SPIE*, to be published.
45. For a comprehensive list of organometallic, O_2 RIE bilayer resist materials please see Y. Ohnishi, M. Suzuki, K. Saigo, Y. Saotome and H. Gokan, *Proc. SPIE, 539*, 62 (1985).
46. G. N. Taylor, unpublished results.
47. F. Watanabe and Y. Ohnishi, *J. Vac. Sci. and Technol.*, *134*, 422 (1986).
48. H. Gokan, Y. Saotome, K. Saigo, F. Watanabe and Y. Ohnishi, in "Polymers for High Technology", M. J. Bowden and S. R. Turner, eds., ACS Symp. Series 346, 358 (1987).
49. J. Shaw, E. Babich, M. Hatzakis and J. Paraszczak, *Solid State Technol.*, *30*, 6, 83 (1987).
50. E. D. Roberts, *J. Electrochem. Soc.*, *120*, 1716 (1973).
51. J. M. Shaw, M. Hatzakis, J. Parasczack and E. Babich, *Microelectronic Eng.*, *3*, 292 (1985); E. Babich, J. Shaw, M. Hatzakis, J. Paraszczak and D. Witman, *Microelectronic Eng.*, *5*, 299 (1986).
52. A. Tanaka, M. Morita and K. Onose, *Jpn. J. Appl. Phys.*, *24*, L112 (1985); M. Morita, A. Tanaka and K. Onose, *J. Vac. Sci. Technol.*, *B4*, 414 (1986).
53. N. Hayashi, T. Ueno, H. Shiraishi, T. Nishida, M. Toriumi and S. Nonogaki, in "Polymers for High Technology", M. J. Bowden and S. R. Turner, eds., ACS Symp. Series 346, 211 (1987).
54. M. Morita, S. Imamura, A. Tanaka and T. Tamamura, *J. Electrochem. Soc.*, *131*, 2402 (1984); T. Tamamura and A. Tanaka in "Polymers for High Technology", M. J. Bowden and S. R. Turner, eds., ACS Symp. Series 346, 67 (1987).

55. C. Rosilio, A. Rosilio and F. Buiguez, *Microelect. Eng., 1*, 197 (1983).
56. R. Brault, R. L. Kubena and R. A. Metzger, *Proc. SPIE, 539*, 70 (1985).
57. S. Fukuyama, Y. Yoneda, M. Miyagawa and K. Nishii, U.S. Patent 4657843 (1987).
58. R. E. Trujillo, *J. Organomet. Chem., 198*, C27 (1980).
59. R. West, L. D. David, P. I. Djurovich, K. L. Stearly, K. S. V. Srinivasan and H. J. Yu, *J. Amer. Chem. Soc., 103*, 7352 (1981).
60. D. C. Hofer, R. D. Miller and C. G. Willson, *Proc. SPIE, 469*, 16 (1984); R. D. Miller, D. Hofer, J. Rabolt, R. Sooriyakumaran, C. G. Willson, G. N. Fickes, J. E. Guillet and J. Moore, in "Polymers for High Technology", ACS Symp. Series 346, 170 (1987).
61. J. M. Zeigler, L. A. Harrah and A. W. Johnson, *Proc. SPIE, 539*, 166 (1985).
62. K. Saigo, F. Watanabe and Y. Ohnishi, *J. Vac. Sci. Technol., B4*, 692 (1986).
63. K. Nate, H. Sugiyama, T. Inoue and M. Ishikawa, Paper presented at Soc. of Polymer Sci. Japan Meeting, Tokyo, Japan, August, 1986.
64. A. S. Gozdz, G. L. Baker, C. Clausner and M. J. Bowden, *Proc. SPIE, 771*, 18 (1987).
65. A. S. Gozdz, C. Carnazza and M. J. Bowden, *Proc. SPIE, 631*, 2 (1986).
66. A. S. Gozdz, *Solid State Technol., 30, 6*, 75 (1987).
67. A. S. Gozdz, D. Dijkkamp, R. Schubert, X. D. Wu, C. L. Klausner and M. J. Bowden in "Polymers for High Technology", M. J. Bowden and S. R. Turner, eds., ACS Symp. Series 346, 334 (1987).
68. T. M. Wolf and G. N. Taylor, unpublished results.
69. M. Nakase, Y. Utsugi and A. Yoshikawa, *J. Vac. Sci. Technol., A3*, 1849 (1985); A. Yoshikawa and Y. Utsugi in "Polymers for High Technology", M. J. Bowden and S. R. Turner, eds., ACS Symp. Series 346, 309 (1987).
70. H. Okamoto, T. Iwayanagi, K. Mochiji, H. Umezaki and T. Kudo, *Appl. Phys. Lett., 49*, 298 (1986).
71. G. N. Taylor and T. M. Wolf, *J. Electrochem. Soc., 127*, 2665 (1980).
72. G. N. Taylor, T. M. Wolf and J. M. Moran, *J. Vac. Sci. Technol., 19*, 872 (1981).
73. G. N. Taylor, T. M. Wolf and L. E. Stillwagon, *Solid State Technol., 27, 2*, 145 (1984).
74. G. N. Taylor, L. E. Stillwagon, T. Venkatesan, *J. Electrochem. Soc., 131*, 1658 (1984).
75. T. N. C. Venkatesan, G. N. Taylor, A. Wagner, B. Wilkens and D. Barr, *J. Vac. Sci. Technol., 19*, 1379 (1981); H. Kuwano, K. Yoshida and S. Yamazaki, *Jpn. J. Appl. Phys., 19*, L615 (1980).
76. G. N. Taylor and T. M. Wolf, unpublished results.
77. D. Follett, K. Weiss, J. A. Moore, A. J. Steckl and W. T. Liu, Proc. Electrochem. Soc. Meeting, Detroit, MI, Oct. 17-21, 1982, Abst. 20, pp 321-322.
78. T. M. Wolf, G. N. Taylor, T. Venkatesan and R. T. Kraetsch, *J. Electrochem. Soc., 131*, 1664 (1984).
79. S. A. MacDonald, H. Ito, H. Hiraoka and C. G. Willson, "Proc. Reg. Tech. Conf., Soc. Plast. Eng., Mid-Hudson Sec.", Ellenville, NY, Oct. 28-30, 1985, pp 177-196.
80. S. A. MacDonald, L. A. Pederson, A. M. Patlach, and C. G. Willson in "Polymers for High Technology", M. J. Bowden and S. R. Turner, eds., ACS Symp. Series 346, 350 (1987).
81. T. B. Brust and S. R. Turner, *Proc. SPIE, 771*, 102 (1987).
82. F. Coopmans and B. Roland, *ibid., 631*, 34 (1986); *633*, 262 (1986).
83. F. Coopmans and B. Roland, *Solid State Technol., 30, 6*, 93 (1987).
84. F. Coopmans, B. Roland and R. Lombaerts, *Microelectronic. Eng., 5*, 291 (1986).
85. B. Roland, R. Lombaerts, C. Jakus and F. Coopmans, *Proc. SPIE, 771*, 69 (1987).
86. F. Coopmans, G. Brasseur, B. Roland, R. Lambaerts and S. J. Till, *ibid., 772*, 159 (1987).
87. J. P. W. Schellekens, R. J. Visser, M. E. Reuhman-Huisken, and L. J. Van Yzendoorn, *ibid., 771*, 111 (1987).
88. M. Rothschild, C. Arnone and D. J. Ehrlich, *J. Vac. Sci. Technol., B4*, 310 (1986).
89. M. Rothschild, C. Arnone and D. J. Ehrlich, *J. Vac. Sci. Technol., B5*, 389 (1987).

SPIN-ON GLASS FILMS IN MULTILEVEL IC INTERCONNECTION

SATISH K. GUPTA
Allied-Signal Inc., Planarization and Diffusion Products, 1090
S. Milpitas Blvd., Milpitas, CA 95035

ABSTRACT

Among the variety of techniques available today for
interlevel dielectric planarization, techniques based on spin-on
glass (SOG) films are relatively attractive because of process
simplicity and minimal equipment requirements. This paper
reviews the materials and processes of SOG technology as it is
applied in dielectric planarization. The SOG planarization
schemes employ SOG films either as a permanent part of the
interlevel insulation layer or as a sacrificial layer in the
'etch-back' techniques. The properties of the various types of
commercially available SOG materials are discussed in relation
to their functional and processing characteristics.

INTRODUCTION

In VLSI circuit process technology, the fabrication of
reliable interconnect structures with practical yields requires
the deposition of metallization layers of uniform thickness and
their subsequent patterning with good line-width control. These
processing goals are difficult to realize unless the substrate
is planarized prior to the metallization step. That is, the
interlevel dielectric must fill the space between the closely
packed, vertical-walled metal lines of the lower interconnect
level so as to produce a smooth topography. Since the
conventional CVD (chemical vapor deposition) oxide layers
provide only a conformal step coverage, a number of techniques
have been developed in recent years which either deposit a
planarizing dielectric layer or effect the planarization of the
conformal CVD layer [1 11].

Dielectric planarization processes based on spin-on glass
(SOG) films are particularly advantageous because of process
simplicity and low equipment costs. In the last few years, a
variety of new SOG materials with significantly improved
properties have become commercially available. Accordingly, a
number of SOG planarization processes have been developed and
are in use today.

DIELECTRIC PLANARIZATION TECHNIQUES

The reported techniques for planarization of IC substrates
can be grouped into two types:

"Physical" Methods

Flow of doped CVD oxides (PSG, BPSG) at high temp.
Bias sputtered quartz (BSQ)
Sputtering of CVD layers
Biased plasma enhanced CVD (PECVD)
Etch-back of thick CVD layers

Mat. Res. Soc. Symp. Proc. Vol. 108. ©1988 Materials Research Society

Spin-On Methods

Polyimide spin-on dielectric
Resist etch-back (REB)
Spin-on glass films

In the "physical" techniques, the sharp step profile produced by a conformally deposited dielectric film is smoothed by
the application of a physical process, such as high temperature
flow (fusion) or sputter/plasma etching. The high temperature
flow technique can be used only over the polysilicon interconnect level, and there too, it is becoming less practical with
increasing use of shallow doped junctions and silicide contact
metallurgies in state-of-the-art IC's. The other physical
techniques make use of accelerated erosion of a dielectric
surface during sputtering or plasma reactive ion etching, when
the surface makes an angle of about 45 degrees with the incident
ions. In practice, such processes require expensive and
elaborate equipment that is difficult to control and maintain,
and have low wafer throughputs. Moreover, there is always a
potential of substrate damage due to radiation or energetic
particle bombardment [12].

The spin-on techniques exploit the inherent ability of spun
on films to planarize the substrate topography. Spin-coated
polyimide films have been extensively evaluated as an interlevel
dielectric layer for many years. While polyimides generally
have good dielectric and mechanical properties, their use in IC
fabrication technology is rather limited. The reasons include:
long term reliability concerns, lack of high temperature and
oxidative stability, and process complexity. The resist etch-
back (REB) technique combines the excellent planarization
capability of hard baked novolac-based positive photoresist
films with the well-characterized, good dielectric properties of
CVD oxide films. The REB process consists in replicating the
planarized profile of a thick resist layer into the underlying
conformal CVD oxide layer by means of plasma etching under
conditions such that the resist and oxide are etched at the same
rate. Although the REB technique is currently in fairly common
use, it suffers from the drawbacks of difficult process control
and reproducibility and the limitation of applicability on
geometries below about 1.5 micron line/space widths.

Spin-on glass planarization techniques combine the
planarization effect of spun-on films with the oxide-like
material characteristics of SOG, resulting in simple and
straightforward processing. As a dielectric layer in multilevel
interconnection structures, SOG films offer the following
advantages over other spin-on materials:

1. High thermal/oxidative stability.
2. Etch characteristics similar to those of CVD oxides.
3. Good adhesion to silicon, oxides, aluminum, etc.
4. Can be doped with phosphorus for Na^+ gettering.
5. Low trace metal contamination levels.

Thus, SOG materials are compatible with the materials and
processes of IC fabrication technology and SOG processes are
easily integrated into existing process flows.

Before the discussion of the SOG planarization schemes, the
basic chemistry and properties of spin-on glass materials are

briefly discussed in the next section.

SOG: THE MATERIAL

SOG Chemistry

Spin-on glass liquids consist of Si-O network polymers dissolved in common organic solvents, such as alcohols, ketones, and esters. The polymers are prepared through the same basic chemistry as that utilized in sol-gel technology. The prototypical reaction is the hydrolysis-condensation reaction of tetraethylorthosilicate (TEOS), $Si(OEt)_4$, given by

$$Si(OEt)_4 + H_2O \longrightarrow EtOH +$$

$$Si(OEt)_3OH + Si(OEt)_2(OH)_2, \text{ etc.} \qquad \text{HYDROLYSIS} \quad (1)$$

$$\equiv SiOH + OHSi \equiv \longrightarrow$$

$$\equiv Si-O-Si \equiv + H_2O \qquad \text{CONDENSATION} \quad (2)$$

$$nSi(OEt)_4 + 0.5n(4-a+b)H_2O \xrightarrow{\text{H}^+}$$

$$[Si(OEt)_a(OH)_bO_{0.5(4-a-b)}]_n + n(4-a)EtOH \qquad \text{OVERALL} \quad (3)$$

$$\text{'Polysilicate'}$$

The nature of the silicate polymer is determined by reaction conditions, such as molar ratio of H_2O to TEOS, pH and concentration of the solution, etc. Unlike in sol-gel technology where the goal is to form dense gels quickly, the reaction conditions in SOG synthesis are chosen such that the polymers are stable toward molecular weight increase for periods of several months. Since gelation time is a strong function of concentration, the equivalent silica (SiO_2) content of most commercial SOG products is typically 10 percent or lower. Therefore, the SOG film thickness is usually limited to a few thousand angstroms.

The properties of SOG materials can be modified by incorporating a substituted alkoxysilane, $RSi(OEt)_3$ where R = methyl or phenyl, or a dopant such as phosphorus or boron during the hydrolysis reaction. Commercially available SOG materials are of four major types:

POLYMER	FILM COMPOSITION
Silicate	$[SiO_2]_n$
Phosphosilicate	$[SiP_xO_y]_n$
Siloxane	$[R_xSiO_y]_n$
Phosphosiloxane	$[R_xSiP_yO_2]_n$

Properties

The material characteristics of SOG films are fundamentally similar to those of sol-gel glasses. There are two important distinctions, however. The first one stems from the fact that an SOG film is always formed on a substrate toward which it exhibits good adhesion. When such a film is dried and cured, the shrinkage can occur only in the z-direction since the film is constrained to remain adhered to the substrate. This results in the build up of tensile stress in the film parallel to the surface. Consequently, SOG films have a low threshold of thickness for cracking. Secondly, in IC fabrication the maximum temperature at which an SOG film can be cured is often limited to about 450°C because of the presence of aluminum interconnects. After such low temperature cures, the SOG film is far from being densified and may contain significant amounts of residual silanols, $\equiv$Si-OH, and absorbed water. Of course, when there is no aluminum metallization on the substrate, the SOG film can be densified at high temperatures, typically at 800-900°C, to obtain a silanol and water-free film. The infrared spectra of a silicate SOG film taken after thermal cures at 425°C and 900°C, shown in Figure 1 illustrate the typical behavior of SOG films with respect to thermal treatments. In the particular case of this silicate SOG film, the elimination of the silanols and water after the 900°C cure is accompanied with a drop in the wet etch (HF) rate of the film to that of thermal SiO_2. This implies complete densification. While all SOG films show complete loss of silanol and water after a high temperature cure, the extent of densification varies, especially among siloxane type SOG's.

The dielectric properties (for example, dielectric constant) of an SOG film are, to a large extent, determined by its silanol/water content [13]. For instance, the dielectric constant of the silicate films after cures at 425°C and 900°C is measured to be about 9 and 4, respectively, consistent with the spectra of Figure 1.

Considerable progress has been made in the last few years in reducing the silanol/water content (after a low temperature cure) as well as in improving the cracking resistance of SOG films. The improvements have been achieved both through the optimization of the hydrolysis conditions and by incorporating Si-R groups [14] in the polymer composition. Typical physical properties of the various types of commercially available SOG materials/films are summarized in Table I.

Table I. General Properties of SOG Materials

SOLUTIONS

Shelf-life (RT)	:	6 months, typical
Viscosity	:	1-3 cp
Filtration	:	0.2 micron
Na^+ contamination	:	<100 ppb

FILMS

Thickness range	:	up to 4-5,000Å
Thickness uniformity	:	2% or better
Pinhole/particulate density	:	<<1/cm^2
Refractive index	:	1.35 - 1.50

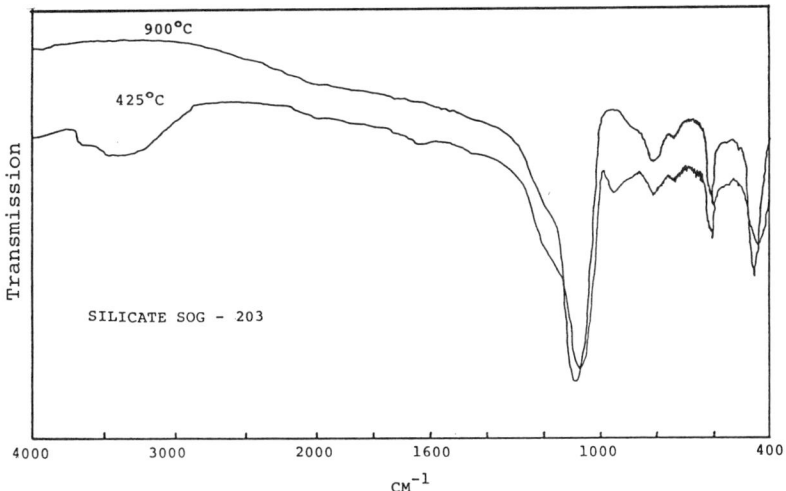

Fig. 1.   IR spectra of SOG as a function of temperature.

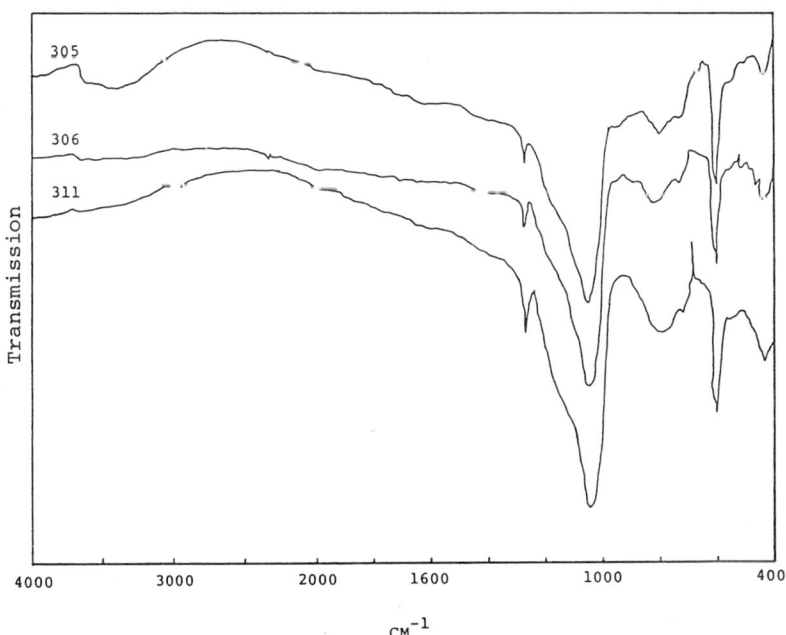

Fig. 2.   IR Spectra of methylsiloxane SOG's (425 C°cure).

Film stress (tensile)   : $1-5\times10^8$ dynes/cm^2
Etching media   :   HF, fluorine plasmas

A selected set of properties relevant to IC fabrication
processing are compared for different SOG types in Table II. In
general, the incorporation of $\equiv$Si-R groups in the Si-O network
improves the cracking behavior as well as the dielectric
characteristics of the SOG film. Methyl groups are more
effective than phenyl groups in lowering the dielectric constant
by reducing the silanol/water content. Within the methyl-
siloxanes, the extent of improvement is a function of the amount
of the methyl character and the manner in which the hydrolysis
reaction is carried out. The infra-red spectra of three methylsiloxane SOG films (after a 425°C cure) are shown in Figure 2.
The SOG's 305 and 306 contain equal but small amounts of the
$\equiv$Si-R groups; the difference is that the SOG 306 is synthesized
under slow hydrolysis conditions. The reduction in the silanol
absorptions at about 3650 cm^{-1} and 950 cm^{-1} for SOG 306 are
clearly seen. The SOG 311 contains a higher amount of the
methyl groups and is also prepared through slow hydrolysis. Its
IR spectrum is virtually free of any silanol absorptions. It
should be noted that the silanol absorptions of all three
methylsiloxane films are considerably diminished compared to
those of the silicate SOG 203 (Figure 1).

SOG PLANARIZATION PROCESSES

The limited film thickness of the available SOG materials,
though not sufficient to allow their use as a stand-alone
interlevel dielectric layer, is adequate for planarizing or
smoothing a wide range of substrate topographies. The effect of
a thin (2000-2500Å thick) SOG film on the step (space) profiles
at two different geometries is shown in Figure 3.

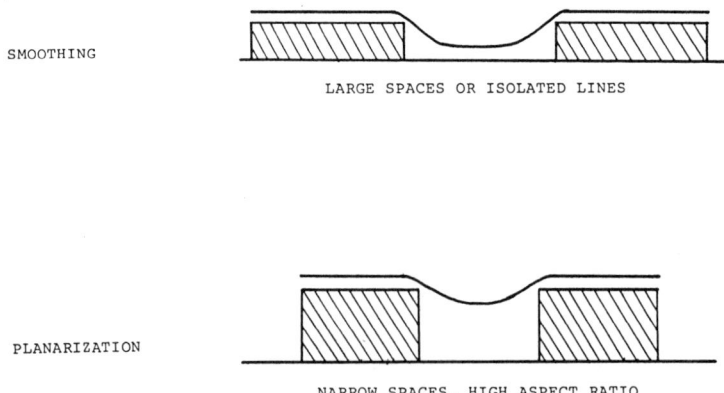

SMOOTHING

LARGE SPACES OR ISOLATED LINES

PLANARIZATION

NARROW SPACES, HIGH ASPECT RATIO

Fig. 3. Smoothing/planarization by thin (2500Å SOG films).

Table II. COMPARISON OF DIFFERENT TYPES OF SOG MATERIALS

MATERIAL	PRODUCT NAME*	CRACKING RESISTANCE	FILM SHRINKAGE	DIELECTRIC CONSTANT	SILANOL/H_2O CONTENT**	THERMAL STABILITY+
Silicate	203	Low	18-20%	9-10	High	Excellent
Phosphosilicate	P-5	Low	18-20%	7-8	Fair	Excellent
Methylsiloxane	305	Medium	16-18%	7-8	Fair	Good
Methylsiloxane	306	Medium	18-20%	<6	Low	Good
Methylsiloxane	311	High	10-12%	<5	Negligible	Good
Phenylsiloxane	204	Medium	12-14%	>9	High	Excellent
Methylphenylsiloxane	208	Medium	10-12%	>9	High	Good
Phosphosiloxane	PC-302	Medium	16-18%	8-9	Fair	Good

*From Allied's ACCUGLASS® Series of SOG products

**Qualitative, from IR spectra

+To 450°C, 60 min. in air

Very little planarization (defined as percent reduction in step height) is obtained over isolated lines or lines separated by 3-4 micron wide spaces. However, the 90 degree angle of steps is reduced to about 45 degrees. This 'smoothing' of the vertical-walled features is quite suitable for conformal deposition of subsequent layers with a high degree of step coverage. At smaller geometries, where the aspect ratio of the space between lines approaches unity, a high degree of planarization is produced by similarly thin SOG films. At such geometries, a mere smoothing effect would not be acceptable. In either case, the SOG thickness on top of the lines is too small to provide interlevel insulation.

Thus, the SOG planarization processes in use today utilize SOG films primarily as a planarizing agent with the bulk of the dielectric insulation functions being provided by CVD oxide layers. In some schemes, SOG films are utilized as a sacrificial planarization layer. Four of the more common schemes for planarization with SOG are [10, 11, 15-19].

1. CVD/SOG Two-Layer Dielectric
2. CVD/SOG/CVD Sandwich Dielectric
3. Partial Etch-Back of SOG in a Sandwich Structure
4. Total Etch-Back of SOG

The two-layer CVD/SOG dielectric structure shown schematically in Figure 4 is the simplest process. Since the SOG layer is in direct contact with the interconnects in these structures, the SOG must have very good dielectric characteristics. At polysilicon level, this can always be insured by carrying out a high temperature (800-900°C) cure of the SOG film. The order in which the SOG and CVD layers are deposited is a function of the geometry and the nature of the underlying dielectric layer. In MOS IC fabrication, the use of a phosphorus-containing SOG is common, especially at polysilicon level, for Na^+ gettering purposes.

The use of a CVD/SOG/CVD sandwich (Figure 5) structure relaxes the requirements on the dielectric properties of the SOG layer. Moreoever, the bottom CVD layer serves to buffer the SOG from the effect of the relatively large thermal expansion of aluminum lines during thermal processing.

Under certain conditions, as will be discussed later, the presence of SOG inside the via holes etched through the composite dielectric can potentially cause poor via contact resistance. In the partial etch-back sandwich process (Figure 6), the problem is avoided by etching back the SOG layer to a point where it clears the top of the interconnect lines. The process is particularly useful at very small geometries where it is difficult to fill the narrow spaces with a good SOG material without cracking or void formation. In such a case an SOG material with low shrinkage characteristics is necessary. The partial etch-back approach allows the use of such a material regardless of its effect on the via contact resistance.

The SOG etch-back process is totally equivalent to the REB process. The use of an SOG as the sacrificial planarization layer offers several advantages, particuarly in the etching step. The etch rates of CVD oxide and SOG are easily matched through simple adjustments of the plasma chemistry. The process control is greatly improved as the plasma micro-loading effects are minimized and the etch chamber is free from organic residues or deposits. In most cases, the wafer throughput in SOG etch-

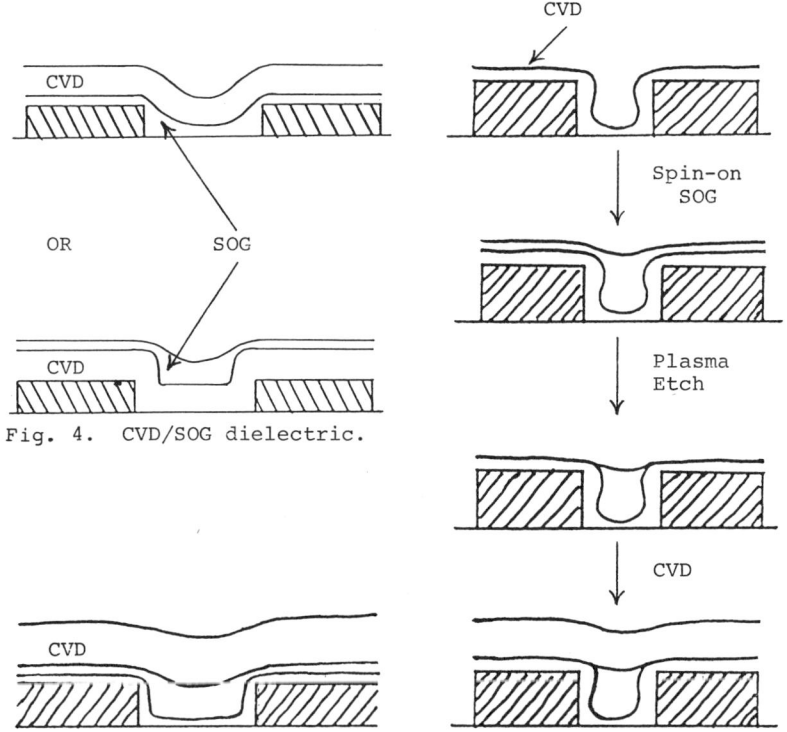

Fig. 4. CVD/SOG dielectric.

Fig. 5. CVD/SOG/CVD sandwich.

Fig. 6. SOG partial etch-back.

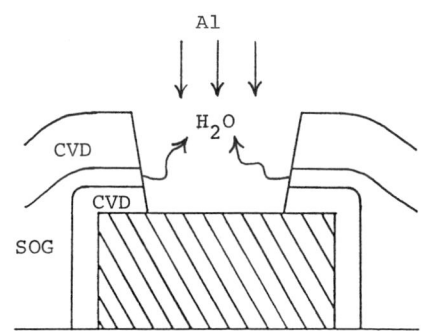

Fig. 7. Outgassing from SOG inside a via.

back processing is significantly higher compared to that of REB. Finally, the thinner SOG planarization layer does not accumulate excessively in low-lying areas of the chip, thus minimizing via-depth variations in the planarized dielectric layer.

The selection of an optimal SOG material and planarization scheme in a given application is dictated by a number of factors including: device geometry, nature of the underlying interconnect (aluminum or polysilicon), post-planarization thermal processes, sensitivity of the device to mobile ion (Na^+) contamination, conformality of the CVD process, and the thermal budget available for SOG cure.

Processing Problems

The most common types of problems encountered in the use of SOG processes include: particulate contamination, cracking of dielectric layer, poor adhesion (delamination), and poor via contacts. The particulate problem is related to the generation of flakes or dust from the SOG residues on various surfaces in the spin-on coater. It is easily eliminated by employing proper spin-on techniques and equipment. In fact, spin-on coaters designed specifically for SOG are now commercially available. Cracking can occur if the SOG layer is too thick or if the composite thickness of CVD/SOG dielectric is too large. The solution to SOG cracking is to reduce the thickness of the SOG layer as far as possible while retaining a usable degree of planarization or smoothing. Adhesion problems are rare but can occur if the dielectric layer under the SOG is highly doped with phosphorus (>5% P). The remedy is either to reduce the phosphorus content or to employ a different planarization scheme which avoids contact between the SOG and P-doped layer. The via contact problem arises from the formation of insulating deposits at the via bottom by the reaction of aluminum with moisture outgassed by SOG during metal deposition [Figure 7]. The out-gassing from SOG can be due to incomplete curing of the SOG or due to the inherent water content of SOG. In most cases, a dehydration step, carried out in-situ in the sputtering apparatus or externally, is effective in eliminating the problem.

REMARKS

Spin-on glass processes for dielectric planarization fill an important need of today's IC interconnect technology. Current SOG materials are significantly better than those available some years back, and are adequate for most applications. However, improved materials are desirable for process simplification as well as to meet the requirements of VLSI and ULSI fabrication. Specifically, materials with better dielectric and low shrinkage characteristics are needed. Apart from modification of the SOG chemistry, the improvements may come from novel low temperature cure techniques. Recent work indicates that cures in N_2 plasmas or in steam ambients significantly improve SOG characteristics [20].

REFERENCES

1. W. E. Armstrong and D. C. Tolliver, J. Electrochem. Soc. 121, 307 (1974).
2. W. Kern and L. Schnable, RCA Review 43, 423 (1982).
3. D. Desbiens, Electrochem. Soc. Ext. Abs. 84(2), 605 (1984).
4. H. Kotani, H. Yakushiji and H. Harada, J. Electrochem. Soc. 130(3), 645 (1983).
5. T. Mogami, M. Morimoto, H. Okabayashi and E. Nagasawa, J. Vac. Sci. Technol. 3(3), 857 (1985).
6. K. Machida and H. Oikawa, J. Vac. Sci. Technol. B 464, 818 (1986).
7. J. S. Mercier, H. M. Naguib, V. Q. Ho and H. Nentwich, J. Electrochem. soc. 132(5), 1219 (1985).
8. A. C. Adams, Solid State Technol. 24, 178 (1981).
9. L. B. Rothman, J. Electrochem. Soc. 130, 1131 (1983).
10. L. B. Vines and S. K. Gupta, Proc. IEEE VLSI Multilevel Interconn. Conf., 506 (1986).
11. J. K. Chu, J. S. Multani, S. K. Mittal, ibid, 474 (1986).
12. Y. Hazuki and T. Moriya, ibid, 121 (1986).
13. S. K. Gupta and R. L. Chin, ACS Symp. Ser 295, 349 (1986).
14. H. Schmidt, Mat. Res. Soc. Symp. Proc. 32, 327 (1984).
15. A. Rey, D. Lafond, J. M. Mirabel, M. C. Tacussel and M. F. Coster, IEEE VLSI Multilevel Interconn. Conf., 491 (1986).
16. N. Parekh, R. Allen, W. Yao and R. Fulks, Electrochem. Soc. Ext. Abs. 86-2, 530 (1986).
17. P. Pai, W. G. Oldham and C. H. Ting, IEEE VLSI Multilevel Interconn. Conf., 364 (1987).
18. R. M. Brewer and R. A. Gasser, Jr., ibid, 376 (1987).
19. M. D. Tui, T. A. Streif, K. E. Schoenberg, R. H. Dorrance and P. P. Proctor, ibid, 385 (1987).
20. K. G. Lubic, J. L. Blodgett and S. K. Gupta, Electrochem. Soc. Ext. Abs. 87-2, 645 (1987).

MATERIALS INTERACTIONS IN THE FIRING
OF COPPER THICK FILM MULTILAYER CERAMICS

WILLIAM BORLAND AND VINCENT P. SIUTA
E. I. Du Pont De Nemours & Co., Inc., Electronics Department,
Wilmington, De 19898

ABSTRACT

In recent years, copper thick film materials have gained rapid acceptance in ceramic multilayer interconnect boards because of their ability to meet advanced packaging requirements at reasonable cost. With high volume production, however, problems such as opens, shorts, blisters and porosity have been experienced. Many of these failures may be attributed to undesirable materials interactions caused by reducing conditions which can be caused by incomplete burnout of thick film organics during firing.

This paper considers the interactions in current copper thick film material systems. Primary emphasis will be placed on chemical interactions that occur in both ideal and non-ideal conditions during firing. Failure modes will be related to chemistry and processing. New systems designed to overcome processing sensitivity will be discussed.

1.0 INTRODUCTION

Copper multilayer interconnect boards consist of copper conductive layers connected by copper vias through isolating layers of ceramic dielectric. The most popular method of manufacture is by a sequential printing and firing operation termed the "thick film process". The first step is to screen print a conductive copper paste composition onto an alumina substrate and then fire it in a nitrogen atmosphere. Next, up to three layers of dielectric, with via openings, are screen printed and separately fired. A copper via-fill composition is printed and fired in order to fill the vias. This sequential process of printing and firing copper, dielectric and via filling is repeated until the required number of conductive and insulating layers has been formed. Most copper multilayer interconnect boards typically have 5 to 7 conductive layers.

The advantage of the sequential multilayer process is that, except for firing in a nitrogen atmosphere, it utilizes simple processing equipment and substrates can be inspected after each step. The main disadvantage is that the process requires many steps. Sometimes two layers of dielectric, or the copper conductor and via-fill, are cofired to reduce the number of process steps. If not properly controlled, however, such process shortcuts can affect the yield and subsequent reliability of the multilayer substrate.

2.0 MATERIALS SYSTEM

Nitrogen-fired thick film multilayer compositions are high viscosity thixotropic pastes containing various functional phases and binders suspended in an organic vehicle.

2.1 Copper Conductors

Copper conductors typically contain a functional phase consisting of 1 to 2 micron copper powder. A binder composed of similar sized oxides and glasses is also present to bond the sintered copper to the substrate. Oxides such as Cu_2O or PbO can be used alone or in combination with the glass powders. The glass serves to wet the substrate and can accelerate sintering of the copper. The glass traditionally is of a lead-bismuth-silicate composition. The organic vehicle consists of an organic resin in a suitable solvent. Thick film screen printing vehicles are typically composed of ethyl-cellulose polymers in a single or mixed solvent system consisting of high boiling alcohols or esters such as butyl-carbitol acetate, terpineol, or dibutyl carbitol.

2.2 Dielectric Composition

Nitrogen fireable dielectric compositions are typically composed of alumino-boro-silicate glasses modified by small amounts of lead, sodium and potassium. Refractory fillers, such as alumina and silica, are also included to control glass flow and subsequent densification at the firing temperature and to provide expansion matching to 96% alumina substrates. The organic vehicle is similar to that employed in copper conductors. Table 1 summarizes typical properties of a copper compatible dielectric.

Table 1

Fired Thickness (μm)	50-55
Dielectric Constant	6-7
Dissipation Factor (%)	< 0.2
Insulation Resistance (Ω)	> 10^{11} at 100 VDC
Voltage Breakdown (VAC/25 μm)	350

3.0 REACTIONS DURING FIRING

Copper material systems are designed for firing in nitrogen in order to minimize oxidation of the copper conductors. The nitrogen contains a low level of oxygen, typically 1 ppm. Firing schedules are simple, rapid and continuous. Belt furnaces are employed with door-to-door firing cycles of approximately one hour. The temperature/time profile generally peaks at 900°C for about 10 minutes. In order to achieve a reliable product, three reactions must occur in the firing part of the fabrication process: burnout of organics, densification and reactive bonding.

3.1 Burnout of Organics

Ethyl cellulose based vehicle systems are used in copper thick film materials because of their superior screen printing characteristics. Air drying of the printed film, typically at 125°C, eliminates solvents by evaporation. Removal of the resin requires higher temperatures.

Burnout of ethyl cellulose is the first and most important step of the copper material system firing process. Brown and Tipper[1] found that the rate of pyrolysis of ethyl cellullose in vacuum is only 0.5%/minute at 300°C. On heating in the absence of oxygen, cleavage at the ethyl groups and dehydrogenation takes place, causing chemical unsaturation, cross-linking and eventual carbonization of the resin.

$$ROC_2H_5 \; ---> \; H_xC_y \; ---> \; C \tag{1}$$

where R represents the cellulose radical.

Complete removal of organics from copper thick film materials during firing requires oxidation of the polymer according to the following reaction.

$$H_xC_y \; (polymer) + O_2 \; (in \; N_2) = H_2O + CO + CO_2 \tag{2}$$

Burnout of organics should take place prior to the softening point of the dielectric glass, typically around 550°C for a 900°C peak firing system. Otherwise, carbonaceous residues from decomposition of the resin will become trapped in the film. As discussed later, this can cause undesirable material interactions which can lead to defects in the multilayer substrate.

Complete burnout of the organics prior to significant flow of the glass phase requires adequate nitrogen flow rates, both into and out of the furnace. Typically, half the flow of gas should enter and leave the burnout section. Good distribution of nitrogen in the burnout zone ensures each part is continually subjected to fresh gas. Sufficient time in the 300-550°C temperature range and a proper balance between belt loading and belt speed is essential. Oxygen or nitrous oxide additions to the burnout zone can also be used to advantage in burnout of the organic[18]

3.2 Densification

In the second stage of the firing process, densification of the dielectric and conductors takes place. The copper conductors sinter by a combination of solid state diffusion and reactive liquid-phase sintering. The dielectric densifies by a viscous flow mechanism.

3.2.1 Solid State Diffusion

Densification of the copper conductor begins at relatively low temperatures when metal particles begin to bond together by "neck" formation via solid state diffusion. The driving force for sintering is the reduction of surface free energy. The temperature ($T°K$) at which sintering takes place is strongly dependent upon the melting point ($Tmp°K$) and particle size of the metal powder. Copper begins to sinter at about 400°C which correlates to $T/Tmp = 0.5$. However, for significant densification, most metals and oxides require a T/Tmp temperature of greater than 0.8, i.e., above 800°C for copper. Since the thick film process does not allow enough time at peak temperature for significant solid state diffusion. low softening point glasses are used as sintering promoters to achieve high levels of densification.

3.2.2 Activated Sintering

Reactive phase sintering of the copper film only occurs when the temperature exceeds the softening point of the glass contained in the conductor. The molten glass phase serves to enhance the sintering of the copper. As the copper sinters, the liquid glass is extruded from the densified film structure. At higher temperatures, copper oxide dissolves in the glass phase which then acts as a transport agent for copper ions. This accelerates sintering of the copper and the formation of a strong reactive bond with the alumina substrate.

3.2.3 Viscous Flow Sintering

Viscous glass flow is the main densification mechanism for thick film dielectric compositions. Kingery[2] showed that densification due to viscous sintering is described by the following expression:

$$\Delta V/V_o = 3\ \Delta L/L_o = \frac{9\gamma t}{4\zeta r} \tag{3}$$

where $\Delta V/V_o = 3\Delta/L_o$ is the volume shrinkage in time t
γ = surface tension of glass
ζ = viscosity of glass
r = particle size of the glass

Equation (3) points out the important variables which control viscous sintering. For example, the viscosity of the glass and, consequently, the rate of densification can change by a factor of 1000 over a 100°C temperature interval[2]. Also, the rate of sintering will be doubled if the glass particle size is decreased from 2 microns to 1 micron.

3.3 Reactive Bonding

Reactive bonding generally only occurs near the peak firing temperature when the glass can significantly flow and wet other surfaces. Wetting is often improved by the addition of fluxing agents, such as Bi_2O_3 or borates, and this action serves to promote reactive bonding. Since the glass phase concentration is minimized to obtain good solderability, reactive bonding is essential to achieve good adhesion. Compounds such as $CuAlO_2$ or $MO.Al_2O_3$ spinel phases[3-7] can form at the glass/alumina interface. These chemically bond the film to the substrate. The glass also continues to extend into the copper structure, anchoring the film by a keying and interlocking action.

4.0 MATERIAL INTERACTIONS

4.1 Copper Dissolution in the Dielectric Glass

Two possible mechanisms for the introduction of copper ions into the glass are copper oxide dissolution and ion exchange. Copper oxide dissolution can occur when the glass is sufficiently fluid to flux reactive oxides. As shown in Fig. 1[8], Cu_2O

rather than CuO should be the predominant copper oxide existing in a conductor fired in nitrogen at 900°C. Consequently, Cu^{+1} ions can be introduced into the dielectric glass at the conductor interface near the peak firing temperatures. This reaction should continue until the solubility limit of the Cu_2O in the glass is reached. Factors controlling the concentration of Cu_2O introduced into the glass are time at peak temperature, the dissolution rate, and the relative diffusion rates of Cu^{+1} and its charge compensating nonbridging oxygen into the dielectric bulk away from the interface.

FIGURE 1

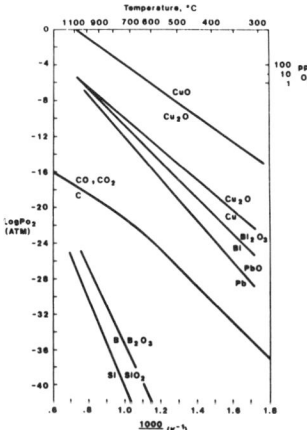

Equilibrium Oxygen Partial Pressure of Metal/Oxides as a Function of Temperature

The other possible mechanism is exchange of Cu^{+1} at the dielectric glass interface with other single valent ions in the glass. Strain and polarization forces favors an exchange between ions with near equivalent sizes. The most likely exchange candidate for Cu^{+} with a radius = 0.96A is Na^{+} with an excellent radius match of 0.97A. Ion exchange is expected to proceed more rapidly at lower temperatures than dissolution, because its activation energy should be much lower. Dissolution may dominate, however, because sluggish kinetics of Na_2O interdiffusion in Cu_2O may deplete Cu_2O at the conductor-dielectric interface and thus limit the exchange mechanism.

Single valent species like Cu^{+1} and Na^{+1} will be relatively mobile at temperatures above the glass transition point. The diffusion of oxide constituents in the glass, such as Cu_2O, away from the conductor interface, will be much slower than the species alone, due to the relative immobility of oxygen. Therefore, to maintain ionic charge balance in the glass, Cu^{+1} is more likely to interdiffuse by simply exchanging sites with equivalent ions of similar size like Na^{+1} rather than simultaneously diffuse with a non-bridging oxygen with its compensating negative charge. If dissolution plays a dominant role, total concentration of single-valent species in the dielectric glass matrix (e.g., Cu^{+1} + Na^{+1}) should increase as the conductor interface is approached. Evidence for this has been found[9], thus lending support for this mechanism.

4.2 Interactions Due to Entrapped Carbon

If the firing process is not carefully controlled, many other complex interactions, other than those discussed, can occur. Firing in nitrogen with rapid heating, inadequate gas flow and/or high belt loading hinders burnout of organics. Organics which are not removed in the burnout zone of the furnace will carbonize rapidly and become entrapped in the film.

Entrapped carbon plays a critical role in triggering unwanted material interactions and failure mechanisms. Carbon inhibits sintering, and this action alone can cause localized areas of unsintered copper leading to conductor open problems[10]. At high temperatures, carbon becomes a powerful reducing agent. It will react with 10 to 25 times its weight in reducible oxides, such as Cu_2O, Bi_2O_3 and PbO. These oxides are reduced to metal with the evolution of CO and CO_2. Such reactions can lead to loss of conductor adhesion, conductor opens, shorting across dielectric layers or blistering problems.

4.3 Thermodynamics of Oxide Instability

Copper conductors generally contain Cu_2O, Bi_2O_3 and PbO as components of the binder system. The stability of an oxide as a function of temperature is dependent on $\Delta F°$, its standard free energy of formation. The equilibrium dissociation pressure Po_2 at temperature T can be calculated from free energy data from the following expression:

$$1/Po_2 = \exp. (-\Delta F°/RT) \tag{4}$$

For our purposes, it is useful to plot Po_2 versus T for the oxides of interest, as given in Fig. 1[8].

From thermodynamic considerations, Bi_2O_3, PbO, Cu_2O etc., will be stable when fired in nitrogen to 900°C, provided the oxygen partial pressure (in N_2) exceeds the dissociation pressure of the oxide. All the oxides present in Fig. 1, except for CuO, are expected to be stable when fired in nitrogen containing 1-10 ppm of oxygen. However, if burnout of the organic phase has not been accomplished fully, entrapped carbon will reduce the oxygen partial pressure in the vicinity of the copper thick film. This will result in reduction of all oxides with a lower (less negative) free energy of formation than CO. In Figure 1, any oxide above the line for the $C/CO/CO_2$ equilibrium is reducible by carbon according to the reaction

$$MO + C = M + CO \tag{5}$$

Furthermore, any H_xC_y and CO atmosphere formed by organic decomposition can result in the following reactions, provided the free energy change for the reaction is negative.

$$3MO + 2HC = 3M + H_2O + 2CO \tag{6}$$

$$MO + CO = M + CO_2 \tag{7}$$

At firing tempratures of 900°C, the kinetics for the above reactions are favorable. Therefore, if firing conditions are not properly controlled to insure good burnout of organics, Cu_2O, Bi_2O_3 and PbO will be reduced to metallic Cu, Bi and Pb with the subsequent evolution of gas.

4.4 Evaporation of Bismuth and Lead

The vapor pressures of bismuth and lead are greater than 10^{-3} mm of Hg at temperatures above 600°C[11]. This is sufficiently high to result in Bi and Pb loss by evaporation.

According to the Langmuir equation[12], the rate of bismuth loss by evaporation is directly proportional to the area of the source. Since large substrate areas are exposed to high temperatures, if any metallic bismuth or lead exists, then the majority of it will be lost by evaporation.

According to LeChatelier's principle, removal of a reaction product will drive the reaction to completion. Therefore, the reduction of easily reducible oxides such as Bi_2O_3 and PbO is accelerated by the loss of bismuth and lead through evaporation. Measurement of bismuth loss above 600°C from Du Pont 9924 copper conductor under typical nitrogen firing conditions is shown in Fig. 2[3].

FIGURE 2

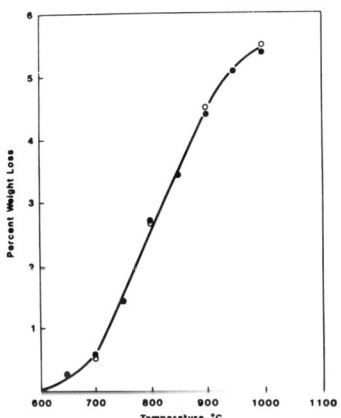

Weight Loss of 9924 Copper Thick Film
on Firing in Nitrogen 10 Minutes at
Peak Temperature

4.5 Eutectic Formation

Phase diagrams for the Cu-Bi and Cu-Pb binary systems are shown in Figs. 3 and 4. Copper forms low melting point eutectics with metallic bismuth and lead. As shown, the eutectic composition is essentially that of pure bismuth and lead, respectively. The solid solubility of copper in bismuth and lead is nil.

294

FIGURE 3

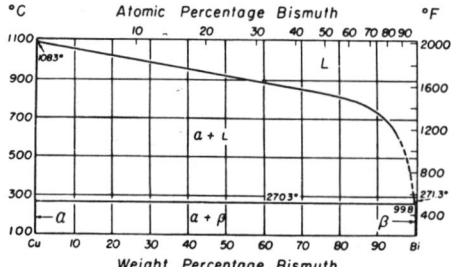

Phase Diagram for the Copper/Bismuth Alloy System

FIGURE 4

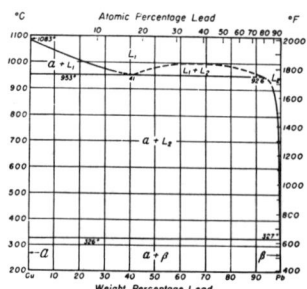

Phase Diagram for the Copper/Lead Alloy System

FIGURE 5

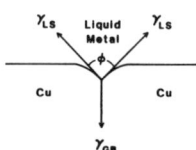

$\gamma_{GB} = 2\gamma_{LS} \cos\phi/2$

where γ_{GB} = Grain Boundary Energy of Cu

γ_{LS} = Liquid/Cu Solid Interfacial Energy

ϕ = Dihedral Angle

Schematic Diagram Illustrating the Relationship Between Interfacial Energy and Dihedral Angle for Molten Metals on Copper

When copper materials are fired under reducing conditions, it is apparent that molten Cu-Bi alloy can form in the copper grain boundaries. On cooling below 270°C, the eutectic alloy composition, consisting of essentially pure bismuth solidifies. The distribution of the bismuth phase in the solid copper and the properties of the copper film are largely determined by surface energy considerations. Figure 5 illustrates the wetting of liquid bismuth and lead on copper. The ratio $\gamma_{GB}/2\gamma_{LS}$ determines the dihedral angle and distribution of the liquid metal phase in copper.

In the case of Cu-Pb, the dihedral angle is about 60° and the copper maintains its ductility. However, in the Cu-Bi case, the dihedral angle is zero, and metallic bismuth forms a thin, continuous film in the copper grain boundaries. This bismuth film is extremely brittle and its presence destroys the ductility of copper. The amount of bismuth required for this catastrophic effect is extremely small (ca. 0.01% Bi)[13]

5.0 PROPOSED FAILURE MECHANISMS

5.1 Opens in Copper Conductors

The development of open circuits, such as that shown in Fig. 6[10,14], on completed copper thick film multilayer interconnect boards after thermal cycling has a dramatic effect on yield and cost. The following observations in the production of multilayer interconnect boards are relevant. 1) The problem was associated with high belt loadings at high production volumes. 2) While running at full capacity a dark grey "soot" built up on the furnace at a rate faster than would be predicted by extrapolation from lower production volumes. Chemical analysis of the furnace "soot" indicated it contained carbon, bismuth and lead. 3) Increased nitrogen flow rates through the furnace did not show significant improvements. 4) Small lot production failed to produce opens[10]

FIGURE 6

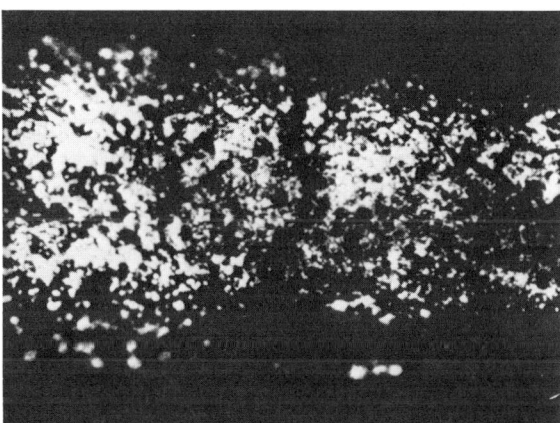

Currently, two mechanisms to explain electrical opens are believed possible. Both are associated with entrapped carbon due to incomplete removal of organics from the burnout zone of the furnace. Toch, et al[10] proposed that presence of elemental carbon inhibits the sintering of copper to such a level that local areas of copper conductor lines essentially remain unsintered. Examination of such failed areas show unsintered particles of copper of a size consistent with the orginal particle sizes. Bacher and Siuta[3] proposed that the entrapped carbon reduces bismuth trioxide to metallic bismuth. This forms a grain boundary film on cooling from 900°C. As previously described, this significantly embrittles the copper film. Failures associated with this mechanism show large grain growth in the area of the open[14]. This is attributed to the accelerated sintering effect of a liquid phase (Cu-Bi- Pb eutectic) at the grain boundary[3]. Failures are always observed to be intergranular in nature.

With regard to these considerations, it is useful to consider the strain induced in the copper due to the thermal coefficient expansion mismatch. Assuming the worst case, that all the differential strain is accommodated in the conductors, the strain ε can be estimated from the following relationship:

$$\varepsilon = (\alpha_{cu} - \alpha_{dielectric})\Delta T \tag{8}$$

where α = coefficient of thermal expansion
ΔT = temperature difference

Since copper has a higher thermal expansion than the dielectric, the copper film will be under tension as it cools from its strain relief temperature of approximately 300°C. In addition, thermal cycling from -75°C to +150°C adds an additional strain. Assuming the strain developed in thermal cycling is superimposed on the strain from cooling, the following calculation can be made:

$$\varepsilon = (16-6)(375) \text{ ppm} \tag{9}$$

$$= 3750 \text{ ppm}$$

The tensile stress in the copper σ_{cu} can be estimated from Hooke's law:

$$\sigma_{cu} = \varepsilon \times E = 3750 \times 10^{-6} \times 16 \times 10^{6} \tag{10}$$
$$= 60,000 \text{ psi}$$

where E is Young's modulus of elasticity.

This is the worst case calculation because α_{cu} (thick film) is less than α_{cu} (bulk) and some of the tensile stress will be relieved by plastic flow in ductile areas of the copper film. A more reasonable value would be in the order of 25,000 psi.

The important point here, however, is that the stress generated is considerable and areas of copper, weakened or embrittled by poor sintering or the presence of bismuth will be subjected to repeated high tensile stress levels on thermal cycling. Since both mechanisms are weakened in tension, the combination of thermal processing and cycling can result in open circuit failures of the copper conductor.

5.2 Shorting Across Dielectric Layers

R. Lund[15] described copper shorts which developed across the dielectric layers in multilayer interconnect substrates. These shorts formed during processing and were not the result of copper migration in the presence of moisture and electric field. Physically, the shorts consisted of thin copper films, spheroidal copper precipitates and rounded copper masses which Lund called protuberances. Fig. 7, taken from Lund's paper, is a SEM photograph of a typical copper protuberance. The resultant short from one of these protuberances is shown in Fig. 8.

FIGURE 7

Copper Short (Protuberance) Extending Upward from the First CMS Metal Layer

FIGURE 8

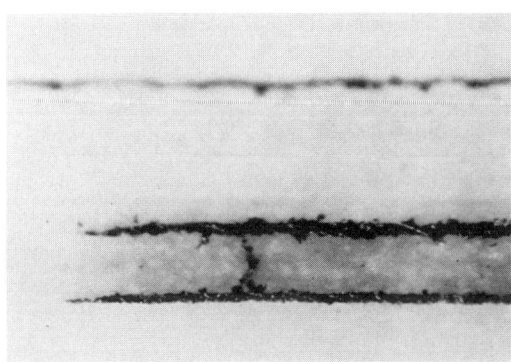

A number of observations in the production of multilayer interconnect boards are considered relevant in the formation of shorts during the firing process. 1) The incidence of shorting increased substantially with hermetic dielectric lots. 2) The problem was worse with thinner dielectric layers. 3) Shorts increased with higher belt loadings and higher production volumes. 4) Poor nitrogen flow rates and patterns were found to be a factor. 5) Shorts often showed up in lower levels of the substrate well after the layer had been fired and successfully tested. 6) The position of shorts was mostly associated with areas of heavy metal deposits such as between ground and power planes or originating from blind vias to lower signal lines. 7) Shorts appeared to increase with the use of oxygen injection in the burnout zone of the furnace.

With regard to these observations, the following two mechanisms are potential causes of shorting problems. Both are dependent upon entrapped carbon associated with heavy deposits of copper metal.

Bacher and Siuta[3] proposed that the formation of copper protuberance shorts could be explained by the simultaneous reduction of Cu_2O (glass) and PbO (glass) by entrapped carbon or CO. As described previously and verified experimentally[16], Cu_2O can dissolve into the dielectric glass up to approximately 2 weight percent. Cu_2O is a constituent of the binder in the copper conductor and forms on the surface of the copper film during firing. Significant amounts of Cu_2O are, therefore, available to diffuse to the conductor-dielectric interface where it will dissolve in the glass. At high temperatures Cu_2O and PbO in the glass of the dielectric can be reduced by C, CO or H_xC_y to metallic $Cu°$ or $Pb°$.

Blushing[9], a rosy discoloration of the dielectric, is caused by copper precipitates which form at the interface of the dielectric layers due to localized reducing conditions, e.g., entrapped carbon between the dielectric layers. Precipitates formed in this manner between dielectric layers are submicron in size and do not occur in sufficient numbers to pose a shorting problem. Reduction of Cu_2O in the glass, however, accounts for the copper films and spheroidal precipitates found by Lund in shorted dielectric areas. They are formed by the same reduction reactions responsible for the blushing phenomenon, except they form in void areas within the dielectric.

The simultaneous precipitation of $Cu°$ and $Pb°$ will form a eutectic Cu-Pb alloy as shown in the phase diagram of Figure 4. Above 327°C, a small amount of liquid phase is present which causes rounding and coalescence of the copper precipitates into protuberances via liquid phase sintering. Furthermore, any reaction of the molten Cu-Pb alloy with the dielectric to form silicates, spinels or other compounds will reduce the interfacial energy and enhance wetting of the dielectric void surface. This is expected to favor the growth of continuous copper shorts across the dielectric layer.

Another mechanism, proposed here by the present authors, also originates with entrapped carbon. In areas of heavy copper deposits such as blind vias depicted in Figure 9, a large amount of Cu_2O from the copper formulation is available for dissolution into the dielectric glass. In cases where hermetic dielectrics are used to build the circuit, the dielectric sinters early in the firing process and traps carbon from incomplete burnout of the vehicle system. Reacting with Cu_2O at high temperatures, the carbon precipitates copper close to the via fill and the resultant carbon monoxide gas creates pressure and seeks to escape by the path of least resistance. Since the dielectric and via fill above the reaction point is densely sintered, the carbon monoxide attempts to escape sideways and downwards. At these temperatures, the dielectric glass is relatively fluid, allowing the gas to create interconnected porosity to its escape route. Since, the copper/dielectric interface is porous, the carbon monoxide will preferentially form an escape route to a signal line. As the carbon monoxide flows through the dielectric, it reacts with Cu_2O at the surface of the pore to create copper precipitates. As Cu_2O is removed from the dielectric at the surface of the pore, more Cu_2O diffuses into the region to maintain concentration equilibrium. The escaping carbon monoxide continues to precipitate copper at the surface of the pore until a continuous film from the via is created. Copper then diffuses from the via to reduce the surface energy of the precipitated film. Eventually enough copper diffuses from the via to form a hard short.

FIGURE 9

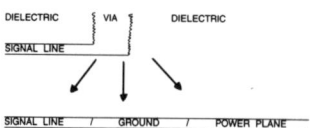

This mechanism provides an explanation of why oxygen addition to the nitrogen in the burnout zone apparently leads to increased shorting tendency. The oxygen oxidizes the surface of any exposed copper and creates a large reservoir of Cu_2O. Added oxygen also accelerates the sintering of copper and the dielectric; therefore promoting carbon entrapment in local areas of heavy copper deposits. As proposed, this situation can lead to growth of copper shorts.

5.2 Blistering and Localized Porosity

Entrapped carbon inhibits sintering of the dielectric and can, therefore, lead to high localized dielectric porosity. When this entrapped carbon reacts with oxygen donors to form gases, blisters can be formed if the escape route of the gas is impeded. Blisters can occur deep in the interconnect board and are formed by gas creating a bubble in the softening glass at high temperatures. Often the blister can extend to the surface during repeated firing of the board.

Usually, the hermeticity and blister density of a copper/dielectric multilayer are directly related. For example, decreasing the peak temperature reduces blistering, but the hermeticity of the dielectric is also reduced[17]. Therefore, process conditions generally involve a balance of hermeticity and blistering conditions.

6.0 NEW MATERIALS CHEMISTRY

Firing related failures of copper multilayer interconnect boards are associated with incomplete burnout of organics. Reliable substrates can be manufactured under carefully controlled process conditions designed to ensure complete burnout of organics in the 300-550°C temperature range[3,10]. However, next generation materials need to be more tolerant of poor processing.

Current generation materials have been improved by the addition of a dedicated via fill composition. The new product eliminates the tendency to form shorts from vias to signal lines through reducing conditions in the furnace. Low surface area, spherical copper powder of closely controlled particle size allows for control of sintering and shrinkage properties. The copper via fill composition sinters later in the firing cycle; therefore allowing burnout of organics to proceed to completion. The material is free of Cu_2O and PbO additives. This reduces the tendency to form copper precipitates, low melting eutectics and high pressure gaseous byproducts. Ceramic additives with low temperature coefficient of expansion, are also present to reduce the mismatch of expansion coefficient with the dielectric.

Next generation materials are currently being designed. Crystallizable glasses with softening points above 700°C are expected to form the basis of the dielectric. Glasses of low viscosity eutectic compositions with rapid crystallization kinetics are needed to ensure good densification and complete devitrification during the firing cycle. The chemistry will be reduction resistant to carbon or carbonaceous gases.

The copper conductor compositions will be free of reducible species such as Bi_2O_3, PbO and Cu_2O and will also be designed to allow for improved burnout of organics. Spherical copper, higher softening point glasses and new vehicle systems with improved burnout properties will also be utilized. By use of these concepts, it is hoped that copper material systems will grow to be the preferred material system of the future.

REFERENCES

1.   W. P. Brown and C. F. H. Tipper, J. Appl. Polymer Sci. 22, 1459-1468, (1978).

2. W. D. Kingery, H. K. Bowen and D. R. Uhlmann, Introduction to Ceramics, 2nd edition, John Wiley & Sons, Inc., New York (1960, 1976), p. 492.

3. R. J. Bacher and V. P. Siuta, in Proc. 36th Electronic Components Conf. (IEEE, New York, 1986).

4. J. R. Larry, R. M. Rosenberg, and R. O. Uhler, IEEE Trans. CHMT-3 (2), 211-225 (1980).

5. R. W. Vest, Ceramic Bulletin 65 (4), 631-636 (1986).

6. P. J. Holmes and R. G. Loasby, Handbook of Thick Film Technology, (Electrochemical Publications, Ltd., Glasgow, 1976).

7. R. G. Loasby, N. Davey and H. Borlow, Solid State Technology, May 1972.

8. R. W. Vest, "The Physical Chemistry of Copper Thick Film Firing", ISHM Symposium, Dallas, TX, 1984.

9. D. P. Button and V. P. Siuta, "Copper Interactions in Thick Film Ceramics: Dielectric Blushing and Performance", Proc. of the Materials Research Society Symposium: Electronic Packaging Materials Science, V. 72, Ed. by K. A. Jackson, Materials Research Soc., Pittsburgh, PA, 1986.

10. P. L. Toch, J. Goodrick and B. A. Shaw in Proc. 36th Electronic Component Conf. (IEEE, New York, 1986)

11. R. E. Honig, "Vapour Pressure Data for the More Common Elements", RCA Review, Vol. 18, No. 2, pp 195-204, June, 1957.

12. Langmuir, Phys. Rev., Vol. 8, 149 (1916).

13. E. Voce and A. P. C. Hallowes, "The Mechanism of Embrittlement of Deoxidized Copper by Bismuth", J. Inst. Metals, Vol. 73, pp 323-376, 1947.

14. M. Schneider, Hughes Aircraft, Unpublished work.

15. R. Lund, "Circuit Shorting Mechanisms in Copper Multilayer Systems", Proc. International Symposium on Microelectronics, ISHM Conference, Anaheim, CA, November 11-14, pp 463-471, 1985.

16. D. E. Pitkanen, J. P. Cummings and J. A. Sortell, "Compatibility of Copper/Dielectric Thick Film Materials", Proc. 1979 ISHM Symposium, Los Angeles, CA, pp 148-156.

17. C. R. S. Needes and A. R. Travis, "The Processing of Thick Film Copper Multilayer Materials in a Fast Firing Furnace", Proc. International Symposium on Microelectronics, ISHM Conference, Anaheim, CA, November 11-14, pp. 454-462, 1985.

18. E. A. Hayduk and B. M. Adams, Proc. 1987 ISHM Conference, p. 569.

MATERIALS AND PROCESSES FOR HIGH FUNCTIONALITY HYBRID CIRCUIT PACKAGES

Donald Jaffe, AT&T Bell Laboratories, 555 Union Boulevard, Allentown, PA 18103

ABSTRACT

Materials and processes used for high functionality hybrid circuit packages based on ceramic substrates are described. Emphasis is on hybrid circuits used for telecommunications and related applications. Examples include packages utilizing thin film technology, thick film technology and combinations of the two technologies. Various thick and thin multilayer approaches to achieve high interconnection density are discussed. These employ a variety of metal conducting systems in conjunction with glasses and/or polymeric materials as dielectric insulating layers.

INTRODUCTION

This paper presents a review of materials and processes for high functionality hybrid circuit packages based on ceramic substrates. The emphasis is on hybrid circuits used for telecommunications and related applications. The topics which will be treated are:

1. Background

2. Overview of high functionality (multilayer) hybrid packages.

3. Specific Thin Film Multilayer: POLYHIC

4. Future Trends

BACKGROUND

A definition of a hybrid integrated circuit and a list of the main characteristics of the two basic types, thick film and thin film are given in Table I.

TABLE I

HYBRID INTEGRATED CIRCUITS (HIC'S)

Definition
  A microelectronic circuit consisting of several components (passive and/or active, integrated or appliqued) on a ceramic substrate.

Basic types

    o Thick film HIC's

— Conductors, resistors and capacitors formed on a ceramic substrate using thick film processing techniques i.e. print, dry and fire.

    o Thin film HIC's

— Conductors, resistors and capacitors formed on a ceramic substrate using thin film processing techniques i.e. vacuum deposition, photolithography and anodization.

OVERVIEW OF HIGH FUNCTIONALITY PACKAGES

The primary role of high functionality hybrid circuit packages is to provide the most cost effective overall solution to a particular electronic application. Often the hybrid is most cost effective when there are critical requirements relative to performances (speed), size and/or weight.

The interconnection hierarchy of a typical telecommunications apparatus mainframe is shown in Figure 1 (Ref. 1).

INTERCONNECTION HEIRARCHY

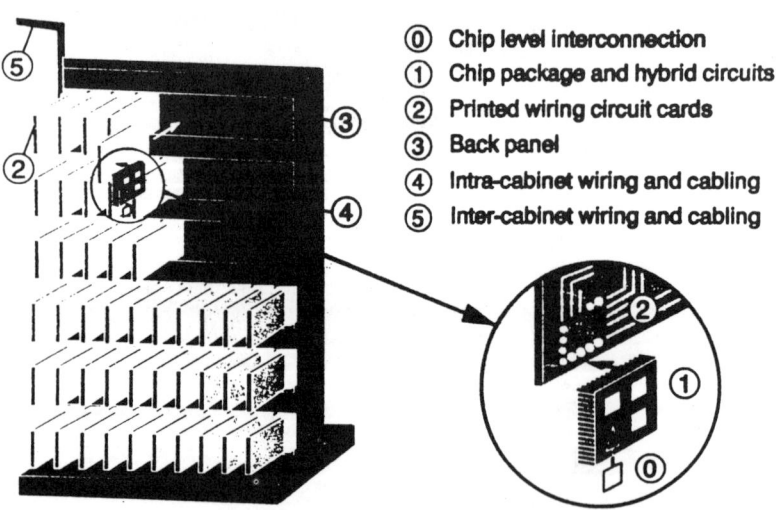

⓪ Chip level interconnection
① Chip package and hybrid circuits (HICS)
② Printed wiring circuit cards
③ Back panel
④ Intra-cabinet wiring and cabling
⑤ Inter-cabinet wiring and cabling

Figure 1

The cost of an interconnection (Ref. 2) is shown as a function of distance from the center of a silicon chip for a typical telecommunication application in Figure 2.

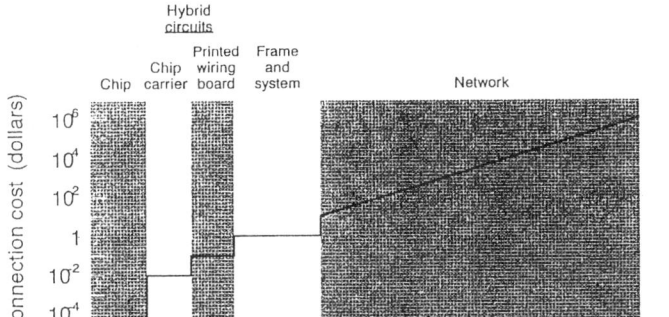

COST OF AN INTERCONNECTION*

* John S. Mayo, "Materials for Information and Communication," Scientific American, October, 1986

Figure 2

It is obvious that cost effectiveness is improved by higher functionality hybrids which have higher interconnection densities.

There are many approaches to achieving high functionality packages. A listing of several multilayer types and their major features is given in Table II.

TABLE II

GENERIC TYPES OF HIGH FUNCTIONALITY (MULTILAYER) HYBRID PACKAGES

Package type	Typical base substrate	Typical metallizations	Typical dielectrics	Typical dielectric constants
Thin film	Alumina (99.5%)	TiPdAu CrCu TiPdCuNiAu	Triazenes Polyimides	2.5 to 3.5
Thick film (sequentially fired)	Alumina (96%)	AuPdAg, Au, Cu	Glass	7-10 4.0 (special)
Thick film (co fired)	None	W, Mo, Cu	Alumina 90-92%	9-10

Combinations of the above

An example of a sequentially fired thick film multilayer can be found in Ref 1. There are many references which describe thick film co fired circuit materials and processes both with refractory and non-refractory metal systems.

SPECIFIC THIN FILM MULTILAYER - POLYHIC

Recent work at AT&T has lead to the development of a new thin film multilayer, high functionality hybrid called POLYHIC. Details are presented in Reference 3. Tables III, IV and V and Figures 3, 4 and 5 are from Reference 3.

The POLYHIC is an extension of the thin film hybrid technology which has been used extensively by AT&T for telecommunications applications. A review of this technology is given in Reference 4. The primary new feature of the POLYHIC is the use of a specially developed, photodefinable triazine based polymeric material as the interlevel insulator. A sample polymer formulation is given in Table III.

TABLE III

SAMPLE POLYMER FORMULATION

Ingredient	Function
Triazine resin	High crosslink density High glass transition temp. Good solvent
Rubber resin	Toughness Adhesion
Acrylated epoxy	Toughness Photosensitivity
N-vinyl pyrrolidone	Hardener
Trimethylolpropane triacrylate	Photo-crosslinker
Glycidoxypropyl-trimethoxy silane	Adhesion
2,2-dimethoxy-2-phenol acetophenone	Photoinitiator

Properties of the cured Polymer are given in Table IV. The glass transition temperature and thermal stability values are compatible with the fabrication and assembly processes used for POLYHICS.

TABLE IV

PROPERTIES OF CURED POLYMER

Property	Value
Glass transition temp.	150°C
Thermal stability	180°C
Thermal conductivity	0.2W/mK
Dielectric constant (ASTM D150)	3.6
Dielectric strength (ASTM D149)	36V/μm (900V/mil)
Volume resistivity (ASTM D257)	>3.7 x 10^{12} ohm-cm
Surface resistivity (ASTM D257)	>4.25 x 10^{12} ohm-cm

The major physical and materials features of POLYHICS are summarized in TABLE V. The Ti Pd Cu Ni Au Conductor System provides high conductivity capabilities (.0025 to .005 ohms per square) which are required for many hybrids for telecommunication applications. Also, the fine line capabilities noted for both the resistors and conductors permit high interconnection densities to be achieved.

TABLE V

PHYSICAL AND MATERIALS FEATURES OF POLYHICS

Feature	Description or value
Substrate	Alumina ceramic
Resistors	Ta N (on ceramic)
Conductors	Ti Pd Cu Ni Au
Linewidths	Resistors (>25μm) Conductors (>50μm)
Vertical interconnections	Vias in polymer (150μm)
Polymer thickness	50μm
External leads	Solder or TC bond
Applique devices	Beam lead Die and wire Solder Tab

306

A drawing of a conceptual POLYHIC is shown in Figure 3.

MULTILAYER POLYMER HIC [CONCEPTUAL]

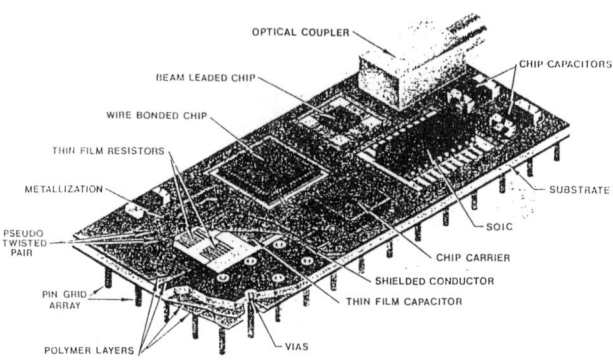

Figure 3

The capability to bring many diverse devices together on a single circuit should be noted.

Figure 4 outlines the major steps used in the fabrication of POLYHICS. As shown in Figure 4, the formation of the vias in the polymer is achieved by photolithographic techniques.

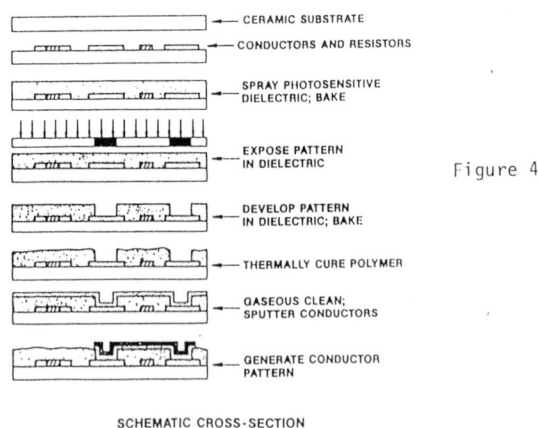

Figure 4

SCHEMATIC CROSS-SECTION
OF POLYHIC FABRICATION PROCESS

A photograph of a completed high functionality POLYHIC is shown in Figure 5. It should be noted that on a hybrid circuit only 1.2 inch x 1.6 inch in size, it is possible to combine 5 silicon chips (two in chip carriers, 3 die/wire bonded) and 50 thin film resistors. As noted, a major reason for this compact size is that all 50 thin film resistors are on the ceramic under the polymer.

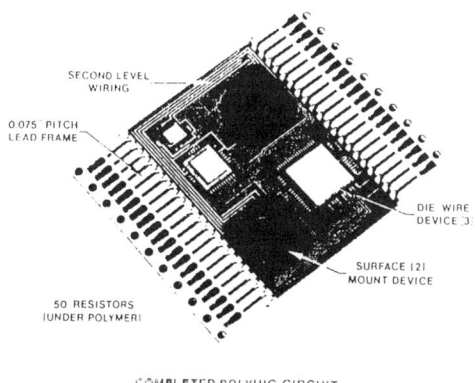

COMPLETED POLYHIC CIRCUIT

Figure 5

FUTURE TRENDS

In my view, the major future trends in high functionality hybrids are:

Finer lines and spaces

More layers.

Higher thermal conductivity substrates
(e.g. Aluminum Nitride)

More and finer pitch external I/O leads
including surface mountable types.

CONCLUSION

High functionality hybrid circuit packages offer demanding challenges in many materials and process areas.

These include:

- Thick Film metals
- Thin film metals
- Ceramics and glasses
- Polymeric films
- Vacuum deposition
- Photolithography
- Electrochemistry
- Screen printing
- Precision furnace firings.

REFERENCES

1. Giffels, C. A. et al, Interconnection Media, AT&T Technical Journal, July/August, 1987, pp. 31-44.

2. Mayo, J. S., Materials for Information and Communication, Scientific American, October, 1986, pp. 59-65.

3. Shiflett, C. C. et al, High-Density Multilayer Hybrid Circuits made with Polymer Insulating Layers (POLYHIC's), Proceedings 1986 ISHM Symposium, Atlanta, GA, October, 1986, pp. 481-486.

4. A History of Engineering and Science in the Bell System, Electronics Technology (1925-1975), F. M. Smits, Editor, Chapter 9.

MATERIALS AND STRUCTURES FOR HIGH DENSITY I/O INTERCONNECTION SYSTEMS

S. HONG, J. C. BRAVMAN, T. P. WEIHS AND O. K. KWON*
Dept. of Materials Science and Engineering, Stanford University, Stanford, CA. 94305
Dept. of Electrical Engineering, Stanford University, Stanford, CA. 94305*

ABSTRACT

In order to examine the use of a compliant cantilever structure as a contact scheme for a Multi-chip Interconnection System (MIS), multi-layer (metals and SiO_2) cantilever beams were fabricated utilizing standard silicon processing and micromachining technologies. The mechanical behavior and electrical characteristics of the beams were investigated in order to establish their optimum dimensions for use in the MIS. During the course of this study, a new mechanical testing method for thin films has also been developed, which makes use of the same cantilever beam structure and a "Nanoindenter." The Young's modulus and yield strength of thermally grown SiO_2 and Au were measured using this technique.

INTRODUCTION

A variety of micromechanical devices have been fabricated during the last 15 years. These include pressure sensors [1], accelerometers [2], gas chromatographs [3], and heat sinks [4]. The three-dimensional structures which are the basic building blocks of micromechanical devices - trenches, cantilevers, etc. - can be precisely fabricated via anisotropic etching of single-crystal silicon. We have investigated the use of one of these structures, the cantilever beam, as the basis for a contact scheme for a Multi-chip Interconnection System (MIS) for VLSI circuitry [5]. In this approach, the integrated circuit (IC) chip is inverted over the beams in order to establish electrical contact between the chip's I/O pads and the top metal surface of the cantilever beams. The system configuration, which makes use of established microcapillary die-attach [6,7] and microchannel heat sink [4] technologies, is schematically illustrated in Fig. 1.

This paper discusses the fabrication, mechanical behavior and electrical characteristics of multilevel cantilever beams as they relate to use in an MIS, as well as a description of a new method for testing the mechanical behavior of thin film materials.

FABRICATION PROCEDURES FOR MULTI-LAYER CANTILEVER BEAMS

The techniques for fabricating cantilevers over a trench are well established and have been extensively used in various micro-sensor applications [1,2]. These techniques exploit the tendency of anisotropic etchants to undercut convex shapes in a mask with respect to the exposed silicon. Since multiple etching of high indexed planes is involved in the undercutting at a convex

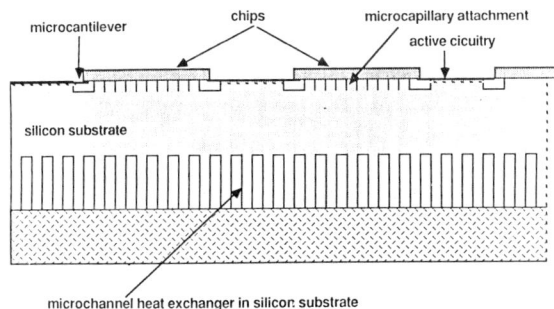

Figure 1. Multi-chip Interconnection System (MIS) configuration (components not to scale)

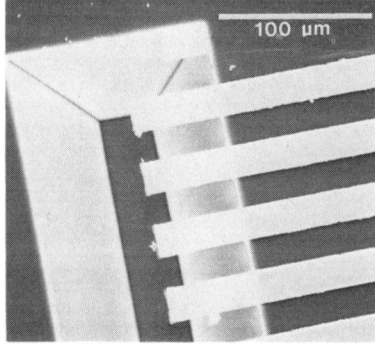

Figure 2. Top view of the multi-layer canti-lever beams.

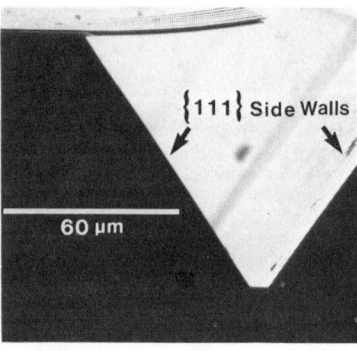

Figure 3. Side view of multi-layer canti-lever beams.

region, undercutting proceeds relatively quickly, with the etch rate higher than the etch rate of {100} plane. After a {111} plane is exposed (e.g., beneath the attached end of a beam) the etch rate drops rapidly. The specific experimental procedures we have developed are as follows:

(1) A layer of wet-thermal SiO_2 1.5 μm thick was grown on a lightly doped P-type {100}Si wafer at 1100°C.

(2) The SiO_2 layer was patterned by standard photolithography processes and a wet etchant of buffered HF. To minimize the undercutting at the concave corners of the final structure, the straight sides of the mask pattern were aligned parallel to the <110> directions of the substrate.

(3) Photoresist was applied and baked at 75°C for 25 minutes. After exposure to a UV light source, it was soaked in a chlorobenzene hardener for 5 minutes in preparation for "lift-off" metal patterning.

(4) Thin films of metals (Cr, Au) with various thicknesses were deposited sequentially on the prepatterned photoresist by evaporation at room temperature, with a deposition rate of ~50 Å per minute. A typical thickness for the Cr adhesion layer was 200 Å. These films were patterned as the photoresist underneath was removed in warm acetone.

(5) Finally, the exposed silicon was etched in an anisotropic etchant, and then rinsed and dried. For EDP (66 w/o ethylenediamine, 23 w/o DI water, 11 w/o pyrocatechol) at 115 °C, the etch rate of {100} silicon is ~50 μ/hr, while that of silicon dioxide is extremely low (~180 Å/hr). Although difficult to measure, the etch rates of Cr and Au in EDP are believed to be comparable to that of SiO_2.

Figures 2 and 3 are SEM photographs of multi-layer cantilever beams which were fabricated using these procedures. The length, width and thickness of the beams in both figures are 60 μm, 20 μm and 2.5 μm, respectively. The spacing between adjacent beams is 20 μm. Thus, the total number of the cantilevers around the periphery of a 1 cm² standard chip site is ~1,000. Figure 4 shows a top view of an MIS structure after mounting an inverted test chip on its site.

DESIGN CONSIDERATIONS FOR THE CANTILEVER BEAMS IN THE MIS

Mechanical Parameters of Thin Films

In order to predict the mechanical behavior of the cantilever beams more precisely, the fundamental mechanical parameters (e.g., Young's modulus, yield stress, etc.) of the constituent thin films must first be determined. Evaluation of these parameters has been carried out using a special sub-micron testing machine, known as a "Nanoindenter." This device measures the depth of penetration of a diamond tip into a material as a function of applied force, just as a conventional hardness tester does, but does so with very high resolution: the maximum resolution of force and displacement change are about $0.25\mu N$ and 4Å, respectively. To determine the elastic parameters of the thin films used in this study (SiO_2 and Au), however, the Nanoindenter was used to measure the *deflection* of a cantilever beam, rather than the depth of penetration, in a simple bending test: see Fig. 5. [The depth of penetration of the indenter tip is quite small compared to the amount of deflection of the beam, and can be accounted for.] The cantilever beams tested here were single layered, and were fabricated with similar but simpler procedures than those used for multi-layer beams (described above). The elastic deflection of a cantilever beam with a rectangular cross section is given by [9]

$$\delta = \frac{Pl^3}{3EI} = \frac{4P}{wE}\left(\frac{l}{t}\right)^3 \tag{1}$$

where P is the force applied, I is the moment of inertia ($I = wt^3/12$ for a rectangular cross section), w is the width, t is the thickness, and l is the effective length of the cantilever beam. Since the deflection varies linearly with the force, the Young's modulus of the beam can be determined from the slope in elastic region. Using plate theory, a small correction was added to equation (1) in order to account for the point loading of the beam [8], although it was found that this correction was small for the beam geometries used in this study.

Figure 6 shows a typical load/deflection curve of an SiO_2 beam obtained with the Nanoindenter. The initial linear section of the data represents only the loading of the indenter tip which is supported by springs. After a small displacement, the diamond tip contacts the SiO_2 beam causing it to deflect. From that point, the elastic spring constants for both the machine and the beam support the total force at a given displacement. To account for this the spring constant for the instrument was subtracted from the total slope in order to calculate the Young's modulus of the beam, which was found to be 65 GPa. Several oxide beams were elastically deflected to depths of about one-half of their effective lengths. The elastic strains corresponding to these deflections were as high as ~1%. We believe such high strength is due to the differences in microstructure between thin film and bulk materials (e.g., the absence of defects, such as cracks and voids). The high strength of thermal oxide provides a good mechanical support in a multi-layer beam structure used in MIS. With the same procedures and analysis, the elastic modulus of gold was found to be 73 GPa. The yield stress of Au was determined by observing the onset of plastic deformation, and was found to be 320 MPa. More detailed experimental procedures and results will be reported separately [8].

The Curvature and Bending Mechanics of Multi-layer Cantilever Beams

In general, multi-layer beams will be curved (see Fig. 3). The curvature arises from two main factors: the difference in thermal expansion coefficients of the metals and the SiO_2, and the intrinsic stresses generated during the deposition process. Thus, the radius of curvature varies considerably with changes in deposition conditions, such as substrate temperature, pressure, deposition rate, the mechanism of deposition method, etc. The bending mechanics for an initially curved beam, along with the analysis of the residual stress state in the films, forms the basis for predicting the maximum elastic deflection of the cantilever beam. This is an important design parameter for the beams when they are used in the MIS packaging scheme.

For an initially curved beam, as in the case of a multi-layer beam, the deflection and resisting force of a cantilever beam can be calculated using Castigliano's theorem [10]. In this theory, the deflection is given by

$$\delta = \frac{\partial U}{\partial P} = \frac{\partial}{\partial P}\left[\int_0^L \frac{M^2}{2EI}ds\right] \tag{2}$$

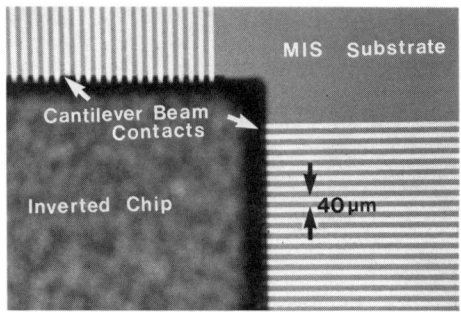

Figure 4. Top view of an MIS structure after
mounting an inverted test chip on its site.

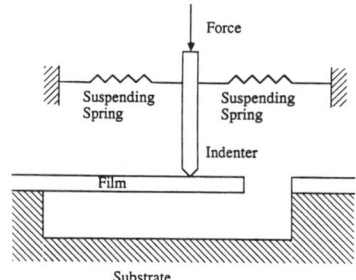

Figure 5. Schematic diagram of a bending
test using a Nanoindenter.

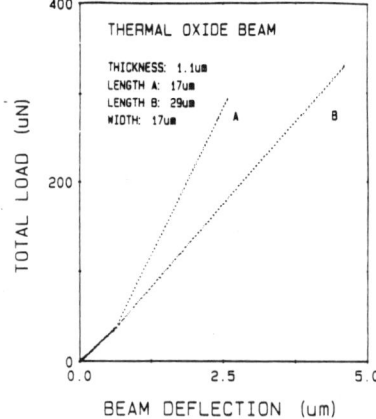

Figure 6. Small deflection/load curves for a
thermal SiO$_2$ beam.

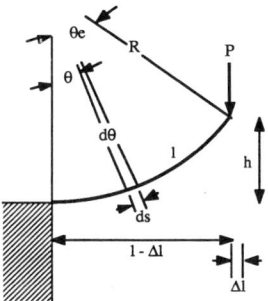

Figure 7. Schematic illustration of the deflection
of an initially curved beam.

where U is the internal strain energy, ds is an incremental arc length (ds = R dθ, see Fig. 7).
The moment, M, varies along the length of a curved beam, and can be expressed in terms of the
angles

$$M(\theta) = PR(\sin\ \theta_e - \sin\ \theta)$$ (3)

where θ_e is the angle corresponding to the full length of a beam. Then the deflection equation
(2) can be written as follows:

$$\delta = \frac{PR^3}{EI}\left[\sin^2\theta_e \cdot \theta_e + 2\sin\ \theta_e \cdot (\cos - 1) + \frac{\theta_e}{2} - \frac{\sin\ 2\theta_e}{4}\right]$$ (4)

In order to determine the initial radius, R, the deflection at the tip is measured by an optical
microscope focusing the edge and tip areas. If we assume that the curvature is uniform along the
length and that it is circular in shape, as shown in Fig. 9, the radius is related to the curved
height, h, at the tip as follows:

$$R = \frac{l}{2\tan^{-1}\left(\frac{h}{l-\Delta l}\right)} \cong \frac{l}{2\tan^{-1}\left(\frac{h}{l}\right)}$$ (5)

where l is the total length of a beam.
For the calculation of the residual stress state in each film, we adopted the general
expressions for the curvature and the stresses in a multi-layer beam, which were developed by P.
H. Townsend et al. [11] :

$$K = \frac{1}{R} = \frac{\sum\limits_{i=1}^{n} E_i\gamma_i\frac{t_i}{2}\left[-\ln d_i + \frac{\sum\limits_{j=1}^{n}E_j t_j \ln d_j}{\sum\limits_{j=1}^{n}E_j t_j}\right]}{\sum\limits_{i=1}^{n}E_i t_i\left[\left(\frac{\pi t}{2} - \frac{t^3}{3}\right) + (t-\pi)\frac{\gamma_i}{2} - \frac{1}{12}(3\gamma_i^2 + t_i^2 + t^2)\right]}$$ (6)

and

$$\sigma_i = E_i \left[-\ln d_i + \frac{\sum_{j=1}^{n} E_j t_j \ln d_j}{\sum_{j=1}^{n} E_j t_j} + (\pi - y)K \right] \tag{7}$$

where K and R are the initial curvature and radius of a beam, d_i is the length of the ith layer in its relaxed state, π is the distance between the neutral axis and the bottom of the beam, y is the distance from the bottom, and t is the total thickness. The coefficients γ_i and β_{ij} are defined as follows:

$$\gamma_i = \sum_{j=1}^{n} \beta_{ij} t_j \tag{8}$$

and

$$\beta_{ij} = -1 \text{ (if } i > j), \ 0 \text{ (if } i = j), \ +1 \text{ (if } i < j) \tag{9}$$

These equations were derived with the assumption that the individual layers are deformed in a purely elastic fashion (to obtain zero resultant force and net moment), and that the normal strain varies continuously throughout the total thickness. The normal stresses in each film thus vary linearly. However, this model considers neither any microscopic strain variations at the interface nor any atomic motion that could cause a local plastic deformation. This implies that the strain and stress distributions predicted by this model could differ from their actual values. It is therefore reasonable to take the average stress value in each layer, and to use these numbers for the deflection and force calculations: the initial residual stresses are additive to the stresses developed in each layer during bending.

Design Considerations

During the bending of a simple cantilever beam, the maximum normal stresses, both in tension and compression, always occurs at the top and bottom of a beam. However, we are only concerned with the stress state on the top layer (conducting layer) in our structure because of the high strength of the underlying SiO_2 film. Au was chosen as a conducting metal because of its high resistance to corrosion and low contact resistance. To calculate the maximum deflection of a curved beam within the elastic limit, the maximum force, P_{max}, should be substituted in Equation (4). The maximum force can now be obtained through following relationships:

$$\sigma_{yield} - \overline{\sigma}_2 = \frac{M_{max}(y - \pi)_{max}}{I_{effective}} = \frac{P_{max} R \sin \theta_s (t - \pi)}{I_{effective}} \tag{10}$$

where $\overline{\sigma}_2$ is the average residual stress in the conducting film, and $I_{effective}$ is the effective moment of inertia for a multi-layer beam.

For typical metal-SiO_2 beams (thickness of Au: 5μm, thickness of SiO_2: 1.5μm, width: 20μm, lengths: 50μm, 100μm, 150μm) maximum deflections were calculated and plotted as a function of the initial radius in Fig. 8. For the application of this structure in the MIS, the beams must be reasonably flexible. In other words, the tolerance of the deflection on bending should be appreciable, say a few microns. This could be achieved by choosing a proper deposition method and by increasing the effective length of a beam. As seen in the same figure, reducing the thickness of the metal layer also leads to similar results. However, an increase in the effective length and/or a reduction of the metal thickness must be considered from the viewpoint of electrical performance. First, the resistance and inductance of a cantilever beam rise as either the length increases or the thickness decreases. Secondly, the maximum upward force generated during chip mounting decreases (see Fig. 9). The contact resistances between the top surface of a cantilever beam (Au) and the pad in a chip (Au in our test die) are expected to increase accordingly. Figure 10 shows the plot of contact resistance as a function of the force applied. These data were generated using the Nanoindenter and a specially designed indenter tip, and were in good agreement with published data [12]. The actual contact area was 15μm x 15μm x 15μm (equi-triangular shape). From this plot, it is clear that contact resistance converges to

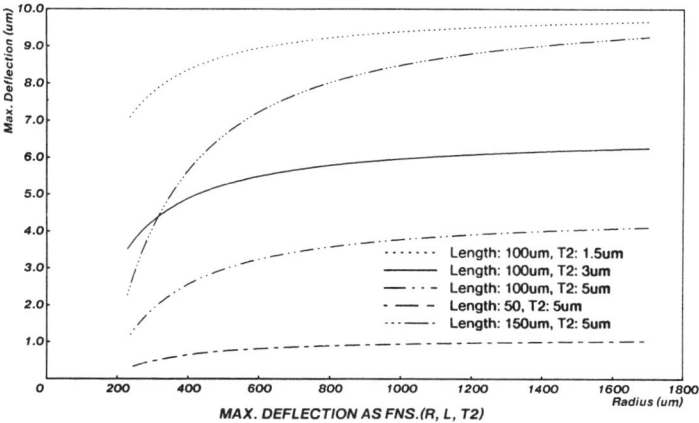

Figure 8. Maximum deflection within elastic limit was calculated as a function of initial radius, beam length, and the metal layer (Au) thickness. The width of the beams and the SiO_2 layer thickness (1.5 μm) are the same for all cases.

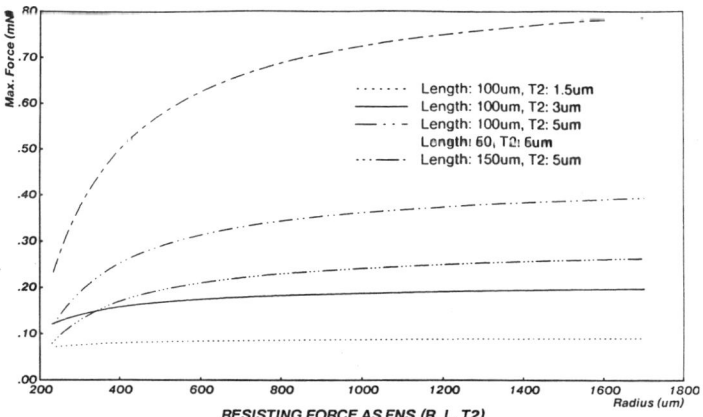

Figure 9. Maximum force corresponding to the maximum deflection within the elastic limit was calculated as a function of initial radius, beam length, and the metal layer (Au) thickness. The width of the beams and the SiO_2 layer thickness (1.5 μm) are the same for all cases.

Table 1. Comparison between the electrical parameters for the cantilever beam structure and for the other conventional packaging techniques.

	Max. lead capacitance* (pF)	Max. lead Inductance*,† (nH)	Max. lead Resistance (Ω)
Cantilever Beams‡	7.34×10^{-3}	0.02	0.01
TAB□	-	0.7	0.01
Ceramic DIP	7	22	1.1
Plastic DIP	4	36	0.1
Pin Grid Array	2	7	0.2

From C.A. Steidel, Chapter 13 in VLSI technology, Ed. S.M. Sze, McGraw-Hill Book Company, New York, 1983

‡ Dimensions: width; 20 μ, length; 100 μ, thickness; 1.5 μ SiO_2 + 5.0μ Au , spacing; 20 μ

* For the ceramic packages, the C and L are for a line 0.5 mm above ground plane.

† Includes bond wire, 1.25 nH.

□ TAB: 1 mil x 3 mil x 50 mil

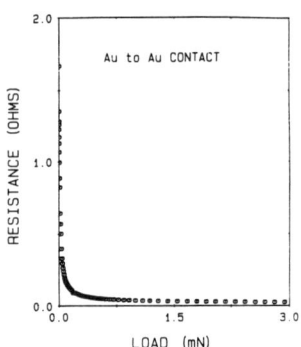

Figure 10. Au to Au contact resistance as a function of applying force. Contact area: 15 μm x 15 μm x 15 μm (equitriangular shape)

below 0.2Ω at a force above 0.3 mN. These values (< 0.2Ω) are adequate for signal transmission in most applications. On the other hand, as the metal thickness increases, the initial curvature and flexibility of a beam tend to decrease. Thus, there is a trade off between the mechanical and the electrical properties of the cantilever structures, and careful selection of the materials, deposition conditions and thicknesses of the beam layers is essential.

The calculations on the capacitance and inductance values of the cantilever beam were carried out using the 3-D computer simulation program [13]. The results show that these values are far less sensitive to the thickness of the conducting layer than is the contact resistance. Therefore, the length and thickness of the beams should be designed such that the contact resistance and flexibility of the beams are optimized first. Table 1 lists the electrical parameters (L, C, R) of cantilever beams along with those for conventional packaging technologies. For a typical cantilever beam (thickness of Au: 5μm, thickness of SiO_2: 1.5μm, width: 20μm, lengths: 100μm), the inductance, capacitance, and resistance values are ~20 pH, $~7.3 \times 10^{-3}$ pF, and 10 mΩ, respectively. The assumption made for these calculations are that power/ground lines are

10% of the total I/O leads. Since the L and C values of this structure are 2-3 orders of magnitude smaller than those for the standard packaging technologies available today, the "ΔI noise" and crosstalk at high frequencies will be reduced accordingly: hence, electrical performance at high frequencies will be significantly improved. The electrical performance can be further enhanced by the proper distribution and design of the power/ground lines.

SUMMARY

In order to examine the use of the cantilever structure as a contact scheme in a multi-chip interconnection system, multi-layer (metals and SiO_2) cantilever beams were fabricated utilizing standard I.C. processing and micromachining technologies. The number of I/O which can be achieved by this structure is ~1000 per 1 cm^2 chip site. The maximum deflection and uprising force of these beams within the elastic limit were calculated as functions of the length, the thickness of the metal layer, and the initial radius of the beam. A typical beam (thickness of gold: 5μm, thickness of SiO_2: 1.5 μm, width: 20μm, length: 100μm) with a moderate initial curvature (radius: 1100 μm) can bend ~4 μm without plastic deformation and produce ~0.4 mN uprising force. Au to Au contact resistance measurements, as a function of force applied, were carried out employing the Nanoindenter. The data show that a force greater than ~0.3 mN produces a contact resistance below 0.2Ω, which should be adequate for signal transmission. Therefore, this approach could be an effective packaging scheme for the high performance I.C. systems in the near future, which require very high speed as well as high I/O pin counts. Lastly, mass production is possible utilizing standard I.C. processing methods.

A new technique for testing the mechanical properties of thin film materials, using a cantilever beam structure and the Nanoindenter, has been developed with other collaborators during the course of this study [8], and was utilized to determine the mechanical parameters of SiO_2 and Au thin film materials. The elastic constants were measured at 65 GPa and 73 GPa for thermally grown SiO_2 and gold, respectively. The yield strength of the gold film was 320 MPa.

ACKNOWLEDGEMENTS

The authors wish to thank to R. F. W. Pease for fruitful discussions. The financial support of the Center for Integrated Systems at Stanford University and of the Semiconductor Research Corporation is gratefully acknowledged.

REFERENCES

[1] Samaun, Ph.D. Dissertation, Stanford University, 1971.
[2] W. D. Frobenius, S. A. Zeitman, M. H. White, D. D. O'Sullivan, and R. G. Hamel, IEEE Trans. Electron Devices, vol. ED-14, p.117,1972.
[3] S. C. Terry, Ph.D. Dissertation, Stanford University, 1975.
[4] D. B. Tuckerman and R. F. W. Pease, IEEE Electron Device Letters, vol. 12, p. 126, May 1981.
[5] R. F. W. Pease, J. C. Bravman, O. K. Kwon, S. Hong, S. C. Douglas, B. W. Langley, and A. F. Paal, Proceedings of 1987 Advanced Research in VLSI Conference, Stanford, March, 1987, pp. 279-292.
[6] D. B. Tuckerman and R. F. W. Pease, Technical Digest of the VLSI Symposium, Sept. 1983.
[7] A. Paal and R. F. W. Pease, IEEE IEMT Symposium Proceeding, pp. 169-172, Sept. 1986.
[8] T. P. Weihs, S. Hong, J. C. Bravman and W. D. Nix, To be published.
[9] S. P. Timoshenko and J. M. Gere, Mechanics of Materials. New York: Van Nostrand, 1972, p. 516.
[10] F. B. Seely and J. O. Smith, Advanced Mechanics of Materials. New York: John Wiley & Sons, 1952, p.442-445.
[11] P. H. Townsend, D. M. Barnett and T. A. Brunner, To be published.
[12] M. Antler, IEEE Circuit and Devices Magazine, vol. 3, #2, pp 8-20, March, 1987
[13] O. K. Kwon and R. F. W. Pease, Proceedings of 1986 IEEE IEMT Symposium, San Francisco, Sept. 1986, pp. 34-39.

A STUDY OF THRESHOLD AND INCUBATION BEHAVIOUR DURING ELECTROMIGRATION IN THIN FILM METALLISATION

C. A. ROSS AND J. E. EVETTS
Department of Materials Science and Metallurgy, University of Cambridge, Pembroke Street, Cambridge CB2 3QZ, UK

ABSTRACT

We have previously described a microstructural model for electromigration in thin film metallisation, based on the concept of a stress-driven backwards flux opposing the forwards electromigration flux. The model can be used to calculate the build-up of stress in a metallisation track subject to arbitrary variations in electromigration flux, and to predict both the location of electromigration damage and incubation times before its appearance. In this paper we present an experimental study of electromigration behaviour in thin aluminium films on a niobium substrate and relate the results obtained to the model.

INTRODUCTION

Electromigration is the transport of the atoms of a metallic conductor under the influence of a high electric current density. It is important in integrated circuits because it can cause failure in the thin film metallisation, which can sustain the necessary high current densities (typically 10^{10} Am^{-2}) because the silicon substrate acts as a heat sink for the Joule heating developed. Electromigration damage consists of voids, which form at places from which conductor atoms are depleted, and hillocks or whiskers which form where atoms accumulate. Integrated circuit metallisation, consisting of aluminium alloyed with silicon and copper, fails either by the coalescence of voids or by short circuits caused by hillocks or whiskers.

In aluminium films, mass flux is found to occur mainly via the grain boundaries[1], and the average 'steady state' ionic velocity v_{ave} over the cross-section is given by

$$v_{ave} = \frac{\rho Z^* e}{kT}(j - j_c)\frac{\delta}{d}D_o e^{-Q/kT} = \alpha(j - j_c) \tag{1}$$

where ρ is the resistivity, $Z^* e$ the effective charge on the ion (e is the electronic charge), k the Boltzmann constant, T the temperature, δ the grain boundary width, d the grain size, j the current density, j_c the threshold current density, D_o the grain boundary diffusion coefficient preexponential and Q the activation energy for grain boundary diffusion. Direct measurement of v_{ave} has been made[2-6] using a geometry consisting of blocks of test metallisation on a refractory underlayer; an electric current in the underlayer is taken up by the test metallisation which drifts on the underlayer at a velocity of v_{ave}.

The threshold has been interpreted[3,4,6] as arising from a stress-driven backwards flux of ions which can balance the forwards electromigration flux if j is sufficiently small. As material is transported from the cathode to the anode end of the stripe, a long range stress gradient builds up which drives a 'mechanodiffusion' flux from regions of compression to regions of tension. Below the threshold, where the electromigration and mechanodiffusion fluxes balance,

$$lj = \frac{\Omega \Delta \sigma}{Z^* e \rho} \tag{2}$$

where l is the stripe length, Ω the atomic volume and $\Delta \sigma$ the stress difference between the ends of the stripe. At threshold, when $j = j_c$, $\Delta \sigma$ represents the maximum stress difference which can be sustained between the ends of the stripe, and is of the order of the yield stress of the film. The current density-threshold length product $(lj)_c$, denoted by β, is constant for a metallisation material and has a slow temperature dependence since the yield stress varies with temperature.

The concept of a stress-driven backflow has been extended[7] to sources of flux divergence within a metallisation track. A divergence dipole, defined as consisting of a source and sink of flux separated by a length l, can arise as a consequence of microstructural or temperature inhomogeneities. It gives rise to a local stress field which drives a backwards flux; this stress field is additional to the long-range stress field existing in the stripe. A metallisation track will contain a distribution of divergence dipoles of different sizes and strengths, and the stress fields developed around all these dipoles are superposed on the long-range stress. The total stress reached at any point determines what type of damage (*i.e.* hillocks or voids) forms, or whether a dipole remains quiescent. This allows us to predict both the incubation times before damage forms and its location, assuming that damage nucleates once specific values of stress or strain are reached in the film. We can also explain why the inevitable dipolar small-scale flux divergences present in metallisation do not give rise to damage, because they are below their threshold conditions. This paper presents evidence in support of this model by comparing experimental results on test metallisation to computer simulations.

EXPERIMENTAL WORK

Samples based on a bilayer metallisation were fabricated using sputter deposition and lithography techniques. Finished samples consisted of variously sized stripes of aluminium on a continuous niobium underlayer. C-plane sapphire (cut parallel to (0001)) of size 12mm x 3mm x 0.5mm was used as the substrate and a layer of niobium followed by a layer of aluminium was deposited using UHV getter sputter deposition[8]. In each case the aluminium and the niobium were of similar thickness, which ranged from 0.07μm to 0.5μm. The films were made into finished samples using a 1:1 lithography system; aluminium was etched in $20H_3PO_4$:$1HNO_3$:$1CH_3COOH$ acid mixture and niobium was plasma etched in CF_4/8%O_2 gas. The films were annealed at 450°C for 30 minutes under 0.5 atm argon, generally before any fabrication was performed.

Testing was carried out on a probe test rig with a heated stage, typically at 200°C. Formation of hillocks and voids was observed *in situ* using optical microscopy. Temperature was determined from a thermocouple and also the resistance of both the sample and test meanders, so self-heating could be assessed. Temperature control is important in these experiments since drift velocity varies exponentially with it, so the temperature during a test was stabilised by mounting the sample on a block of copper mantained at constant temperature by a heating element. Film thickness was measured using a Talysurf profilometer and by interferometry.

The film microstructure was assessed by measuring resistance ratios to 77K and also by using X-ray diffraction. As-deposited aluminium films had a zone 1/zone T microstructure[9] consisting of a dense array of fibrous grains with lengths of the order of 60 nm perpendicular to the film, and aspect ratios of approximately 2, with a preferred (111) orientation in the plane of the film. Annealing leads to grain growth[10] so that the final grain size is of the order of the film thickness.

RESULTS

An example of a test structure is given in Fig. 1(a), which shows drift in a row of annealed aluminium stripes of different lengths, including one which is below the threshold length. The variation of drift length with the original stripe length is clearly visible. Fig. 1(b) is a tilted scanning electron micrograph of another sample which was tested unannealed, and Fig. 1(c) is a thinner annealed sample showing hillocks which have nucleated in rows parallel to the current flow.

(a) Drift velocity as a function of time

Equation (2) gives the critical current density j_c for a uniform block of metallisation. As the cathode end of the stripe drifts at v_{ave} and material accumulates as hillocks at the anode end, the overall stripe length l decreases and the backflux increases, so the stripe velocity decreases. The variation of stripe length l with time t is given by

$$j\alpha t = (l_o - l) - l_c \ln\left(\frac{l - l_c}{l_o - l_c}\right) \tag{3}$$

where l_o is the initial length and l_c the threshold length corresponding to a current density of j, *i.e.* $l_c = \beta/j$. It is seen that l approaches l_c at long times.

Fig. 2 shows the drifted length $(l_o - l)$ plotted against time for four aluminium stripes of different initial lengths, tested in series at 200°C and a current density of 1.1×10^{10} Am^{-2}. The aluminium thickness was 0.07 μm and the width 40 μm. The drift velocities tend to zero as the stripe lengths decrease towards l_c. It should be noted that the velocities obtained are higher than literature values[3] by approximately one order of magnitude, due presumably to the very fine grain size in these films. A test on a thicker (0.35 μm) sample gave a drift velocity lower by a factor of 10 than that seen in Fig. 2.

The value of l_c can be obtained from a plot of stripe velocity against $1/l$, which has intercepts of $1/l_c$ and αj on the axes. Fig. 3 shows this plot for a stripe of original length $l_o = 150$ μm; this gives $l_c = 21 \pm 3$ μm, $\alpha = (4.4 \pm 0.3)$ 10^{-19} A^{-1}m^{3}s^{-1} and $\beta = (2.4 \pm 0.4)$ 10^5 Am^{-1}, in

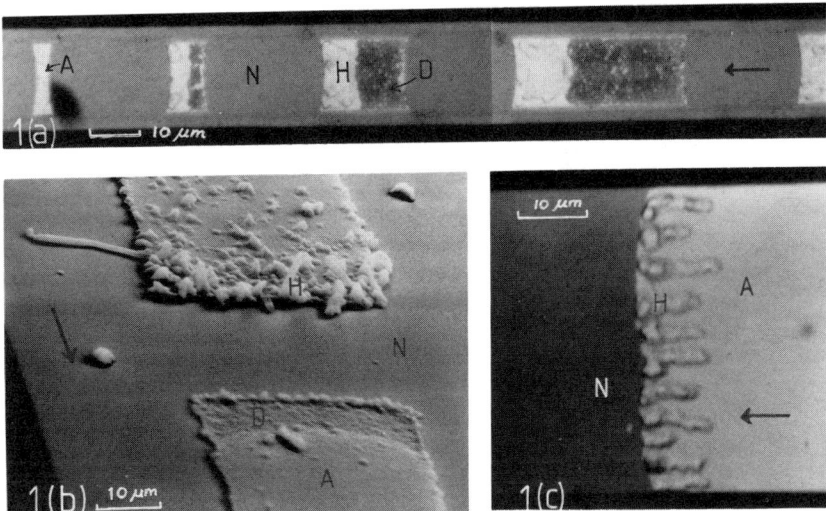

Figure 1. Test structures. In each case, A marks the aluminium, N the niobium underlayer, D the depleted region and H the hillocked region. The arrow shows the direction of current flow.

(a) Optical micrograph of a 0.35 μm annealed sample after 60 hours at 1.5 10^{10} Am^{-2} and $T \approx 150°$C. Note that the shortest stripe has not drifted: l_o is below the threshold length.

(b) Tilted scanning electron micrograph of a 0.35 μm unannealed sample tested for 10 hours at 5 10^{10} Am^{-2} and 150°C. A whisker can be seen growing horizontally from the edge of the stripe.

(c) Optical micrograph of a 0.07 μm annealed sample tested for 2 hours at 10^{10} Am^{-2} and 200°C. Hillocks have grown in rows parallel to the current flow.

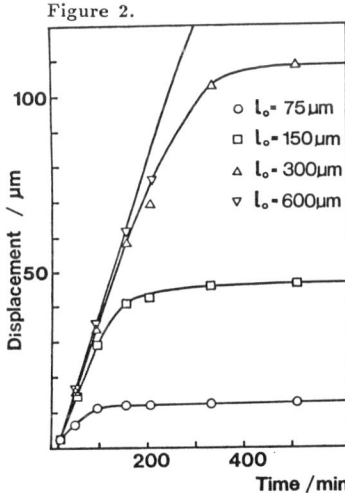

Figure 2. Edge displacement as a function of time for four stripes of different l_o tested at 1.1 10^{10} Am^{-2} and 200°C.

Figure 3.

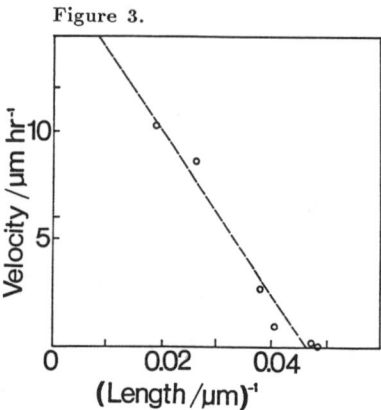

Figure 3. Edge velocity of the stripe of $l_o=150\mu$m from Fig. 2 plotted against $1/l$.

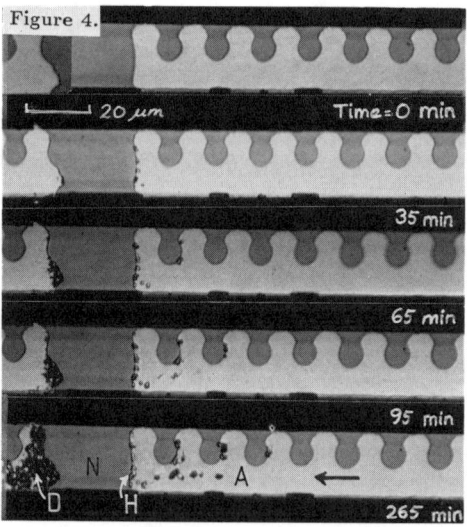

Figure 4. Optical micrographs of part of
a notched stripe after different times at 9 10^9
Am^{-2} and 200°C. Annotations are as in Fig. 1.
Notice the sequential nucleation of the hillocks
at the notches in the anode end of the right-
hand stripe, and drift of the cathode end of the
left-hand stripe.

agreement with a literature value[3]. The longer stripes tend towards a larger asymptotic length because the large amount of material which builds up at the anode end reduces the current density and thus increases the effective l_c.

(b) Behaviour of artificially introduced flux divergences

The damage which forms at a site of flux divergence depends not only on the strength and size of the dipole but on its position with respect to the long-range stress gradient. This has been demonstrated for divergences in the form of artificial voids introduced in a stripe[11].

The effect of artificial divergences in the form of notches in the aluminium has been studied experimentally. Fig. 4 shows a notched aluminium stripe after different testing times at $9.3 \ 10^9 \ Am^{-2}$ and $200°C$. The notches act as sites of flux divergence because the flux-carrying ability of the stripe is reduced by approximately 10% as part of the current is diverted into the niobium underlayer. (The current density at the notch is higher so the notch region is expected to be hotter, leading to an enhanced electromigration flux. However, the high thermal conductivity of the sapphire substrate reduces these temperature fluctuations so the reduction in flux-carrying ability is the dominant factor.) Hillocks are therefore observed at the start of the notch; as the stress builds up in the stripe the notches reach the hillock nucleation condition sequentially. Hillocks were seen at the anode end within 10 min, before forming at the notches, then formed earliest at the notches near the anode end.

(c) Incubation times and the response to reversed current

An incubation time before the start of testing and the start of drift of a stripe has been observed[5] and related to the build-up of stress in the film. Incubation times before the formation of hillocks and voids were determined by *in situ* observation of samples under test. In general, hillocks were observed before the displacement of the cathode end began, for instance in the sample of Fig. 1, hillocks were seen to form within 5 minutes of current stressing at $1.1 \ 10^{10} \ Am^{-2}$, though there was an incubation of 30 minutes before displacement of the cathode end started. However, the incubation for visible edge displacement is longer than that for void formation; material had been depleted from the cathode end before edge displacement was observed. The cathode end had a slanted profile extending over a few microns after 5-10 min test, as revealed by dark-field optical microscopy.

Tests on reversed and pulsed currents can be compared directly to model predictions. A $1020 \ \mu m$ long, $0.35 \ \mu m$ thick and $20 \ \mu m$ wide aluminium stripe was tested at $200°C$ at a current density of $3 \ 10^9 \ Am^{-2}$ until hillock nucleation was observed. Hillocks formed between 30 and 40 min after current stressing was started; at this point edge displacement of the cathode end of the stripe had not begun. The current was then reversed and hillocks formed at the original cathode end 48 to 58 min later. This demonstrates that stressing in one direction reduces the compressive stress at the cathode end so the formation of hillocks there is delayed when the current is reversed. This result will be discussed further below.

DISCUSSION

A computer model to calculate the evolution of stress and strain in a one-dimensional metallisation track containing specific sites of flux divergence has previously been described[7]. A block of test metallisation of length l corresponds to a divergence dipole of length l; smaller length dipoles can be superposed on this to represent sites of flux divergence due to structural or temperature inhomogeneities.

It is important to note that the model describes the build-up of stress and strain before the nucleation of damage, *i.e.* in a sub-threshold configuration. Once damage has nucleated, the subsequent behaviour of the metallisation will depend on the values of stress and strain at the sites of damage formation, and on the perturbing effects on temperature and current density of changes in cross-section, caused by, for instance, a $2 \ \mu m$ high hillock growing from a $0.5 \ \mu m$ thick film. Damage formation can thus lead to extra sources of flux divergence and the development of further damage; we have observed that hillocks often grow in rows parallel to the current flow, indicating preferential nucleation of hillocks behind each other with respect to the electron flow (Fig. 1(c)).

An example of a model calculation is shown in Fig. 5, in which the stress build-up is shown for a stripe of length $100 \ \mu m$ containing divergence dipoles of length $20 \ \mu m$ separated by distances of $10 \ \mu m$. The parameters used are typical of aluminium tested at $200°C$ and $10^{10} \ Am^{-2}$ and the internal dipoles represent regions in the stripe in which the flux is halved. After 3000 s the development of the long-range stress gradient between the ends of the stripe has started, with superposed short-range stress fluctuations resulting from the smaller dipoles. The overall flux has

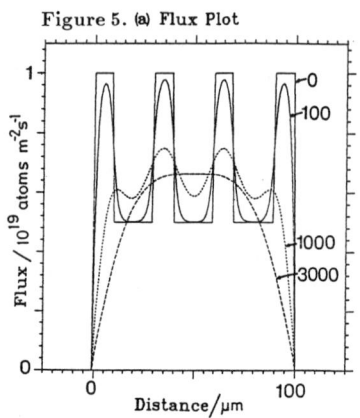

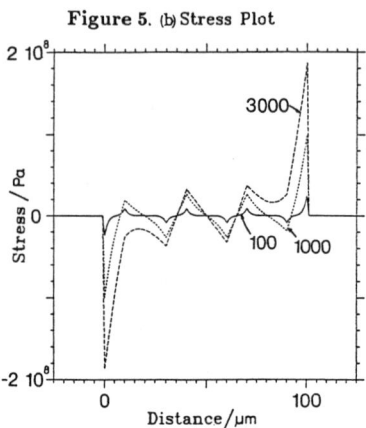

Figure 5. Computer simulation of a 100μm stripe containing 20μm regions of reduced flux-carrying ability. The flux and stress are shown at different times; an oscillating stress pattern develops as the flux smooths out. The numbers represent times in seconds.

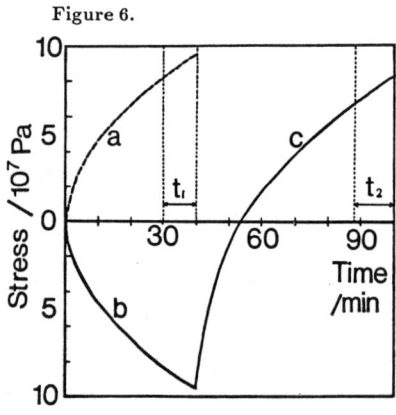

Figure 6. Computer simulated stress build-up in a stripe subject to 3×10^9 Am^{-2} at 200°C. (a) shows the stress at the anode end and (b) the stress at the cathode end. In the time interval t_1, hillocks nucleated at the anode end. On reversing the current, the stress at the original cathode end follows (c), and hillocks nucleated in the interval t_2. The stresses reached within t_1 and t_2 are similar.

been smoothed by the addition of a backwards mechanodiffusion flux, so the flux variations caused by the small dipoles are no longer discernable. The nature of damage forming at a particular site of flux divergence is seen to depend on the position of the divergence in the long-range stress gradient developed in the stripe, because this will determine whether or not the hillock or void nucleation conditions are satisfied more easily. Note that in this simulation the stress which builds up has not been limited, so the entire configuration of flux will become quiescent eventually. In the experiment of Fig. 4 the internal dipoles could not be made quiescent without reaching hillock nucleation conditions so hillocks were observed.

Incubation times before damage formation can be predicted from the model. Fig. 6 shows how the stress at the end of the 1020 μm stripe described above changes with time during 40 mins forward current and 60 min reverse current. After annealing at 450° C, the biaxial stress in the aluminium at the test temperature would be expected to be tensile, of approximately 5 10^8 Pa, but this level of stress will be relaxed by creep to 10^7 Pa or less[12]. With appropriate values of aluminium parameters, the compressive stress at the anode end of the stripe after 30-40 min testing is 8-10 10^7 Pa, which represents the model hillock nucleation stress. Reversing the current for 48-58 min gives a compressive stress at the other end of 7-8 10^7 Pa, a similar value, and at that point hillocks had formed there.

CONCLUSIONS

We have discussed the electromigration behaviour of aluminium stripes on a refractory substrate in terms of a microstructural model. The importance of the model is its prediction of the evolution of stress and strain in a microstructure subject to an inhomogeneous electromigration flux; the location of electromigration damage and the incubation time before it forms can be calculated, and we have shown that model values compare favourably with experimental results. This has implications for integrated circuit design, since the effects of inhomogeneities (such as changes in cross-section or microstructure when a track passes over a feature on the substrate) can be modelled, and the circumstances under which electromigration damage would occur can be predicted.

ACKNOWLEDGEMENTS

This work was supported by the Science and Engineering Research Council and the General Electric Company Hirst Research Centre. The authors would like to thank R.E. Somekh for film deposition, A.L. Greer, A.P. Schwarzenberger, Z. Barber and R.E. Somekh for constructive discussion, and P.V. Evans for advice on heat treatment.

REFERENCES

1. F.M. d'Heurle and P.S. Ho, Electromigration in thin films, Chapter 8 of *Thin Films - Interdiffusion and Reactions* ed. J.M. Poate, K.N. Tu and J.W. Mayer (John Wiley and Sons Inc., New York, 1978)
2. I.A. Blech and E. Kinsbron, *Thin Solid Films* **25** 327 (1975)
3. I.A. Blech, *J. Appl. Phys.* **47**(4) 1203 (1976) and *Erratum, J. Appl. Phys.* **48**(6) 2648 (1977)
4. I.A. Blech and Conyers Herring, *Appl. Phys. Lett.* **29**(3) 131 (1976)
5. E. Kinsbron, I.A. Blech and Y. Komem, *Thin Solid Films* **46** 139 (1977)
6. I.A. Blech and K.L. Tai, *Appl. Phys. Lett.* **36**(8) 387 (1977)
7. C.A. Ross and J.E. Evetts, *Scripta Metall.* **21**(8) 1077 (1987)
8. R.E. Somekh and C.S. Baxter, *J. Crys. Growth* **76** 119 (1986)
9. J.A. Thornton, *Ann. Rev. Mater. Sci.* **7** 239 (1977)
10. S.S. Iyer and C.Y. Wong, *J. Appl. Phys.* **57**(10) 4594 (1985)
11. A.P. Schwarzenberger and A.L. Greer, presented at the 1987 MRS Fall Meeting, Boston, MA, 1987 (published in these Proceedings, Paper I4.20)
12. H.J. Frost and M.F. Ashby, *Deformation-Mechanism Maps* (Pergamon Press, 1982)

ASYMMETRIES IN THE FORMATION OF ELECTROMIGRATION DAMAGE AROUND DIVERGENCE DIPOLES IN A METALLIZATION TRACK

A.P. SCHWARZENBERGER AND A.L. GREER
University of Cambridge , Department of Materials Science and Metallurgy,
Pembroke Street, Cambridge, CB2 3QZ, U.K.

ABSTRACT

Electromigration damage mechanisms around lithographically introduced notches in unannealed and annealed test tracks are studied.

INTRODUCTION AND THEORY

Electromigration is mass transport driven by electric current. It is significant only at current densities greater than about $1\times10^{10}Am^{-2}$. Continuing miniaturization of electronic devices with corresponding increase in the current density in the metallization tracks, has led to the reappearance of electromigration (previously controlled by alloying) as a significant failure mechanism.

Simple theory [1] predicts that the atomic flux $\mathbf{J}$ is linear with the electric field $\mathbf{E}$, and is given by

$$\mathbf{J} = D \left(-\nabla N + \frac{N}{kT} z^* e\mathbf{E} \right) \qquad (1)$$

where D is the effective diffusivity, N the vacancy number density, T the absolute temperature, e the electronic charge and k Boltzmann's constant. For aluminium at normal temperatures we expect that the effective charge $z^* \approx 10$ and that grain-boundary diffusion will dominate.

Electromigration of itself will result only in mass transport, not in any damage to metallization tracks. Damage occurs only when there is a divergence in the electromigration flux $(\nabla \cdot \mathbf{J})$, given by

$$\nabla \cdot \mathbf{J} = D \left\{ -\nabla^2 N + \left(\frac{N}{kT} \right) z^* e\mathbf{E} \left(\frac{\nabla N}{N} + \frac{\nabla z^*}{z^*} - \frac{\nabla T}{T} \right) \right\} + \left(\frac{\nabla D}{D} \right) \mathbf{J}. \qquad (2)$$

Variations in N, z^*, T, and D are possible contributors to the total flux divergence. The main effect of T is through its effect on D, which may also be affected by changes in microstructure, mainly grain size. It is important to note that because the atomic flux is in strict linear proportion to the electron flux there will be no flux divergence arising directly from non-uniform current density.

A flux divergence will cause damage only when the resulting biaxial stress exceeds a critical value. For hillocks and whiskers we take this threshold to be the compressive yield stress [2,3]. The appropriate value for voids is discussed later in this paper.

It has been found that a significant component of damage accumulation is the growth and migration of voids [4]. In this work we study the mechanisms involved by making use of lithographic techniques to introduce controlled notches and voids into test structures. To investigate the effects of microstructure, both unannealed and annealed samples have been used. The results are interpreted in terms of the local microstructure and stress.

EXPERIMENTAL RESULTS

Pure aluminium, typically 0.5µm thick, was deposited by UHV DC magnetron sputtering onto oxidized silicon substrates. Sputtering conditions were such that the as-deposited films are in compression. Annealing to relieve stresses and coarsen the grain structure was at 450°C for 30 minutes in flowing Ar (to prevent oxidation).

Current stressing of annealed and unannealed samples was performed in air at selected nominal temperatures between 100°C and 200°C. Current densities were in the range 1 to $3\times10^{10}Am^{-2}$ causing, for example, a temperature rise of 12°C (for a current density of $1.6\times10^{10}Am^{-2}$) in the thin film test track with respect to the contact pads, which are assumed to be at, or near, the nominal oven temperature.

For an unannealed sample, hillocks are observed after current stressing near the positive terminal (mostly at the sides of the track) and voids near the negative terminal. This confirms that z^* for Al is positive, giving material accumulation where the temperature gradient (near the colder contact pads) is antiparallel to the potential gradient. However, at a notch (defined lithographically in the test track) the voids and hillocks form the other way around, i.e. with hillocks nearer the negative terminal. For example, voids are seen at A in fig. 1b whilst hillocks are seen at D. Also, voids and hillocks are observed at the extreme corners of the notch (at B and C on fig. 1b), unexpectedly where the current density must be lowest.

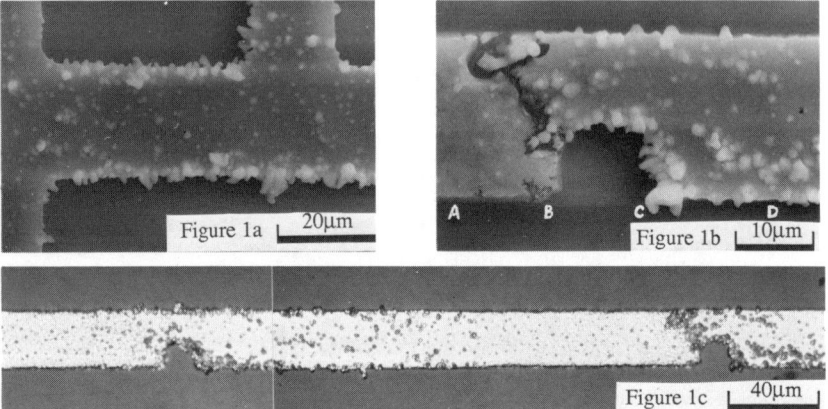

Figure 1a | 20μm |

Figure 1b | 10μm |

Figure 1c | 40μm |

Figure 1: Notches in an unannealed sample stressed at $1.8 \times 10^{10} Am^{-2}$ for 245 hours at 125 °C. In all photographs the electron flow is from right to left.
a) Scanning electron micrograph of the area of the track near the negative terminal.
b) Scanning electron micrograph of a notch.
c) Optical micrograph of the central region of the track, containing two notches.

When two notches are placed adjacent on a track, it is found that there is a sharp transition from voids to hillocks between them. For the pair of notches shown in fig. 1c, which were closer to the positive terminal, the transition is somewhat offset from the midpoint in the direction of the negative terminal.

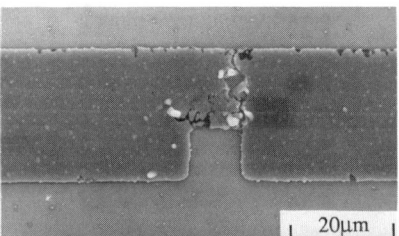

20μm

Figure 2: A notch in an annealed sample stressed at $1.8 \times 10^{10} Am^{-2}$ for 36 hours at 165 °C. The electron flow is from right to left.

When samples are annealed before current stressing, thermal hillocks are seen. Subsequent current stressing gives a markedly different result from the unannealed films. Very few hillocks are now seen, damage is mostly by voiding at the sides of the track. The voids and hillocks on either side of the notches are no longer seen (fig. 2). The effect of the notch is now to produce damage (mostly voiding) where the highest current density occurs. No damage is seen at the extreme corners of the notch.

DISCUSSION

The temperature difference between the test track and the bond pads provides a "divergence dipole" the length of the test track, as manifested by the hillocks and voids near the bond pads. As the atom flux evolves towards a steady state, the dipole gives rise to a stress gradient along the track [2]. Small scale microstructural divergence-dipoles (caused e.g. by grain boundary configuration) along the track may lead to perturbation of the stress, reaching either the hillock or void nucleation stress. This is most likely near the positive terminal for hillocks and near the negative terminal for voids.

Lithographically introducing a notch provides an additional significant divergence dipole, at a position of our choice in the test track. The notches used in this experiment approximately halved the track width but this does not, of itself, give a flux divergence. However, the current density will be doubled at the notch, giving four times the Joule heating per unit area, and so approximately four times the temperature rise compared to the rest of the test track. Experiment gave a 12°C temperature difference between track and oven indicating that the notch will be about 50°C hotter than the oven. This temperature rise gives a higher local diffusivity and resistivity, both of which would give voids towards the negative terminal – the opposite of what is observed in fig. 1b. There must be a local reduction in the diffusivity at the notch to get the observed effect. Knowing this we can interpret the positions at which hillocks and voids form by estimating the stress field in the film necessary to have a steady state atomic flux.

We approach this problem by considering the stress field that would be required to give the same steady state atomic flux, but in the absence of electric current. This is illustrated in fig. 3a. The line 'A' represents the steady state stress condition in a track with no notches. The zig-zag line 'B' represents the steady state stress on the 'front edge' of the track (as defined in fig. 3c), and curve 'C' represents the stress on the back edge of the track. The greater stress gradient in the notch is necessary both because of the lower diffusivity there and because of the higher flux density. The difference in stress, e.g. between points 'D' and 'E' on the two edges of the track is necessary to drive atoms around the notch. Note also that the stress gradient away from the notches is lower than in a track with no notches, giving a smaller total atomic flux due to the constriction at the notch.

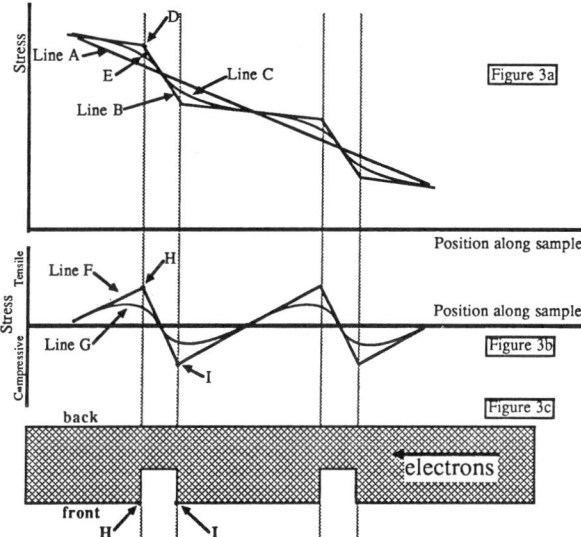

Figure 3: Schematics showing the stress distibution around a pair of notches. Details in text.

In fact the atomic flux is largely driven by the electron flux. So to determine the actual stress field for steady state atomic flux, we subtract from the stress field of fig. 3a the stress field which would give the total force on the atom due to the electric potential gradient (i.e. that due to the combined 'wind' and 'electrostatic' forces), so that the actual stress has decayed to zero away from the notches. Although the potential-gradient-equivalent stress field will also be disturbed by the presence of the notch, the disturbance is much less than that on the overall stress field which is largely affected by the local decrease in diffusivity. The actual stress field is shown schematically in fig. 3b where lines 'F' and 'G' represent the actual stress of the sample on front and back edges. This explains the voids on the negative terminal side of the notches as seen. It also shows that the maximum tensile stress is at a position such as 'H' and the maximum compressive stress is at a position such as 'I' (in figs. 3b and 3c) and damage would therefore be expected there although the current density is at a minimum. Fig. 3b also shows that a changeover from voids to hillocks is expected midway between the two notches. In fact the notches seen in fig. 1c are offset towards the positive terminal, and so the changeover is displaced to the right.

Since grain boundaries are the major diffusion path in aluminium at the temperatures used in this experiment, the lower diffusivity D at the notch could have arisen from a local increase of the grain size d, as we can consider $D=\delta D_{gb}/d$, where δ is the effective grain boundary width. The unannealed sample shown in fig. 1 was tested at a nominal temperature of 125°C which will give a test track temperature of 140°C and a notch temperature of up to 185°C. Comparison with grain growth studies [5] shows that 1 hour at 450°C (a 'complete' anneal) is equivalent to 2990 hours at 140°C and 390 hours at 185°C. Hillocks and voids had appeared within the first 120 hours of current stressing. Clearly in this time, the area around the notch will have annealed significantly, whilst the rest of the track will not have done.The grain size is expected to increase from 0.05μm to 0.5μm on annealing, giving a ten-fold reduction in the effective diffusivity. Another possible contribution, meriting further investigation is that electromigration of itself enhances grain growth. The mechanism of this enhancement would be the pressure due to the electromigration-induced atomic flux on the grain boundaries. This could move a boundary to a less stable configuration capable of rapid coarsening.

Considering the annealed samples, the compressive stress in the as-deposited film will be almost completely relieved at the annealing temperature. However, on cooling to the test temperature, the aluminium will be left in biaxial tension. The main form of damage seen is therefore in the form of voids. Now the effect of the notch is only to concentrate the current. Whilst there will still be the same temperature rise at the notch, this will not result in any further annealing, but will increase resistivity and diffusivity locally. These increases produced a partial reversal of the damage pattern. The effects seen in unannealed samples have disappeared and the annealed track fails by voiding at the position of highest J with only slight damage elsewhere.

Clearly the damage seen is very different in annealed and unannealed test tracks. Although we expect mainly voids and hillocks respectively, it is noticeable that in both cases damage has occurred mainly at the edges of the track, indicating a similar mechanism. We believe that this is due to small notch-like defects along the sides of the tracks acting in a similar manner to the larger lithographically introduced notches. These small notches are probably an artefact of the processing, especially from the etching stage, and may be unavoidable (e.g. due to grain boundaries).

Notches can be considered as half-voids. Preliminary results with unannealed tracks with lithographically introduced voids show a similar effect to notches. Whilst both notches and hillocks are observed to from in test tracks, it is apparent from many studies that more hillocks are seen than voids, although mass conservation must apply. It is suggested that the nucleation stress for voids is a zero rather than a tensile stress. This is supported by observations that general thinning of individual grains or the whole track can occur. Voids can also be mobile [4], particularly in multilevel metallization schemes and so extension of this work to voids may yield relevant results.

SUMMARY AND CONCLUSIONS

Qualtitative prediction of damage patterns in test tracks is readily achieved by considering the stress state necessary for steady state atom flow. This has been applied to notches in unannealed and annealed tracks, and demonstrates the importance of microstructural changes and configuration of track edges on damage initiation. The analysis can be applied equally to voids.

ACKNOWLEDGEMENTS

APS is supported by an SERC 'CASE' studentship with Plessey Research Caswell Ltd. The authors thank Prof. D.Hull for the provision of laboratory facilities, C.A.Ross and J.E.Evetts for many helpful discussions and R.E.Somekh for help with sample preparation.

REFERENCES

1 F.M. d'Heurle , and P.S. Ho, in Thin films - interdiffusion and reactions, (edited by J.M. Poate, K.N. Tu and J.W. Mayer), Wiley, New York, 1978
2 C.A. Ross and J.E. Evetts, Scripta Metall. **21** 1077 (1987)
3 C.A. Ross and J.E. Evetts, presented at the 1987 MRS Fall meeting, Boston, MA, 1987 (Published in these proceedings, paper I7.5)
4 R.W. Thomas and D.W. Calabrese, 21st Annual Proceedings on Reliability Physics, pp1-9 (1983)
5 S.S. Iyer and C.Y. Wong, J. Appl. Phys., **57** 4594 (1987)

AEM STUDY OF THIN AND THICK FILM METALLIZATION ON AlN SUBSTRATES

Alistair D.Westwood & Michael R.Notis,
Department of Materials Science and Engineering,
Whitaker Lab #5, Lehigh University, Bethlehem, PA 18015

Abstract

Microstructural characterization of thin film (Au-Pt-Ti) and thick film (Mo-Mn) metallization on AlN substrates has been performed using Transmission Electron Microscopy (TEM), Analytical Electron Microscopy (AEM), Convergent Beam Electron Diffraction (CBED) and Auger Electron Spectroscopy (AES). The reaction mechanisms for both types of metallization methods are proposed. In particular, the microchemical and morphological nature of grain boundary penetration and precipitation within the AlN near the metallization interface has been examined.

Introduction

Aluminum nitride (AlN) has emerged as a favourable substrate material for high performance substrates in hybrid circuits because of its potentially high thermal conductivity, $\approx 250 Wm^{-1}K^{-1}$[1]. Prior to the need for high thermally conducting substrates, the packaging industry almost exclusively used alumina ceramics with varying amounts of glassy phase in them. The thermal conductivity of these materials is $\approx 30 Wm^{-1}K^{-1}$[2].

These debased alumina substrates are easily metallised using the Mo-Mn process[3-7]. Glass migrates from the alumina into the metallizing layer forming a dense metal/glass composite. For metallizing to pure alumina substrates a glass phase is added to the Mo-Mn paste; the manganese glass penetrates the grain boundaries of the substrate and also forms a glass matrix interpenetrating the molybdenum metallization layer[3].

The successful use of AlN substrates in hybrid circuits requires that they be easily metallized. An understanding of the different bonding mechanisms which occur, as a result of the variety of metals and metallizing processes used, is important in producing a metallized layer with good adhesion properties. An important factor which appears not to have been addressed thus far is that the bonding processes used for metallizing AlN should ideally have a minimal effect on the thermal conductivity of the substrate.

All metallization studies of AlN published[2,8-12] to date have been phenomenological and have evaluated the metallization process by adhesion testing. No significant attempt has been made to microscopically analyse the metal ceramic interface and bonding mechanisms in thick or thin film configurations. A wide variety of materials and processes have been used[8,9]: co-firing of W, direct bonding of Cu, frited and fritless Ag/Pd, MoMn, Ni and RuO_2/glass resistor pastes and molten metal metallization pastes containing Ag, Cu, Ti and Sn have all been used for thick films. Thin films of Ti, Pt, Pd, Au, Ta_2N, NiCr, Mo and Ti-AgCu have been deposited by sputtering or evaporation.

The penetration of a glassy phase down the grain boundaries as in pure alumina, or the presence of a glassy phase at the grain boundaries as in debased alumina substrates would be detrimental to the thermal conductivity of the substrate because the glassy phase is a poor thermal conductor. It has been shown[10,11] that directly bonded copper metallization of AlN requires

that the AlN surface be oxidized which will decrease the thermal conductivity
at the interface region. These two forms of metallizing AlN must be carefully
controlled if its potential as a high thermal conductivity substrate is to be
maximized. Future work should concentrate on minimizing the effect that the
bonding mechanism has on thermal conductivity.

Materials and Experimental

Thin and thick film metallized substrates were obtained from Hitachi.
TEM and CBED[13,14] studies were conducted using the Philips EM400T
microscope operated at 120kV; AEM analysis was undertaken using Scanning
Transmission Electron Microscopy (STEM) and Energy Dispersive Spectroscopy of
X-Rays (EDS). AES was carried out on a Physical Electronics instrument.

Results and Discussion

Thin Film Metallization

The thin film metallization, comprised of gold, platinium and titanium
was ≈1μm thick. A cross-section through the thin film/ceramic interface can
be seen in Fig.1. TEM and AEM investigations revealed smooth planar
interphase interfaces with no visible grain boundary penetration into the
ceramic substrate. A Ti-rich microcrystalline layer ≈130nm thick was present
at the ceramic interface and which contained measureable amounts of Pt and
Al[15]. The Au-Pt layer on top of this Ti-rich layer was ≈900nm thick. AEM
showed Ti to be present in both the AlN and Au-Pt layers, in close proximity
of both these interfaces, demonstrating diffusion across both interfaces.

AES produced a better indication of the concentration gradient through
the deposited layers upto the metal/ceramic interface, Fig.2. The
microcrystalline Ti-rich layer was found to also contain nitrogen and oxygen.

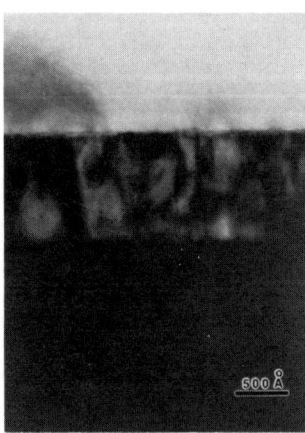

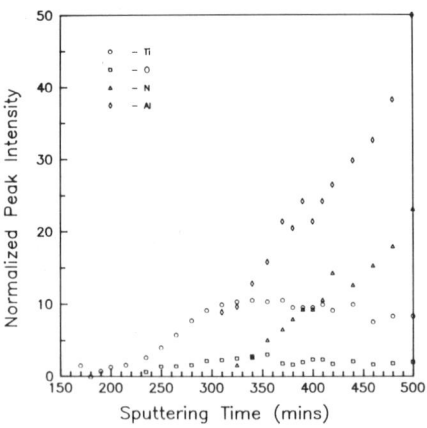

Fig.1. Micrograph of AlN/Ti
reaction layer/Au-Pt interfaces
in thin film metallized Hitachi
substrate.

Fig.2. AES composition profile across
Au-Pt/Ti reaction layer/AlN interfaces.

Nitrogen was present in the microcrystalline phase but not in the Au-Pt layer. The oxygen was associated with the Ti and was also found to have diffused into the Au-Pt layer. Al was observed to have diffused across the microcrystalline Ti-rich layer and into the Au-Pt layer. Pt was detected in the Ti-rich microcrystalline phase. Ti was found in the AlN, similar to AEM results[15]. The leveling out of the Al, N, O and Ti at the metal/ceramic interface at ≈400mins is believed to indicate compound formation Fig.2.

The bonding mechanism appears to involve interdiffusion across the metal/ceramic interface, consistant with previous published work which indicates that Al and N diffuse out of and Ti diffuses into the ceramic substrate[12]. Diffusion, reaction and compound formation result from post-deposition heat treatment.

Nicolet[16] has considered the thermodynamic basis for reaction between thin metal films and alumina substrates; a similar approach can be used to predict the nature of reaction products formed by thin metal film/AlN substrate interaction. Thermodynamic calculations indicate that the Gibbs free energy G_{298} for TiN (-73.6 kcal) is greater than that for AlN (-70.3 kcal).

Table 1 - Gibbs Free energy of relevent reaction products.
Reaction Products

Gibbs Free Energy	AlN	TiN	TiAl	TiAl$_3$	Al$_2$O$_3$	TiO	Ti$_2$O$_3$
G_{298}, kcal	-70.3	-73.6	-17.4	-34.0	-381.0	-115.9	-108.4

Therefore TiN would be expected to form at the Ti-metal/AlN ceramic interface. However, this would result in a high concentration of free elemental Al present at the interface which could form Al$_2$O$_3$ (-381.0 kcal) if oxygen were present, or TiAl$_3$ (-34.0 kcal) if excess free Ti were present. The reaction layer which formed at the Ti-metal/AlN ceramic interface is believed to consist of TiN plus Al$_2$O$_3$ and TiAl$_3$. Brow et al[17] have studied the reaction of Ti containing brazes and Ti on AlN; good agreement is found between the results of Brow et al and those proposed in this paper. Further experiments to identify the nature of this reaction layer and determine the kinetics of the reaction are now being carried out.

Thick Film Metallization

The Mo-Mn metallization layer on the Hitachi substrates was found to be ≈10-12μm thick and was generally consistant with the Mo-Mn metallurgy described in the literature[5-9]; in addition a significant amount of W was also found to be present. However, TEM/AEM studies showed the reaction at the interface to be far more complex than for alumina thick film reactions. Fig.3 shows a schematic view summarizing the observed microstructural features and microchemical variations through the metallized layer into the AlN ceramic.

The Mo-W layer was covered with a layer of crystalline Ni which in turn had a cap of microcrystalline Ni. It appears that the Mo and W powders sintered together to form a continuous upper layer; the W concentration is somewhat higher near the Ni interface and lower near the glass interface. The alumino-silicate glass also contains appreciable Mn and K, penetrates the regions of interconnected porosity present within the Mo-W metallization layer, Fig.4; protrusions from the metallization layer also extend into the glass region. Interpenetration of both layers is therefore extensive. A glassy region exists just below the Mo-Mn metallization, followed by a thin microcrystalline region adjacent to the AlN ceramic substrate. Crystalline regions were present, Fig.4(region B), in the glass layer. CBED measurements indicate the crystal structure to be orthorhombic. AEM measurements made in region B indicate the Al and Si peaks to be in the ratio of ≈3:2[15]. The pertinent Cliff-Lorimer ratio[18] was measured to be k_{AlSi}=1.12; this material is therefore identified as 3:2 mullite.

The major bonding mechanism appears to be by grain boundary penetration into the substrate, and with the formation of a complicated sequence of precipitates, Fig.5, along the grain boundary. Two different precipitate morphologies were present, Fig.5, a more common amorphous form, Fig.5(region A), and a crystalline form, Fig.5(region B). The thin microcrystalline region at the glass/ceramic interface contains Al, Si and Mn,(indicated by arrows in)Fig.6. The presence of the manganese is believed to play a similar role in metallizing AlN as in metallizing high purity alumina[7,8]. Mn was detected in the grain boundary penetrating phases closer to the interface but was not detected at increasing penetration depths. The various crystal structures and compositions of these precipitates have been extremely difficult to identify and measure.

THICK FILM METALLIZATION- Ni Mo Mn

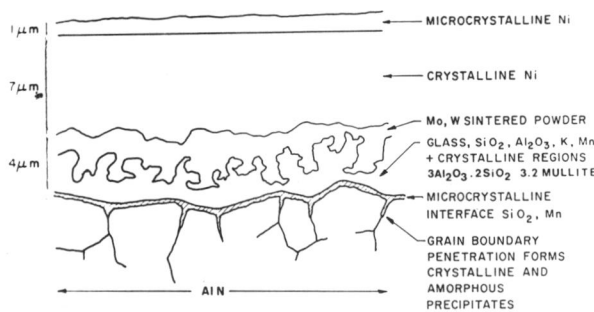

Fig.3. Schematic Cross-section through thick film metallization on Hitachi AlN substrate indicating proposed chemical and structural features at the interface.

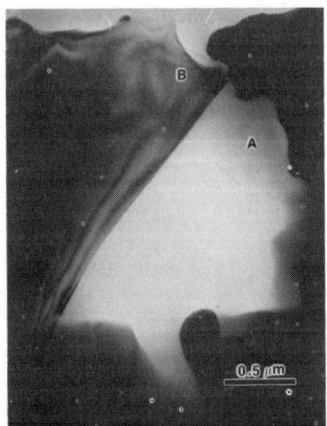

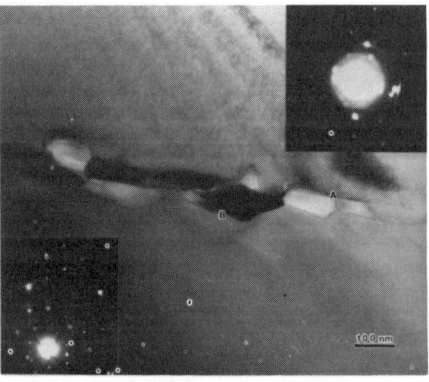

Fig.4. Glass frit region between Mo powder and AlN. Amorphous glass phase, region (A), crystalline phase, region (B), was found to be 3.2 Mullite, black areas are Mo.

Fig.5. Precipitation morphologies in thick film metallization. Penetration of amorphous and crystalline precipitates down AlN grain boundaries. Top right amorphous pattern from precipitate (A), bottom left crystalline pattern from precipitate (B).

The grain boundary penetration extended from the glass/ceramic interface region ≈20μm into the AlN, Fig.7. Below this depth, Ca and O were not detected by AEM, consistent with previous AEM measurements on the bulk AlN substrate material[15].

The adhesion between the metal layers and the ceramic is via a mechanical keying of the Mo into the glass phase and vice versa. The Mn-rich glass also mechanically keys itself to the AlN by grain boundary penetration and precipitation. There is the possibility of dissolution at the AlN interface by the Al-Si-Mn glass phase, enhanced by the solubility of transition metals in AlN[19,20].

The uses of AlN substrates for heat sinks requires that they be metallized; however, no evaluation has been carried out as to how this metallization sequence affects the thermal conductivity. It would be expected that grain boundary penetration and complete encapsulation of surface AlN grains by precipitation of low conductivity phases would severely affect the thermal conductivity behaviour. The solid state bonding encountered in thin film metallizations or those metallization systems avoiding a glassy phase would not be as detrimental to property degradation.

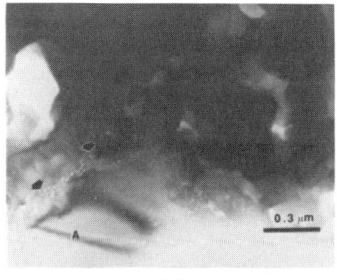

(a)

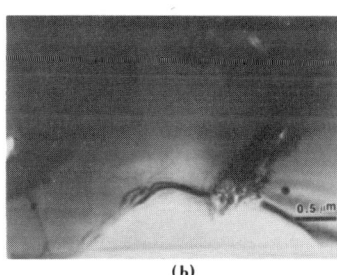

(b)

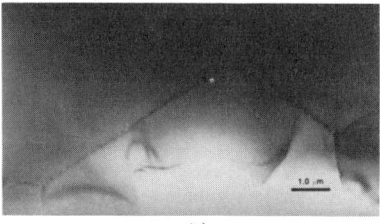

(a)

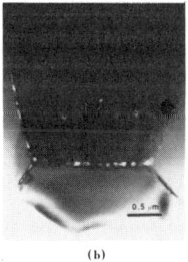

(b)

Fig.6a. Al, Si, Mn microcrystalline region indicated by the arrows at glass/AlN interface, region A is AlN. Above microcrystalline region is the glass frit and Mo particles.
b. Different grain boundary penetration morphologies. Wide and narrow grain boundary penetration is visible.

Fig.7. Extensive penetration of precipitates down AlN grain boundaries.
b. Encapsulation of surface AlN grain by precipitates.

336

Conclusions

1. Ti thin film deposited onto AlN substrates produce a reaction layer during post-deposition heat treatment presumed to be predominantely TiN and $TiAl_3$. This reaction layer is responsible for adhesion in this system.
2. Adhesion produced during Mo-Mn thick film metallization is due to the mechanical keying of the sintered metallized layer into the glass phase and the glass phase also mechanically keying itself to the AlN by grain boundary penetration plus chemical bonding via precipitation.

Acknowledgements

The authors gratefully wish to acknowledge the funding support of the Semiconductor Research Corporation (SRC).

References

[1] G.A.Slack and T.F.McNelly, J. Crystal Growth. 42, 560-565 (1977).
[2] W.Werdecker and F.Aldinger, IEEE Trans. Components Hybrids, Manuf. Technol., CHMT-7 (4), 399-404, Dec (1984).
[3] M.E.Twentyman, J. Mat. Sci. 10, 765-776 (1975).
[4] M.E.Twentyman and P.Popper, J. Mat. Sci. 10, 777-790 (1975).
[5] M.E.Twentyman and P.Popper, J. Mat. Sci. 10, 791-798 (1975).
[6] D.M.Mattox and H.D.Smith, Am. Ceram. Soc. Bull. 64 (10), 1363-1367 (1985).
[7] K.White and D.P.Kramer, Mat. Sci. Eng. 75, 207-213 (1985).
[8] N.Kuamoto, H.Taniguchi and I.Aso, IEEE Trans. Components, Hybrids, Manuf. Technol., CHMT-9 (4), 386-390, Dec (1986).
[9] W.Werdecker, Proceedings ISHM 1985, Stressa/Italy.
[10] P.Kluge-weiss and J.Gobrecht, Mat. Res. Soc. Symp. Proc. 40, 399-404 (1985).
[11] N.Iwase, K.Anzai, K.Shinozaki, O.Hirao, T.Dinh Thanh and Y.Sugiura, IEEE Trans. Components, Hybrids, Manuf. Technol., CHMT-8 (2), 253-258, June (1985).
[12] Y.Kurokawa, H.Hamaguchi, Y.Shimada, K.Utsumi, T.Kamata and S.Noguchi, Proceedings form the 36th Electronic Components Conference (IEEE), 412-418 (1986).
[13] B.F.Buxton, J.A.Eades, J.W.Steeds and G.M.Rackham, Phil. Trans. Roy. Soc. A281, 171-194 (1976).
[14] J.W.Steeds and R.Vincent, J. Appl. Cryst. 16, 317-324 (1983).
[15] A.D.Westwood and M.R.Notis, Proceedings of the International Symposium on Ceramic Substrates and Packages, to be published in Advances in Ceramics, (1988).
[16] X.A.Zhao, E.Kolawa and M.A.Nicolet, J. Vac. Sci. Technol. A4 (6), 3139-3141 (1986).
[17] R.K.Brow, R.E.Loehman, A.P.Tomsia and J.A.Pask, Proceedings of the International Symposium on Ceramic Substrates and Packages, to be published in Advances in Ceramics, (1988).
[18] D.B.Williams, Practical Analytical Electron Microscopy in Materials Science, (Philips Electronic Instruments, 1984), p.71.
[19] J.C.Schuster and H.Nowotny, J. Mat Sci. 20, 2787-2793 (1985).
[20] J.C.Schuster, J. Am. Ceram. Soc. 68 (12), C329-C330 (1985).

ALUMINUM NITRIDE FIBERS
FROM A THERMOPLASTIC ORGANOALUMINUM PRECURSOR

JOHN D. BOLT AND FRED N. TEBBE
E. I. du Pont de Nemours & Co., Experimental Station, E262,
Wilmington, DE 19898

ABSTRACT

A new organoaluminum polymer $(EtAlNH)_n(Et_2AlNH_2)_m \cdot AlEt_3$ derived from triethylaluminum and ammonia, is thermoplastic at elevated temperatures and a glassy solid at ambient temperature. As a thermoplastic it can be processed in certain shapes, solidified, cured and transformed to dense aluminum nitride with retention of its shape. Aluminum nitride fibers are prepared by melt spinning the polymer, pyrolyzing in ammonia and at high temperature in nitrogen. The AlN microstructure forms as very fine particles at 400-600°C, coarsens at higher temperature, and densifies at 1600-1800°C into polycrystalline AlN with submicron grains. Mechanical strength, thermal expansion and dielectric constant are consistent with bulk ceramic values. Initial thermal conductivity deduced from composite measurements is 82 W/m°K in fibers containing 0.5 to 1.0 percent oxygen.

The high thermal conductivity of aluminum nitride (AlN) [1] and its close thermal expansion match to silicon make it a potentially important material for use in microelectronic packaging. Recent AlN research has focused on forming monolithic bodies for replacing alumina and beryllium oxide substrates. However, we recently reported alumina fiber-polymer composites which favorably combine the thermal properties of ceramics with the dielectric and processing advantages of polymers [2]. We predicted that by incorporating AlN fibers in such composites, properties could be tailored to give in-plane thermal conductivity exceeding alumina, in-plane thermal expansion close to silicon, and out-of-plane dielectric constant typical of glass-epoxy. Our AlN fibers are prepared from a thermoplastic organoaluminum preceramic polymer [3-5]. The polymer is melt spun, cured and pyrolyzed in ammonia, and sintered at high temperature. We report properties of the fibers relevant to electronic applications.

Organoaluminum Polymers

Early studies of the reaction of trimethylaluminum with ammonia and conversion of the products were reported by Weiberg [6]. More recently Interrante and others have used this and related systems for forming films [7-10] and powders of AlN [7,11]. In general, alkylaluminums and ammonia react in a 1:1 ratio with stepwise loss of alkane from the initial adduct as illustrated for the ethyl derivative:

$$\text{AlEt}_3 + \text{NH}_3 \longrightarrow \text{Et}_3\text{AlNH}_3 \xrightarrow[-\text{C}_2\text{H}_6]{60°} (\text{Et}_2\text{AlNH}_2)_3 \xrightarrow[-\text{C}_2\text{H}_6]{160°} \underset{\text{I}}{(\text{EtAlNH})_n} \longrightarrow$$

$$\xrightarrow[-\text{C}_2\text{H}_6]{270-350°/\text{NH}_3} \underset{\text{amorphous}}{\text{AlN}} \xrightarrow{400-1000°} \underset{\text{crystalline}}{\text{AlN}}$$

An important characteristic of the infusible polymer (EtAlNH)$_n$, I, is that it transforms to AlN with retention of shape, although some shrinkage occurs as the organic portion is lost and the AlN crystallizes. However, I is infusible and sparingly soluble, making formation of fibers, fine powders or other desirable shapes difficult. Nevertheless organoaluminums are a route to high purity AlN [7]. Purity is controlled by distilling the starting materials, giving total metal ion contamination below 40 ppm [4]. Oxygen contamination is known to be a major factor in limiting thermal conductivity [1]. By careful attention to dry-box conditions during polymer synthesis and fiber preparation we have prepared fibers containing less than 0.1 % O. Ammonia is used in the final conversion to AlN to insure the removal of carbonaceous by-products and to produce a white product.

The solubility of inorganic compounds such as aluminum chloride is greatly enhanced by complexation with alkyl aluminums. By analogy we reasoned that the solubility of I could be similarly enhanced. Indeed, when (Et$_2$AlNH$_2$)$_3$ is heated in the presence of small amounts of AlEt$_3$, the product is not only soluble in common organic solvents, it is a fluid liquid at synthesis temperature (circa 165°C), thermoplastic above 100°C and a glassy solid at room temperature. Details of the polymerization have been given elsewhere [3-4]. We monitor the polymerization reaction by the quantity of ethane produced, stopping the reaction before a full equivalent evolves. The ^{1}H-NMR spectrum of the polymer contains resonances assignable to (Et$_2$AlNH$_2$)$_3$ and other resonances which are tentatively assigned to ET$_2$AlNH$_2$ incorporated in (ETALNH)$_n$ matrices. This information is used to formulate the polymer as (EtAlNH)$_n$(Et$_2$AlNH$_2$)$_m$·AlEt$_3$. A typical polymerization gives n:m of about 6:1.

$$(\text{Et}_2\text{AlNH}_2)_3 \xrightarrow[-\text{C}_2\text{H}_6]{\substack{0.06 \text{ AlEt}_3 \\ 160°C}} \underset{\text{II}}{(\text{EtAlNH})_n(\text{Et}_2\text{AlNH}_2)_m\cdot\text{AlET}_3} \xrightarrow{\text{NH}_3} \text{I}^*$$

$$\text{I}^* \xrightarrow{\text{NH}_3} \text{AlN}$$

Ammonia and II react below the glass transition temperature with conversion of the polymer to an infusible solid, I*. The morphology of the cured polymer, I*, is retained at high temperatures.

In contrast to preceramic polymers based on silicon from which low molecular weight products evolve during pyrolysis, the polymerization and subsequent conversion to AlN are quantitative in this organoaluminum system.

Fiber Spinning

Conventional melt spinning technology is used with provision for the reactive nature of organoaluminums. The polymer, II, is conveniently melt spun from a small piston-driven spinning cell in a nitrogen filled dry-box. Spinning conditions are determined from the glass transition temperature of the polymer. Polymers are typically spun from a 150 μm diameter spinneret hole at 115°C and wound-up at 300 m/min. Many combinations of cross-sectional shapes can be extruded. For determining basic fiber properties, fibers are removed from take-up bobbins as skeins and processed batchwise. Figure 1 shows a sample of uncured fibers.

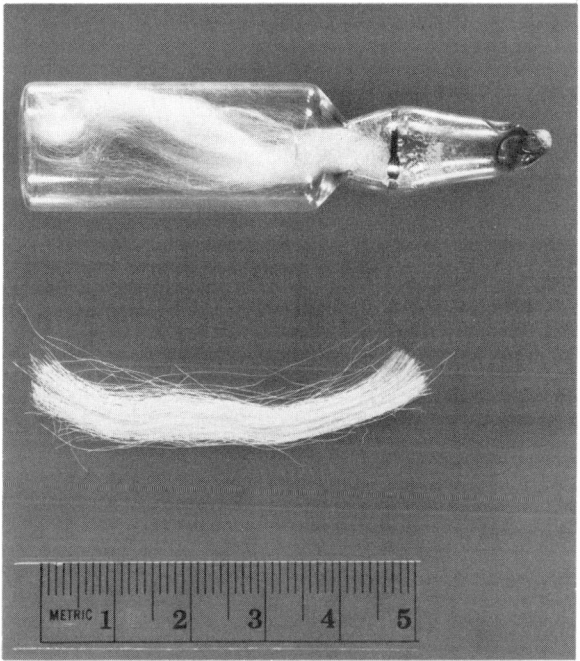

Fig. 1. As-spun organoaluminum fibers (sealed vial) and sintered AlN fibers.

Pyrolysis and Sintering

Fibers are pyrolyzed under a stream of ammonia in quartz tubes. As the temperature is raised, fibers cure and convert into AlN below 400°C. Typically, the fibers are further heated in ammonia to 1000°C before subsequent sintering in a graphite furnace in nitrogen. At 400°C the x-ray diffraction pattern contains very broad peaks suggestive of microcrystalline AlN, Figure 2. By 600°C, a distinct AlN diffraction pattern is visible. At successive higher temperatures the diffraction pattern transforms smoothly into sharp AlN peaks.

The microstructure after 400°C is a very fine, closely packed particulate with surface area in excess of 700 m²/g. The particle density, measured by He pycnometry, is initially quite low and comparable to the polymer density. From a plot of density and surface area after progressively higher heat treatment we see that most particle densification occurs below 1200°C and that particle sintering to a dense fiber occurs between 1600 and 1800°C (Fig. 3). Micrographs reveal the fine particulate microstructure at low temperatures which coarsens as temperature increases and eventually sinters to dense submicron grains, Figure 4.

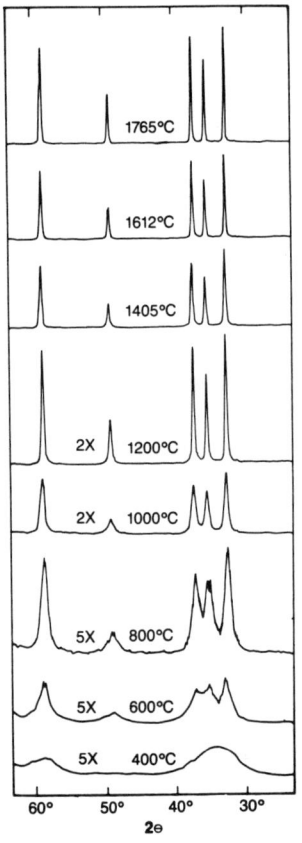

Fig. 2. X-ray diffraction pattern of AlN fibers after different pyrolysis and sintering temperatures. (CuKα)

Fiber Properties

Physical properties of the AlN fibers are summarized and compared to literature values for bulk AlN and to α-alumina fibers (Fiber FP) in Table 1. Coefficient of thermal expansion (CTE) was measured in a thermal mechanical analyzer on ribbons of AlN prepared by polymer extrusion through a slot, pyrolysis and sintering to 1820°C.. The CTE of 4.6 ppm/°C (50-400°C) is about that expected from bulk AlN values [12].

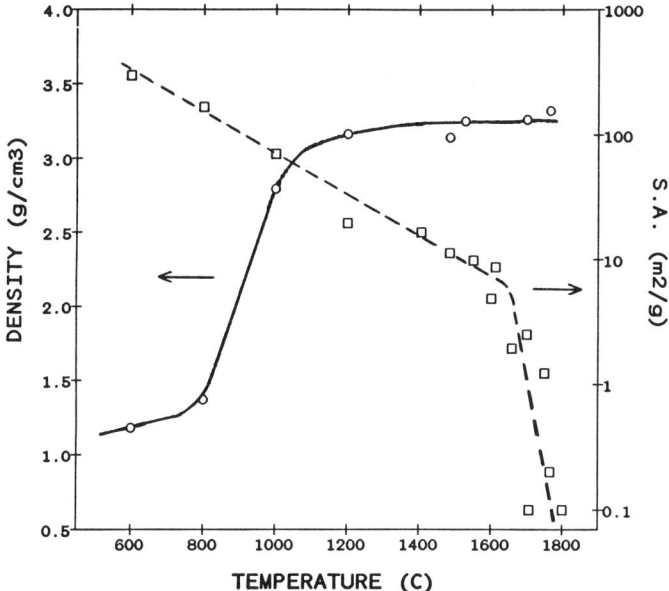

Fig. 3. Surface area (N_2 BET) and density (He pycnometry) of AlN fibers after various heat treatments.

TABLE 1: PROPERTIES OF ALUMINUM NITRIDE FIBERS

	AlN Fibers	Bulk AlN (lit. value)	$\alpha\text{-Al}_2\text{O}_3$ Fibers (Fiber FP)
Thermal Conductivity, W/m°K at room temperature	82 (1.0-0.5%)	40-285 (3.6-0.03%)[a]	28
Coefficient of Thermal Expansion, ppm/°C (50-400°C)	4.6	4.5	7.3
Dielectric Constant at 1 MHz	8.2	8.8	10.5
Tensile Strength, MPa (6.4 mm gauge)	340	(440)[b]	1500
Density, g/cm³	3.14 - 3.32	3.25	3.9
Grain Size, μm	0.3 - 0.8	3 - 5	0.4
Diameter, μm	8 - 25	-	20

a) G. A. Slack, J. Phys. Chem. Solids, 34(1973) 321 and ref [1].
b) 3-point bend, M. Billy and J. Mexmain, Fachberichte, 118 (1985) 245.

342

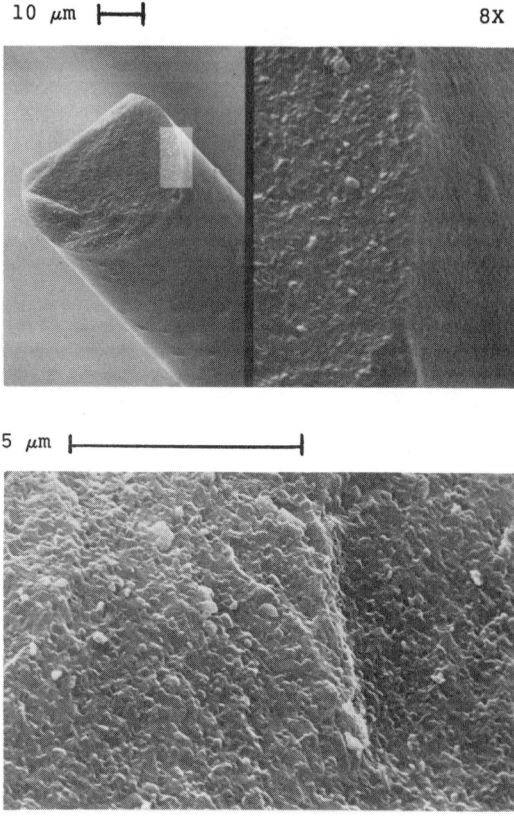

10 μm ⊢──┤   8X

5 μm ⊢────────────┤

Fig. 4.   SEM micrographs of AlN fibers.

        Fiber tensile strength was measured on single filaments
at 6.4 mm gauge length. Typical samples averaged about 340 Mpa
breaking strength. This value is typical of bulk AlN bending
strength from sintered or hot pressed bars [12-13]. However,
the strength is low for fibers of such small dimensions and fine
grain size. We believe the strength can be significantly
improved; surface flaws are observed and probably limit
strength.

        Dielectric constants were estimated from uniaxial
composites. With fibers aligned with the electric field, the
dielectric constant obeys a linear (parallel) rule of mixing
[2].   Figure 5 plots the 1 MHz data for AlN fiber composites in
an epoxy matrix. Extrapolation to 100 volume percent fiber gives
8.2, within experimental error of literature values of 8.8.
Similarly, the dielectric constant of α-alumina fibers was found
to be 10.5 at 1 MHz in good agreement with the literature

[14]. The value for AlN fibers may be low due to porosity in the fibers or composites.

The thermal conductivity of a uniaxial AlN fiber-epoxy composite was measured at 27 W/m°K at room temperature in the fiber direction. With a fiber loading of 30-33 volume percent and a linear dependence on fiber loading [2], we calculate the fiber thermal conductivity as 82 W/m°K. Contribution of the epoxy to thermal conductivity is negligible. Oxygen in the AlN lattice creates aluminum vacancies which limit the thermal conductivity [1]. Based on measured oxygen contents of fibers of between 0.5 and 1.0 percent, the thermal conductivity agrees with expected values from the literature. With the capability to make AlN fibers with considerably lower oxygen content, we can expect significant improvements in thermal conductivity. Even at 82 W/m°K, the thermal conductivity of these fibers is three times that of α-alumina fibers [2].

Acknowledgements

We wish to thank Uma Chowdhry and Daniel Button for discussions which led to this work. David Onn and co-workers at the University of Delaware made thermal conductivity measurements. Dielectric measurements were made by Mas Subramanian.

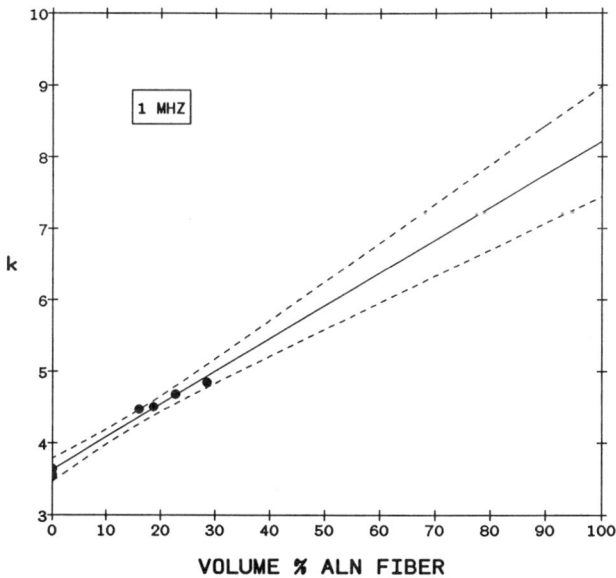

Fig. 5. Dielectric constants of uniaxial AlN fiber composites measured parallel to fibers. ----90% confidence limit.

REFERENCES

1. G. A. Slack, R. A. Tanzilla, R. O. Pohl, and
   J. W. Vandersande, J. Phys. Chem. Solids, <u>48</u>, 641 (1987).
2. D. P. Button, B. A. Yost, R. H. French, W. Y. Hsu,
   J. D. Bolt, M. A. Subramanian, H.-M. Zhang, R. E. Geidd,
   A. J. Whitaker, and D. G. Onn, submitted to <u>Advances in
   Ceramics</u>, edited by H. M. O'Bryan, K. Niwa, W. Young, and
   M. F. Yan.
3. F. N. Tebbe, U.S. Patent No. 4 696 968 (October, 1987).
4. F. N. Tebbe, J. D. Bolt, R. J. Young, O. R. Van Buskirk,
   W. Mahler, G. S. Reddy, and U. Chowdhry, submitted to
   <u>Advances in Ceramics</u>, edited by H. M. O'Bryan, K. Niwa,
   W. Young, and M. F. Yan.
5. J. D. Bolt and F. N. Tebbe, submitted to <u>Advances in
   Ceramics</u>, edited by H. M. O'Bryan, K. Niwa, W. Young, and
   M. F. Yan.
6. E. Wiberg, reported in G. Bahr, FIAT Review of German
   Science, 24, Inorganic Chemistry, Pt.2, W. Klemm, ed.
   (1948), p. 55.
7. L. V. Interrante, L. E. Carpenter II, C. Whitmarsh, W. Lee,
   M. Garbauskas, and G. A. Slack in <u>Better Ceramics Through
   Chemistry 2</u>, edited by C. J. Brinker, D. E. Clark, and
   D. R. Ulrich (Mater. Res. Soc. Proc. 73, Palo Alto, CA,
   1986) pp. 359.
8. U. Rensch and G. Eichorn, Phys. Stat. Sol. (a) <u>77</u>, 195
   (1983).
9. M. Morita, N. Uesugi, S. Isogai, K. Tsubouchi, and
   N. Mikoshiba, Jap. J. Appl. Phys., <u>20</u>, 17 (1981).
10. H. M. Manasevit, F. M. Erdmann, and W. I. Simpson,
    J. Electrochem. Soc., <u>118</u>, 1864 (1971).
11. N. Maeda and H Harada, Japanese pat. Application public.
    Tokuko 54-13 439 (1979).
12. K. M. Taylor and C. Lenie, J. Electrochem. Soc., <u>107</u>, 308
    (1960).
13. M. Billy and J. Mexmain, Sprechsaal, <u>118</u>, 245 (1985).
14. W. H. Gitzen, <u>Alumina as a Ceramic Material</u> (Am. Ceram.
    Soc., Columbus, Ohio, 1970) p.78-84.

Integrated Circuit Packaging

MECHANICAL INTERCONNECTS TO SILICON INTEGRATED CIRCUITS

CHARLES A. STEIDEL
Intel Corporation, 145 S. 79th Street, Chandler, AZ 85226

ABSTRACT

There are three predominant interconnection technologies in use today for silicon integrated circuits: wire bonding; TAB (tape automated bonding); and C4 (controlled collapse chip connection). This paper briefly reviews each of these technologies for their strengths and weaknesses but focuses especially on wire bonding for single chip VLSI applications.

Although wire bonding has been in widespread use ever since the invention of the small scale integrated circuit, the technique is still applicable for today's much larger and denser chips. With leadcounts up to 250 or even higher, many challenges are presented for equipment accuracy and speed and in package design, where novel techniques often are required to prevent the package from being the limiting factor in chip interconnection. These challenges are discussed in some detail.

The paper concludes with a discussion of the future direction for mechanical connections as influenced by the technical and cost requirements of both the chip and the system in which it resides.

Introduction

Three major approaches have been used to mechanically interconnect to silicon: wire bonding, TAB (tape automated bonding), and C4 (controlled collapse chip connection or flip chip). Figures 1 and 2 show the metallurgical structures at the chip bond pads and the mechanical connection schemes, respectively, for each of the connection technologies.

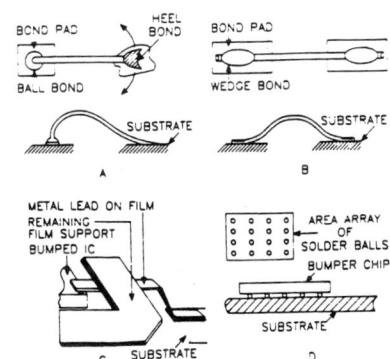

SCHEMATICS OF THREE INTERCONNECT TECHNOLOGIES

Figure 1

Schematics of three inter-connect technologies showing
A) Ball bond   B) Wedge bond
C) TAB and D) C4.

348

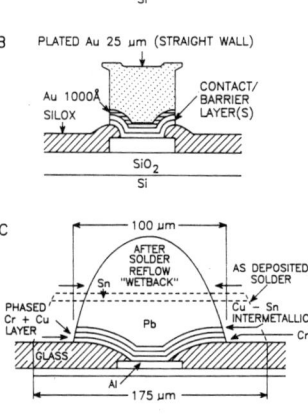

A   SILOX (CVD) 1–1.5 μm

B   PLATED Au 25 μm (STRAIGHT WALL)

C

Figure 2

Bond pad structures for three
interconnect technologies showing   A) Aluminum bond pad,
B) TAB bond pad and C) C4 bond
pad[1].

Wire bonding is the most broadly applied technique of the
three finding use in both single chip devices and in hybrid
circuits. TAB has been used in high volume for plastic molded
small scale circuits, in moderate volume for consumer-like
products (calculators, watches, etc.) and in limited but
growing volume as an advanced interconnect scheme. C4 is used
on a small scale by many hybrid manufacturers but is best
known as the interconnect scheme used in IBM main frame
computers. The future mechanical interconnect schemes to
silicon will be principally based on these three approaches or
some combination of these technologies which have been
described.

This paper outlines the challenges for silicon interconnects over the next several years and then compares and
contrasts the three technologies to meet these challenges.
Wire bonding will be used as the base case technology.

Silicon Interconnect Challenge

The following discussion will be restricted to leading
edge silicon devices and more particularly to logic or microprocessor devices, since they will pace the silicon challenges
in areas such as higher lead count, higher power, and higher
speed. The following attributes of each technology need to be
understood in order to make the interconnect decision:

Density/Electrical Performance
Thermal Performance
Replaceability/Testability
Flexibility
Reliability
Cost

Density/Electrical Performance

Careful consideration must be given to the silicon connections to support the shrinking silicon design rules which will progress from 1um in 1988 to perhaps 0.5um in five years. For practical reasons, the maximum silicon size will not grow much beyond 1.25 cm on an edge so as the number of gates on this size silicon increase dramatically and thus the number of required external connections (assuming a random logic-like part), the density of connections to the silicon will grow dramatically.

The number of interconnections to these chips is forecast to increase to the 400 to 600 range for a random logic part. Today's edge connection density to the silicon is about 150um, pad to pad. With a 1.25 cm chip, 400 to 600 I/O's will require a density of 125um to about 85um, respectively. Further, as the chip interconnect density is shrinking so then is the required density at the package level for a fixed length of interconnect between the silicon and its package.

Because of the ever increasing frequency of operation of such chips, the length of the interconnects need to be shorter rather than longer to minimize inductive noise in power and ground and crosstalk between the signal leads. But in many cases, the standard package technologies, e.g. leadframes for plastic packages, cannot achieve a small enough pitch to match the silicon. Two options exist: either increase the silicon pitch or find novel ways to meet the package pitch and still maintain the interconnection lengths as short as possible. Figure 3 suggests that increasing the silicon pitch by adding non-active silicon is expensive. In a later section, we cover one now-standard, package innovation to achieve finer package pitch and keep the critical interconnection length fixed.

As system performance approaches the 25 - 50 MHz range over the next few years these electrical issues may become so severe that single chip packages in a printed circuit board environment will not be viable. Based on this expectation, most system's companies are hard at work developing multichip technology.

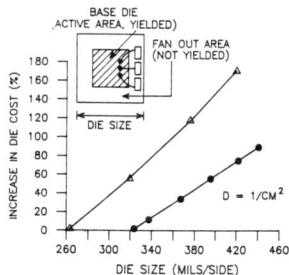

Figure 3

Cost increase to expand die for interconnect as a function of die size - for starting die sizes of 260 and 320 mils.

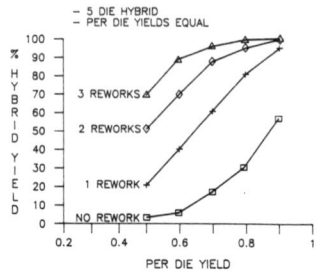

HYBRID YIELD AS A FUNCTION
OF PER DIE YIELD AND REWORK

Figure 4

Hybrid yield as a function
of per die yield and re-
work for a 5 die hybrid.

Thermal Performance

These same higher density chips will dissipate much more
power because of the increased gate count. CMOS technology
may increase to the 10 watt range while bipolar chips may
consume up to 50 watts. Clearly, better thermal performance
at the package and system levels will be required even to
maintain today's junction temperatures, however, the need to
get the heat out of the chip will be much greater because the
allowed junction temperatures must decrease. The need for
lower junction temperature is driven by the fact that as the
silicon design rules shrink, electro-migration is exaserbated
and must be compensated by lowering the chip temperature.
Thus, the mechanical interconnect to the silicon must be
compatible with achieving significantly better thermal
performance than can be delivered today.

Replaceability/Testability

A third consideration is replaceability/testability of a
chip. This is not usually a concern for single chip packages,
but for multichip packages, the final per chip yield must be
extremely high to attain a reasonable finished package yield
(see Figure 4). The high per chip yield can be attained two
ways; the chip can be pre-tested so well that only mechanical
assembly defects (usually low) affect its yield or it can be
easy to repair (replace) with a good device. If burn-in is
required to achieve satisfactory module reliability, the same
strategies apply although the replacement method is more
complex because unless the device has been pre-burned in
another burn-in cycle for the whole module will be required.
Ideally, both pre-testability and pre-burnability, plus ease
of replacement are desirable. The different interconnect
technologies have very different capabilities in these two
areas. We assume that all technologies have equal testability
at the wafer level.

Flexibility

While many future chips will be customized for their particular package (and thus so will their mechanical interconnects), some will need to be more flexible for use in both single chip and in multichip applications, and need to have the right cost/performance for both. The single chip package will usually need to be less costly on a per chip basis.

Reliability

Different use applications have different reliability requirements. A silicon device used in a controlled office environment will see both less varying thermo-mechanical stress than under an automobile hood and less exposure of the silicon interconnects to noxious fumes and humidity. Some aspects of the more severe environment can be handled with hermetic packaging, but generally the trend is to non-hermetic packaging to lower cost. Hermeticity at the chip surface and at its mechanical interconnects is often stated as a future requirement. The bare aluminum bond pad is considered to be the reliability weak link in a wire bonded plastic package because of the poor corrosion resistance of aluminum metal.

Cost

Finally, the cost to interconnect the silicon must be minimized and still meet all other requirements. These interconnect costs usually include individual costs associated with labor, depreciation for assembly equipment, any special treatment of the wafer (extra layers) to make a chip connectable, the actual interconnect itself (wire, TAB tape, etc.), and the extra complexity of the package to accommodate the interconnect.

Wire Bonding

What especially differentiates wire bonding from TAB and C4 is that the bonding is done directly to the standard Al metallurgy one bond at a time. Further, wire bonding is the entrenched interconnect technology and commercial equipment is the most readily available for this technology.

The two principle wire bonding technologies are gold ball and aluminum wedge (although aluminum ball and gold wedge can be done). Aluminum wire is usually not considered to be reliable for non-hermetic packages. Wire bonding is performed after die attachment to a substrate (e.g. leadframe or ceramic). The die attached device has a natural heat flow path from the junctions to the package and then to the environment. Since one bond is made at a time by a computer controlled movable head, wire bonding is flexible because the path between the chip connect and the package can be customized by programming the bonder (See Figure 1a and b).

The challenges for wire bonding arise from the very essence of how it is performed. Since it is one bond at a time, it is relatively slow and expensive (state-of-the-art is about 8 wires/sec.); for example, about 70-400 wire devices can be assembled per hour with a machine cost of $120K. Since wire bonding is serial, the yield for every bond must be very high to yield a high final yield (analogous to our earlier hybrid argument). To achieve 98% device yield with 400 wires will require 0.99995 yield per wire. We have demonstrated this per wire yield in volume production for much lower leadcount devices but in future devices the complexity of bonding will increase as the bond pad spacing on the silicon shrinks. Based on the industry learning curve and continued improvements in bond control, high yield on 400 lead devices with reduced geometries should be achievable in the 1992 time frame.

A major challenge for wire bonding is achieving the densities required at the chip level (85um to 125 um) and at the package. The minimum density at the chip depends on the size of a reliable bond, the accuracy with which the bond is placed on the bond pad and the critical clearance between the bonding tool and adjacent wires. A thorough analysis of the machine related variables was recently published [2]. For a 25um diameter wire the authors conclude based on machine capabilities that ball bonding will be possible on 75 to 87.5um wide pads while wedge bonding will be possible on 62.5 to 80um pads. These translate to about 100um and 75um centers, respectively as better bonding tools are developed. Since aluminum wire is not reliable for plastic packages, gold wedge bonding must be further developed especially for leadframe-like product. It appears that the required densities at the chip level are achievable, but there are further considerations.

As the chip pad space shrinks so must the package lead space shrink if the wire lengths are to be maintained as discussed previously. Again, the motivation to not substantially increase the wire length is to minimize the inductance of the connections, especially for power and ground. Table 1 shows the inductances of a multiple interconnects for wire, TAB and C4 [3]. Since the effective ground inductance goes as the inductance of a single connection divided by the number of such connections, the implication is that wire and TAB would need more chip connections allocated to ground than C4 for the same performance. If the wire length is allowed to fan out to a larger package pitch, the inductance problem is only worsened and the wire is more susceptible to damage. The wire length is limited by its ability to support its own weight, by its resistance to vibration and in molded packages by wire sweep which can lead to short circuits between connections.

Several design innovations have been introduced in recent years to solve the pad pitch/short interconnect dilemma. Figure 5 shows a two tier package bonding scheme where the pitch in the package can be twice that at the chip and, by connecting the grounds to the lower shelf, the inductance problem can be minimized. Similar concepts can be applied to molded plastic packages to improve their electrical performance.

Table 1 [3]

	Cross-section	Length	Inductance*
Wire bond	1 mil diameter	50 mil	0.9 nH
TAB	1 mil x 3 mil	50 mil	0.7 nH
Solder ball	4 mil x 4 mil	2 mil	< 0.05 nH

* 25% of I/O are power/ground

BOND AND PACKAGE STRUCTURE
TO MINIMIZE INTERCONNECT LENGTH

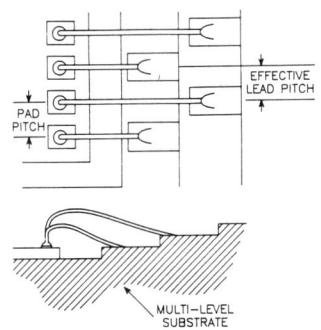

Figure 5

Bond pad and package structure
to minimize interconnect length.

TAB

The standard TAB process today requires forming bumps on the wafer to which planar leads are bonded (See Figure 2b). The bumps are typically formed by sputtering an adhesive layer such as Ti-W plus a thin layer of gold over the entire wafer surface. The areas at the bond pads are plated up to 15 to 25um. Gold is usually used for the bumping process. The cost of the bumping operation must be borne by the good die in the wafer.

The entire TAB process has been thoroughly discussed in the literature; a summary may be found in Ref. [4]. Compared to wire bonding, several hundred chips can be assembled per hour with a machine cost comparable to that of a wire bonder. TAB is not as flexible as wire bonding because a custom tape must be made for each device which has different bond pad locations. For very large die with many leads there is still considerable controversy over whether the bonds should be made by thermo-compression bonding of the gold plated leads to the gold bumps or by eutectically reacting the bump with tin which may be plated on the tape leads or on the bump [5]. The thermo-compression bond requires large deformations. To avoid cracking the silicon, extreme care must be taken in maintaining planarity of the tool and the bumps. Gold-tin eutectic bonding requires much less deformation but there is the constant concern for tin whiskers which could be a severe reliability problem [5]. Straight walled bumps and the lesser deformation of eutectic bonding will ease getting to very fine pad pitches.

TAB has already been well documented at less than 100um centers and 50um centers have been reported [6]. TAB also has sealed bond pads, that is the aluminum is completely blocked from the environment, for better corrosion protection. The TAB device may be die bonded face up to achieve good thermal performance or lead bonded face down to shorten the inter-connect to the substrate and it is the only technology of the three where test and burn-in of the silicon can be done easily before committing to a multichip module. This is accomplished through handling the tape bonded device as a component and performing the usual electrical test, at frequency and temperature, and burn-in before committing the device to a module. The TAB bonded device is more robust than its wire bonded counter part and all bonds can be inspected. TAB bonded devices can be replaced on a hybrid but like wire bonded devices this is a somewhat messy operation and questions of reliability degradation are often raised.

Many exotic tape structures have also been proposed which include power and ground planes which can serve to lower the inductance of the long interconnects to the package. Further, a tape lead can be considerably longer than a wire bond and still deliver the same inductance (See Table 1) and stiffness. Because of this capability, TAB has been proposed as the interconnect technology for fine pitch plastic packages which have limitations on the pitch of the leadframe fingers.

Figure 6 shows how the length of the interconnect varies with the pitch of the leadframe and the pitch on the die. Since we previously showed that adding extra fan out to the silicon to accommodate a greater package pitch was expensive, TAB may well help accommodate both minimum silicon sizes and limitations in package pitch. The extra cost of bumping a device and making a tape, however, creates great incentive for innovative package designs which can still accommodate wire bonding.

Figure 6   INTERCONNECT LENGTH VERSUS
EFFECTIVE LEAD PITCH

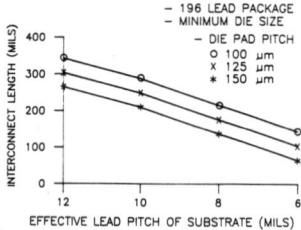

Interconnect length versus
effective lead pitch of
substrate.

C4

The C4, solder ball technology is well described in the literature; a summary may be found in Ref. [4]. Like TAB, extra wafer processing is required to create the solder balls, the bond pads are sealed and the cost must be borne by the good die in the wafer. As with TAB, all bonds to a substrate can be made simultaneously. The rate and cost are better even than TAB since all that is required to achieve a bond is accurate placement of the die on a substrate and reflow of the solder connection. Unlike TAB, there is no easy way to test and burn-in a C4 part before committing to a multichip module nor can the bonds be easily inspected, except possibly by x-rays.

The overwhelming motivations for C4 technology are the dense area array connection capability to the chip and the ease of replaceability. Replaceability follows from the soft, easily melted connection between the chip and substrate. Because C4 stands out for replaceability and TAB for testability they are the clear technology choices for complex, multichip modules.

The C4 chip is of course mounted active side down and is connected to the substrate only through the solder balls. This structure inherently has lower thermal capability than a die bonded TAB or wire bonded device, at least in terms of heat remove through the substrate. However, many mainframe multichip packages are structured to remove heat through a cooling system brought in contact with the chip backside, where the chips, TAB or C4 is mounted active side down [7].

The classic example of this approach is the IBM, TCM module [7]. In order to make contact to the back of the chips an elaborate and only moderately efficient, flexible connection scheme was devised. While successful for today's needs, the cost and technical limitations are significant.

Since solder balls are easily fatigued in thermo-mechanical stressing, care must be used in physical design of the solder ball array. When the silicon is mounted on Al_2O_3 ceramic, the mismatch in expansion coefficients between the Al_2O_3 and silicon must be compensated by restricting the distance to the furthest ball from the center of the chip, since the stress on the ball increases as this distance increases. While large C4 devices can still be used on Al_2O_3, the solder ball array is restricted toward the center area of the die which aggravates the thermal management (uniformity of heat removal) of the device and forces more dense wiring in the substrate to fan out the array. For these reasons, silicon substrates have been proposed as ideal for C4 technology since with matched expansion coefficients between chip and substrate, solder ball positions would be unrestricted thus relieving the fan out problem and allowing more uniform heat removal.

Given all of the above pros and cons for C4 and TAB, there is still not an overwhelming consensus on which connection technology is preferred for multichip packages or whether Al_2O_3 with copper-polyimide interconnects or silicon with copper-polyimide or even silicon with SiO_2 and aluminum wiring is the preferred substrate technology for multichip packages. C4 is probably more costly for single chip packages than either TAB or wire bonding because of the requirement to have a low expansion substrate such as ceramic or silicon.

Summary and Future Trends

We have seen that all three interconnection technologies have definite strengths and weaknesses. Wire bonding will be pushed to its capability limits for single chip packages with TAB coming in as these limits are reached. With multichip modules there are trade-offs between replaceability and pretestability for C4 and TAB, respectively and elaborate and expensive cooling techniques required for C4 because of its face down mounting. At best, these cooling schemes will be limited for C4. In recognition of these problems, still other interconnection schemes have been proposed which can achieve very fine pitch at the die and excellent thermal contact of the die face with the cooling media. Figure 7 shows a sketch of the micro cantilever interconnect scheme used in conjunction with the micro channel cooling scheme [8]. We will see many more of those novel proposals over the next few years as the industry searches for the optimum interconnect scheme which maximizes performance at reasonable cost.

Figure 7

Microchannel heat exchanges and microcantilever sub-connect scheme [8].

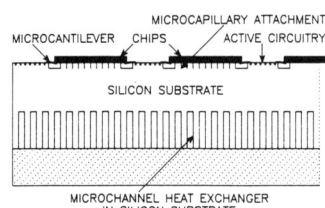

References

[1]   P.A. Totta and R. P. Sopher, "SLT device metallurgy and its monolithic extension", **IBM J. Res. Dev.**, **13**, 1969, p.226.

[2]   Gautan N. Shah, Lee R. Levine and Dipak I. Patel, "Advances in wire bonding technology for high lead count, high density devices", in **Proc. 37th Electronic Components Conf.**, 1987, pp.33-39.

[3]   Charles J. Barlett, John M. Segelken and Nicholas A. Teneketges, "Multi-chip packaging design for VLSI based systems", in **Proc. 37th Electronic Components Conf.**, 1987, pp.518-525.

[4]   Charles A. Steidel, "Assembly techniques and packaging", in S.M. Sze, Ed., **VLSI Technology, McGraw-Hill**, 1983.

[5]   T. Kawanobe, K. Miyamoto and H. Hirano, "Tape automated bonding process for high lead count LSI", in **Proc. Electronic Components Conf.**, 1983, pp.221-226.

[6]   "Tab for VLSI coming from Fairchild", **Electronics** Aug. 20, 1987, p32.

[7]   Abrahm Bar-Cohen, "Thermal management of air - and liquid cooled multichip modules", **IEEE Trans. Components, Hybrids, Manuf. Technol., vol CHMT-10, no.2,** 1987, pp.159-175.

[8]   R.F.W. Pease, J.C. Braman, O.K. Kwon, S. Hong, S.C. Douglas, B.W. Longley and A.F. Poal, "High Performance Packaging for VLSI Systems", **Advanced Research in VLSI "New Technologies For New Architectures"**, Stanford University, Stanford, C.A. March 23-25, 1987.

DIFFUSIONAL PROCESSES IN METALLURGICAL INTERCONNECTIONS

S. K. Kang * and C. G. Woychik **
* IBM T.J. Watson Research Center, Yorktown Heights, NY
** IBM, System Technology Division, Endicott, NY

Abstract

Diffusional processes, either solid- or liquid-state, are essential in promoting reliable metallurgical interconnections used in many electronic packaging applications. The same phenomena often degrade the integrity and reliability of the interconnections as well. In this talk, both examples are presented to illustrate how the diffusional processes are important in many interconnection technologies. Thermocompression bonding, eutectic bonding, soldering and liquid-phase bonding are chosen for this purpose. Several issues related to soldering, such as dissolution, intermetallic formation and microstructure of solder joints, will also be discussed in terms of diffusional processes.

INTRODUCTION

In many VSLI electronic packages, various interconnection technologies often dictate the utility of semiconductor devices, and also they influence critically the reliability, or longevity of whole electronic systems. The interconnections in electronic devices serve as electrical linkages between individual components, or thermal conduction path to the outside world, or mechanical support of the system. Because of complicated processing as well as the demanded multiple functions, the interconnections themselves are well recognized as vulnerable sites of premature failure or long-term degradation.

Examples of the interconnection technologies used in electronic packaging include wire bonding of IC devices, flip-chip bonding, surface mount soldering of IC packages, die bonding, pin brazing to ceramic substrates, tape automated bonding. The joining process is often directly dependent on diffusion-controlled reactions. Diffusional processes can be either in the solid state, such as for wire bonding, or in the liquid state, such as for soldering or flip-chip bonding. The diffusional reaction is responsible in the formation of intermetallic phases and Kirkendall porosity, dissolution of metals, microstructural instability, and others.

The diffusional processes occurring during bonding are rather rapid, since they are either in the liquid or in solid state at an elevated temperature. Although diffusional processes are essential to joint formation, it is also important to minimize the extent of certain metallurgical reactions. For example, formation of intermetallics or dissolution of metals into solder joints may have negative effects on joint integrity. In contrast, the diffusional processes during the functional life of the interconnections occur in the solid state comparatively slowly. However, these processes may be deleterious to the reliability of the interconnections.

In this paper, a few examples of the interconnection technologies will be discussed by examining the underlying mechanisms in the formation of successful interconnections. In addition, several issues are presented where diffusional processes are also responsible for the degradation or failure of the interconnections.

METALLURGICAL INTERCONNECTIONS

The interconnection technologies may be classified simply according to the physical state of the material at interfaces to be joined; solid-state bonding and liquid-phase bonding.

Solid-State Bonding

Solid-state joining involves atomic bonding and/or interdiffusion achieved by mechanically bringing appropriately prepared mating surfaces into intimate atomic contact (1). Removal or dispersion of surface films and asperities is necessary to obtain the required compressive contact. Surface films can be removed mechanically or thermally with increasing temperature by dissolution into the base metal(1).

Wire bonding of IC devices is a common example of a solid-state type interconnection, which is formed essentially by a thermocompression bonding process. In the early days, bonding of a gold wire to an aluminum pad did cause many reliability problems, mainly due to the formation of a deleterious intermetallic phase, known as 'purple plague' (2-7). Due to the high temperature, formation of this intermetallic phase was extensive during wire bonding, having a thickness of 0.5 to 1 μm (6). Further growth of the intermetallic phases is anticipated during the service of the components, which induces formation of Kirkendall porosities (5-7). Often, the porosities are connected to form cracks along the interface, leading to a catastrophic increase in electrical resistance and mechanical failure of the joints. The improvement of wire bonding technology is owed to the introduction of ultrasonic or thermosonic energy, which breaks up the thin aluminum oxide layer just prior to bonding. This has reduced the bonding temperature well below 300 C, and thereby the extent of the intermetallic formation is greatly suppressed. Because the intermetallic formation is simply a diffusion-controlled process, its growth could be minimized by providing a less favorable condition for diffusion, and the extent of the reaction could be estimated by using the experimental data obtained from bulk diffusion couples(4). However, the growth kinetics could be highly dependent upon materials processing and microstructure. For example, gold wire to aluminum pad bonding may be quite different from aluminum wire to gold pad bonding because wires are wrought materials and pad materials are vapor deposited, thereby having different grain sizes and associated diffusion paths.

Another example of thermocompression bonding used in VLSI electronic packaging is tape automated bonding (TAB), (8-13). This interconnection technique is a mass bonding scheme, while wire bonding is a sequential one. Depending upon the combination of tape metallurgy and bump structure on chip, TAB has been demonstrated to be either solid state or liquid phase bonding. For example, a common practice of TAB uses tin-plated-copper beams bonded to gold bumps on the chip (9-12), which is liquid-phase bonding since they form a gold-tin eutectic phase by melting tin during thermocompression bonding. However, TAB is a solid-state thermocompression bonding, when bare copper or gold plated beams are bonded to gold bumps on chip (13-15).

Liquid-Phase Bonding

Liquid-phase bonding involves methods wherein only the filler metal is in the liquid state during the joining process. Generally, flux materials are used to clean the surface chemically by dissolving metal oxides and to protect it from further contamination and oxidation. In liquid-phase bonding, the filler metal wets the parent metal surface and then is drawn into the joint by capillary action until the volume between components to be joined is completely filled. Surface tension forces provide the principal driving force for liquid-phase bonding (1,16).

Wave soldering is a common technology widely used in assembly of printed circuit boards. This technology utilizes the pin-in-hole mounting scheme. Molten solder is pumped up to form a wave, applied underneath the boards, filling up the joint gaps by capillary force, and then is readily solidified to form solder joints. The resultant microstructure of solder joints is greatly dependent upon the process parameters, such as solidification rate. Solder joints have a cast microstructure, which is less understood than the wrought microstructure in terms of the structure-property relationships of materials science.

Recently, surface mount technology (SMT) is increasingly practiced to replace the conventional pin-in-hole (PIH) technology (17). SMT requires a whole new solder technology, which involves new formulations of solder paste, screening of solder paste on printed circuit boards, component placement, new reflow methods such as vapor phase or infra-red reflow. The processes occurring during the reflow soldering in SMT are rather complicated. For example, solder paste first melts, volatile components in the flux are liberated, and at the same time the molten solder coalesces, after which solidification of the solder begins. Because of all the simultaneous reactions in SMT soldering, a greater process control is required as compared to PIH interconnections formed by wave soldering.

Another example of reflow soldering is found where high I/0 chips are attached to ceramic modules by the process called "controlled collapse chip connection (C-4)" or flip-chip connection (18). Here, the C-4 solder bumps made of low Sn, Pb-solder are evaporated onto silicon wafers through metal masks in a vacuum system. The C-4 solder joints are formed by reflowing the chip-and-module assembly in a controlled atmosphere. Whether solder joints are made by wave soldering or reflow soldering, many diffusional reactions occur during joint formation such as dissolution of base metals to solder, intermetallic formation at the interface, precipitation of intermetallics in solder joints, etc. These reactions are easily influenced by the soldering process. For example, the reflow time, (the time of the solder existing in the liquid state) determines the amount of metal dissolution, intermetallic formation and precipitation of intermetallics in solder joints.

In attaching power transistors or IC chips to a heat sink, or in brazing I/0 pins to ceramic modules, the eutectic bonding method is widely used (19,20). This is also classified as a liquid-phase bonding. A preform of Au-Sn or Au-Si eutectic alloys is inserted between the materials to be joined and heated at temperature above the eutectic temperature. Upon solidification, a strong bond is formed between the joining components.

ISSUES AFFECTING JOINT RELIABILITY

Chemical, physical and microstructural changes occurring in the interconnections due to diffusional processes introduce many issues affecting the joint reliability such as corrosion, oxidation, thermal fatigue, creep deformation, growth of intermetallics and voids, crack nucleation and growth, and microstructural instabilities.

Here, two examples are given which describe the microstructural instabilities in solder joints. The first one is the growth of intermetallics
and related problems. The formation of intermetallics at the soldering
interface is almost instantaneous as shown in Ni-Sn system (21). A parallel
reaction, often not disturbed by the presence of the intermetallics, is the
dissolution of the parent metal to molten solder (22,23). For most soldering
interfaces, these reactions are often regarded as enhancements for
solderability. However, when the reactions are excessive, it creates a reliability problem such as those encountered in soldering gold-plated surfaces. The gold layer in contact with molten solder dissolves rapidly to form
Au-Sn intermetallic precipitation, which can lead to solder embrittlement
later during thermal cycle testing (24). When the Cu-Sn or Ni-Sn intermetallics form at the interface, growth can continue upon further thermal exposures. Here two phenomena are expected; degradation in mechanical
properties of solder joints due to the increased thickness of the brittle
intermetallics (25), and depletion of tin content near the joint interfaces
where the intermetallics grow inwards as shown in Figure 1. The comparatively
soft, Pb-rich layer next to the brittle intermetallics is shown to be an easy
path for crack propagation during thermal cycling.

The second example of the microstructural instabilities was obtained from
fatigue testing of the Sn-Pb eutectic solder. The initially lamellar
eutectic microstructure of the solder as shown in Figure 2, has changed into
an equi-axed grain structure such as in Figure 3, after fatigue testing of
100 cycles at 1 percent strain and 1 cpm frequency. This change is due to
dynamic recrystallization during fatigue deformation at room temperature.
Similar microstructural changes have been observed in other solders (26).
The fine microstructure of the solder is generally known to be beneficial
to the mechanical ductility and fatigue resistance of a solder joint (24,
27). The microstructural instability observed here may have an important
bearing on explaining the long-term mechanical behavior of solder joints.
For example, in predicting a fatigue life of solder joints by using an analytical model, it is generally assumed that the microstructure and therefore
the properties of solder are not changing. In order to have a better prediction, it is necessary to take into account the microstructurel/property
changes in modelling.

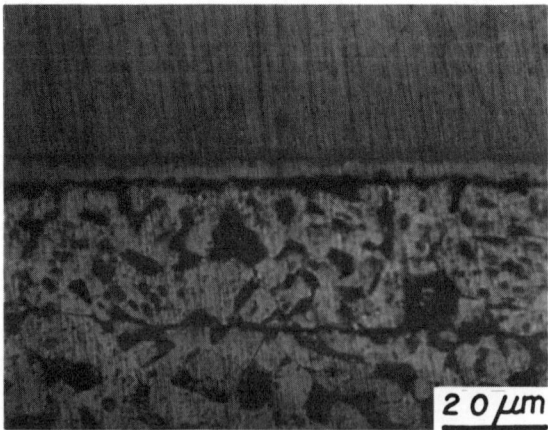

Figure 1.   Solder joints annealed at 150C, 300 h, showing the
            growth of intermetallics as well as the Sn depleted
            region next to the intermetallic layer.

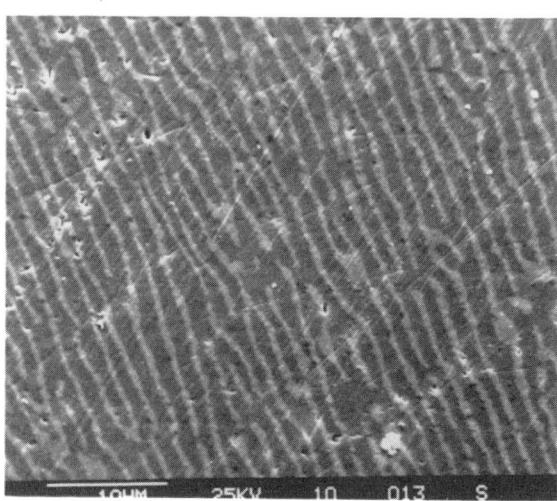

Figure 2.   SEM micrograph of Sn-Pb eutectic solder showing its
            lamellar structure with an average spacing of 2-3
            micrometers.

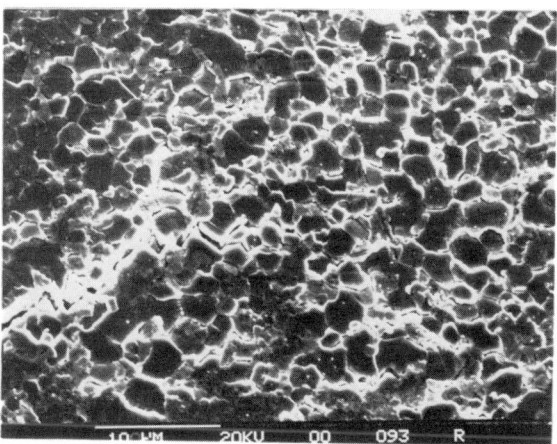

Figure 3.   SEM micrograph of Sn-Pb eutectic solder fatigue tested
            to 100 cycles with 1 % strain, 1 cpm, at room temp.
            Fine, equi-axed grains are the result of dynamic
            recrystallization occurred during the fatigue testing.

In summary, common interconnection technologies used in electronic packaging applications were reviewed in light of how the diffusional processes are important in forming successful interconnections as well as in determining a long-term reliability of the joints. It is clear that the extent of diffusional reactions should be controlled during joint formation to achieve maximum joint integrity. However, the reactions should be minimized further to reduce concerns regarding the long-term reliability. This idea has been successfully used to limit any excessive diffusional reactions in the thin film structures of IC devices by introducing so-called 'diffusion barriers' (28).

REFERENCES

1. C. Bauer and G. Lessmann : Ann.Rev.Mat.Sci., Vol.6, p.361,(1976).
2. I. Blech and H. Sello: J.Electrochem.Soc., Vol.113, p.1052,(1966).
3. M. Kashiwabara and S. Hattori: Rev. Elec. Comm. Lab., Vol.17, No.9, p.1001, (1969).
4. E. Philofsky: Solid St. Elec., Vol.13, p.1391, (1970).
5. J. Newsome, R. Oswald and W. Rodrigues de Miranda: IEEE Proc. 14th Ann.Conf. Reliability Physics, Las Vegas, Nev., p.64, (1976).
6. W. Gerling: IEEE Electr. Comp. Conf., p.13, (1984).
7. G. Clatterbaugh, J. Weiner and H. Charles,Jr.: IEEE Trans. Comp.Hybrid. & Manuf.Tech., Vol.CHMT-7, No.4, p.349, (1984).
8. G. Dehaine and K. Kurzweil: Solid St. Tech., p.46, (1975).
9. M. Hayakawa, T. Maeda, M. Kumura, R. Holly, and T. Gielow: Solid St. Tech., p.52, (1979).
10. T. A. Scharr: Int. J. Hybrid Electrn., Vol.6, No.1, p.561, (1983).
11. T. Kawanobe, K. Miyamoto, and M. Hirano: IEEE Proc. Electr. Comp., p.221, (1983).
12. L. A. DelMonte: Proc. IEEE Electr. Comp. Conf., p.21, (1973).
13. F. A. Lindberg: ASTM STP 850, (D.C.Gupta, ed.), p.512, (1984).
14. A. S. Rose, F. E. Scheline, and T. V. Sikina: Proc. 27th Annual Elec. Comp. Conf., Arlington, VA, p.130, (1977).
15. A. Keizer and D. Brown: Solid St. Tech., p.59, March, (1978).
16. R. J. Klein Wassink: "Soldering in Electronics", Electrochemical Pub. Ltd., Scotland, (1984).
17. ISHM Technical Monograph: "Surface Mount Technology", The Int. Soc. for Hybrid Microelectronics, Silver Spring, MD., (1984).
18. L. F. Miller: IBM J. Res. Develop., p.239, May, (1969).
19. R. K. Shukla and N. P. Mencinger: Solid St. Tech., p.67, (1985).
20. N. G. Ainslie, J. E. Krzanowski, and P. H. Palmateer: U.S.Patent No.4,418,857, Dec.6, (1983).
21. S. K. Kang and V. Ramachandran: Scrpt. Met., Vol.14, p.421,(1980)
22. W. G. Bader: TMS-AIME Fall Meeting, St. Louis, MO, Oct.15-19, (1978)
23. S. K. Kang: Met. Trans., Vol.12B, p.620, (1981).
24. J. K. Lake and R. N. Wild: 28th Nat.SAMPE Symp., Anaheim, CA,(1983).
25. D. S. Dunn, T. F. Marinis, W. M. Sherry and C. J. Williams: MRS Symp. Proc. Vol.40, p.129, (1985).
26. V. Raman and T. C. Reiley: Scrpt. Met., Vol.20, p.1343, (1986).
27. T. C. Reiley and D. Y. Shih: "Package-to-Board Interconnections" in Microelectronic Packaging Handbook (ed.R.Tummala, E.Rymaszewski) to be published.
28. J. R. Shappirio: Solid St. Tech., p.161, (1985).

ESCA AND SIMS STUDY OF THE TETRABROMOBISPHENOL A FLAME RETARDANT
DIFFUSION TO THE METAL-ENCAPSULATING RESIN INTERFACE

ALBERTO TORRISI*, ANGELO CAVALLARO**, ANTONIO PERNICIARO**, GIUSEPPE FERLA**,
SALVATORE PIGNATARO*.
*Dipartimento di Scienze Chimiche, Viale A. Doria 6, 95125 Catania, Italy.
**SGS Microelettronica, Stradale di Primosole 50, 95100 Catania, Italy.

ABSTRACT

Epoxy resins of novolac type were molded to real copper and nickel plated
copper frames in plant conditions. Three resins, formulated with different amount
of tetrabromobisphenol A flame retardant, were used. The samples were stored
at 200°C and then mechanically fractured at given time intervals. The surfaces
obtained in the fracture have been analyzed by ESCA and Ga^+ excited static and
imaging SIMS. The kinetics of migration of the flame retardant to the resin-
metal interfacial zone have been followed. The process was found to be diffusion
controlled. Evidence of a redox reaction between the copper oxides found on the
surface of the copper frame and the tetrabromobisphenol A was obtained. A copper
diffusion through the Ni plating was observed in the case of the nickel plated
frames. This phenomenon is most important at the point on the frame where the
bromine concentration is higher.

INTRODUCTION

In a recent paper [1] different copper surfaces have been molded to a commercial epoxy resin of novolac type. The obtained samples have been stored at
high temperature in the range 150-200°C. These systems have been mechanically
fractured at given time intervals and the surface obtained in the fracture have
been analyzed by ESCA. In this way it was possible to gain information on the
kinetics of migration to the resin-metal interface of the bromine contained in
the tetrabromobisphenol A (TBBPA) flame retardant additive of the resin. In
particular two hypothesis were made. The first was that the migration process
was a diffusion controlled one. This hypothesis was made on the basis of the
observed linear dependence of the bromine concentration at the interfacial zone
with the square root of the storage time. The second hypothesis was that a
chemical reaction occurred at the interface between the TBBPA and the oxides
overlaying the copper frame. This hypothesis was made on the basis of the
observation that the copper oxides disappear from the interfacial zone with the
storage time. The above study was restricted to one type of frame (Cu) and to a
given formulation of the used resin. In this paper the previous work has been
extended to include a different metal frame (Ni plated Cu) and resins having
various content of the brominated flame retardant. This in order to: a) confirm
the above hypothesis; b) have a deeper insight on the influence of the flame
retardant on the resin-metal interfacial chemistry. This work is a part of a
general programme intended to apply surface analysis tools to a fabrication
problems within the semiconductor industry [2-7].

EXPERIMENTAL

The resins used were novolac B type amine cured and were supplied by Hysol
Corporation. Three different type of resins were formulated. The first was with-

out TBBPA, the second was containing a standard amount of this additive (1Br) and the third a double amount of flame retardant to respect the standard one (2Br). The CuBr and CuBr$_2$ pure compounds were supplied by Janssen and were analyzed as pellets. The Cu and Ni electroplated Cu frames from Auge before their use were chemically etched with an acid treatment followed by a degreasing procedure and an electropolishing process. The molding process was done in the plant apparatus. The molding conditions were: T=175°C, P=70kg/cm^2, post curing temperature also 175°C. The high temperature storage (HTS) was done at 150, 175 and 200°C till 1000 hours. Before ESCA and SIMS analysis, the resin molded on the back of the metal frame - in order to ensure a mechanical graft between the metal and the resin itself - was taken off. The interfaces obtained in the rupture of the nominal metal-resin bond were analyzed by using an ESCA ES 300 KRATOS electron spectrometer operating in FRR mode. The radiation used was the Al K$_\alpha$ (1486.6 eV) under a residual pressure of 4 x 10^{-8} torr. The calibration of the spectrometer was achieved according to the recently proposed procedure [5, 8]. The concentration of bromine atoms at the interfaces was obtained from the ESCA spectra by using well known relations [5,6] assuming the C, O, Cu, Ni, Br and Si are the main species that contribute to the surface compositions. The SIMS spectra and images were taken in a SIMSLAB from VG. The primary ions was obtained from a liquid gallium source operating at 10 keV. The primary ion current was kept at low level (typically 10 nanoA). The ion images were obtained setting the target bias for the maximum ion count. For the spectra the target bias was set in order to obtain the maximum count in defocused mode of the quadrupole (zero resolution).

RESULTS AND DISCUSSION

Several samples, prepared by molding epoxy resins containing different amount of TBBPA flame retardant to a copper surface, were stored at high temperature. At given time intervals one of these samples was taken and mechanically fractured. the metal side of the fracture was analyzed by ESCA in order to obtain the percentage of the Br concentration in this point of the system. Fig.1 shows the plot of this concentration vs. the square root of the time of storage at 200°C after the molding. Two different regression lines are obtained. That referring to the systems having a 2Br content in the initial formulation of the resin lies above that of the 1Br containing systems. Both lines obey a simple rate law of the form: $Y = A \sqrt{X} + B$ where X is the time of storage, Y is the Br concentration (%), A is the slope and B is the value of Y when the storage time is zero. This observed time dependence of the migration phenomenon confirms the previous [1] hypothesis of a diffusion controlled migration of the flame retardant (or of a molecular fragment containing Br) to the interfacial zone. The amount of Br arriving at the interfacial zone increases, on the other hand, with the storage temperature and with the amount of Br nominally present in the bulk of the resin, these effects being shown in Fig.2. The second hypothesis made in ref.1 regarded the redox reaction mentioned in the introduction. Also this hypothesis seems to be confirmed by the present study. In particular the study of the interfacial chemistry of the Cu-resin interface shows that the interfacial Cu is oxidized according to the sequence Cu-Cu$_2$O-CuO. Fig.3 shows the Cu$_{2p3/2}$ ESCA peaks for the metal side of the fracture of the metal-resin systems broken after storage at 175°C for various times. For the sample with an amount of flame retardant equal to 2Br, the oxidation to CuO, observed in the sample with only 1Br (left side of the fig.) is strongly retarded. Similar evidence is obtained comparing the results with 0, 1 and 2Br. Fig.4 reports the $Cu_{2p3/2}$ and the Cu_{LMM} Auger peaks for the metal side of the fracture of the Cu-resin systems broken after storage at 200°C for 1000 hours. The figure shows that for the sample molded with a resin without flame retardant (0Br) the storage causes massive oxidation to CuO, the shapes and Binding Energies (B.E.) of the observed main and shake-up ESCA peaks as well as Auger band being those reported [9] for this system. In samples with 1Br and 2Br the formation of Cu^{++} chemical species is strongly reduced. This in agreement with: a) the reduction of relative intensity of the shake-up ESCA peaks; b) the appearance of an ESCA peak at B.E. characteristic of Cu° and Cu^+ species [9]; c) the modification of the Auger band envelope showing shoulders in position characteristic of Cu° (918.8 eV) [9] and Cu^+ (916.6 eV) [9]. All of this support a redox reaction mentioned in the introduction. On the other hand,

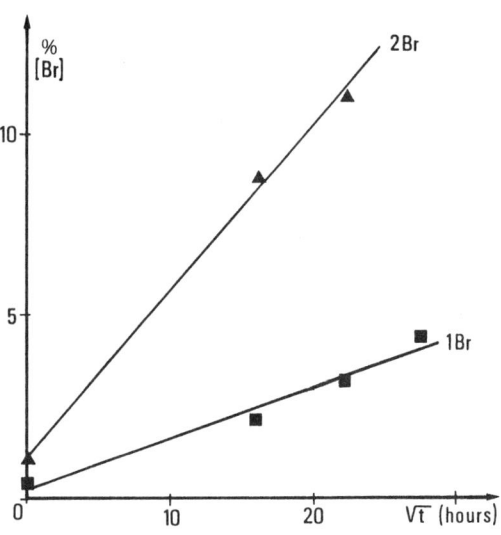

Fig.1. Percentage of bromine atoms on the metal side of the fracture of the Cu-resin joint vs. the square root of the time of storage at 200°C.

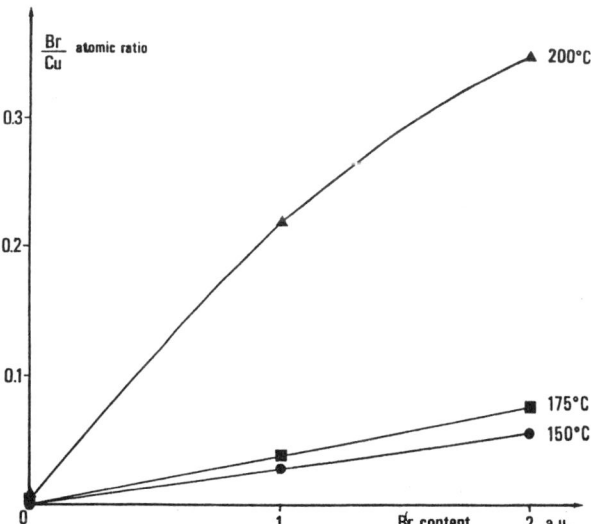

Fig.2. Br/Cu atomic ratio on the metal side of the fracture of the Cu-resin joint vs. the bromine content in the bulk of the resin after 1000 hours of storage at high temperature.

368

this is in agreement with the well known [10] anti-oxidant power of the phenolic moieties of the TBBPA which in turn should act giving quinonic moieties. Further evidence of the strong interaction which occurs between the Cu and the TBBPA is given by the study made on the system having Ni plated Cu frames. Fig.5 shows typical Cu^+, Ni^+, Br^-, O^- SIMS images of the metal side of the fracture of the metal-resin system broken after HTS. This figure demonstrates that copper diffuses through the Ni layer, this diffusion being mainly driven by the bromine itself of a Br containing fragment. This is substantiated by the fact that bromine is found mainly where copper comes on top of nickel. In addition the Cu/Ni atomic ratio on this face increases with the storage time much more rapidly for the systems containing

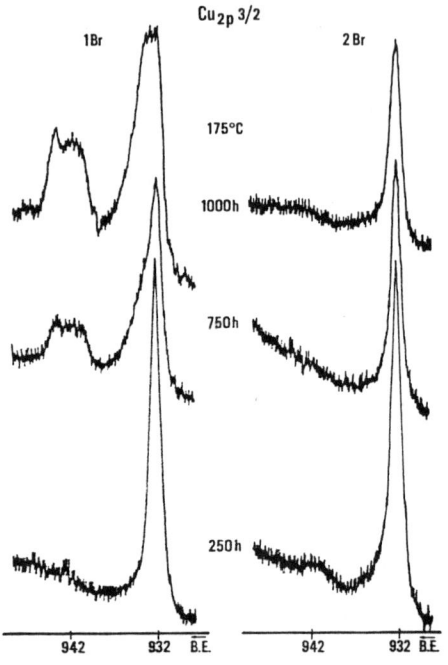

Fig.3. $Cu_{2p3/2}$ ESCA peaks for the metal side of the fracture of the Cu-resin joint stored at 175°C for various time. The resin used were formulated with 1Br and 2Br content of flame retardant.

flame retardant with respect to that without this additive (fig.6). A deeper insight into the chemical nature of the copper and bromine at the interface is obtained comparing ESCA and SIMS data with those of CuBr and $CuBr_2$ pure systems. The strong intensity of the ESCA shoulder at ≈ 933.0 eV in the 2Br system reported in fig.4 would be in agreement with the formation of CuBr, whose B.E. obtained in the present work is just 933.0 eV. SIMS appears to support this indication suggesting CuBr as a possible product of the reaction between the diffused Cu ant the TBBPA. However, the obtained SIMS spectra do not allow us to reach a clear cut conclusion on this point.

ACKNOWLEDGEMENTS

The authors thank the Hysol Corporation for having provided the used resins and for the useful information. CNR Grant (P.F. "Materiali e Dispositivi per l'Elettronica a Stato Solido") is gratefully ackowledged.

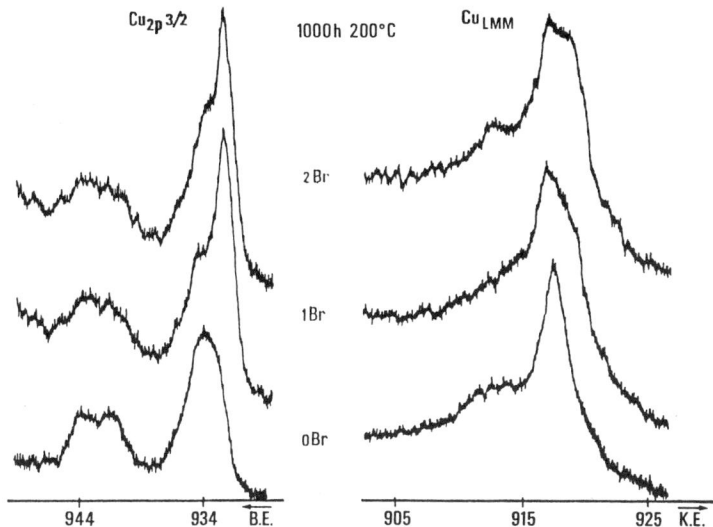

Fig.4. $Cu_{2p3/2}$ ESCA and Cu_{LMM} Auger peaks showing the effect of TBBPA in retarding the oxidation of copper frame

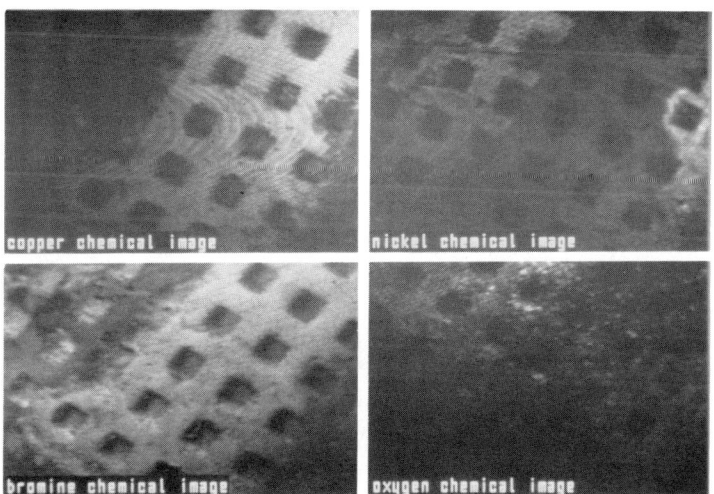

Fig.5. SIMS ion images of the Cu^+, Ni^+, Br^-, O^- on the metal side of the fracture of Ni-resin joint broken after 500 hours of storage at 200°C. the resin used was the 2Br one.

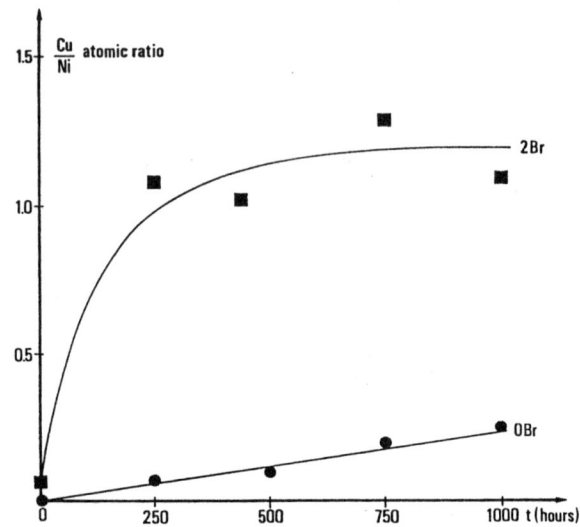

Fig.6. Cu/Ni atomic ratio on the metal side of the fracture
of Ni-resin joint broken after storage (200°C) at
various time

REFERENCES

1. A. Torrisi and S. Pignataro, Surf. Interf. Anal. 9, 441 (1986).

2. A. Torrisi, S. Pignataro, G. Nocerino, Appl. Surf. Sci. 13, 389 (1982).

3. A. Torrisi, G. Marletta, O. Puglisi, S. Pignataro, Surf. Interf. Anal. 5, 161 (1983).

4. S. Pignataro, A. Torrisi, G. Ferla, Surf. Interf. Anal. 7, 129 (1985).

5. S. Pignataro, A. Torrisi, O. Puglisi, A. Cavallaro, A. Perniciaro, G. Ferla, Appl. Surf. Sci. 25, 127 (1986).

6. A. Torrisi, A. Cavallaro, A. Licciardello, A. Perniciaro, G. Ferla, Surf. Interf. Anal. 10, 306 (1987).

7. P. Vasquez, O. Puglisi, A. Licciardello, G.W. Arnold, A. Patti, S. Pignataro, Surf. Interf. Anal. 10, 327 (1987).

8. M.T. Anthony and M.P. Seah, Surf. Interf. Anal. 6, 107 (1984).

9. D. Briggs and M.P. Seah, Practical Surface Analysis, (John Wiley & Sons, Chichester, 1983), p. 496 and references therein.

10. J. Pospisil, in Advances in Polymer Science vol. 36, edited by K. Dusek (Springer Verlag, Berlin, 1980) p. 69.

ENERGY-FREE BONDING OF MATERIALS WITH FINE CONTROLLED SURFACES IN ULTRAHIGH VACUUM

Y.KASHIBA,* K.MACHIDA,* T.OKUDA,* W.SHIMADA,* AND S.NAKATA**
*Mitsubishi Electric Corp., Amagasaki, Hyogo, Japan
**Osaka University, Suita, Osaka, Japan

ABSTRACT

   Bonding phenomena with clean and smooth surfaces at room temperature (293 K) in ultrahigh vacuum has been studied. The metal surfaces, finished to required smoothness in nm level, were bombarded by Ar ions to remove surface contaminants. Then, two cleaned surfaces were put together. Interaction of silver and copper was examined by SEM, AES, and TEM. In order to observe the fundamental phenomena, bonding was attempted between a silver disk with a flat surface and a copper sphere 80 μm in diameter. A bonding area of about 40 μm^2 was confirmed under slight load at 293 K. Strong bonds were achieved and the fracture took place almost completely in silver bulk. Furthermore, it was found that the bonding area obtained by the process was much larger than the one derived from conventional plastic theory.

INTRODUCTION

   In general, welding processes need high energy to achieve bonding. But, the extreme high temperature and pressure easily destroy the unique functional characteristics of semiconductors and other advanced materials. To avoid this deterioration, the development of a bonding process that can be achieved at lower temperature and pressure is required.
   Conventional solid state bonding processes (diffusion welding, pressure welding, etc.) are used for advanced materials, because bonding can be achieved with relatively low energy. For these processes, it has been shown that the surface properties, dominated by oxide layer, surface roughness, and hardness, are very important in getting high quality bonding.[1,2]
   From earlier studies about adhesion in space environment [3,4], it is expected that a low energy bonding should be possible with highly controlled surfaces of materials.
   Fundamental bonding phenomena at room temperature under slight load has been studied. The new bonding process developed here would be useful for connection technology of advanced devices.

APPARATUS

   The experimental set-up for the bonding is shown in Fig.1. This system is composed of three vacuum chambers: preparation, bonding and in-situ AES analysis. Each chamber has vacuum pumps and is closed by a gate valve, so this system enables the bonding chamber to maintain constant ultrahigh vacuum (pressure on the order of 10^{-9} Pa).
   The specimens are mounted on holders and moved by magnet transfer between chambers. The preparation chamber has an entry port and specimens are set up on the jig. Cleaning of specimen surfaces is done with an ion gun in this chamber. In the bonding chamber, specimens on holders are attached to upper and lower jigs which are bellows sealed and can be moved up and down in a vertical direction. In-situ. AES analysis can be carried out on specimen surfaces and fractured surfaces.

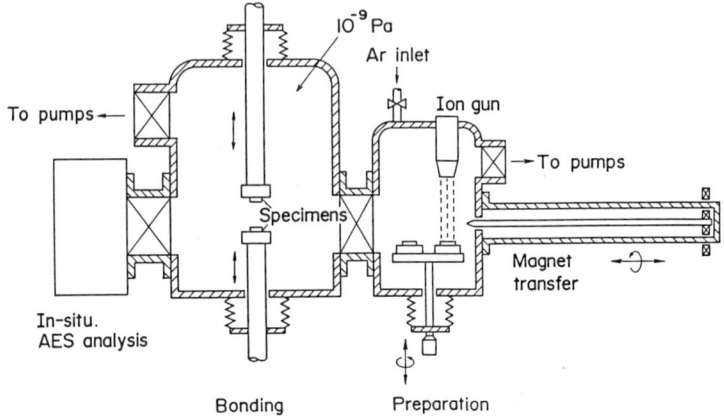

Fig.1. Schematic drawing of the bonding apparatus.

EXPERIMENTAL PROCEDURE

Silver of 99.99% purity and copper of 99.999% purity were used. For preexamination, silver and copper disks of 2 mm in diameter were bonded. To research fundamental bonding phenomena, the contact between silver with a flat surface and copper with a spherical surface was examined. The surfaces of silver and copper disks were diamond turned and were finished to peak-to-valley roughness of about 0.02 μm. The end of copper wire 25 μm in diameter was melted to form a copper sphere 80 μm in diameter by arc discharge in Ar atmosphere.

The bonding process is as follows. At first, specimens are set into the preparation chamber. The chamber, baked out for 8 hours during pump-down procedure, achieve a vacuum of 3×10^{-6} Pa. Next, Ar gas is introduced into the chamber to 6×10^{-3} Pa. Ar ions can be accelerated to high speeds by an electric field and they then remove the layer of contamination. In general, the higher the beam energy, the better the cleaning efficiency, but at excessively high energy, the ions may roughen the smooth surface. In this study, 1 kV acceleration voltage is used for 5 min. to clean the surface. The removal rate is about 100 Å/min. for silver. Then, specimens are transferred to the bonding chamber and put together.

RESULTS AND DISCUSSION

First, in order to confirm the bondability at 293 K, silver and copper disks were bonded. The Auger spectra from the copper surfaces are shown in Fig.2. As shown in the upper spectrum, contaminants of oxygen and carbon are detected on the diamond turned surface before cleaning, and bonding can not be achieved. On the other hand, after cleaning, the surface changes to pure silver. Tensile strength of 220 N was achieved. In this bonding, specimens were exposed for 20 sec. in 5×10^{-6} Pa, and 200 N load was applied for 30 sec. to get apparent contact overall. Cross-sectional TEM image of the bond is shown in Fig.3. The interface is slightly tilted in this figure, but bonding is achieved successfully between silver and copper.

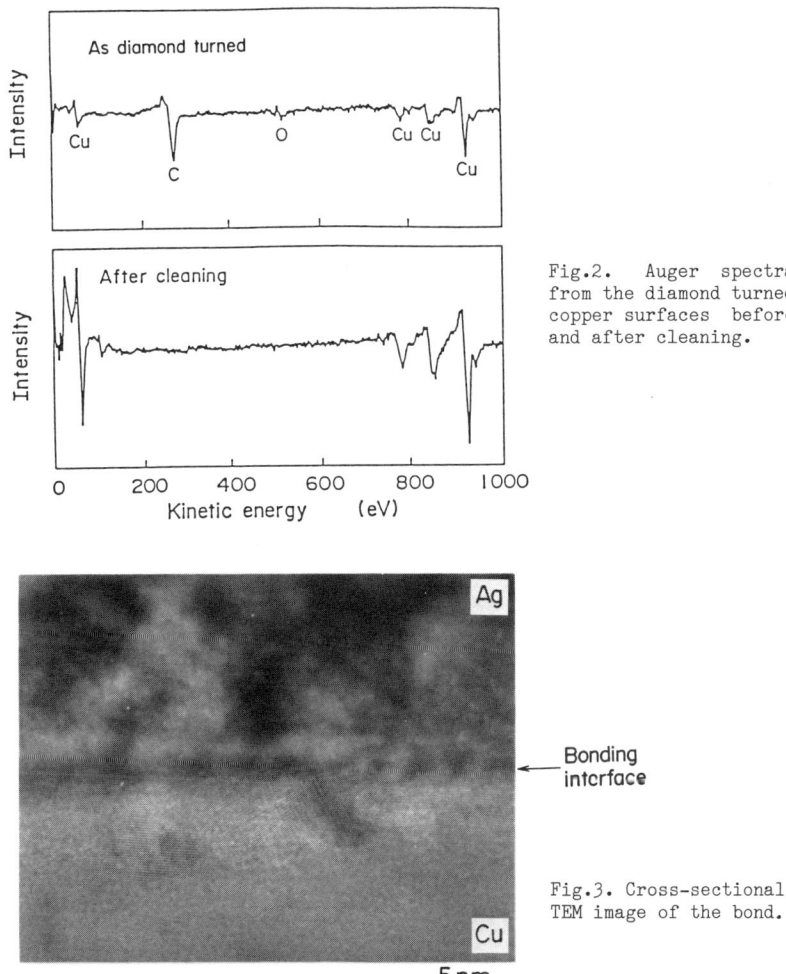

Fig.2. Auger spectra from the diamond turned copper surfaces before and after cleaning.

Fig.3. Cross-sectional TEM image of the bond.

Next, to examine the fundamental bonding phenomena, silver with flat surface and copper with spherical surface were bonded under only 40 ± 10 μN load for 10 sec. The bond is shown in Fig.4. Bonding is achieved without macroscopic deformation. Then, the bond is fractured to evaluate the properties of interface. Fractured surfaces and Auger spectra from those surfaces are shown in Fig.5. It can be seen that the fractured surfaces (a) have many dimples and they show good matching between the silver side and the copper side. Auger spectra (b) were taken from those surfaces. Auger spectrum from the silver side shows only silver peaks. On the other hand, from the copper side, the fractured surface is almost all silver and only a slight amount of copper. These results show that the bond is strong and the fracture takes place almost completely in the silver bulk.

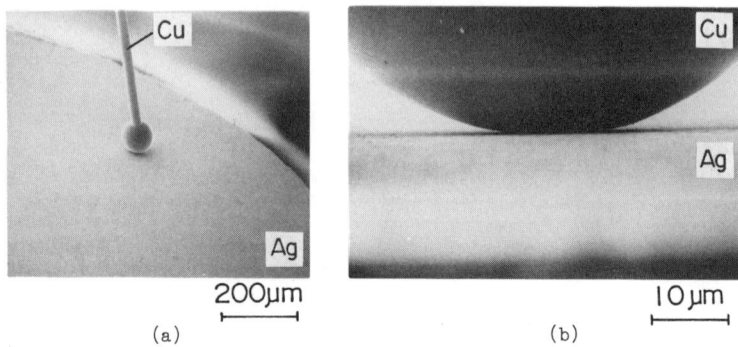

Fig.4. Appearance of the bond.
(a) oblique view   (b) magnification of (a)

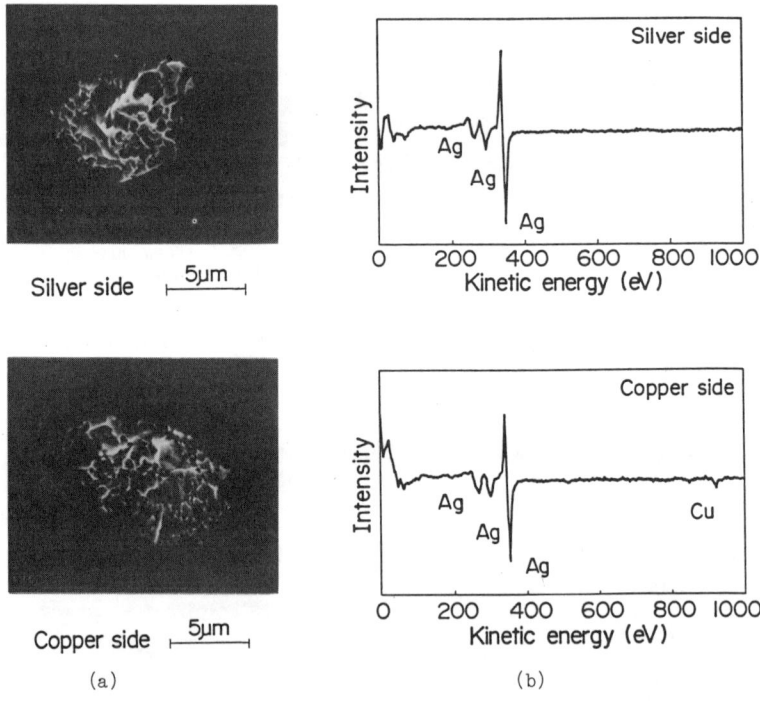

Fig.5. Fractgraphs of the bond (a) and Auger spectra from
the fractured surfaces (b).

The bonding area, measured from the fractured surface, is about 40 μm², so the bonding is achieved at only about 0.9 MPa pressure and this means that the bonding is achieved under the elastic condition.

According to Hertz's equation, the region of elastic contact with sphere and half space will be bounded by a circle of radius r, where

$$r=\left\{\frac{3}{4}\left(\frac{1-\nu_1^2}{E_1}+\frac{1-\nu_2^2}{E_2}\right)F\cdot R\right\}^{1/3} \tag{1}$$

where F is the applied load, R is the radius of sphere, E_1, E_2 are Young's moduli and ν_1, ν_2 are Poisson's ratios for the sphere and the half space. As the load is increased, the mean pressure $Pm(=F/\pi r^2)$ increases until it reaches the value of critical point of elastic deformation. After this point, deformation conforms to the plastic formula, where

$$Pm = C \cdot \sigma y \tag{2}$$

where σy is the elastic limit of the softer metal, C is a coefficient value which increases from 1.1 to 3.0 as the load increases.[5]

Relation between applied load F and contact area S is obtained from these equations and shown in Fig.6. In this figure, solid lines are drawn on the assumption that R=40 μm, E_1=12.98×10¹⁰N/m², E_2 =8.27×10¹⁰ N/m² , ν_1=0.343,ν_2=0.367, σy=40 MPa. Plastic deformation occurs at a area larger than 0.013 μm², and applied load of a few thousands μN is necessary to get an area of 40 μm². This result is not in agreement with measured values, that is, much larger contact area is obtained in experiment.

The state of contact in ultrahigh vacuum was compared with that in air. Cross-sectional outlines of contact between a flat surface and a sphere are shown in Fig.7. The specimen bonded in ultrahigh vacuum is half-polished and the values of G and D are measured. Contact of copper sphere and transparent sapphire in air is evaluated by observing the Newton rings which appear every additional 0.28 μm of D value. Furthermore, the theoretical results are shown by solid and broken curves. The result obtained in air is closely in agreement with the theory but the result in ultrahigh vacuum is not in agreement.

This disagreement seems to be caused by interaction force or atomic migration between clean surfaces in ultrahigh vacuum and it is important to clarify the mechanism of this new bonding process.

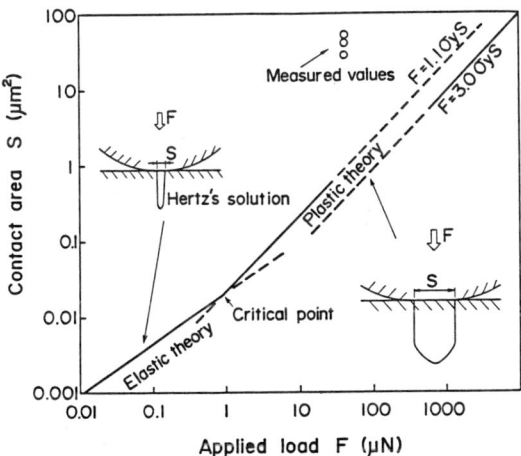

Fig.6. Relation between applied load and contact area derived from both elastic and plastic theories.

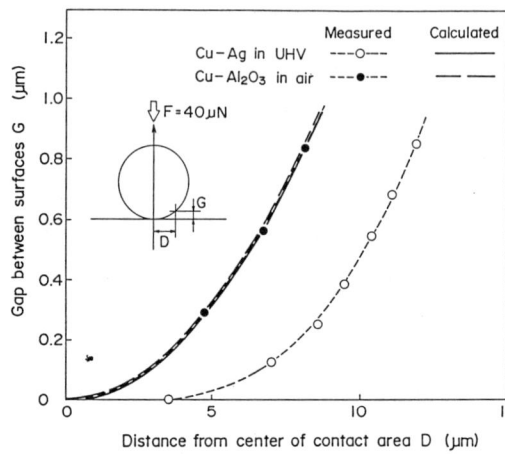

Fig.7. Cross-sectional outlines of contact compared with in ultra-high vacuum and in air.

CONCLUSIONS

A silver disk with flat surface and a copper sphere 80μm in diameter were cleaned and put together at room temperature in ultrahigh vacuum, and characteristics of bonds were examined. It was clarified that strong bonds could be achieved at 293 K under slight load. The bonding area obtained by the process was much larger than the one derived from plastic theory.

It was proved that the new bonding process could be applied to micro connection technology for electronics devices.

REFERENCES

1. M.Kobayashi, J.Hirota, N.Watanabe, and K.Machida, in Thin Films-Interfaces and Phenomena, edited by R.J.Nemanich, P.S.Ho, and S.S.Lau, (Mater.Res.Soc. Proc. 54, 1986) pp.793-798.
2. Z.A.Munir, Welding Journal, 62, 333s, (1983)
3. D.H.Buckley, ASTM Special Technical Publication, 431, 248, (1976)
4. Gilbert H. Walker and Beverley W. Lewis, Metallugical Transactions 2, 2189, (1971)
5. F.P.Bowden and D.Tabor, Friction and Lubrication of Solids, (Oxford University Press, London, 1950) ,pp.10-14.

LATERAL DIFFUSION OF $NiAl_3$ AND Ni_2Al_3

JOYCE.C. LIU AND J.W. MAYER
Department of Materials Science and Engineering, Cornell
University, Ithaca, NY 14583

ABSTRACT

Phase formation of $NiAl_3$ and Ni_2Al_3 were studied individually in lateral diffusion couples A and B and simultaneously in sample C. The lateral growth of certain phases was controlled by varying the thickness ratios of Al to Ni at the source and thin film area. Parabolic dependence of the phase width on annealing time was found for $NiAl_3$ and Ni_2Al_3 in the one-phase and two-phase growth at 425°C.

INTRODUCTION

There are four intermetallic phases, $NiAl_3$, Ni_2Al_3, $NiAl$ and Ni_3Al, in the Al-Ni system as shown in Fig 1[1]. In conventional bulk diffusion studies, all four equilibrium phases were formed after a long anneal at 600°C[2]. The reacted zones extended from several to hundreds of microns. However, only one phase grew at a given time in the thin film reaction, and the reacted region was in the order of 100 nm[3]. In order to study the difference of phase growth between thin film and bulk couples, lateral diffusion couples were developed for metal-silicon system[4]. The lateral diffusion couples consist of a thin film with a source deposited on top, where multi-phase growth can take place and the reacted width is about several microns. The difference between thin film and lateral diffusion couples for Al-Ni system is shown in Fig. 2.

(a) Thin Film Couple

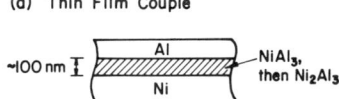

(b) Lateral Diffusion Couple

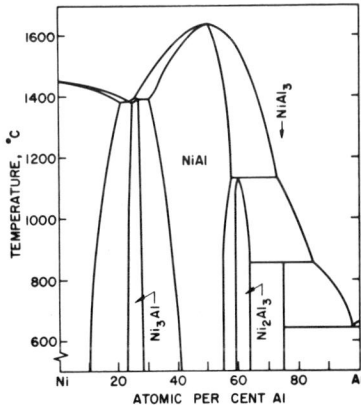

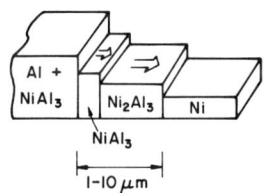

Fig. 1. Al-Ni phase diagram.

Fig. 2. Schematic diagrams of thin film and lateral diffusion couples.

Mat. Res. Soc. Symp. Proc. Vol. 108. ©1988 Materials Research Society

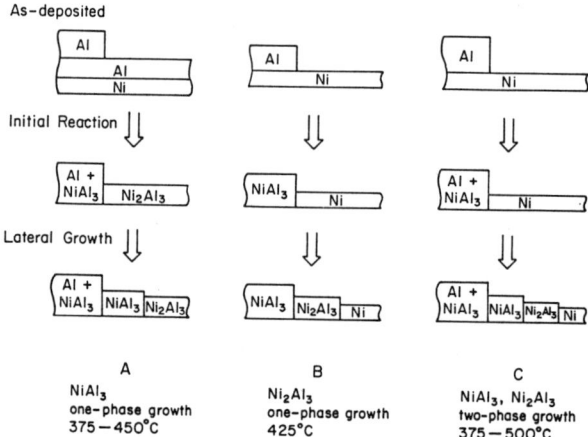

Fig. 3 Schematic diagrams of sample A, B and C at their as-deposited, initial reaction and lateral growth stages.

Table 1. Thickness of Al and Ni at source and thin film area

	Al source(nm)	Ni or Al/Ni film(nm)	Al/Ni at. ratio source	thin film
A	260	105/43	5.6	1.6
B	275	59	3.07	
C	420	52	5.3	

In this report, three types of lateral diffusion couples were designed to study the phase formation of $NiAl_3$ and Ni_2Al_3 individually and simultaneously. The schematic diagrams of the three samples at their as-deposited stages are given in Fig. 3, where sample A is composed of $125 \times 125 \mu m^2$ Al islands on an Al/Ni bilayer and sample B and C consist of Al islands on a single Ni thin layer. The thickness of the Al islands, Al and Ni thin films are given in Table 1. The resulting kinetic data at 425°C, was analyzed in detail.

RESULTS AND DISCUSSION

Phase Formation

It is known that the end products of thin film reactions are determined by the supply of components as well as annealing temperature and time. In the study of Al/Ni thin film reaction[3], $NiAl_3$ was the initially formed phase. The $NiAl_3$ phase continued to grow until all the Al was consumed, then the next phase started to form. The final compound was eventually governed by the thickness ratio of Al to Ni. Since lateral growth of phases take place after the completion of vertical reactions in the lateral diffusion couples, the formation of certain phases can be investigated by controlling the Al/Ni thickness ratios at the source and thin film areas.

In sample A, the atomic ratio of Al to Ni at the source and thin film regions was 5.6:1 and 1.6:1 respectively, so that initial reaction produced $NiAl_3$ plus extra Al at the source and

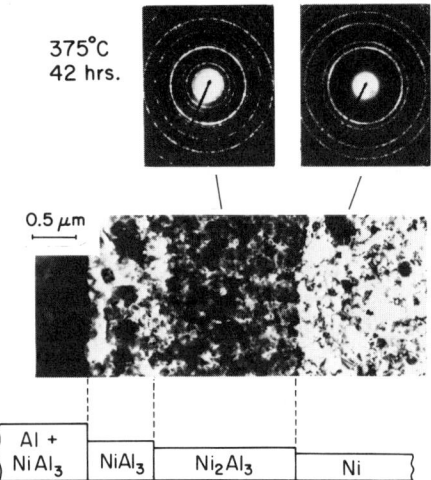

375°C
42 hrs.

0.5 μm

| Al + NiAl₃ | NiAl₃ | Ni₂Al₃ | Ni |

Fig. 4. TEM micrograph of sample C annealed at 425°C for 42 hrs.

Ni_2Al_3 at the thin film area. During further annealing, Al atoms diffused out from the source through the $NiAl_3$ and reacted with the previously formed Ni_2Al_3. Therefore, $NiAl_3$ grew laterally, i.e., the $NiAl_3/Ni_2Al_3$ interface moved in the same direction as the Al flux and the source/$NiAl_3$ interface moved in the opposite direction.

The atomic ratio of Al to Ni (3.04:1) at the source region of sample B determined the fact that there was almost no remaining pure Al at the source area after the initial reaction was completed. Ni_2Al_3 can only grow laterally by decomposing the pre-formed $NiAl_3$ at the source and diffusing Al atoms through the newly formed Ni_2Al_3. Compared with sample A and B, there was semi-infinite amount of Al (in the source) and Ni (in the thin film) for lateral growth of phases in sample C. First, the initially formed $NiAl_3$ grew laterally. Since the Al flux in $NiAl_3$ was reduced as the width of $NiAl_3$ increased, the next phase, Ni_2Al_3, then formed at growth fornt and grew simultaneously with $NiAl_3$. Schematic diagrams of the initial and lateral rection stages for the three samples were shown in Fig. 3. The microstructure of a sample C annealed at 375°C for 42 hrs. was shown in Fig 4.

Kinetics

The width of the growing phases, X, was plotted versus the square root of annealing time, $t^{1/2}$, for $NiAl_3$ and Ni_2Al_3 in sample A, B and C at 425°C in Fig. 5. Some aspects shown in Fig. 5 are: (1) the parabolic dependence of phase width on annealing time; (2) the different slopes, $(X/t^{1/2})$, for $NiAl_3$ growing between saturated neighboring phases(sample A) and growing with Ni_2Al_3 simultaneously(sample C); (3) in sample C, Ni_2Al_3 formed after $NiAl_3$ was about 0.5μm wide.

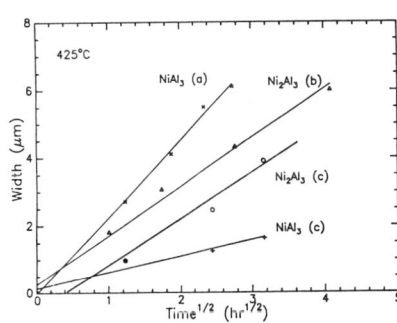

425°C

NiAl₃ (a)   Ni₂Al₃ (b)

Ni₂Al₃ (c)

NiAl₃ (c)

Width (μm)

Time$^{1/2}$ (hr$^{1/2}$)

The parabolic dependence was found for $NiAl_3$ in one-phase growth processes and for $NiAl_3$ and Ni_2Al_3 in two-phase growth processes from bulk diffusion couples[2]. The parabolic dependence of

Fig. 5. Lateral growth kinetics of $NiAl_3$ and Ni_2Al_3 at 425°C in sample A, B and C.

phase growth implies a diffusion-controlled phase growth, whereas interface reaction-controlled process would lead to a linear dependence[5]. The difference of growth rates in one-phase and two-phase growth processes for $NiAl_3$ can be explained by the kinetics of multi-phase growth. Based on the sample geometries and experimental results, equations were derived[6], where it showed the growth rate of $NiAl_3$ in an one-phase growth process was determined by the Al flux in this phase only, but the rate was reduced by the Al flux in its adjacent phase when two phases growing simultaneously.

Diffusivities

Al was the dominant moving species in the lateral growth of $NiAl_3$ and Ni_2Al_3 at 425°C. Therefore, the diffusivities of Al in $NiAl_3$ and Ni_2Al_3 can be estimated from the growth constants, X^2/t, in one-phase growth processes. In terms of the potential change[7], the diffusion coefficient of Al in $NiAl_3$ was estimated to be 9.8×10^{-13} cm^2/sec. The diffusivity of Al in Ni_2Al_3, 1.6×10^{-11} cm^2/sec, was evaluated from the concentration gradient.

CONCLUSION

The growth of $NiAl_3$ and Ni_2Al_3 was studied individually in sample A and B, and simultaneously in sample C, which had different configurations. Al was the dominant moving species in the growth, and $NiAl_3$ was the initial phase. Kinetic results at 425°C showed a parabolic dependence of phase growth on annealing time, which implied a diffusion-controlled process. Hence, the diffusivities of Al were calculated.

ACKNOWLEDGEMENTS

The authors acknowledge the Cornell Materials Science Center for the use of the facilities. The research at Cornell was supported in part by the National Science Foundation through Cornell Materials Science Center.

REFERENCES

[1] M. Hansen, Constitution of Binary Alloys(McGraw-Hill, New York, 1958) p.118.

[2] M.M.P. Janssen and G.D. Rieck, Trans. TMS-AIME 239, 1372 (1967).

[3] E.G. Colgan, M. Nastasi, and J.W. Mayer, J. Appl. Phys. 58, 4125 (1985).

[4] L.R. Zheng, L.S. Hung, and J.W. Mayer, J. Vac. Sci. Technol. A1, 758 (1983).

[5] U. Gösele and K.N. Tu, J. Appl. Phys. 53, 3252 (1982).

[6] J.C. Liu, J.W. Mayer, and J.C. Barbour, (submitted to J. Appl. Phys.).

[7] U. Gösele, K.N. Tu, and R.D. Thompson, J. Appl. Phys. 53, 8759 (1982)

DEFORMATION BEHAVIOR OF LEADLESS 60% Pb-40% Sn SOLDER JOINTS

RAVICHANDRAN SUBRAHMANYAN, DONALD STONE and CHE-YU LI
Department of Materials Science and Engineering,
Cornell University, Ithaca, NY 14853

ABSTRACT

Room temperature deformation data of leadless solder joints
are reported. The joints were sheared under cyclic, displacement
controlled loading at frequencies between 0.001 and 0.01 Hz. A
microplastic model was utilized to simulate the stress-strain
loops, which demonstrated a pronounced Bauschinger effect. The
implications of microplasticity on fatigue life of solder joints
are discussed. This phenomenon must be taken into account in an
accurate prediction of solder deformation at low strain ranges.

INTRODUCTION

Failure life during low cycle creep fatigue is sensitive to
stress and nonelastic strain range. The first step, therefore, in
predicting the fatigue life of a solder joint under service conditions is to estimate the stresses and strains it encounters during a thermal or power cycle. Already, a model has been proposed
for solder joint configurations of simple geometry [1]. This
model makes use of the flow properties of the solder, the elastic
behavior of the soldered assembly and solder joint, and the expansion properties of the soldered assembly. The model has had success in predicting the stresses and strains undergone by solder
joints under thermal cycle conditions when the nonelastic strain
ranges were relatively large (~4%). However, for the smaller
strain ranges of less than 1% often encountered in actual service,
the model has not been tested. In particular, the flow laws used
for characterizing plasticity of metals generally provide a rather
poor description of deformation at low strain ranges, where microplasticity can be important [2]. At low strain ranges, microplasticity results in wider stress-strain loops than would be predicted based solely on plastic flow behavior.
    This article extends the base of data for cyclic deformation
of solders to low strain ranges and incorporates low strain range
effects into the description of solder flow behavior. The conse-
quences of introducing these effects into a fatigue correlation
are discussed.

EXPERIMENTAL WORK

Fatigue specimens were manufactured by soldering together 0.5
mm-thick ceramic substrates by using 0.625 mm diameter 60% Pb-40%
Sn solder balls. Each specimen contained four solder joints on
circular, 1 mm diameter metal pads. Some ceramic substrates were
plated with Ti-Ni-Pd pads and others were deposited with Cr-Cu-Au
pads to act as base metallizations for the solders. No significant difference in solder flow behavior was detected amoung specimens of different metallization.
    Reflow was performed in air under a controlled temperature vs
time profile. The soldered assemblies were held at 218°C for 35

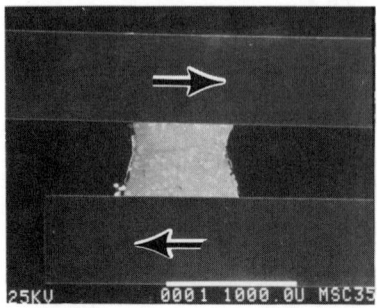

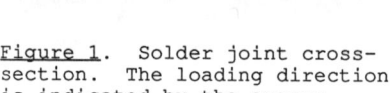

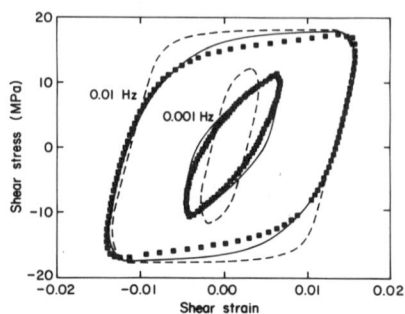

Figure 1. Solder joint cross-
section. The loading direction
is indicated by the arrows.

Figure 2. Stress-strain data
and model simulations. The
solid curves incorporate microplasticity.

seconds before air-cooling to 25°C. In all, the specimens spent
150 seconds above the eutectic temperature. Solder joint shape
was controlled by fixing the positions of the two ceramic pieces
during soldering. The cross section of a typical solder joint is
shown in Figure 1.

Specimens were cyclically deformed in shear under displacement
controlled loading as shown in Figure 1. A triangular waveform
was used with frequencies ranging between 0.001 Hz and 0.01 Hz.
The load and displacement of the soldered assembly were monitored
continuously with an inline load cell and a capacitive displacement gauge with sensitivity of 25 nm. The reported stresses and
strains were calculated based on load and displacement measurements and height and nominal area of the solder joint.

RESULTS AND DISCUSSION

Typical stress-strain curves are shown as data points in
Figure 2.. The low yield stress observed after each displacement
reversal can be regarded as a manifestation of the Bauschinger
effect [3].

To simulate the curves, we have adapted a mechanical model
originally used by Stone, et al. [1] to describe thermal displace-
ment-controlled loading of solder joints. The shear stress in the
solder joint may be obtained by integrating the equation:

$$\dot{\tau} = S\, G_s\, [\dot{\epsilon}_0 - \dot{\epsilon}_n] \qquad (1)$$

where S is the combined machine-solder joint stiffness, G_s is the
shear modulus of the solder, $\dot{\epsilon}_0$ is the applied displacement rate
divided by the solder joint height, and $\dot{\epsilon}_n$ is the nonelastic strain
rate of the solder.

The room temperature plastic flow properties of these solder
joints have been measured and are reported in the form of load relaxation curves in Ref. [4]. Based on these properties and Eq. 1,

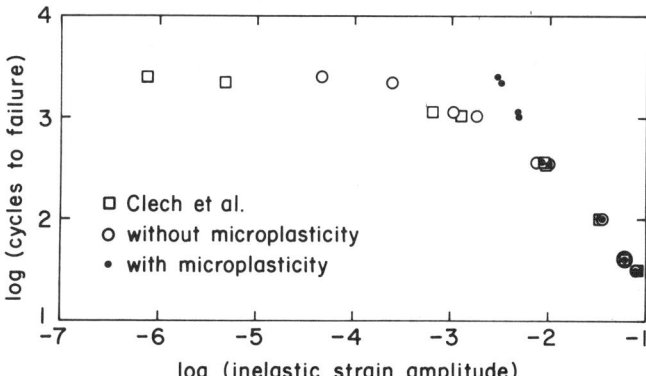

Figure 3. Number of cycles to failure (N_f), measured by Sherry and Hall [7], versus inelastic strain range ($\Delta\varepsilon_n$), calculated by Clech and Augis [8] and in this work. N_f is also plotted versus inelastic strain range ($\Delta\varepsilon_n$), calculated based on model incorporating microplasticity.

simulations of the stress-strain loops are shown as dashed curves in Figure 2.

It is seen that, upon displacement reversal the specimen appears to yield at a much lower level of stress than predicted based on the plastic flow model. This, the Bauschinger effect can be attributed to the presence of unrelaxed internal stresses [5,6].

Due to the complexity of the solder microstructure, a number of parallel deformation processes may cause the effect. For the present purposes, without regard to the specific mechanisms involved, a phenomenological model of microplasticity has been adapted from Jackson, et al. to simulate the behavior of the solder joints. Details of the model are found in Ref. [2]. The model is able to simulate the observed behavior quite well, as shown in Figure 2. Parameters used in the simulation are $M_1 = 3\,G_s$, $M_2 = 0.33\,G_s$, and $m_2 = 3$, where M_1, M_2, and m_2 are defined in Ref. [2]. The plastic flow law used is described in Ref. [1].

An important consequence of microplasticity is that the solder is much "softer" at low levels of displacement than predicted based on an analysis that fails to include microplasticity. For instance, the measured strain range of the 0.01 Hz curve in Figure 2 agrees well with predictions of both models, whereas only the model that incorporates microplasticity can be used to predict the strain range of the 0.001 Hz curve. The smaller the displacement range is, the larger is the discrepency between elastic-plastic analysis and the actual material behavior.

Inelastic strain range ($\Delta\varepsilon_n$) is often correlated with cycles to failure (N_f). Recently, Clech and Augis [9] attempted this correlation for the thermal fatigue data of Sherry and Hall [7]. The inelastic strain ranges were estimated based on a model similar to ours but without microplasticity. The $N_f-\Delta\varepsilon_n$ data, replotted in Figure 3, shows that the trend of the data obtained at low thermal displacements corresponding to low strain ranges deviates significantly from that of the high thermal displacements. It should be noted that the deviation at low strain ranges is not that expected based on isothermal fatigue data at the corresponding strain ranges. We have recalculated the strain ranges with and without microplasticity. Without microplasticity, our predictions agree with that of Clech and Augis [8] quantitatively at high strain ranges and qualitatively at low strain ranges, where slight differences in the models can cause big effects. On the other hand, the trend in the data when microplasticity is taken into account is similar to that observed in isothermal fatigue.

CONCLUSIONS

Isothermal, displacement controlled fatigue data of solder joints demonstrate a pronounced Bauschinger effect caused mainly by microplasticity of the solder. Microplasticity and macroscopic strain locallization must be taken into account in predicting solder joint response to small thermal displacements, around and below the 1% strain range.

ACKNOWLEDGMENTS

The authors are thankful to P. M. Hall of AT&T Bell Labs, Allentown, PA for kindly providing the Ti-Ni-Au metallized substrates. They also thank J.-P. Clech for sending information about his calculations and V. Raman for his stimulating discussions. The Cr-Cu-Au metallizations were deposited at the Technical Operations Laboratory at Cornell, funded by the NSF. This work was performed in part at the National Nanofabrication Facility supported by the NSF under Grant ECS-8619049. This work was supported by the Semiconductor Research Corporation.

REFERENCES

1. D. Stone, S.-P. Hannula, and C.-Y. Li, Proc. 35th Electronics Components Conference, IEEE, New York NY, 1985, pp.46-51.
2. M. S. Jackson, C. W. Cho, P. Alexopoulos, and Che-Yu Li, J.Eng. Mat. Tech., 103, 314 (1981)
3. F.A. McClintock and A.S. Argon, Mechanical Behavior of Materials, (Addison-Wesley, Reading MA., 1968), pp. 184-186.
4. D. Stone, H. Wilson, R. Subrahmanyan, and C.-Y. Li, Proc. National Electronic Packaging Conference East, June, 1986 pp. 175-180.
5. R. Sowerby and D.K. Uko, Mat. Sci. Engr. 41, 43 (1979).
6. H. Mughrabi, in Constitutive Equations in Plasticity, edited by A. S. Argon (MIT Press, Cambridge, Mass., 1975) p. 199
7. W. M. Sherry and P. M. Hall, Proc. 3rd International Conference in Interconnection Technology in Electronics, Fellbach, W. Germany, Feb. 18-20, 1986, pp 47-81
8. J.-P. Clech and J.A. Augis, in Proceedings, 7th Annual International Electronics Packaging Society Conference, Boston MA, 1987.

THERMODYNAMICS OF WETTING BY LIQUID METALS

F. G. YOST and A. D. ROMIG, JR.
Sandia National Laboratories
Albuquerque, NM 87185

ABSTRACT

The wetting of metal surfaces by molten solder is usually considered to be driven solely by an interfacial energy imbalance. The effect of chemical reactions on the wetting process is neglected, although the growth of an intermetallic layer in the wetted interface is commonly observed. In this work, the energy release during the incremental advance of a spreading solder droplet due to the interfacial energy imbalance and the formation of the intermetallic layer is calculated. The free energy of formation, ΔG, of the intermetallic layer is shown to be an important driving force for solder wetting.

This approach to wetting has been applied to three systems, Cu-Sn, Cu-Sb and Cu-Cd. Liquid Sn, Sb and Cd react with solid Cu to form $Cu_6Sn_5(\eta)$, $Cu_2Sb(\gamma)$ and $CuCd_3(\epsilon)$, respectively. The free energy of formation, ΔG, for these intermetallic compounds is unknown experimentally, but can be calculated from the phase diagrams and other solution data using classical thermodynamics. These thermochemical calculations yield $\Delta G(\eta) = 465-3.09T$ for Cu-Sn, $\Delta G(\gamma) = -2500+0.54T$ for Cu-Sb and $\Delta G(\epsilon) = -825+0.44T$ for Cu-Cd (cal/mole). These relations were evaluated at the respective melting temperatures and compared with the interfacial energy exchange. In all three cases the ΔG contribution was approximately two orders of magnitude larger than the interfacial energy exchange making it the dominant driving force for wetting kinetics.

INTRODUCTION

Since the original work of Thomas Young in 1805 [1], a substantial amount of progress has been made in the understanding of the wettability of solids. Three recent reviews [2-4] cited a total of 505 references on the subject of wetting. The familiar Young expression for the equilibrium wetting angle has been derived in many ways and is valid for many weakly interacting systems such as polymers on steel surfaces. For metal/metal systems wetting often results in the formation of intermetallic compounds. Several kilocalories per mole of compound are lost by the system in a typical intermetallic reaction. For example, in the technologically important system Cu-Sn, the heat of formation of the compound Cu_3Sn is measured to be approximately 1800 calories per mole [5]. This suggests that the surface tension imbalance which drives the wetting process in weakly interacting systems may be an insignificant energy loss in reacting systems. Earlier work [6,7] has suggested this, but no formal treatment of wetting in reacting metal systems is known to the authors. In this work, a pure solid metal, A, will be assumed wetted by a pure liquid metal, B, resulting in the formation of an intermetallic compound, C, having the stoichiometry A_xB_y. The energy lost to the system due to the incremental advance of the wetting front will be calculated. The free energy of formation of compounds in three model systems is then calculated with thermodynamic data in the literature. The magnitude of this driving force is then shown to be much larger than that due to the surface tension imbalance.

ANALYSIS OF REACTIVE WETTING

Consider a clean, oxide free, pure metal substrate, A, on which there is a spreading droplet of oxide-free metal B. (Analyses of wetting have consistently neglected the important role of flux, especially in technologically important systems. It is neglected here also to simplify the analysis in an effort to emphasize the importance of reaction free energies.) The two metals react during spreading thus forming an intermetallic compound C, according to the reaction

$$xA + yB \rightarrow A_x B_y$$

As the circular drop spreads an incremental distance, dr, energy is lost to the environment in the amount dE. The process is assumed to yield the above mentioned compound having uniform thickness, ℓ_C, and to consume a uniform thickness of A amounting to ℓ_A as shown in Figure 1. Over a considerably longer period of time, C grows in thickness due to solid state diffusion, but this is neglected here. During the wetting advance, dn_A moles of A and dn_B moles of B are transferred into the compound C, yielding dn_C additional moles of C. These molar amounts are related by

$$\frac{dn_A}{x} = \frac{dn_B}{y} = dn_C$$

and since A and B are pure

$$dn_A = \frac{\rho_A}{M_A} dV_A \qquad \text{and} \qquad dn_B = \frac{\rho_B}{M_B} dV_B \qquad (1)$$

where ρ is density, M is molecular weight and V is volume, therefore,

$$\frac{\rho_A}{xM_A} dV_A = \frac{\rho_B}{yM_B} dV_B. \qquad (2)$$

The energy lost by the system in this incremental spreading is

$$dE = (\mu_A^C - \mu_A)dn_A + (\mu_B^C - \mu_B)dn_B + 2\pi r(\gamma_{AC} + \gamma_{BC} - \gamma_A + \gamma_B Cos\phi)dr$$

where μ_A is the chemical potential of metal A, μ_A^C is the chemical potential of metal A in compound C, γ_{AC} is the energy of the AC interface, γ_{BC} is the energy of the BC interface, γ_A is the surface energy of A, γ_B is the surface energy of B and ϕ is the wetting angle. This can be rewritten with equations 1 and 2 as

$$dE = \{x(\mu_A^C - \mu_A) + y(\mu_B^C - \mu_B)\} \frac{\rho_A}{xM_A} dV_A + 2\pi r \Gamma(\phi)dr \qquad (3)$$

where

$$\Gamma(\phi) = \gamma_B(\cos\phi - \cos\phi_e) \quad \text{and} \quad \cos\phi_e = \frac{\gamma_A - \gamma_{AC} - \gamma_{BC}}{\gamma_B} \qquad (4)$$

where ϕ_e, the equilibrium wetting angle, is always less than ϕ. Furthermore, from Figure 1,

$$dV_A = 2\pi r \ell_A dr$$

and upon rearrangement equation 3 becomes

$$\frac{1}{2\pi r}\frac{dE}{dr} = \{x(\mu_A^C - \mu_A) + y(\mu_B^C - \mu_B)\}\frac{\rho_A \ell_A}{x M_A} + \Gamma(\phi) . \qquad (5)$$

The term in the bracket is recognized as the Gibbs free energy of formation, ΔG, of the compound C. Therefore, equation 5 can be written

$$\frac{1}{2\pi r}\frac{dE}{dr} = \sigma + \Gamma(\phi), \text{ where} \qquad \sigma = \frac{\rho_A \ell_A}{x M_A}\Delta G \qquad (6)$$

and dE/dr is the driving force for wetting. In the following, the terms σ and Γ will be evaluated and compared.

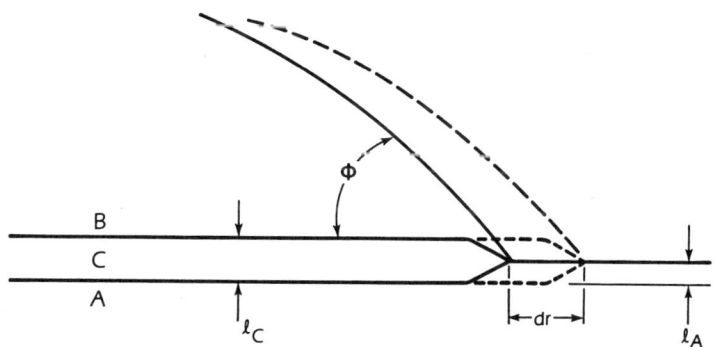

Figure 1
Reactive wetting of metal A by liquid metal B resulting in compound C.

CALCULATION OF REACTIVE ENERGIES OF INTERMETALLIC COMPOUND FORMATION

The Gibbs free energy of formation of intermetallic compounds, ΔG, can be calculated from the phase diagram and other relevant thermodynamic information using classical solution theory. The procedure used for all of the systems of interest here (Cu-Sn, Cu-Sb and Cu-Cd) is identical. As an example of the computational scheme, consider the determination of the free energy of formation of η in the Cu-Sn system.

When liquid Sn reacts with solid Cu, the intermetallic phase which forms is,

$$6Cu + 5Sn = Cu_6Sn_5 \ (\eta)$$

To calculate the free energies of formation (ΔG) of this compound, the free energy of mixing (ΔG_ℓ^{mix}) versus composition curve is needed at several temperatures where the liquid and the intermetallic compound coexist. To calculate the ΔG_ℓ^{mix} versus composition curve at the temperature of interest, the following information is required: (1) The Cu-Sn phase diagram - Figure 2 [5], (2) ΔS, ΔH and ΔG of mixing for Cu-Sn [5] (given at 1400 K in reference), (3) Integral enthalpies and entropies for Cu-Sn as a function of composition [5] (given at 1400 K in reference), and (4) The free energy of melting for Cu [8]. A change in standard state will be required to examine the interaction between solid Cu and liquid Sn, since the thermodynamic information given in the literature [5] is for liquid alloys.

$$\Delta G^o_{melt} = 3100 - 2.286 \ T \ cal/mole. \tag{7}$$

From these data, the free energy of mixing at any temperature is calculated by

$$\Delta G_\ell^{mix} = \Delta H_\ell^{mix} - T\Delta S_\ell^{mix} + X_{Cu}\Delta G^o_{melt}(Cu), \tag{8}$$

where ΔH_ℓ^{mix} and ΔS_ℓ^{mix} are assumed to be independent of temperature. From the free energy of mixing curves, the free energy of phase formation can be calculated directly.

For the case of Cu-Sn, the temperature dependent free energy of formation for the η phase is determined by common tangent construction from the free energy of mixing curve at 600 and 688 K, where the η + L phases coexist.

With the data and procedure described above, the free energy of mixing curve for Cu-Sn is calculated at 688 K, as shown in Figure 3. The L and η phases coexist at this temperature, with compositions (atom fraction Sn) of 0.867 and 0.435, respectively (note that the ϵ phase, Cu_3Sn, also exists at the 688 invariant, as shown in Figure 3). The partial molar enthalpy and entropy of the liquid alloys are given in the literature [5] at 1400 K for liquid Cu-Sn alloys at 0.1 atom fraction intervals [5]. ΔS and ΔH for both Sn and Cu are determined from the data by linear interpolation at the composition of interest (0.867 Sn). The partial molar enthalpies at 1400 K are ΔH_{Sn} = 91 cal/mole and ΔH_{Cu} = 15 cal/mole. The partial molar entropies at 1400 K are ΔS_{Sn} = 0.38 cal/mole/K and ΔS_{Cu} = 6.29 cal/mole/K. From the common tangency point on the mixing curve at the composition of the liquid (0.867 Sn), ΔG_{Sn} and ΔG_{Cu} are determined. Once again assuming temperature independence,

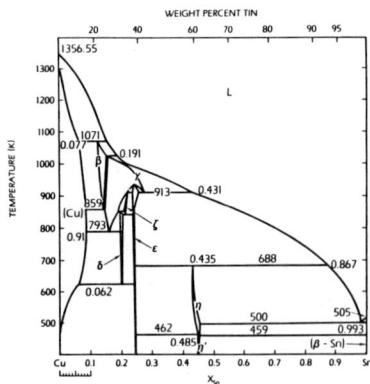

Figure 2
Cu-Sn phase diagram [5].

Figure 3

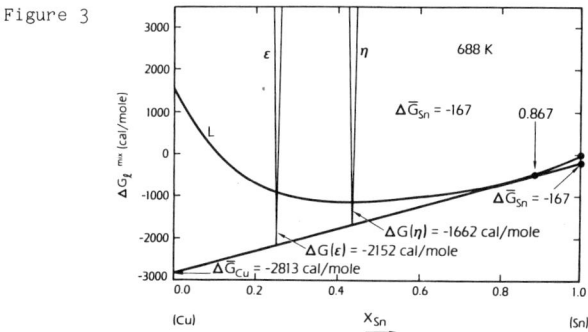

Calculated free energy of mixing curve for Cu-Sn at 688K. Gibbs free energies of formation for ε and η are also shown.

$$\Delta \overline{G}_{Sn} = \Delta \overline{H}_{Sn} - T\Delta \overline{S}_{Sn} \qquad (9a)$$

and

$$\Delta \overline{G}_{Cu} = \Delta \overline{H}_{Cu} - T\Delta \overline{S}_{Cu} + \Delta G^{o}_{melt}(Cu). \qquad (9b)$$

The standard state change for Cu is again required, since $\Delta \overline{H}$ and $\Delta \overline{S}$ are for the liquid standard state, but the reaction of interest is between solid Cu and liquid Sn. The partial molar free energies for a liquid with composition $X_{Sn} = 0.867$ at 688 K are calculated to be $\Delta \overline{G}_{Sn} = -167$ cal/mole and $\Delta \overline{G}_{Cu} = -2813$ cal/mole.

The free energy of formation of a phase is then calculated by,

$$\Delta G = X_{Cu} \Delta \overline{G}_{Cu} + X_{Sn} \Delta \overline{G}_{Sn}. \qquad (10)$$

For the η phase, $X_{Sn} = 0.435$, which yields $\Delta G(\eta) = -1662$ cal/mole. The free energies of formation of this intermetallic compound is illustrated graphically in Figure 3 ($\Delta G(\varepsilon)$ is also shown).

Similar calculation are performed at 600 and 913 K and the results yield a temperature dependent expression for the free energy of formation of the η phase,

$$\Delta G(\eta) = 465 + 3.09T \quad cal/mole. \qquad (11)$$

The identical approach was used to calculate the free energies of formation for intermetallic compounds in Cu-Sb and Cu-Cd utilizing thermodynamic data available in the literature [5,8]. The results are summarized in Table I.

RESULTS AND CONCLUSIONS

The results of the above analysis provide a means of estimating ΔG at any flow temperature. In order to compare the two driving force terms, σ and $\Gamma(\phi)$, in equation 6, it is assumed that wetting occurred at the melting temperatures of Sn, Sb and Cd. The melting temperatures, T_m, and the ΔG's evaluated at these temperatures are given in Table I. Assuming that in each case the reaction consumes approximately 1 μm of the copper substrate

Table I
Summary of Data and Calculations

Compound	$\Delta G(T)$ cal/mole	T_m(K) (metal B)	$\Delta G(T_m)$ cal/mole	$\sigma(T_m)$ ergs/cm^2	$\gamma(T_m)$ ergs/cm^2
Cu_6Sn_5	465 - 3.09T	505	-1096	-6026	550
Cu_3Sn	-3000 + 1.23T	505	-2377	-26700	550
Cu_2Sb	-2500 + 0.54T	903	-2015	-33740	380
$CuCd_3$	-825 + 0.44T	594	-562	-18800	590
Cu_5Cd_8	-1430 + 0.97T	594	-854	-5713	590

(ℓ_A=1 μm) the quantity,

$$\frac{\rho_A \ell_A}{M_A} \approx 1.4 \times 10^{-5} \text{ moles/cm}^2.$$

The driving force, σ, can be calculated with the appropriate ΔG from Table I and from the stoichiometric number, x. These results are also shown in Table I. To estimate $\Gamma(\phi)$ it is realized that the Cos terms in equation 4 can be no larger than 1.0, and therefore, $\Gamma(\phi) \leq \gamma_B$. The surface energy of the liquid metal, γ_B, at its melting point [9] is shown in the last column of Table I. For the three cases considered here, the contribution to the driving force for wetting which results from the reaction between liquid metal and solid substrate far exceeds that due to surface energy imbalance, i.e. $\sigma \gg \Gamma(\phi)$.

In this work a simplified model of the wetting process has been assumed. Simultaneous intermetallic reaction and associated free energy was included in the process energy imbalance and then shown to be the dominant driving force for wetting. The technologically important wetting processes are complicated by multicomponent liquids, fluxes and oxidized surfaces, non-planar compound layers and alloy substrates. All of these factors make for a more difficult thermodynamic analysis. However, reaction energies will continue to dominate the overall driving force for wetting.

ACKNOWLEDGEMENT

This work performed at Sandia National Laboratories supported by the U.S. Department of Energy under Contract Number DE-AC04-76DP00789.

REFERENCES

1. T. Young, Trans. R. Soc., London 94, 65 (1805).
2. J. V. Naidich, Prog. Surf. and Memb. Sci. 14 353 (1981).
3. P. G. deGennes, Rev. Mod. Phys. 57(3), 827 (1985).
4. F. Delannay, L. Froyen and A. Deruyttere, J. Mater. Sci. 22, 1 (1987).
5. R. Hultgren, P. D. Desai, D. T. Hawkins, M. Gleiser and K. K. Kelley, Selected Values of the Thermodynamic Properties of Binary Alloys, (American Society for Metals, Metals Park, OH, 1973), p. 566, 746, 795.
6. G. L. J. Bailey and H. C. Watkins, J. Inst. Metals 80, 57 (1951-52).
7. R. J. Klein Wassink, J. Inst. Metals 95, 38 (1967).
8. O. Kubaschewski, E. L. Evans and C. B. Alcock, Metallurgical Thermochemistry, (Pergamon Press, New York, 1967), p. 366.
9. L. E. Murr, Interfacial Phenomena in Metals and Alloys, (Addison-Wesley Publishing Company, Reading, Massachusetts, 1975), p. 101-106.

REACTION DIFFUSION IN THE Au-Cu-Pb-Sn SYSTEM AT LOW TEMPERATURE

J.F. ROEDER, M.R. NOTIS, and J.I. GOLDSTEIN
Dept. of Materials Science and Engineering
Lehigh University, Bethlehem, PA 18015

ABSTRACT

The microstructure of a Cu_3Au v. Pb-5wt%Sn diffusion couple was examined in the as-dipped and annealed conditions. The as-dipped couple showed no reaction between the solder and the base metal. The couple developed intermediate layers of two intermetallic phases upon annealing for 289 hours at 169°C. These phases were identified as phi and eta' which are contained in the Au-Cu-Sn ternary system. No phases containing significant amounts of Pb formed, indicating that even at low concentrations in a Pb-Sn solder, Sn is the active chemical specie in the solder for the Au-Cu-Pb-Sn system. The microhardness of the eta' phase was found to be equal to that of Cu_3Sn, which forms in the absence of Au in the Cu base metal. However, the the phi phase was harder, and therefore the presence of Au in the base metal may adversely affect the thermal fatigue resistance of the solder joint.

INTRODUCTION

An understanding of the nature of intermetallic phase formation in solder joints for electronic applications is of fundamental importance to fabricating reliable connections. Due to their brittle nature, intermetallic compounds may not respond favorably to stresses induced by thermal cycling [1]. This can lead to joint failure. These compounds can form both during the soldering process by liquid-solid reaction and during service at elevated temperatures [2]. This paper examines the nature of intermetallic phases formed with an Au-Cu alloy as the base metal and a high lead Pb-Sn solder. Thus, the quaternary system of interest is Au-Cu-Pb-Sn.

Of the six binary subsystems, only three form intermetallic compounds: Au-Sn [3], Au-Pb [4], and Cu-Sn [5]. Neither the Au-Pb-Sn [6] nor the Cu-Pb-Sn [7] systems contain ternary phases. The Au-Cu-Pb system remains uninvestigated. However, the Au-Cu-Sn system has been shown to contain several ternary phases [8,9]. Creydt and Fichter [8] identified the following compounds in the Au-Cu-Sn system: $AuCu_5Sn_5$, $Au_2Cu_4Sn_5$, and $Au_3Cu_3Sn_5$. These phases formed between 110 and 212°C in both Au-Cu-Sn and Au-Ag-Cu-Pb-Sn diffusion couples made by electroplating Cu billets. Recently, we have found two other ternary compounds in this system: chi (at ≈ 33Au-33Cu-33Sn, at%) and phi (at ≈ 25Au-55Cu-20Sn) as well as significant solubility of Au in the eta (Cu_6Sn_5) phase and Cu in the delta (AuSn) phase [9]. Keller has also reported the compound Cu_6AuSn_5 which formed in a Mod I soldered terminal connection [10], although the actual composition of the phase was not measured [11].

In this paper, the results of quaternary diffusion between Cu_3Au and a Pb-5wt%Sn solder are interpreted and compared to diffusion between Cu_3Au v. pure Sn [9] and Cu v. Pb-5wt%Sn [12].

EXPERIMENTAL

The Cu_3Au alloy (50.9 wt% Au) was made from 99.999% Au and Cu by induction melting in an inert atmosphere. The ingot was deformed and annealed in vacuum at 900°C for 48 hours to homogenize the ingot and to insure a large grain size (1-3mm). A 9mm diameter by 5mm thick disc was cut from the ingot and polished on one side through 1/4 μm diamond paste. This side was covered with an activated rosin flux (Superior flux #99), suspended face-up in a tungsten wire basket, dipped into a Pb-5wt%Sn solder bath at ≈350°C for 5 seconds, and then

water quenched. The Pb and Sn were also 99.999% pure. The as-dipped billet
was cut into two pieces. One half was examined metallographically in the
as-dipped condition. The other half was annealed in vacuum at 169°C ± 2° for
289 hours. Very careful metallographic preparation was necessary to avoid
polishing artifacts. The specimens were ground through 1 µm SiC papers using
kerosene and paraffin as a lubricant. This was followed by 1 µm diamond paste
on napless cloth, 0.3 µm alumina on napped cloth, and finally colloidal silica
with 3% EDTA solution added. Electron metallography was performed on a JEOL
733 electron probe microanalyzer (EPMA)operated at an 18kV accelerating voltage
and a 20 nA beam current. Wavelength dispersive spectroscopy (WDS) was used
for x-ray mapping and quantitative analysis. Pure element standards were used
for Au, Cu, and Sn. An NBS Pb-Si-O glass was used as a standard for Pb.

The microhardness of intermetallic phases formed in the couple was
determined using a LECO M-400 FT hardness tester with a diamond pyramid
indenter. Due to the limited spatial extent of the intermetallic layers in the
couples, the hardness measurements of the same phases were made in annealed
ternary alloys. The phase compositions in the alloys closely matched those in
the couples. The indentation load, 10gf, was chosen to keep the size of the
indentation small enough in a particular phase so that its stress field would
be unaffected by neighboring phases. In all cases, there was at least one
indentation width on either side of the indentation used for measuring the
hardness of a particular phase.

RESULTS AND DISCUSSION

A backscattered electron (BSE) image of the as-dipped diffusion couple
appears in Figure 1. Tin rich precipitates are evident in the solder. There
was no visual evidence for the formation of a reaction product at the bond
interface. However, a WDS x-ray map for Sn did indicate some enrichment at the
interface. Therefore it is possible that a very thin reaction layer was
present at the interface. Marcotte and Schroeder [12] found a 16 µm thick
Cu₃Sn layer at the interface in a Pb-5wt%Sn v. Cu diffusion couple held in the
liquid state at 350°C for 2 hours. The presence of a layer in the as-dipped
couple would not affect the interpretation of the annealed couple results.

A BSE image of the annealed diffusion couple appears in Figure 2a. X-ray
maps from the same area are shown in Figures 2b-e. Note that two intermediate
layers have formed as well as isolated precipitates in the Pb-5wt%Sn solder
near the interface. In all cases the phases formed contain Au, Cu, and Sn but
practically no Pb. The compositions of the layers in the solder couple are
given in Table I. The wider layer is a true ternary phase identified
previously as the phi phase [9]. The composition for the narrow layer must be
considered semi-quantitative as the analysis totals were quite high (≈ 106%).
This was attributed to a significant amount of surface relief at the interface.
However, the composition matches that of the eta' phase [9] closely and
therefore the narrow layer is identified as eta'. The isolated precipitates
were too small to measure their composition directly. It is assumed that they
are also eta' based upon their BSE contrast. In contrast to the quaternary
results, Marcotte and Schroeder [12] found only one intermediate layer, Cu₃Sn,
in a ternary Cu v. Pb-5wt%Sn couple annealed at 160°C for 1000 hours. The
presence of Au in the base metal appears to favor the stability of the phi and
eta' phases over the binary Cu₃Sn phase.

The morphology of the diffusion zone in the solder couple closely
resembles that in a Cu₃Au v. Sn diffusion couple [9] annealed in the
solid-state at 170°C ± 2° for 256 hours. As in the solder couple, the same two
intermediate layers were present: phi and eta'. No isolated precipitates were
present in the pure Sn couple. In this couple, which does not contain Pb, the
eta' layer was the same width as the phi layer. The eta' layer was thinner
than the phi layer in the lead containing couple. This is attributed to the
larger Sn concentration gradient between the end members in the Cu₃Au v. Sn
couple than in the solder couple. The important result is that Sn appears to

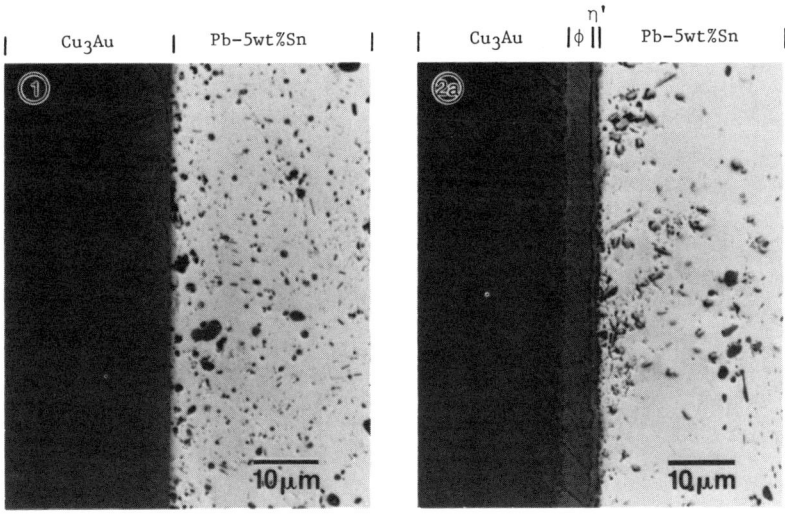

| Cu₃Au | Pb-5wt%Sn |

η'

| Cu₃Au | φ | Pb-5wt%Sn |

Figure 1 - BSE image of Cu₃Au v. Pb-5wt%Sn diffusion couple in as-dipped condition.

Figure 2a - BSE image of Cu₃Au v. Pb-5wt%Sn diffusion couple annealed at 169 C for 289 hours.

Figures 2b-e - WDS x-ray maps from same area shown in Figure 2a: b) Au, c) Cu, d) Pb, e) Sn.

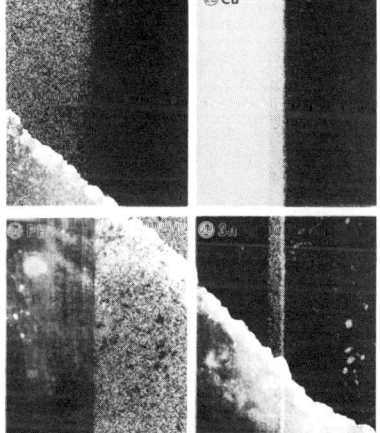

be the active diffusing specie in the solder. Even though the Sn content in the solder end member was only 5 wt% compared with pure Sn in the second couple, the same two phases formed. In addition, no phases formed which contained significant amounts of Pb despite the presence of Au-Pb compounds in that binary system and the rich source of Pb in the solder.

The microhardness of the intermediate phases and Cu₃Sn are given in Table II. The phi phase was found to be harder than the binary Cu₃Sn phase. Eta' has approximately the same hardness as Cu₃Sn. These results indicate that the phi phase is more likely to fail in a brittle manner than either the Cu₃Sn or the eta' phase. Thus, the presence of 51 wt% Au in the Cu base metal could lead to a less favorable microstructure from a mechanical viewpoint in a Pb-Sn solder joint in service at elevated temperature.

Table I - Phase compositions of
intermetallic layers formed in
a Cu₃Au v. Pb-5wt%Sn diffusion
couple annealed at 169°C for
289 hrs.

Phase	Composition (atomic %)			
	Au	Cu	Pb	Sn
Phi	25.4	52.5	0.02	22.0
Eta'	17.7	33.3	0.68	48.3

Table II - Vickers hardness of
intermediate phases (10 gf load).

Phase	Hardness	Standard Deviation
Phi	396	9
Eta'	318	14
Cu₃Sn	320	11

CONCLUSIONS

1. Solid-state diffusion between Cu₃Au v. Pb-5wt%Sn resulted in the formation
of two intermetallic compounds in the Au-Cu-Sn ternary system: phi and eta'.

2. No phases formed containing significant amounts of Pb despite the presence
of a very Pb-rich end member, indicating that Sn is the chemically active
component in the solder for the system investigated.

3. The Au-Cu-Sn ternary phi phase was found to be considerably harder than
either the eta' phase or the binary Cu₃Sn phase which forms at similar
temperatures in the absence of Au in the Cu base metal. This indicates that the
presence of Au may be unfavorable with respect to the mechanical properties of
the solder joint.

ACKNOWLEDGEMENTS:

     The authors acknowledge the support of the National Science Foundation
under grant #DMR 8023955. The authors also gratefully acknowledge Engelhard
Industries, Inc. for supplying on loan the precious metal used in this study.

REFERENCES

1. P. Dehaven, Mat. Res. Soc. Symp. Proc. 40, 123-128 (1985).
2. C.A. Mackay and S.W. Levine, IEEE Trans. Comp. Hybr., and Mfg. Tech.
CHMT-9, 195-201 (1986).
3. H. Okamoto and T.B. Massalski, Bull. Alloy Ph. Dia. 5, 492-501 (1984).
4. H. Okamoto and T.B. Massalski, Bull. Alloy Ph. Dia. 5, 276-284 (1984).
5. Bulletin of Alloy Phase Diagrams 1, 87-89 (1979).
6. Y.A. Chang, J.P. Neumann, A. Mikula, and D. Goldberg, in INCRA Monograph
IV: The Metallurgy of Copper. Phase Diagrams and Thermodynamic Properties of
Ternary Copper-Metal Systems (1979) pp. 631-642.
7. G. Humpston and B.L. Davies, Mat. Sci. and Tech. 1, 433-441 (1985).
8. M. Creydt and R. Fichter, Metall 25, 1124-1127 (1971).
9. J.F. Roeder, M.R. Notis, and J.I. Goldstein, Diffusion and Defect Data, in
press.
10. H.N. Keller, IEEE Trans. Comp. Hybr. and Mfg. Tech. CHMT-9, 433 (1986).
11. H.N. Keller (private communication).
12. V.C. Marcotte and K. Schroeder, in Alloy Phase Diagrams, edited by L.H.
Bennett, T.B. Massalski, and B.C. Giessen, (Mat. Res. Soc. Symp. Proc. 19,
Boston, Ma. 1983) pp. 403-410.

MICROSTRUCTURE DEVELOPMENT IN THIN FILMS DURING DEPOSITION FROM THE VAPOR PHASE

D. GOYAL*, A.H. KING* AND J.C. BILELLO**
*Department of Materials Science and Engineering, State University of New York, Stony Brook NY 11794-2275.
**School of Engineering and Computer Science, California State University, Fullerton CA 92634.

ABSTRACT

Transmission electron microscope observations of the structures of single-phase metallic thin films have been made. Films prepared by sputtering and by thermal evaporation have been used and the structures have been evaluated quantitatively as a function of film thickness. It is found that the grain size in all films increases linearly with the film thickness, indicating that grain growth occurs during film deposition.

INTRODUCTION

The mechanical and electrical properties of thin films can be substantially affected by their structures, and it is well recognized that a thin metallic film can be amorphous, polycrystalline or single crystal (epitaxial), depending upon the deposition process and composition. In particular the deposition rate and the substrate temperature have important influences upon the eventual structure. It has variously been suggested that the grain structure of a polycrystalline film is determined by the nucleation density or by recrystallization effects, both of which are temperature dependent. We have conducted a series of experiments upon thin films deposited at constant rates using thermal evaporation or ion sputtering, in order to assess what mechanisms are responsible for the development of their final structures. It will also be shown that the mechanisms that determine the film structures also have important effects upon the stresses developed in the films during deposition.

EXPERIMENTAL

In the present studies, chromium, nickel and tungsten films of varying thicknesses were deposited onto silicon single crystal wafers of (111) and (100) surface orientations. The substrates were cleaned only with non-corrosive solvents before use, so they were coated with native oxide layers. The substrates were nominally at room temperature for all depositions. Chromium and nickel films were deposited by thermal evaporation while tungsten films were deposited by ion sputtering. A deposition rate of 0.15nm per second was used for all of the coatings.

Transmission electron microscopy specimens were prepared using a modified design of the apparatus suggested by Booker and Stickler (1962). Disc specimens were trepanned from the whole wafers and chemical jet polished from the silicon side, in order to yield thin specimens consisting of the substrate and the coating.

The average grain size was determined from the transmission electron micrographs using the mean linear intercept method. These calculations were based on values obtained from several readings, using a line length of 1 cm on micrographs taken at 100,000X magnification.

RESULTS

For every system that we studied, it was found that the grain size that we measured was dependent upon the thickness of the film that was used. The grain sizes increased linearly with the film thickness, as is shown in Fig.1 and Fig.2, which represent thermally evaporated and sputtered films, respectively. It should be noted that for this experiment, the mean grain diameter was always much less than the film thickness, so that effects such as the Mullins limit on grain size (2) are not important. It is apparently the case that the film structures were not columnar grained, since we observed moire fringes in most specimens, Fig.3, indicating the overlap of crystals with nearly parallel planes.

These results clearly indicate that grain growth occurs during the deposition of thin films either by thermal evaporation or by ion sputtering. We may extrapolate our results back to zero film thickness, then, to obtain the mean separation between film nuclei. When we do this, we notice that the nucleation density for thermally evaporated films is identical for both (100) and (111) substrates, although the grain growth rates are different. On the other hand, for the sputtered films, the growth rates are identical for (100) and (111) substrates, but the nucleation densities are different.

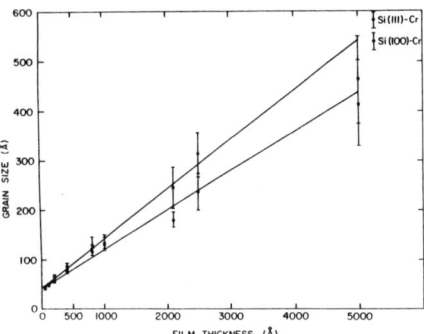

Fig.1a. Grain size as a function of film thickness for thermally evaporated chromium on (111) and (100) silicon substrates.

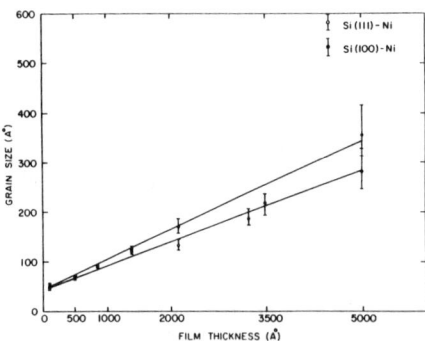

Fig.1b. Grain size as a function of film thickness for thermally evaporated nickel on (111) and (100) silicon substrates.

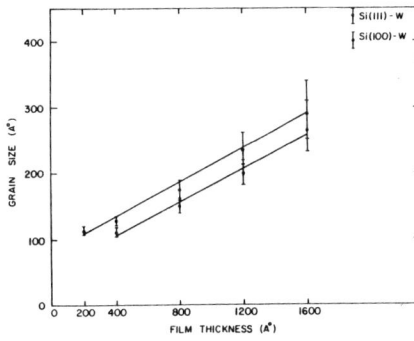

Fig.2. Grain size as a function of film thickness for sputtered tungsten on (111) and (100) silicon substrates.

DISCUSSION

Grain growth of the type found in this study has also been observed by several other authors (3-6).

The causes of the grain growth during thin film depostion are not immediately apparent, since grain growth is a thermally activated process, and we were using nominally room temperature substrates and high melting point materials. However, there are two sources of heat available to the specimen during the deposition of a thin film: The first is the enthalpy contained in the evaporated material, including the latent heat of solidification of the film. The second source of heat is radiant heat from the material source, operating at high temperature, a few centimeters from the substrate. During each film deposition we monitored the substrate temperature by means of a small K-type thermocouple in contact with the back side of the substrate wafer. For thermal evaporations, the temperature rise was found to be about 0.2°/nm for chromium and 0.4°/nm for nickel. The temperature rise for the sputtered tungsten specimens was approximately 0.1°/nm. These results are consistent with a significant enthalpic contribution to the temperature rise, since the evaporation source for chromium is necessarily hotter than that for nickel, and so provides a greater radiant heat contribution, but the temperature rise for the nickel films was found to be larger. Since our thermocouples were placed on the backs of the substrates, we may infer that the temperature rise in the film itself is much larger than that which we measured. The differences between the grain growth rates on (100) and (111) substrates may be attributed to the different thicknesses of the native oxides that are found on the two different orientations, because the oxide provides an effective thermal barrier, preventing conductive heat losses to the substrate.

Since the grain growth of the specimens is a thermally activated process, it is to be expected that the same process will lead to the relaxation of grown-in film stresses, whatever their source might be. Typically, it is found that evaporated films are formed in a state of tensile stress, while sputtered films appear to be compressively loaded (7), but these statements refer to the average state of stress for the film as a whole. When films are formed with a tensile internal stress for reasons such as constrained thermal contraction, each new layer of material added to the film will relieve the stress in the underlying layer, by means of recrystallizing it or permitting grain growth, but the new layer will be formed with the intrinsic stress present. This reasoning leads us to conclude that the grown-in stress is not homogeneous through the thin film, but is largest at the top surface and smallest (or even conceivably reversed) at the bottom surface. Evidence for such stress distributions is found when thin films delaminate from their substrates as shown in Figs. 4 and 5, which show evaporated films (tensile surface stress) curling upward, and sputtered films (compressive surface stress) curling downward, respectively.

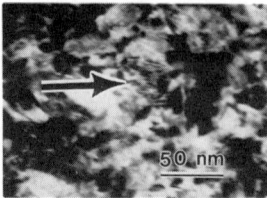

Fig.3. Bright field transmission electron micrograph showing Moire fringes in a 100nm chromium film formed on a (100) silicon substrate. This indicates grain overlap.

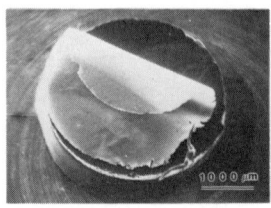

Fig.4. Scanning electron micrograph showing curling of a 320nm evaporated nickel film on a (100) silicon substrate. The sign of the curl indicates that the top surface is in a state of higher tensile stress than the bottom surface of the film.

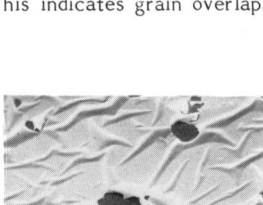

Fig.5. Scanning electron micrograph showing curling of a 160nm sputtered tungsten film on a (100) silicon substrate. The sign of the curl indicates that the top surface is in a state of higher compressive stress than the bottom surface of the film.

CONCLUSIONS

The intrinsic stresses and microstructures of thin films deposited onto substrates are considerably modified by an auto-annealing process facilitated by the deposition of heat with the material itself. Models for the generation of film stresses need to take this into account.

ACKNOWLEDGMENT

This work was supported by the Army Research Office, under contract number DAAG2984K0168.

REFERENCES

1. G.R. Booker and R. Stickler, Brit. J. Appl. Phys. 13, 446 (1962).
2. W.W. Mullins, Acta Met. 6, 414 (1958).
3. K.L. Chopra, Thin Film Phenomena (McGraw Hill Book Company, New York, 1969), p.183.
4. A.S. Nowick and S. Mader, in Basic Problems in Thin Film Physics, edited by R. Niedermayer and H. Mayer (Vandenhoeck & Ruprecht, Goettingen, 1969), p. 212.
5. E.I. Alessandrini, M.O. Aboelfotoh, R.B. Laibowitz and J.A. Lacey in Thin Films: The Relationship of Structure to Properties, edited by C.R. Aita and K.S. Sreeharsha (Mater. Res. Soc. Proc. 47, Pittsburgh, PA 1985), pp. 27-34.
6. R. Berger and H.K. Pulker, in Thin Film Technologies, edited by J. Roland Jacobson (Proceedings of S.P.I.E. 401, Bellingham, WA, 1983), pp. 69-72.
7. D. Goyal, W. Ng, A.H. King and J.C. Bilello, this volume.

STRESSES IN ANISOTROPIC THIN FILMS BONDED TO STIFF SUBSTRATES

JOHN C. LAMBROPOULOS AND SHIH-MING WAN
Department of Mechanical Engineering, University of Rochester, Rochester, NY 14627

ABSTRACT

Numerical techniques are used to calculate the stress concentrations arising near the interface of a single-crystalline film which is bonded to a stiff substrate. The film has cubic elastic symmetry, and it is characterized by the anisotropic constants A and H which show the deviation of the material from elastic isotropy. The normal to the film-substrate interface is taken to be along the 100, 111 or 110 directions. The inhomogeneous stresses near the free edge and the uniform stresses far from the free edge are calculated, and the effects of cubic elastic anisotropy and of film growth direction are established for material parameters typical of metallic and semiconducting films.

INTRODUCTION

It is well known that thin films grown on substrates by various techniques are in a state of internal stress which arises as a result of the deposition process (intrinsic stress) or as a result of differential thermal mismatch between film and substrate at a temperature different than the deposition temperature (thermal stress) [1]. Large values of the internal stress may lead to failure by delamination from the film's free edge [2], by buckling [3] or by cracking along the interface [4].

Stresses near the vicinity of the free edge of an isotropic film were calculated by Aleck [5], and later by Zeyfang [6] and Blech and Levi [7], who established that close to the free edge large peeling and shear stresses develop. Williams [8] showed that for an isotropic film bonded to a rigid substrate stress singularities develop near the point where the interface meets the free edge. Hein and Erdogan [9] calculated the stress singularity for varying stiffness between film and substrate. They showed that when the film is much stiffer the singularity is $-\frac{1}{2}$ for large values of the film material angle. When the substrate is much stiffer, the singularity is generally milder [9]. Approximate solutions for the stresses along the interface have been presented by Suhir [10], and by Yang and Freund [11], who have approximated the variation of the stresses through the thickness of the film.

In all works cited above, it has been assumed that the film and substrate are elastically isotropic. Still, situations often arise where the films are not isotropic. Examples are epitaxial films which may have a considerable amount of anisotropy [12], and magnetic films in which the material properties normal to the interface differ from the properties parallel to it [13].

It is the objective of this note to establish the effect of elastic anisotropy on the stresses near and far from the free edge in thin films of cubic symmetry which are bonded to stiff substrates.

ANISOTROPIC ELASTIC STRESSES

The geometry of the film and the substrate is shown inserted in Fig. 1. The z direction is normal to the film-substrate interface, and it is assumed that plane-strain conditions exist along the y direction. The substrate is modelled as being rigid and rigidly bonded to the film whose constitutive relation in the coordinate system of the crystal is

$$\underline{\sigma} = \underline{C}(\underline{\varepsilon} - \varepsilon^{T}\underline{I}) \tag{1}$$

where $\underline{\sigma}, \underline{\varepsilon}$ denote the stress and strain tensors, respectively, $\underline{I}$ is the identity tensor, and $\underline{C}$ is the 4-th order tensor of elastic stiffness with the usual symmetries and the non-zero entries

$$
\begin{aligned}
C_{1111} &= C_{2222} = C_{3333} = C_{11} \\
C_{1122} &= C_{2233} = C_{3311} = C_{12} \\
C_{1212} &= C_{2323} = C_{3131} = C_{44}
\end{aligned}
\qquad (2)
$$

The quantity ε^T denotes a misfit strain and it may be due to differential thermal mismatch or due to the intrinsic stress.

It is convenient to define the anisotropic factors A and H by [14]

$$
A = 2C_{44}/(C_{11}-C_{12}) \quad , \quad H = 2C_{44}-(C_{11}-C_{12}) \qquad (3a)
$$

For an isotropic material A=1 and H=0. An equivalent Poisson ratio v is also defined

$$
v = C_{12}/(C_{11}+C_{12}) \qquad (3b)
$$

For typical metals [14] and semiconductors (elemental or III-V compounds) [15] A lies between 1.5 and 4, and v between 0.2 and 0.45.

The orientation of the interface normal n is taken along the 100, 111 or 110 directions. Knowledge of n allows equation (1) to be expressed in the coordinate system of the film which is inserted in Fig.1.

The far field uniform stresses are found by noting that far from the free edge

$$
\varepsilon_{xx} = \varepsilon_{yy} = \varepsilon_{xy} = 0 \quad , \quad \sigma_{zz} = \sigma_{zx} = \sigma_{zy} = 0 \qquad (4)
$$

The only non-zero stresses are, in general, σ_{xx}, σ_{yy} and σ_{xy}. For n=100 or 111 $\sigma_{xx} = \sigma_{yy}$ and $\sigma_{xy} = 0$, whereas for n=110 $\sigma_{xx} - \sigma_{yy} \neq 0$ and $\sigma_{xy} \neq 0$. The insert in Fig. 1 shows the far field σ_{xx} for n=110 or 111 for various values of the anisotropic parameter A. For n=100 it can be easily shown that the far field stresses do not depend on C_{44} or A, but only on C_{11} and C_{12}. Thus for arbitrary A with n=100 the far field stresses are equal to the n=111 or 110 values with A=1.

To find the stresses distribution near the free edge, finite element techniques were used employing 4-node bilinear isoparametric elements. The half-length of the film (parallel to the interface) was taken to be five times the thickness of the film. The Poisson ratio was taken to be v=0.2, and A was equal to 1 (isotropic), 2 or 4. For n=100 the plane-strain condition was enforced along the 010 direction; For n=110 the plane-strain condition was along 001, and for n=111 the plane-strain condition was along $1\bar{2}1$.

Figures 1 and 2 show the stress distribution near the free edge and along the interface (z=0.008h) for n=111. Figures 3 and 4 show the same results for n=100. The insert in Fig. 1 and Fig. 3 shows the levels of the uniform far field stresses σ_{xx} for n=110 and 111. In order to clearly show the effect of anisotropy and the effect of the interface normal n, the stresses are measured in units such that the far field stress $\sigma_{xx}=1$ for an isotropic material with the same Poisson ratio.

DISCUSSION

It is seen from Figures 1-4 that the main effect of elastic anisotropy is to increase the stresses both in the vicinity of the free edge, as well as in the part of the film which is uniformly stressed. Furthermore, the extent of the interface which is subjected to large shear stresses is seen to increase considerably as A increases, especially for the case n=111 (Fig. 2). Although the peeling stress σ_{zz} has diminished considerably by the time x=0.1 or 0.2 h,

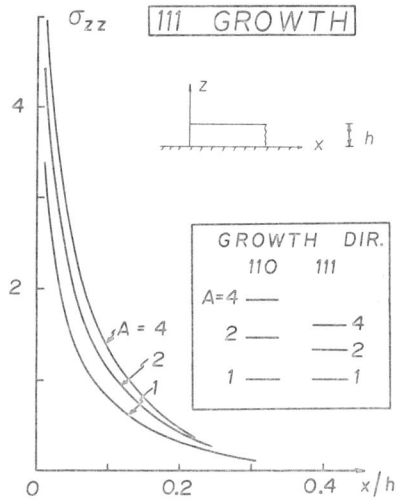

Figure 1: The peeling stress σ_{zz} for n=111. The insert shows the far field uniform σ_{xx}. Units are such that far field uniform $\sigma_{xx}=1$ for isotropic material (A=1).

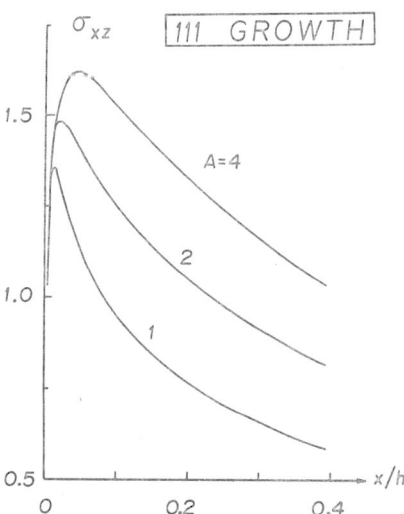

Figure 2: The shear stress σ_{xz} for n=111. Units are such that far field uniform $\sigma_{xx}=1$ for isotropic material (A=1).

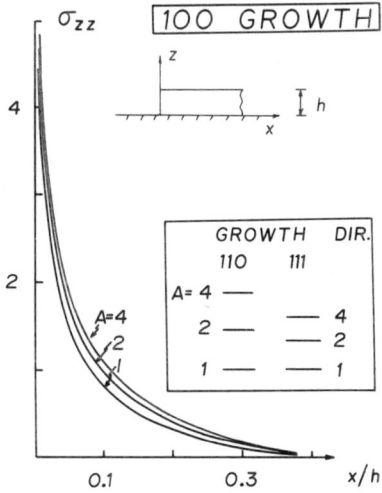

Figure 3: The peeling stress σ_{zz} for n=100. The insert shows the far field uniform σ_{xx}. Units are such that far field uniform $\sigma_{xx}=1$ for isotropic material (A=1).

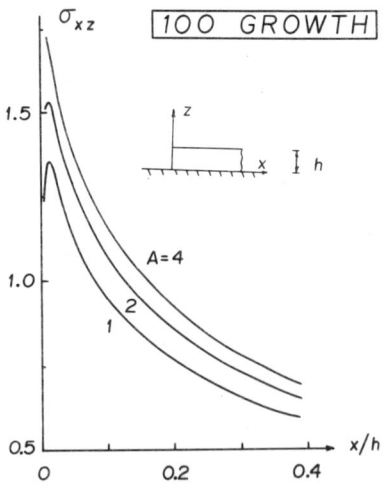

Figure 4: The shear stress σ_{xz} for n=100. Units are such that far field uniform $\sigma_{xx}=1$ for isotropic material (A=1).

the levels of the shear stress still remain close to the far field value of the internal stress. When compared to the isotropic results (A=1), it is seen that anisotropy does not change the qualitative features of the stress distribution along the interface.

Concerning the effect of the interface orientation n on the resulting stress distribution along the interface, the numerical results showed that the stresses near the free edge with n=110 were within 5% or less of the stresses with n=111 for the same value of A. Thus, it was concluded that the effect of anisotropy is essentially the same when n=111 or 110. On the other hand, when n=100 the numerical results show that the effect of A on the peeling stress σ_{zz} is much smaller than when n=111 or 110; The effect of A on the shear stress σ_{xz} is stronger, but still weaker than the case with n=111 or 110. A more detailed analysis of the effect of anisotropy is forthcoming.

Several assumptions had to be made in this work, such as the isotropy of the misfit strain ϵ^T, or that the substrate is rigid, or that the assumption of plane-strain is valid (thus reducing the problem from three to two dimensions), or that the film remains linearly elastic at the high levels of stress present near the free edge, whereas plastic deformation is expected to relieve the high stresses by a variety of deformation mechanisms, such as dislocation flow or diffusional creep depending on the level of the temperature and on the level of the far field uniform stress [12]. A careful examination of these assumptions is currently under way.

ACKNOWLEDGEMENT

This work has been supported by the ONR under Grant N00014-87-K-0488 and in part by the DOE Office of Inertial Fusion under Agreement No. DE-FCOB-85DP 40200 and by the Laser Fusion Feasibility Project at the Laboratory for Laser Energetics which has the following sponsors: Empire State Electric Energy Research Corporation, GE Company, NY State Energy Research and Development Authority, Ontario Hydro, and the University of Rochester, and the Office of Naval Research.

REFERENCES

1. R.W. Hoffman, in Physics of Thin Films, edited by G. Hass and R.E. Thun (Academic Press, New York, 1966). p. 211.
2. J.W. Hutchinson, M.E. Mear and J.R. Rice, Report MECH-95, Harvard University, 1987.
3. A.G. Evans and J.W. Hutchinson, Int. J. Solids Structures, 20, 455 (1984).
4. D.B. Marshall and A.G. Evans, J. Appl. Phys., 56, 2632 (1984).
5. B.J. Aleck, J. Appl. Mech., 16, 118 (1949).
6. R. Zeyfang, Solid State Electronics, 14, 1035 (1971).
7. I.A. Blech and A.A. Levi, J. Appl. Mech., 48, 442 (1981).
8. M.L. Williams, J. Appl. Mech., 19, 526 (1952).
9. V.L. Hein and F. Erdogan, Int. J. Fract. Mech., 7, 317 (1971).
10. E. Suhir, J. Appl. Mech., 53, 657 (1986).
11. W. Yang and L.B. Freund, Brown University, 1986 (to be published).
12. M. Murakami, in CRC Critical Reviews in Solid State and Materials Science, vol. 11, edited by D.E. Schuele and R.W. Hoffman (CRC Press, Boca Raton, 1984), p.317.
13. J.K. Howard, J. Vac. Sci. Technol., A4, 1 (1986).
14. J.P. Hirth and J. Lothe, Theory of Dislocations, 2nd edition,(John Wiley, New York, 1982), p. 42, 430-431.
15. J.C. Lambropoulos, J. Crystal Growth, 80, 245 (1987).

PART VII

GaAs and Optoelectronic
Packaging Materials

INTEGRATED OPTOELECTRONIC MATERIALS AND CIRCUITS
FOR OPTICAL INTERCONNECTS

L. D. Hutcheson
Raynet Corporation, 181 Constitution Dr., Menlo Park, California 94025

ABSTRACT

Conventional interconnect and switching technology is rapidly becoming a critical issue in the realization of systems using high speed silicon and GaAs based technologies. In recent years clock speeds and on-chip density for VLSI/VHSIC technology has made packaging these high speed chips extremely difficult. A strong case can be made for using optical interconnects for on-chip/on-wafer, chip-to-chip and board-to-board high speed communications. GaAs Integrated Optoelectronic Circuits (IOC's) are being developed in a number of laboratories for performing Input/Output functions at all levels. In this paper integrated optoelectronic materials, electronics and optoelectronic devices are presented. IOC's are examined from the standpoint of what it takes to fabricate the devices and what performance can be expected.

INTRODUCTION

The throughput of data and signal processors is being pushed to ever-increasing limits. The development of faster, more complex silicon integrated circuits (ICs) and the use of parallel processing are largely responsible for this improved performance. At the same time, it has been necessary to improve the electrical packaging and interconnect technology in order not to compromise the speed of the IC. In an attempt to develop even faster circuits, major research programs have been started in gallium arsenide (GaAs) electronics, and to a lesser extent indium phosphide (InP). One of the goals of these programs is to develop circuits having gigahertz clock rates. A few MSI-level GaAs parts have already been demonstrated with clock rates above 1 GHz.

One of the problems that must be faced when designing a processor to operate at these higher speeds is the extreme difficulty of transmitting data at gigabit/second (Gbit/s) rates. The performance of electrical interconnects is adversely affected by increases in capacitance and reflections due to impedance mismatches. Multilevel board technology is being developed to address this problem for chip-to-chip interconnects at hundreds of megahertz. The interlevel vias, however, are electrical discontinuities which become increasingly more troublesome as frequency increases.

One solution may be the use of optical interconnects to transmit the data. The optical fiber, integrated optical waveguides and freespace all provide an excellent transmission medium, while optical sources and detectors have been demonstrated at operating frequencies above 10 GHz. In addition to the transmission medium, high speed optical interconnects also benefits from freedom of capacitive loading effects, immunity to mutual interference effects and the flexibility to utilize the third dimension allowing more efficient utilization of space [1]. Other potential benefits include reduced system power, increased fanout capability, decreased complexity, reduced pinout count, smaller volume, increased density, and new architectures not previously possible [2].

One major advantage of GaAs and InP over silicon in high speed circuits is that III-V materials can emit light. It therefore becomes possible to integrate optical emitters on the IC to perform the I/O (input/output) functions. Integrated optoelectronic circuits (IOCs) are now being developed in a number of laboratories around the world. Interested parties in the United States include major computer manufacturers, defense contractors, communications companies, and a number of universities. It has been announced that Japan is establishing a research and development company for IOCs to be staffed by several different companies.

When designing an optical interconnect, it must be remembered that the performance is dependent on the optical components as well as the electronics. Therefore, in order to see a significant improvement in the performance of an IOC over a hybrid circuit, the processing sequence must not compromise either the electronics or the optoelectronic components (e.g. lasers and detectors). This means that the circuit must accommodate the materials and processes needed for the electronics as well as those of the optoelectronics. This is where the similarities in the development of integrated optoelectronic components differ greatly from those in the development of silicon integrated circuits. The drive to increase the level of complexity of silicon ICs is based on considerations of cost, performance, density and ease of packaging. All parts of the silicon ICs are made with the same materials and processing sequence. For IOCs, however, the materials and processes for the optoelectronics and the electronics are vastly different and the process for integration is more complex still.

The obvious question, then is: Why bother integrating the devices? There is no single answer to this question. It is actually application dependent. There are many applications where optical interconnects have higher performance than electrical interconnects. The major reasons for integrating the devices, then, fall into three areas: lower parasitics means higher performance, optical integration increases density and fewer parts eases the packaging task. It is not clear whether there will be a reduction in cost for producing integrated optoelectronic chips over an IC connected to a discrete optical device, particularly if the IOC has MSI level circuitry. This is because the integration of the optical components adds processing steps to the electronics, thereby increasing the cost per circuit and decreasing the circuit yield. In addition, the crystal growth on a substrate for an integrated optoelectronic circuit costs just as much as one for a wafer to be made into discrete parts, but the integrated wafer will produce relatively few optical components, making each optical component more expensive. So instead of trying to justify the components from a cost standpoint, they should be examined on the basis of what can be done with this technology that cannot be done any other way.

In this paper the components that make up an IOC - materials, electronics, and optoelectronics - are presented. The parameters that are important to a designer of interconnects (e.g., bandwidth, power, density and bit error rate) are described. Other operating characteristics, such as temperature sensitivity, are discussed to provide an appreciation of what must be considered when the optical interconnect is taken out of the laboratory and designed into a system. The present status of IOCs is described for both optical transmitters and optical receivers. Finally, a few examples are given of the expected performance of IOCs and their impact on the system.

For a review and status of optical interconnect development the reader is referred to reference [2].

MATERIALS FOR INTEGRATED OPTOELECTRONIC CIRCUITS

An IOC begins with a GaAs substrate and requires the growth of epitaxial layers to fabricate the optoelectronic components. These materials must provide a method for the electronics to be compatible with the optoelectronics. In production, this means that the substrate must be at least 3 inches in diameter and the epitaxial growth system must be able to handle the 3 inch substrate. Further requirements are placed on both the substrate and the epitaxial layers by the devices that are to be fabricated. These are described below.

Substrates

The requirements of the starting GaAs substrate are dependent on the electronics and the optoelectronics. The electronic technologies which are fabricated using selective ion implantation require uniform, high-resistivity substrates which maintain their properties after the implant anneal steps. The circuits which require epitaxial layers are not as dependent on the electrical properties of the substrate, but do depend on the density of defects. Optoelectronic components, especially lasers, are very susceptible to defects in the substrate which propagate up through the active region. These defects have proved to be one of the major causes of short-lived lasers.

There are two basic methods for growing GaAs substrates: horizontal Bridgeman (HB) and liquid-encapsulated Czochralsky (LEC). Most people today choose to use 3 inch LEC material which have resistivities in the range of 1×10^8 ohm-cm. These resistive properties are maintained fairly well during implant annealing. The major problem with the quality of the substrates is the defect density. Present 3 inch round LEC wafers have an etch pit density on the order of 10^3 to 10^4 cm^{-2}. Nonuniformities in the threshold voltage of FETs and defects in the epitaxial growth have been attributed to these defects.

Recently "zero-defect" LEC material has been reported to be available from Sumitomo. These wafers are fabricated by doping with indium, which ties up any defects and thus prevents them from propagating. Because the indium is isoelectronic, it does not cause the same electrical problems as chrome during implant anneals. The major drawback to this material is its cost. Each wafer is approximately three times as expensive as a standard LEC wafer. Only about 10 good In-doped wafers, as opposed to 80 standard wafers, can be obtained from a boule of GaAs. It is possible that the success of integrated optoelectronics depends on the development of In-doped material. The reason is that the laser lifetime may not be long enough on the high defect standard LEC material [3]. Therefore, low defect density is a must.

Epitaxial Growth

The requirements for the growth of the epitaxial layers on the substrate are also dependent on the type of component to be fabricated. The electronics technologies requiring epitaxial growth generally need low background carrier concentrations (around 1×10^{13} cm^{-3}) and the ability to control doping accurately. Lasers are not as dependent on electrical properties, but need high photoluminescence efficiency. Waveguide structures need low carrier concentration (low capacitance) and excellent morphology (low scattering). Quantum-well lasers and MODFET-type electronics also require extremely sharp interfaces (on the order of a few angstroms) between layers of GaAs and AlGaAs.

There are two epitaxial growth techniques that can be considered for use in the production of integrated optoelectronic circuits: molecular beam epitaxy (MBE) and metal organic chemical vapor deposition (MOCVD). Liquid phase epitaxy (LPE) has been used for years to manufacture high quality lasers and light emitting diodes. The substrates used in LPE, however, are limited in size to approximately 2 in.^{2}, which is not compatible with GaAs electronics. MBE [4] has been the standard technique for the development of MODFET structures. The system can achieve the low background carrier concentrations, layer thicknesses, and sharp interfaces needed for low threshold lasers [5] and optoelectronic circuits [6]. There are two drawbacks to MBE as a production process for IOCs. The first is the limited throughput of the machines. MBE can grow on only one 3 inch wafer at a time, and the growth rate is approximately 1 um/hr. Therefore, the growth of a laser structure would take up the better part of an 8 hour shift. There are some experimental MBE machines being designed to grow on multiple 3 inch wafers simultaneously. The second problem is the defect density of the grown layers. A better than average MBE machine will produce a 2 μm-thick layer with approximately 1000 defects /cm^2. This is higher than can be tolerated for VLSI-level circuits or for waveguide structures. Some pioneering work has been done to reduce the level of defects to less than 100 cm^{-2}, but this technique is far from production.

MOCVD has the advantage that it can be used for multiple wafer growths and the growth rate is higher than MBE. The layers can be made with very low defect levels, and the photoluminescent efficiency is extremely high. MOCVD is currently being used in production by a number of laser diode manufacturers. The major drawback to MOCVD is its background carrier concentration. Typical MOCVD layers have a background level of around 1 x 10^{15} cm^{-3}. This is not acceptable for either MODFET electronics or waveguide structures. The major limitation appears to be in the purity of the starting metal organic sources.

ELECTRONICS FOR INTEGRATION

For integrated optoelectronic ICs to meet the needs of high speed processors, it is imperative that the IOCs be fabricated in a way that guarantees an adequate supply of chips. This will occur only if the IOCs are fabricated using a standard GaAs production process to which the extra steps for the optoelectronics have been added. It is very doubtful that a new GaAs electronics process will be developed solely for the purpose of allowing optoelectronic components to be integrated.

Figure 1 shows the cross sections of the most common transistors fabricated in GaAs; Fig. 1(a) is the depletion-mode MESFET (metal-semiconductor field-effect transistor), Fig. 1(b) shows the enhancement-mode MESFET and Fig. 1(c) is the JFET (junction FET). Each of these technologies is fabricated using undoped, semi-insulating (10^8 ohm-cm) GaAs as a starting substrate [3]. The channel and contact regions are formed by selective ion implantations which are activated using a high temperature annealing step.

Until now, nearly all GaAs IOCs have been fabricated using depletion-mode MESFETs. Very few demonstrations, however, have used the process as it exists on a production line. The researchers have chosen instead to use a fabrication sequence which is more conducive to experimentation and which does not require the same restrictions on substrate size and uniformity.

A detailed discussion of GaAs processing and high speed GaAs IC demonstrations will not be presented here. For the interested reader a review of the production process and state-of-the-art demonstrations for depletion-mode, enhancement-mode, JFETs and MODFET can be found in reference [3].

INTEGRATED OPTOELECTRONIC DETECTORS AND RECEIVERS

Semiconductor optical detectors are two-terminal devices that convert optical inputs into electrical carriers. By connecting the detector to an appropriate circuit, the electrical carriers are collected and the signal is amplified to levels adequate to drive a digital IC. The detector and amplifier must be designed as a unit. The integration of the detector with the amplifier provides a significant improvement in bandwidth and sensitivity over a hybrid circuit [3]. The reason for the improvement is that the capacitance at the connection of the detector to the amplifier can be made as low as 0.2 pF for a monolithic circuit as opposed to > 0.5 pF for a discrete detector/amplifier pair. The significance of this can be seen in Figure 2 where the detected optical power versus bit rate for several values of input capacitance is plotted [7] for a BER = 10^{-9}. The best experimental results for a hybrid receiver are shown in the figure as closed circles [8] and is consistent with a total capacitance between 0.6 and 0.8 pF. At 1 Gbit/s there is a 5 dBm increase in receiver sensitivity between a hybrid and an integrated receiver. Alternately, this three or fourfold decrease in capacitance would permit the detector/amplifier to be operated at nearly three or four times the bit rate with no degradation in accuracy. From this figure one can also see that the effect of capacitance is even more dramatic at higher bit rates. This is a strong argument for integration.

Figure 2 also shows that avalanche photodiodes (APD) have much higher sensitivity than pin structures. However, APDs require maintaining gain control at high reverse biases (typically 60 V or more) over a wide range of temperatures due to the temperature sensitivity of the avalanche gain. This can significantly increase complexity and cost.

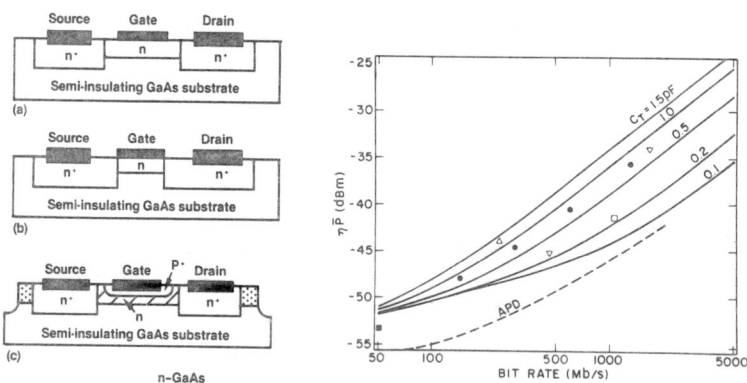

Figure 1. Schematic cross-sections of GaAs field effect transistors: (a) Depletion-mode MESFET; (b) Self-aligned-gate enhandement-mode MESFET; (c) Junction FET.

Figure 2. Plot of sensitivity at 10^{-9} BER versus bit rate as a function of node capacitance. (From Ref. 7)

Requirements for Integration

If a photoconductor is to be integrated on a circuit with more than a few electronic components, it must meet a number of criteria. First, the detector must be compatible with the electronics processing. Therefore, it must be processed on a production line and be compatible with the substrates used for the electronics. For GaAs, this means a 3-inch round semi-insulating substrate. Second, the material for the detector cannot interfere with the electronics. Thus if epitaxial material growth is needed it must be excluded from the regions in which the electronics will be fabricated and the transition to the epitaxial region must be smooth enough to permit fine-line photolithography. Third, any process needed to fabricate the detector cannot degrade the performance of the electronics. For example, a very high temperature step may cause unacceptable surface damage. Fourth, the detector and electronics must be adequately electrically isolated from each other on the substrate. Finally, the detector must meet specifications after integration.

Although most of these sound obvious, they are not easily accomplished. If the devices are fabricated on semi-insulating substrates, a method must be found to make electrical contact to both sides of the detector other than via the substrate. The photoconductor and back-to-back Schottky detectors have both contacts at the surface. However, for both the p-i-n and avalanche detector one contact is below the surface. A process step must be added to make contact to this area. One method for solving this problem is shown in Figure 3 where the layers for the p-i-n detector was grown in a well [9]. After the epitaxial layers were removed from the area in which the electronics were to be fabricated, the n^+ region was exposed at the edge of the well, as shown in Figure 3, permitting easy access for contacting. Growing in a well is one step closer toward a monolithic receiver.

Detector/Amplifier Demonstrations

The first and simplest detector/amplifiers reported were p-i-n/FETs fabricated at Bell Labs in InGaAs/InP [10]. Both the detector and FET were fabricated in epitaxial layers grown by LPE. One such example is shown in Figure 4 where a transimpedance amplifier is integrated with a p-i-n detector [11]. The layers for the detector and amplifier were selectively grown in the two regions by a two-step MOCVD process. As in the structure shown in Figure 3 the three p-i-n layers were grown first in a preetched 7 um deep well. Next the FET layers were grown, forming a nearly planar surface for photolithography. The pn junction was formed by a Zn diffusion into the lightly doped GaAs absorption region. The technique of selectively growing the p-i-n and FET layers allows for the independent optimization of both circuit segments. For example, high-transconductance FETs require thin (<0.5 μm) n-type channels with high impurity concentration. The transimpedance amplifier consists of

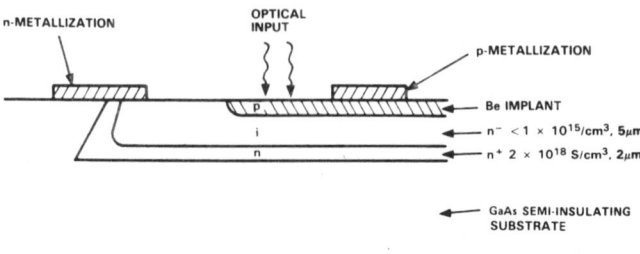

Figure 3. Schematic cross-section of a p-i-n
detector grown in a well on a semi-
insulating substrate. (From Ref. 9)

six GaAs MESFETs. The amplifier consists of two stages with a transimpedance of 1 k-ohm. The output impedance is 50 ohm, and the circuit had a 400 ps rise and fall time.

A third example fabricated by Honeywell researchers is GaAs receiver [12], consisting of a detector, amplifier, and a 1:4 GaAs demultiplexer operating at 1 GHz clock rates. The detector is an interdigitated back-to-back Schottky diode with 1 μm lines and 3 μm spaces fabricated directly on the semi-insulating substrate. The circuit consists of depletion-mode MESFETs fabricated using selective ion implantation on a 3 inch processing line. This is truly a production compatible part.

INTEGRATED OPTOELECTRONIC TRANSMITTERS

The fabrication of semiconductor laser diodes requires crystal growth and processing steps similar to IC manufacturing in order to define the electrical and optical cavity in the two dimensions perpendicular to the direction of light propagation. The length of the cavity is defined by partial mirrors which are formed by cleaving the semiconductor along parallel crystal planes. The integration of the laser diode with the associated electronics is much more difficult from a materials and processing standpoint than the integration of the detector with its associated electronics [3]. There are three major reasons for this: (1) lasers require a multilayered heterostructure up to 7 μm thick; (2) they need two high quality parallel mirrors separated by on the order of 200 μm; and (3) a method is needed that can achieve electrical and optical confinement in the lateral dimension.

Requirements for Integration

The requirements for monolithically integrating a laser diode with high speed electronics fall into three categories: compatibility, required components and performance. The decisions that must be made to obtain a working component will be application dependent. In general, the materials and processing for the laser and the electronics will be different. In addition, one of the major requirements for monolithic integration is that the epitaxial material grown for the laser covers only specific regions of the substrate while the rest of the surface consists of high quality planar material for the fabrication of the electronics. As a second requirement, the epitaxial layers for the laser must be grown on a semi-insulating substrate. It is also necessary that the resulting substrate after processing of the laser be highly planar to permit fine line lithography for the electronics. Finally, it is necessary to isolate the laser from the electronics both electrically and optically.

Optical Transmitter Demonstrations

The first monolithic laser/electronics demonstration was reported by researchers at Cal Tech [13]. This component consisted of an AlGaAs laser which was integrated with a GaAs Gunn oscillator. Since that time, there have been numerous demonstrations of different laser and electronic devices in both GaAs and InP material systems. The demonstrations were limited to single lasers, grown by LPE, integrated with a single transistor. The laser mirrors were formed by cleaving. These parts demonstrated the feasibility of integrating lasers with electronics, but they were far from being practical.

In more recent years, advances have been made in the areas of crystal growth, circuit design, and complexity which brings these components closer to production. As discussed in the earlier section on material growth both MBE and MOCVD are capable of growing on 3 inch round substrates which are standard for current production of GaAs ICs.

Demonstrations have also been reported which show an increase in the complexity of the electronics, an on-chip mirror and power monitor. Figure 5 shows an optoelectronic IC with the laser grown in a well etched in the semi-insulating substrate and has a multi-quantum-well active region [14]. The back facet was formed by reactive ion etching and resulted in a laser threshold of 40 mA. The electronics were formed by selective ion implantation into the semi-insulating substrate. The

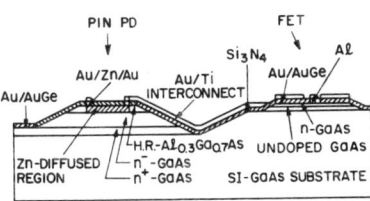

Figure 4. Cross-section of a monolithic optical receiver fabricated in GaAs/AlGaAs. (From Ref. 11)

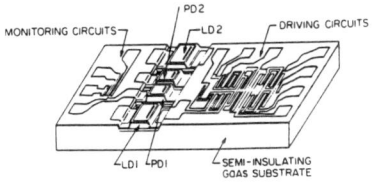

Figure 5. Integrated optoelectronic laser transmitter fabricated with a quantum well laser and selective ion implantation. (From Ref. 14)

circuit design uses input buffers as well as a differential drive. It was demonstrated at modulation rates up to 2 Gbit/s using nonreturn-to-zero format.

A more sophisticated and complex demonstration consisted of a TJS laser integrated with a 4:1 multiplexer [15]. The chip was approximately 1.8 x 1.8 mm^2. The 4:1 multiplexer (MUX) and its associated circuitry is formed by selective ion implantation and contains 36 NOR gates (approximately 150 D-mode MESFETs). The TJS laser was grown by LPE in an etched well and the rear facet was formed with an undercut mirror process. This particular chip was tested at speeds up to 160 MHz.

EXPECTATIONS

Integrated receivers and transmitters are still in their infancy. The greatest immediate challenge is still one of materials and processing compatibility between the optoelectronic and electronic components. Once this accomplished, the question is still how to get these circuits into applications. This section will contain some speculations and projections for integrated optoelectronics and how we might expect infusion of this technology into future applications.

First, the IOCs must be fabricated with one of the higher speed, lower power GaAs IC technologies such as E-mode MESFET or MODFET. Depletion mode MESFET is a good technology to use when fabricating demonstration units, but the power required for the circuit is too high to make it attractive for high speed applications. Not only would the system power be high, but the heat generated would be detrimental to the laser threshold.

Second, the 1.5 V swing for D-mode causes problems for the receiver. First, a p-i-n detector produces a relatively small voltage swing, on the order of 10 to 30 mV requiring an amplifier gain of approximately 100. Because GaAs FETs have a relatively low gain, many stages of amplification are required to produce a GaAs level signal. Thus a lot of power is burned in the amplifier. It also leads to excessive latency. The large voltage swing at the output of the amplifier may also couple back into the detector, causing excessive noise. This coupling can occur either through the power and ground lines or directly through the semi-insulating substrate.

Third, the threshold of the laser diode must be brought down. The development of quantum-well lasers may be the answer to this problem. Figure 6 shows a multiple-quantum-well laser (MQW) having a ridge waveguide structure that is being developed for monolithic integration [16]. The MQW laser structure is grown by MOCVD and consists of five 100-Å GaAs wells separated by four 40-Å $Al_{0.2}Ga_{0.8}As$ barriers. The ridge waveguide is ion milled, having a width of 5 um. The milling is adjusted to stop approximately 0.2 um above the active region. Room temperature CW threshold currents as low as 12 mA and a differential quantum efficiency of 60% has been measured for devices with a cavity length of 200 um.

In the future one can expect even higher levels of optoelectronic integration such as a complete transceiver [17] (transmitter and receiver integrated on a single substrate). One possible implementation of a transceiver chip is shown in Figure 7 [2,17]. The chip has an integrated laser, laser driver, 4:1 MUX, photodetector, preamplifier, post-amplifier, 1:4 DMUX, and a customized gate array to perform some signal processing functions. A slightly different implementation of a transceiver chip is described in reference [16]. The primary difference between the two implementations is the one shown in Figure 7 has an edge emitting laser and uses a 4:1/1:4 MUX/DMUX and the latter version uses a surface emitting laser and a 8:1/1:8 MUX/DMUX.

As can be seen from the discussion in this paper significant progress has been made in IOC materials, processing and device demonstrations. Even though we have seen this progress and a lot more development is needed the electronic circuit designers must feel comfortable designing with IOCs and implementing them for relieving I/O problems. This still requires a great deal of work.

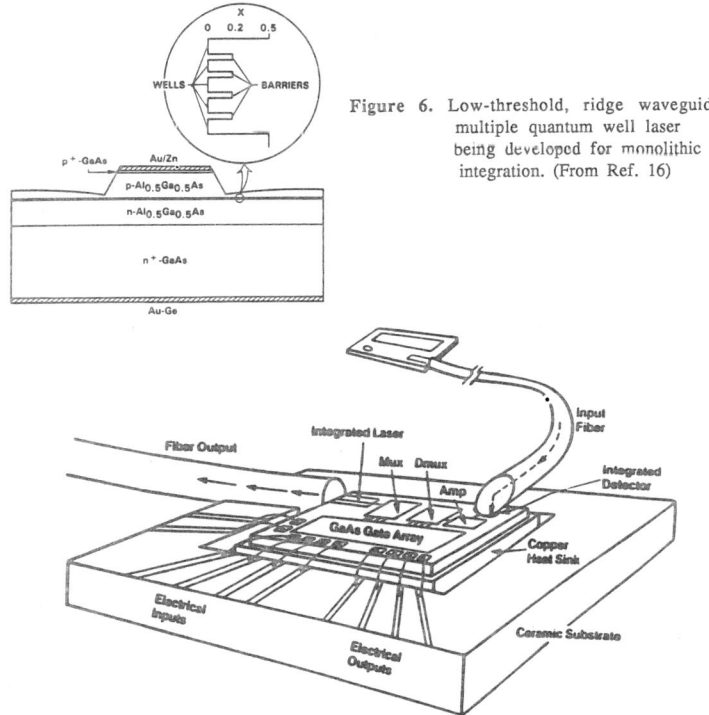

Figure 6. Low-threshold, ridge waveguide multiple quantum well laser being developed for monolithic integration. (From Ref. 16)

Figure 7. Monolithic optoelectronic transceiver chip to be used in an optical interconnect application.

REFERENCES

1. J.W. Goodman, F.J. Leonberger, S.Y. Kung and R.A. Athale, Proc. IEEE, 72, p. 850 (1984).

2. L.D. Hutcheson, P.R. Haugen and A. Husain, IEEE Spectrum, p. 30 March (1987).

3. L.D. Hutcheson, ed., Integrated Optical Circuits and Components;Design and Applications, Marcel Dekker, New York, (1987).

4. A. Y. Cho, Thin Solid Films, 100, p. 291 (1983).

5. W.T. Tsang, Appl. Phys. Lett., 40, p. 217 (1982).

6. T. Sanada, J. Yamakaski, O. Wada, T. Fujii, T. Sakurai and N. Sasaki, Appl. Phys. Lett., 44, p. 325 (1984).

7. S.R. Forrest, J. Light. Tech., LT-3, p. 1248 (1985).

8. P.P. Smythe, P.J. Chidgey, M.C. Brian, B.R. White, R.C. Hooper and D.R. Smith, Tĕch. Digest10th European Conf. Opt. Fibre Comm., Stuttgart, Germany, paper 11B-5 (1984).

9. R.M. Kolbas, J.K. Abrokwah, J.K. Carney, D.H. Bradshaw, B.R. Elmer and J.R. Biard, Appl. Phys. Lett., 43, p. 821 (1983).

10. R.F. Leheny, M.A. Nahary, M.A. Pollack, A.A. Bellman, E.D. Beebe, H.C. DeWinter and R.J. Martin, Elect. Lett., 16, p. 353 (1980).

11. M. Makiuchi, O. Wada, S. Miura, H. Hamaguchi, H. Hachida, K. Nakai, H. Horimatsu and T. Sakurai, Tech. Digest of IEDM, p. 862 (1984).

12. S. Ray and M.B. Walton, Proc. IEEE Microwave and Millimeter IC Symposium, Baltimore (1986).

13. C.P. Lee, S. Margalit, I. Ury and A. Yariv, Appl. Phys. Lett., 32, p.574 (1986).

14. H. Nakano, S. Yamashita, T. Tanaka, N. Hirao and N. Naeda, J. Light. Tech., LT-4, p.574 (1986).

15. J. K. Carney, M. J. Helix and R.M. Kolbas, Tech. Digest GaAs IC Symposium, Phoenix, p. 48 (1983).

16. M.K. Kilcoyne, D. Kasemset, R. Asatourian and S. Beccue, Proc. SPIE, Vol. 625 (1986).

17. L.D. Hutcheson, Technical Digest of Conference on Lasers and Electro-Optics, San Francisco (1986).

PACKAGING OF COMPONENTS FOR OPTICAL FIBER TECHNOLOGY

S. SRIRAM AND R.L. HOLMAN
Amphenol Fiber Optic Products, 1925 Ohio St., Lisle, IL 60532

ABSTRACT

A number of promising techniques used to interconnect two or more optical fibers and optical fibers with guided wave optical devices are reviewed. Such waveguide-to-waveguide interconnections impose severe optical and mechanical design constraints. Components used in single mode infrared optical systems use fibers of five to ten microns in core diameter. This creates the need for submicron alignment tolerance between the waveguides. Such severe tolerance requirements are satisfied, in the case of single mode fiber connectors, through the use of precision ceramic ferrules. Achieving the same performance, when interconnecting fibers and waveguide devices, is somewhat less straightforward. Several approaches are currently under study. One such approach uses precisely etched silicon V-grooves. Another approach uses self-aligned grooves or slots that are formed in the substrate material through ion-milling or laser-assisted etching. Components used in multimode fiber systems use fiber with fifty to one thousand micron core diameters, and are therefore more forgiving to misalignment. Despite this, careful design is required to achieve satisfactory performance. The misalignment penalties associated with a number of interconnect situations are illustrated and discussed in terms of alignment tolerances and fabrication difficulties.

INTRODUCTION

The many types of passive and active components needed in optical fiber systems must be ruggedly packaged and simply interconnected. Their performance is expected to remain essentially constant even when operating in fairly harsh fluctuating environmental conditions. In most cases, this implies that optical waveguides must be placed and then held in micron and often, submicron proximity with other optical waveguides. This requirement is relatively straightforward in the laboratory; but when applied to the field, it becomes the primary obstacle to the successful device operation.

The simplest and most mature case is that of single-mode and multimode fiber optic connector components. These well-known fiber optic component packages are now largely taken for granted. One common version [1] uses precision ceramic ferrules to capture and secure optical fibers. The fibers are secured with epoxy in the precise hole, or bore, pre-machined down the ferrule's axis. Great effort is taken to insure that the bore is concentric with the ferrule's outer diameter. Interconnection and alignment of two such ferrules is then achieved within a third cylindrical package known as an adaptor. The adaptor consists of a precise ceramic cylinder that captures and aligns the ferrules to be interconnected. When properly constructed, the axis of each fiber is made to precisely coincide with the central axis of both the ferrule bores and the adaptor. If dimensional tolerances are carefully controlled, an interconnection loss of less than 0.5 dB can be achieved [1]. This low-loss interconnection can be held over a wide temperature range, if the fibers are made to be in physical contact. Springs within the connector component package, or the use of screw threads combined with judiciously chosen dimensions and tolerances, can maintain fiber-to-fiber physical contact without fiber damage and performance degradation.

Interconnecting more than two fibers intensifies the demand on package precision. Linear arrays of etched V-grooves have been fabricated routinely in silicon substrates using methods developed for silicon microelectronics [2], fiber optics [3,4], and integrated optics [5]. Fiber-to-fiber locations in a single array have been achieved with submicron accuracy as expected of a photolithographic fabrication. Yet, despite the greater ease with which silicon arrays can be produced, silicon arrays have proven more difficult to align with each other than their single-fiber ceramic ferrule counterparts.

The packaging challenges become greater still as fiber interconnection with the more complex component architectures found in semiconductor lasers or planar waveguide circuits is considered. In these cases, the interaction of dissimilar packaging materials can frustrate performance stability due to ambient temperature fluctuations. Moreover, techniques using simple cylindrical symmetry used for fiber positioning and alignment are no longer applicable.

In this paper, we will review some of the current issues and promising approaches associated with successful packaging of advanced fiber optic components. We will restrict our discussion to the problem of input and output fiber coupling to simple waveguide circuit elements, such as those made in the electrooptic material lithium niobate. We feel this is a particularly good example for several reasons. First, lithium niobate devices are under active consideration in a number of promising electro-optical sensor and communication applications [6]. Second, while the circuit elements built on glass and semiconductors may not necessarily be identical to those built on lithium niobate, the size and layout are quite common among these different materials. Finally, the technology developed for lithium niobate can be readily extended to other optical and optoelectronic materials such as glass and semiconductors.

FIBER COUPLING TO OPTICAL DEVICES

The technology of interconnecting an optical fiber to a waveguide circuit is commonly referred to as pigtailing. Waveguides are formed near the surface of a polished lithium niobate substrate by solid state diffusion so that the guided wave is confined to within a 3 to 5 micron layer. In order to couple these guides optically to a fiber pigtail, the end-faces of the crystal and the fiber must be polished flat and perpendicular, and each end-face and must be free of defects. The fiber axis must be carefully aligned to that of the waveguide, and the fiber then secured to the crystal with a bonding agent. It is this bonding step that presents the greatest challenge. Most bonding agents, whether epoxy or solder, change their dimensions asymmetrically over time as they harden. As a result, the waveguide-to-waveguide alignment is disturbed, decreasing coupling efficiency. In addition, a small gap can sometimes exist between the fiber and the waveguide leading to Fabry-Perot resonances. This can contribute as much as a 30% variation in coupling efficiency with change in ambient temperature. Such difficulties escalate greatly when two or more fibers have to be bonded to the crystal in close proximity; the center-to-center spacing being less than 300 microns. Methods for holding and placing one fiber next to another without disturbing the alignment are not straightforward. Moreover, dimensional asymmetries that can occur during bonding agent hardening usually frustrates the simultaneous attachment of fiber arrays, unless special grooved fixturing is provided.

Fiber coupling to waveguides in anisotropic crystals such as lithium niobate pose additional complexity. The polarization direction of the input light must be oriented so that it is along the c-axis of the lithium niobate crystal. Such a requirement is satisfied by launching linearly polarized light into a pre-oriented polarization maintaining (PM) fiber. The PM fiber has the property such that a linearly polarized input light maintains its state of polarization as it travels through the fiber. The standard circular core fiber does not ordinarily hold a single polarization constant with time and is therefore not suitable for these applications.

Although several methods have been demonstrated in the laboratory, additional work is still required to reduce the time, effort, cost and repeatability of fiber attachment. Despite the difficulties, a few single input and output fiber pigtailed lithium niobate integrated optical circuit products have been commercialized [7]. A photograph of one such prototype, a lithium niobate phase modulator, is shown in Figure 1. The fiber is captured in a precision ceramic ferrule to provide the required mechanical strength to the structure during alignment and attachment operations. Positioning and angular alignment are achieved with micromanipulators while monitoring optical performance. Ferrule attachment to the waveguide substrate is accomplished with a UV curing epoxy. The total optical insertion loss for the device shown in Figure 1 has been as low as 5 dB, with 2.5 dB attributable to input and output coupling inefficiencies.

Future improvements are possible by increasing the efficiencies at the optical fiber and crystal interface. To do this, we must first establish an appreciation of the physics of the optical coupling interface. The physics at the interface can be understood by the use of the overlap integral calculations. A mathematical description of the overlap integral is given in the following section and will clarify some of the issues in interconnecting two optical devices. The materials and geometry issues are less easily quantified, and will be developed by way of example. We will consider the use of precision ceramic ferrules and micromachined silicon V-grooves. Since lithium niobate devices are used primarily because of its excellent electrooptic properties, the electronics aspects of packaging will also be discussed.

FIGURE 1. PHOTOGRAPH OF A PIGTAILED AND PACKAGED PHASE MODULATOR

Efficiency of Optical Coupling and the Overlap Integral

In a fiber optic system light propagates through many different optical waveguide devices. The overall throughput efficiency of the system is maximized when the interface between any two devices is designed such that the greatest energy transfer takes place between them. A theoretical estimate of the transfer of energy from one device to the other is obtained through the use of the overlap integral. An estimate of both the input and output spatial field distributions must be made before the overlap integral can be calculated.

Expressions for the optical field can be derived for several waveguide systems. The spatial fields of interest are of a single mode fiber and, for example, a channel waveguide in lithium niobate. The fields of a fiber are accurately represented by Bessel functions or a Gaussian. The fields of diffused channel waveguide require series representations of, for example, parabolic cylinder functions.

Theoretical Approach

When light propagating in a waveguide impinges upon a boundary between two different sections normal to its direction of propagation, some energy may be reflected, thus reducing the amount of light transmitted into the second guide. The reflection may result from a difference in effective indices of the sections, or a difference in the spatial distribution of the fields in the two guides, or both. The junction may be analyzed using a method due to Marcuse [8]. The boundary shown in Figure 2 contains both the guided and radiated modes propagating in the forward and reverse directions.

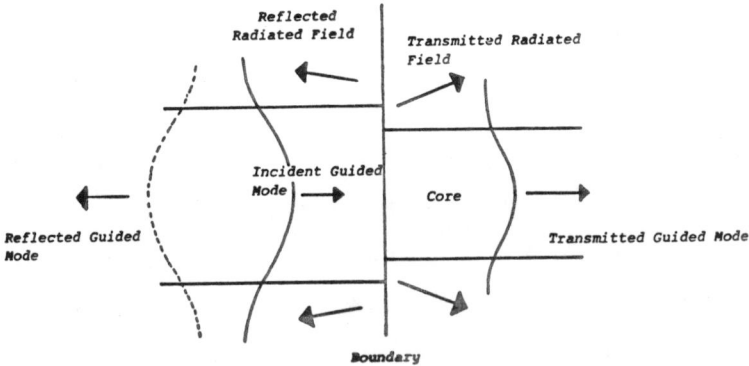

FIGURE 2. BOUNDARY AT THE FIBER - WAVEGUIDE INTERFACE

The transfer of energy from one optical field to the other can be obtained by using the "overlap integral" given by Equation 1. The overlap integral is a measure of the coupling efficiency between two guides and is only to be used as an optimization tool. For two similar single mode guides, the square of this quantity describes the power transferred between the two guides at each interface, neglecting the loss due to Fresnel reflection. The expression in Equation 1 breaks down for greatly dissimilar guides.

$$\kappa = \frac{\int\limits_{-\infty}^{\infty}\int E_1(x,y)\ E_2(x,y)\ dx\ dy}{[\int\limits_{-\infty}^{\infty}\int E_1 E_1{}^* dxdy\ \int\limits_{-\infty}^{\infty}\int E_2 E_2{}^* dxdy]^{\frac{1}{2}}} \tag{1}$$

where E_1 and E_2 are the electric field distributions for the modes propagating in guides 1 and 2.

The overlap integral may be calculated using the experimentally determined optical fields. The near-field of the optical waveguide is imaged onto a camera sensitive to the light at the wavelength of operation. Careful attention to focussing ensures that the profile is a true representation of the waveguide's electric field distribution within the limits imposed by diffraction of the focussing lens. The field profile is sampled at 64,000 points, and the data thus obtained are recorded. A similar process is carried out for the second waveguide. The overlap integral for the two-waveguide system is evaluated using a point-by-point multiplication and normalization.

If one of the waveguides is displaced relative to its optimum position, the fraction of power launched into the second guide will be:

$$n(\Delta x) = \frac{[\int\limits_{-\infty}^{\infty}\int E_1(x+\Delta x,y)\ E_2(x,y)dxdy]^2}{[\int\limits_{-\infty}^{\infty}\int E_1 E_1{}^* dxdy\ \int\limits_{-\infty}^{\infty}\int E_2 E_2{}^* dxdy]} \tag{2}$$

Exact expressions for two similar single mode fibers can be obtained by approximating the optical fields as Bessel functions. The overlap integral for successive waveguide displacements enables the loss associated with a transverse misalignment to be determined. Typical calculated loss results for two single mode fiber waveguides are shown in Figure 3.

The actual field profile associated with titanium diffused lithium niobate waveguides is asymmetric in the direction normal to the air:crystal interface and is, in general, elliptical. Thus, perfect coupling to a circular core fiber is not possible.

Since the diffused waveguide's mode profile is asymmetric, the function describing the effect of misalignment between the fiber and the waveguide depends on the direction of movement. Figure 4 illustrates the experimentally determined excess loss as a function of fiber:waveguide displacement for single mode systems at both 0.85 and 1.3 micron wavelengths. These results are analyzed in greater detail in the following sections to determine the necessary tolerances in the alignment and those of the precision componentry used.

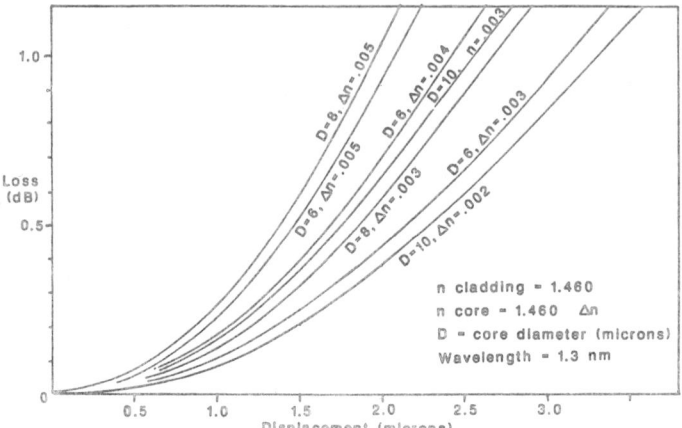

FIGURE 3. CALCULATED COUPLING EFFICIENCY BETWEEN TWO SINGLE MODE FIBERS

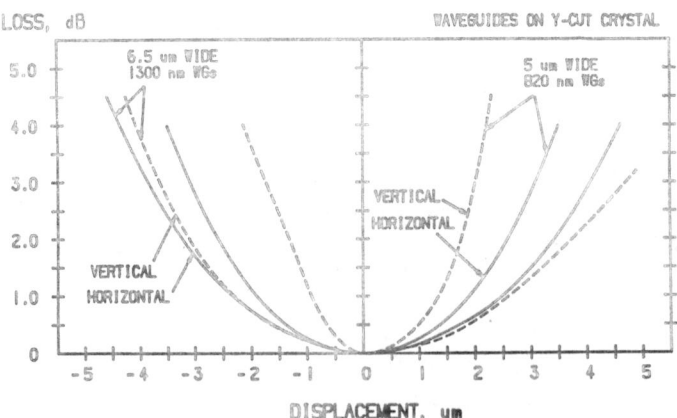

FIGURE 4. EXPERIMENTALLY DETERMINED COUPLING EFFICIENCY BETWEEN
FIBER AND WAVEGUIDE IN LITHIUM NIOBATE

Precision Required to Connect Optical Fibers to IOCs

As already described, lithium niobate-based devices require polarization maintaining (PM) fiber for successful operation. Although such fiber is commercially available, its geometric tolerances are not as tight as conventional single mode fiber. Consequently, pigtailing with PM fiber is more difficult and must accommodate both current tolerances and the required angular control of the polarization axes.

Angular alignment accuracy determines the polarization crosstalk in the waveguide. Mechanical orientation of the fiber axis is capable of providing the required angular alignment within 1 degree. This yields a 35 dB crosstalk rejection.

The alignment tolerances required in attaching single mode fibers to a diffused lithium niobate waveguide are summarized in Figure 4. For devices designed to operate at 820 nm wavelength, submicron alignment tolerances must be maintained. A misalignment of about one micron in the plane perpendicular to the waveguide substrate leads to an excess loss of 1 dB, increasing by 3 dB for every additional one micron in misalignment. Alignment requirements are less stringent in the plane parallel to the waveguide substrate. Alignment tolerances are further relaxed for devices designed to operate at 1300 nm wavelength. In this case, misalignment of up to two microns can be tolerated in the plane perpendicular to the waveguide.

The total insertion loss for a pigtailed integrated optical device is given by the sum of the waveguide propagation loss, twice the modal mismatch loss, the Fresnel reflection loss, and any loss associated with non-optimal fiber positioning. Additional loss can result during the fiber bonding process due to the drift caused by shrinkage during the curing. Adhesives that exhibit minimal shrinkage during cure and a stable opto-mechanical system are needed to perform successful fiber pigtailing. Conventional adhesives have not proven to be adequate in this regard, and a more stable bonding scheme is under investigation.

Single Fiber Attachment

When only a single fiber pigtail has to be attached to one side of a waveguide circuit, a precision ceramic ferrule is used to capture the fiber. Thus a rigid and a more easily handled structure is obtained. Ceramic ferrules were developed by the fiber optic connector industry as alignment structures for use in single mode fiber connectors. Since the fiber connector industry has an annual requirement for these ceramic ferrules in the hundreds of thousands, these components are economically priced and can therefore be used for other alignment purposes.

In the case of fiber pigtailing, a fiber pigtail is cleaved and epoxied in a ceramic ferrule whose bore diameter corresponds to that of the fiber (approx. 125um). Standard connector polishing techniques are used to prepare the fiber's end-face, with polishing times typically about two minutes. The fiber protrudes slightly from the end-face of the ceramic ferrule. The fiber end-faces are inspected for optical quality using a microscope. For best results, the core region of the fiber should be optically flat with a mirror finish and free from visible defects.

Light is coupled to this pigtail from a laser operating at 1.3 micron wavelength. The fiber pigtail is positioned close to the input of the waveguide circuit. Light from the output of the waveguide substrate is focussed onto a photodetector via a 10x microscope objective. The ferrule is then manipulated using a three axis piezoelectric stage as shown in Figure 5, until a maximum in the waveguide output is observed. A small amount of UV curing epoxy is then applied between the fiber and the

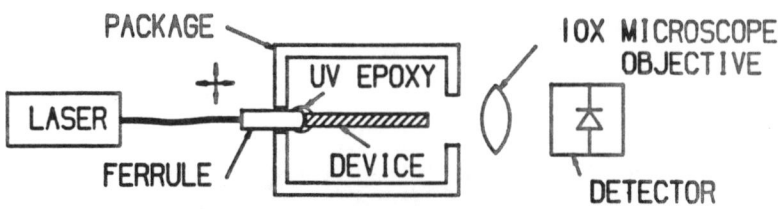

A. POSITIONING OF FIRST FIBER

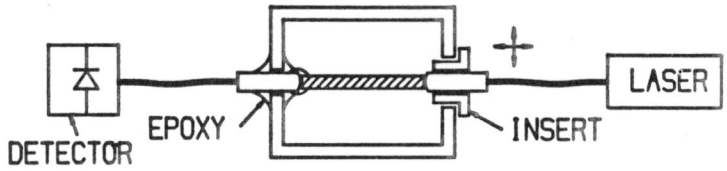

B. POSITIONING OF SECOND FIBER

FIGURE 5. ALIGNMENT OF FIBER TO A WAVEGUIDE

waveguide, and the gap between them is varied until a maximum in the waveguide output is obtained. The adhesive is then cured. Conventional epoxy is introduced into the gap between the ferrule and the end-face to provide greater final attachment strength than obtained by the use of the UV material alone. The subassembly is installed in the package, and the process repeated for the fiber pigtail at the waveguide output, yielding the final assembly shown in Figure 1.

Multiple Fiber Attachment Using Grooved Array

When two or more fiber pigtails have to be attached to one side of a waveguide circuit, requirements for relatively close channel-to-channel spacings do not permit the use of the ceramic ferrules. Instead, the same advantages can be achieved by capturing the fibers in a silicon V-groove array. The silicon V-groove technology was initially developed by the microelectronic industry [2]. Precisely etched parallel grooves can be etched with angled sidewalls. These sidewalls support the fiber within the groove.

Silicon V-grooves are fabricated by anisotropic chemical etching of (100) or (110) oriented silicon substrates depending on the shape of the V-grooves required. Typically, (100) oriented silicon is used to generate either trapezoidal or V-shaped grooves. The fibers are positioned and captured in the grooves and their end-faces polished. Thus, an array of fibers with accurate separation between each fiber is obtained. Aligning the first and the last fiber in the array to their respective waveguide targets will ensure the alignment of all the fibers in the array to their respective waveguides, provided the V-grooves and the fiber geometries are perfect. While laboratory demonstration of simultaneous multi-fiber attachment using silicon V-groove arrays have been successful [9,10], the approach has some drawbacks. Thermal expansion differences among the silicon, glass, and the lithium niobate substrate material can cause degradation in the device performance due to the temperature variations. This technique, however, has the potential to be developed into a practical production technique for the multiple fiber waveguide devices.

More advanced pigtailing approaches are also under investigation that can overcome the difficulties associated with the use of silicon. This includes locating grooves on the waveguide substrate itself by ion-milling [11] or laser etching [12]. Integrating the alignment structure with the waveguide circuit can greatly simplify the fiber pigtailing process.

Such an integrated alignment technique reported in the literature [11] uses ion-milling to generate fiber positioning slots in lithium niobate. Ion-milled slots about 20 microns wide and 10 microns deep are first milled in lithium niobate. Then, titanium is deposited and the titanium stripe is accurately defined using photolithography. Thus the optical waveguide formed by subsequent diffusion is automatically aligned with the ion-milled slot in the substrate. The fiber pigtail is etched in HF to match the slot width, cleaved and then dropped into the lithium niobate slot. There is no need for micromanipulation of the fiber to maximize the throughput. The depth of the slot and the outer diameter of the fiber must be carefully controlled to achieve satisfactory results.

Difficulties arise due to the need to chemically etch the fiber to match the slot width generated by ion-milling. Optical fibers, when etched in HF, may show greater fragility over their lifetime. Trace amounts of the etchant may contaminate other parts of the packaged device. Forming a deep slot of the order of 60 microns is necessary to overcome the need to etch the fiber. Lithium niobate mills at a rate of 300A/min [13] which makes it impractical to obtain the required slot depth.

A similar approach using laser etching of lithium niobate to generate grooves was reported by Ashby and Brannon [12]. The sample was covered with KF and irradiated with an Excimer laser operating at 248 nm or a dye laser operating at 270 nm. The results obtained have so far been less satisfactory for this application.

Multiple Fiber Attachment Using Active Alignment

The technique of capturing fibers within precision components is not the only promising approach for fiber attachment. Early attempts have been made to extend the single fiber attachment process to multiple fibers through computer control [14]. The approach for single fiber attachment described in the previous section depends on capturing the fiber inside a ceramic ferrule prior to alignment and bonding to the substrate material. The large outside diameter (approx. 2.5 mm) of the ceramic ferrule makes it impractical to extend this approach to multiple fiber alignment, since the waveguide circuits in the substrate are typically separated by 250 microns.

A promising approach is to align and attach individual fibers which have been stripped of their plastic jacket material leaving each fiber core and cladding intact. This approach requires special tooling to handle and move one fiber at a time without disturbing the position of previously placed fibers. Ordinary fiber grabbing hardware is not capable of maneuvering within the tight fiber spacings encountered in many multiple fiber device applications. Conventional epoxy bonding is not applicable since even minutely dispensed UV epoxies can flow over considerably larger surface areas, damaging adjacent fiber and waveguide channels. While these epoxies may set up in seconds, most adhesives do not develop full strength for minutes to hours. Careful surface preparation, using special bonding agents, of the fiber end-face, the fiber outer surface and the substrate surface adjacent to the waveguide

circuit is required. This bonding agent must:
* Instantaneously bond the fiber to the waveguide substrate on command;
* Confine the bonding agent to an area less than the fiber's end-face, so as not to damage adjacent channels;
* Show pull strengths between 5 and 25 grams.

Alignment times obtained with this method can be as short as 30 ms per fiber [14,15] after initially positioning the first fiber to within +/- 10 microns of the optimum to the corresponding waveguide circuit. The fiber is pre-positioned close enough to the target waveguide to provide a sufficient optical feedback signal. The required level of pre-positioning can be obtained without difficulty by using mechanical machine references, and is only an issue for the first fiber, as all other relative fiber locations are known to the computer in advance [14].

This approach also requires the use of a supporting substrate system, or submount, to hold the waveguide substrate during pigtailing and to provide strain relief for the fibers after bonding [14]. The submount is chosen to match the thermal expansion coefficients of the waveguide substrate. In principle, even grooved supporting substrates can be used as templates to guide the fiber pigtails, provided the grooves are perfectly registered with the target waveguides.

Given instantaneous bonding of one fiber at a time, the attractiveness of this approach as compared to passive alignment schemes hinges on the incremental time required to align each fiber. Optimizing fiber-waveguide alignment using piezoelectric micromanipulators, optical feedback and a phase locked loop has been shown to be quite practical.

ELECTRONIC ASPECTS

While the fiber pigtailing is the most challenging issue in the packaging of optical fiber components, the electrical interconnection is equally important when electrooptic operation is desired. Lithium niobate IOCs require the application of external voltages to the device's electrodes. There are several ways to achieve this.

For electronic signals up to a few GHz, lumped electrodes as shown in Figure 6 are adequate. Two SMA electronic connectors apply the external voltage across the waveguide's electrodes. The center contact of each SMA connector is wire-bonded to the bonding pads on the device.

For electronic signals faster than a few GHz, lumped electrodes are inadequate, and alternate approaches are required [16]. Traveling wave microwave stripline electrodes shown in Figure 7 are capable of accepting electronic signals up to 40 GHz. The bandwidth is limited by the velocity mismatch between the optical and the electrical fields, as opposed to the RC time constant of a lumped electrode. Optical devices with such traveling wave electrodes have been sold with bandwidths of 8 GHz. These devices have used high frequency electronic SMA connectors to launch the input signal to a microstrip line fabricated on an alumina substrate. The microstrip line carries the signal to the input port of the traveling wave electrode on the waveguide substrate. After traversing the traveling wave electrode, the signal exits at the output port and continues onto a second microstrip line which connects to a second high speed SMA connector. This second SMA connector can be tied to external circuits for impedance matching or other requirements.

CONCLUSIONS

Major strides have been made in the packaging of fiber optic components, particularly in the areas of fiber pigtailing and electrical

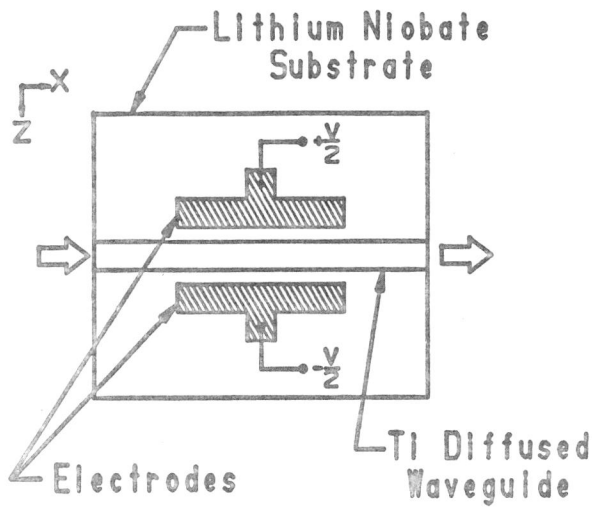

FIGURE 6. SCHEMATIC OF A LOW SPEED LUMPED CIRCUIT MODULATOR

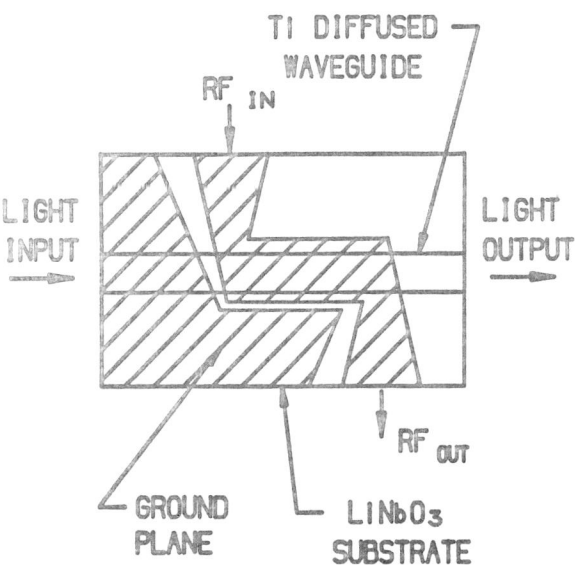

FIGURE 7. SCHEMATIC OF A HIGH SPEED TRAVELING WAVE MODULATOR

interconnection. Except for the simplest of optical devices, such as, connectors, photodetectors and CD lasers, fiber optic packaging technology has not yet reached the level of large scale manufacturing. Fiber pigtailed sources and detectors are beginning to benefit from advanced materials and automated processes. Simple waveguide modulators have been commercialized in prototype quantities, and both producers and users are gaining valuable experience. Major effort will be required to satisfy overall performance requirements in terms of thermal management, shock, vibration and radiation effects. These packaging related issues are now under active laboratory investigation around the world. We expect that as these issues become resolved, acceptance of the technology will spread, and larger-scale commercialization will become a reality.

REFERENCES

1. G.J. Sellers et al., FOC/LAN Proceedings, Orlando, FL 1986.
2. R.M. Finne and D.L. Klein, J. Electrochem. Soc. Solid State Sci. 114, 965 (1967).
3. P.C. Chang, S. Sriram, and A. Wey, SPIE Proceedings 837, San Diego, CA, 1987.
4. C. M. Miller, Bell System Technical Journal 4, 1228 (1986); Journal of Lightwave Tech. 4, 1228 (1986).
5. J. T. Boyd and S. Sriram, Applied Optics 17, 895 (1978).
6. G.J. Sellers and S.Sriram, Laser Focus, September 1986.
7. Amphenol Data Sheet, Lini-Guide Phase Modulator, Amphenol Fiber Optic Products, Lisle, IL 60532.
8. D. Marcuse, Bell System Technical Journal 49, 272 (1970).
9. H.P. Hsu and A.F. Milton, Electronics Lett. 12, 404 (1976).
10. E.J. Murphy and T.C. Rice, IEEE J. Quantum Electronics QE-22, 928 (1986).
11. I. Andonovic et al., Proc. of the 1st ECIO, pp. 8-10 (1981).
12. C.I.H. Ashby and P.J. Brannon, Applied Phys. Lett. 49, 475 (1986).
13. Bei Zhiang, S. Forouhar, S.Y. Huang, and W.C. Chang, J. Lightwave Tech. LT-2, 528 (1984).
14. R.L. Holman, 37th Electronic Components Conference, May 1987.
15. J. Goodwin, SPIE Proceedings 703, Cambridge, MA, 1986.
16. M. Izutsu et al., IEEE J. Quantum Electronics QE-13, 287 (1977).

LITHIUM NIOBATE PACKAGING CHALLENGES

E. J. MURPHY, R. J. HOLMES, R. B. JANDER, A. W. SCHELLING
AT&T Bell Laboratories, 555 Union Boulevard, Allentown, PA 18103

ABSTRACT

The use of lithium niobate integrated optic devices outside of the research laboratory is predicated on the development of a sound packaging method. We present a discussion of the many issues that face the development of a viable, robust packaging technology. We emphasize the interaction of lithium niobate's physical properties with available packaging materials and technologies. The broad range of properties (i.e. electro-optic, piezo-electric, pyro-electric, photorefractive...) that make lithium niobate an interesting material in many device applications also make it a packaging challenge. The package design, materials and packaging technologies must isolate the device from its environment so that lithium niobate's properties do not adversely affect the device performance.

INTRODUCTION

Over the past decade a myriad of optical devices has been conceived, manufactured and put into service. Lasers, LED's, photodetectors, optical isolators, splitters, couplers and opto-mechanical switches are currently in use. These devices provide the basis for current optical communication systems. Integrated optic devices will play an important role in future generation fiber systems by supplanting some of the existing devices and by providing new functionality.

Lithium niobate is currently the material of choice for integrated optics because of its relatively advanced state of development. Lithium niobate waveguide technology has evolved from the laboratory to prototype manufacturing and small volume commercial sales in the last few years. However, most of the work to date has dealt with device physics, materials science and fabrication technology; very little attention has been focused on packaging this complex material. Package design and packaging technology must now be developed for lithium niobate devices. This should be done with specific applications in mind because with each potential application, new considerations can act to constrain the package design. Device cost, performance, reliability, and environmental sensitivity are the usual parameters. In the relatively controlled environment of the central office, high performance and reliability are usually required, and higher costs are tolerated. In the local loop, devices must be inexpensive and capable of tolerating extremes of temperature and humidity.

Figure 1 illustrates a typical lithium niobate waveguide device. Optical waveguides are formed by diffusion of titanium strips into a flat substrate. Waveguides which are closely spaced interact and the guided light couples between the two guides. This coupling is critically dependent on the refractive index difference between the guides. A voltage applied to surface electrodes induces a changes in the refractive index difference through the linear electro-optic effect and alters the coupling properties. With proper waveguide and electrode design, the phase, amplitude or spatial position of the guided light can be altered. Optical phase modulators, amplitude modulators, polarization converters, 2x2-, 4x4-, 1x16-, and 8x8-, electro-optic switches have been demonstrated in lithium niobate. [1-5]

A package for these devices must be capable of supporting the crystal, protecting it from environmental effects and delivering the optical signals and controlling voltages. In many applications, optical devices are required to operate at high frequencies so the package must be capable of handling very high speed (>1 GHz) drive voltages. In other applications, there is a need to package optical sources, detectors and/or signal processing electronics with the device. As will become apparent in the next section, lithium niobate is a complex material. Its complexity coupled with device requirements presents many challenges to the package designer.

LITHIUM NIOBATE'S PROPERTIES

Lithium niobate is highly valued for optical waveguiding devices because of its large electro-optic coefficients (Table 1) that allow devices to operate at reasonable voltages. The crystal structure which produces the large electro-optic response also creates difficulties for the designers, fabricators and packagers of these devices. Lithium niobate is a ferroelectric, piezoelectric, pyroelectric and photoelastic single crystal material which exhibits anisotropic responses to physical stimuli. These various properties are interrelated and must be considered when producing a device. The large device size (typically 0.5 to 1 mm thick, 20 to 60 mm long, 5 to 10 mm wide) can amplify the influence of these properties on device performance.

In the next few paragraphs, we will discuss the crystal structure and the influence of several of its properties on device performance. It is important to note during the following discussion that the operation of an ideal device varies strongly with slight changes in refractive index. Therefore, subtle changes to the device or device environment are important. Because of large anisotropies in many of the properties, the choice of crystal orientation will influence the magnitude of the effects.

The crystal structure of lithium niobate is that of a rhombohedrally distorted perovskite, similar to the mineral ilmenite ($FeTiO_3$). While conventionally written as $LiNbO_3$, it is in fact not stoichiometric but contains percent levels of cation defects. [11, 12] The crystals are grown by the Czochralski method from a congruent melt at 48.45 mole% Li_2O. [13] Because of the large defect concentrations and because the crystal is thermodynamically metastable at room temperature, stresses on the crystal can impart large changes to the structure. For instance, the direction of spontaneous polarization can be reversed by simply scratching the crystal surface. [14] Lithium niobate is a brittle material, has natural cleavage planes and has a Mho Hardness of about 5, (approximately that of soft glass). [15] Thus it must be handled carefully during fabrication and packaging.

Temperature variations can affect device performance in several ways. Table 2 lists the (anisotropic) thermal expansion coefficients. Because of the large physical size of the device, slight differences in thermal expansion between the crystal and the package can generate significant stresses. These can couple via the piezoelectric (Table 3) and electro-optic effects to generate changes in device performance. (In fact, a similar effect can occur even without temperature changes if the package material flexes and applies a torque to the crystal.) The crystal's low thermal conductivity can combine with local heating during operation or packaging of the device to cause stresses which will alter the optical properties. Such local stresses are particularly important because they are likely to be applied in the active region of a device.

The pyroelectric response (Table 4) of the crystal to a change in temperature is perhaps the most serious impediment to stable performance in certain devices. The fully oxidized crystal is an electrical insulator at room temperature. Because of the crystal's pyroelectricity, temperature changes of a few degrees can generate large bound charges. Resultant changes in the surface charge can change the electro-optic response. [19] Non-uniform pyroelectrically-induced fields developed in the crystal can couple with the applied field and degrade device performance. In addition, the rapid discharge of excessive surface charge can induce physical damage in the crystal.

TABLE I

LINEAR ELECTRO-OPTIC COEFFICIENTS
(10^{-12} V/m)

parameter	λ (nm)	r(51)	r(22)	r(13)	r(33)	ref.
r(T)	633	32	6.81	10.0	32.2	6, 7
	1150		5.4			8
	3390		3.1			8
r(S)	633	28	3.4	8.6	30.8	9
	3390	23	3.1	6.5	28	10

TABLE 2

THERMAL EXPANSION COEFFICIENTS

$\alpha(33) = 7.5 \times 10^{-6}$ $\beta(33) = -7.7 \times 10^{-9}$ ref. 16

$\alpha(11) = 15.0 \times 10^{-6}$ $\beta(11) = 5.5 \times 10^{-9}$

TABLE 3

PIEZOELECTRIC STRAIN COEFFICIENTS
(10^{-11} C/N)

d(15)	d(22)	d(31)	d(33)	ref.
6.92	2.08	-0.085	0.60	6
0.8	2.1	-0.1	.06	34
7.4	2.1	-0.087	1.6	17

TABLE 4

PYROELECTRIC COEFFICIENT

$p(3) = -4 \times 10^{-5} \text{ cm}^{-2} \, {}^{\circ}\text{C}^{-1}$ ref. 23

TABLE 5

PHOTOELASTIC STRAIN COEFFICIENTS (AT CONSTANT FIELD)

p(11)	p(33)	p(44)	p(12)	p(13)	p(14)	p(31)	p(41)	ref.
0.036	0.066		0.072	0.135		0.178	0.155	24
0.025	0.068		0.079	0.132	0.1	0.168	0.158	25
0.034	0.060	0.30	0.072	0.139	0.66	0.178	0.154	26
-0.02	0.07	0.12	0.08	0.13	-0.08	0.17	-0.15	27

The last effect of temperature we will discuss involves phase transitions that occur near 100° Centigrade [20-22]. The diagram in Figure 2 shows the niobium-oxygen octohedra and the angular relationships. The alpha and beta angles, which correspond to a rhombohedral indexing of the crystal, change nonlinearly as a function of temperature. There are several important implications. First, this phenomenon must be considered when packaging devices for high temperature applications. Secondly, the influence of processing and packaging temperatures on the material must be considered. Finally, accelerated thermal aging of devices to establish reliability criteria cannot necessarily be linearly extrapolated.

Humidity levels may also affect device performance and reliability. All lithium niobate crystals contain water (protons) which is incorporated during growth and poling. In early work on the growth of ferroelectric oxides [23], the crystals were poled in a water vapor atmosphere. This was done to facilitate creation of a single domain crystal. The intrinsic water may play a role in the time dependent response to applied electric fields in an operating device. Such behavior can be explained with reference to Figure 3. There are two positions available for a proton within the octohedral cage of oxygen anions. The low energy position is in the trigonal plane of three anions; the high energy position is in the near center of the octohedral cage. Protons initially in the high energy position will eventually relax to the low energy position and this change in charge distribution will lower opposition to an applied electric field. In addition to these protons, protons will be incorporated from the water vapor atmosphere sometimes used during the diffusion of titanium into the lattice. These mobile protons can induce microdomain wall motion when the crystal is under a bias field. They can also contribute to the time dependence of electric fields within the crystal by drifting to oppose an applied voltage.

Lithium niobate's photoelasticity (Table 5) is the potential source of a subtle device-related problem because strains can induce changes in the refractive index. A cross section of a typical device is shown in Figure 4. The surface electrodes above the waveguides are used to apply the controlling voltages. An intermediate dielectric layer is often used to decrease the coupling of optical energy from the waveguides into the overlaid metal. Occasionally it is necessary to etch a gap in the dielectric layer (as shown) to eliminate ionic drift under the high applied fields. The metal and the dielectric are usually deposited at elevated temperatures and their thermal expansion coefficients differ from the substrate's. Thus there is a resultant strain in the layers that creates stress regions in the gap in the waveguide area. The resultant change in the refractive index (from the photoelastic effect) can have a magnitude greater than the electro-optic change due to the applied voltage. In fact, several authors have reported waveguides produced by this effect. [28, 29] If the strain varies or relaxes with time or temperature, the characteristics of the device will change during its lifetime.

The last phenomenon we will discuss here is the photorefractive effect. [30-32] This is a process by which absorption of optical energy changes the refractive index. More "optical damage"is likely to result from the power levels of the guided light but care should be taken to exclude room light from the package lest it have a long-term effect on reliability. The cross-section of the photorefractive effect appears to be small enough at $\lambda = 1.3$ and 1.5 μm to preclude optical damage at moderate power levels from these typical operating wavelengths.

PACKAGE EXAMPLE

Despite these difficulties, it is possible to design a suitable package for lithium niobate. A generic package concept is shown in Figure 5. The crystal is bonded to a support pedestal which may contain a thermoelectric heater/cooler to isolate the device from ambient temperature changes. Electrical signals enter the package through pins which may be hermetically sealed. Wirebonds feed the electrical signals to the device. Where high speed operation is critical, packages must be designed to incorporate transmission lines through the package wall. Integrated circuits which convert standard logic inputs into appropriate drive voltages might also be incorporated inside the package.

Two fiber ribbons are shown entering the package through the end wall. Outside the package there is strain relief and a bend radius limiter. Inside the package, a vertical displacement generates a slight S-bend in the fibers to provide protection against differential thermal expansion.

The major difference between packaging of optical and electronic devices is the attachment of optical fibers. [33] This is perhaps the single most difficult assembly task. Attachment of single mode fibers requires alignment tolerances that are at least an order of magnitude smaller than those normally encountered in electronic packaging. Fiber bending restrictions also impact the design. The fiber-lithium niobate problem is more difficult than the common fiber-laser problem because many fibers must be passed through two opposite walls. Bonding of fibers to the substrate is also a major issue. The method of holding the fibers during attachment operations and the bonding material must be chosen with package reliability in mind.

SUMMARY

Despite lithium niobate's sensitivity to external stimuli, reliable device performance can be obtained as evidenced by devices that are commercially available and by devices that have operated continuously for long periods of time. However, we must strive for improvements. As much as possible, we need to take advantage of packaging concepts and materials that have been developed for electronic components. Apart from the reliability, cost and performance trade-offs mentioned in the introduction, a package designer should consider the manufacturing line where workable assembly and testing sequences must be put in place. This is an important next step in lithium niobate packaging because it is likely that testing and packaging operations will be the dominant contributor to device cost.

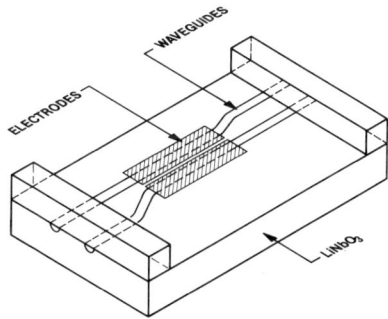

FIGURE 1

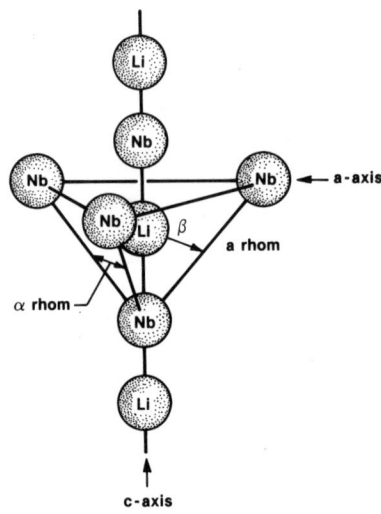

**RHOMBOHEDRAL-HEXAGONAL
CATION ARRANGEMENTS**

FIGURE 2

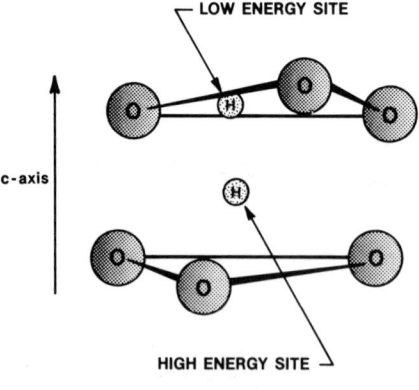

FIGURE 3

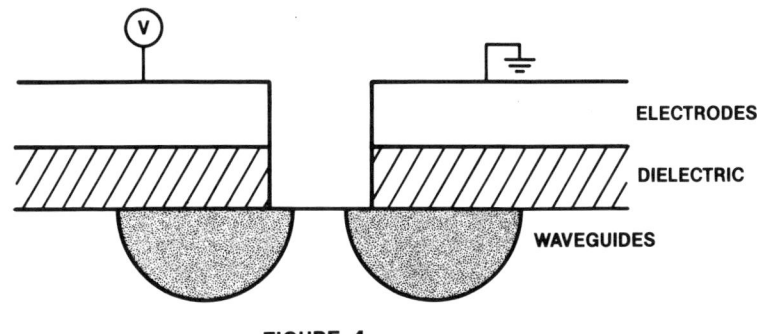

FIGURE 4

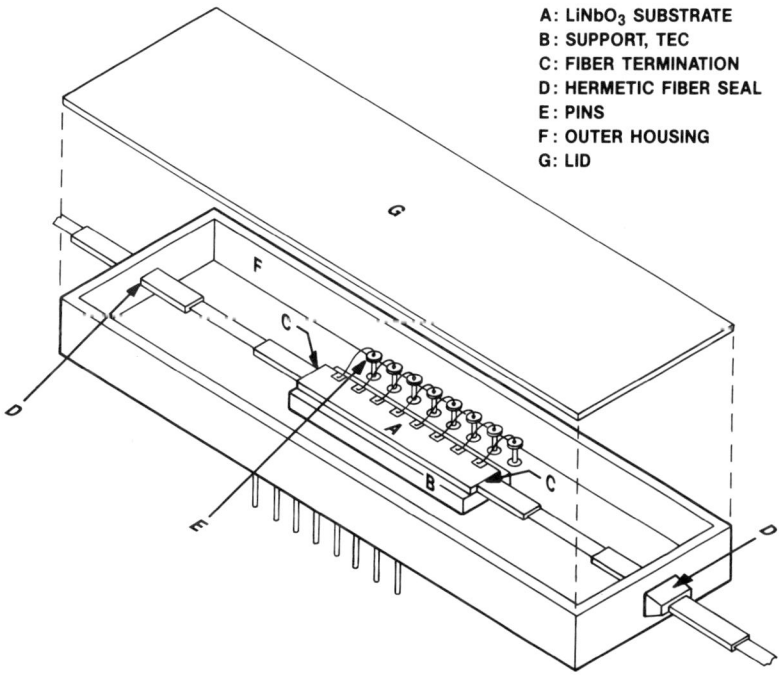

A: LiNbO₃ SUBSTRATE
B: SUPPORT, TEC
C: FIBER TERMINATION
D: HERMETIC FIBER SEAL
E: PINS
F: OUTER HOUSING
G: LID

LITHIUM NIOBATE DEVICE PACKAGE EXAMPLE

FIGURE 5

REFERENCES

[1] R. C. Alferness, IEEE J. Quantum Electron., QE-17, 946 (1981).

[2] G. A. Bogert, E. J. Murphy, R. T. Ku, J. Lightwave Tech., LT-4, 1542 (186).

[3] M. J. Wale, P. J. .Duthie, C. J. Groves-Kirkby, I. Bennion, Conf. on Lasers and Electro-Optics, p. 198 (1987).

[4] P. Granestrand, B. Stoltz, L. Thylen, K. Bergvall, Electronic Lett., 22, 816 (1986).

[5] J. E. Watson, M. A. Milbrodt, T. C. Rice, J. Lightwave Tech., LT-4, 1717 (1986).

[6] J. D. Zook, D. Chen, G. N. Otto, Appl. Phys. Lett. 11, 159-161 (1967).

[7] E. Bernal, G. D. Chen, and T. C. Lee, Phys. Lett. 21, 259-260 (1966).

[8] P. H. Smakula and P. C. Claspy, Trans. Metall. Soc. AIME 239, 421-424 (1967).

[9] E. H. Turner, Appl. Phys. Lett. 8, 303-304 (1966).

[10] E. H. Turner, J. Opt. Soc. Am. 56, 1426 (1966).

[11] E. K. Chang, A. Mehta, and D. M. Smyth, NATO Advanced Study Workshop, Sept. 1986.

[12] S. C. Abrahams and P. Marsh, Acta Cryst. B42, 61 (1986).

[13] C. D. Brandle, P. K. Gallagher, and H. M. O'Bryan, J. Am. Cer. Soc. 68, 493 (1985).

[14] S. Miyazawa, J. Appl. Phys. 50, 4599 (1979).

[15] A. Rauber, "Chemistry and Physics of Lithium Niobate", Current Topics in Materials Science, ed. E. Kaldis, North-Holland Publishing C., Amsterdam, 481 (1978).

[16] Y. S. Kim and R. T. Smith, J. Appl. Phys. 40, 4637 (1969).

[17] T. Yamada, N. Niizeki, and H. Toyoda, Jpn. J. Appl. Phys. 6, 151-155 (1967); Y. Cho and K. Yamanouchi, J. Appl. Phys. 61(3), 875 (1987).

[18] A. Savage, J. Appl. Phys. 37, 3071 (1966).

[19] P. Skeath, C. H. Bulmer, S. C. Hiser, W. K. Burns, Appl. Phys. Lett., 49, 1221 (1986).

[20] I. G. Ismailade, V. I. Nesterenko, F. A. Mirishi, Sov. Phys. Cryst. 13, 25, (1968).

[21] Y. Wang, H. Shen, Z. Xu, and H. Zhon, Ferroelectrics Lett. 6, 1-6 (1986).

[22] H. Engelmann, N. Kramer, H. Yuanfu, L. Rongchuan, and U. Gonsis, Ferroelectrics 69, 217-222 (186).

[23] H. Levinstein and C. D. Capio, J. Appl. Phys. 38 (7), 2761 (1967).

[24] R. W. Dixon, J. Appl. Phys. 38, 5149-5153 (1967).

[25] V. V. Lemanov, O. V. Shakin, and G. A. Smolenskii, Sov. Phys. - Sol. State. 13, 426-428 (1971).

[26] R. J. O'Brien, G. J. Rosasco, and A. Weber, J. Opt. Soc. Am. 60, 716 (1970).

[27] G. A. Coquin, as cited in D. A. Pinnow: Handbook of Lasers with Selected Data on Optical Technology, ed. R. J. Pressley, CRC Press, 482 (1971).

[28] P. A. Kirkby, P. R. Selway, L. D. Westbrook, J. Appl. Phys. 50, 4567 (1979).

[29] A. R. Nelson, Appl. Opt. 19, 3423 (1980).

[30] A. Ashkin, G. D. Boyd, J. M. Dziedzic, R. J. Smith, A. A. Ballmann, J. J. Levinstein, and K. Nassau, Appl. Phys. Lett. 9, 72 (1966).

[31] A. M. Glass, D. von der Linde, D. H. Auston, and T. J. Negran, J. Electron Mat. 4, 915 (1975).

[32] R. S. Weis and T. K. Gaylord, Appl. Phys. A 37, 191-203 (1985).

[33] E. J. Murphy, J. Lightwave Tech. to be published.

[34] A. W. Warner, M. Onoe, G. A. Coquin, J. ACoust. Soc. Am., 42, 1223, (1967).

CERAMIC OPTICAL PACKAGE: MATERIAL REQUIREMENTS
AND GUIDELINES FOR MATERIAL SELECTION

M. F. YAN AND W. W. RHODES
AT&T Bell Laboratories, Murray Hill, NJ 07974

ABSTRACT

Recently Lightwave Device Packaging Department at AT&T Bell Laboratories has demonstrated that ceramic materials can provide cost effective and high quality packages to house optical and electronic components for lightwave communication applications. In this paper we examine the material requirements for optical packages. We also study the material properties of metals and ceramics with a potential application in optical packages. In particular, we review hermeticity, thermal conductivity, thermal expansion coefficient, dielectric constant, electrical resistivity, sintering temperature and mechanical strength of these materials. Our study will provide a data base and useful guidelines for designers to make uniformed decisions on material selection for optical package. We also review the mixing rules to predict the resultant property of a composite from the known attributes of its constituents and the use of new composite materials will provide a new degree of flexibility in the optical package design.

INTRODUCTION

Recently there has been a significant demand for ceramic packages for lightwave communication applications. It has been demonstrated that certain ceramic materials can provide cost effective and high quality packages to house light sources (e.g. lasers and LED's) and other electronic components and to provide interconnections with optical fibers and other electronic systems.[1-3]

For optical package applications, ceramic and metal are usually co-fired to form a composite to provide a hermetic sealing and electrical and optical interconnection for optical fibers, light sources, integrated circuits and other electrical components. Figure 1 shows an Astrotec Lightwave Transmitter Package developed by Lightwave Device Packaging Department at AT&T Bell Laboratories.[1-3] This package consists of three ceramic layers co-fired with tungsten metallization to form a hermetic housing for a 1.3 μm multi-frequency laser transmitter operating over a wide range of bit rates from 1 Kb/s up to 1 Gb/s. By the use of a thermoelectric heat pump inside the ceramic package the laser transmitter can operate at 20°C over an ambient temperature range of −65 to +85°C. The opening for lense and single-mode optical fiber access was formed in the ceramic prior to sintering. The ceramic material in the package structure was 90-94% Al_2O_3 and the ceramic structure was approximately 2.41 inch long, 1.67 inch wide and 0.29 inch thick. The stable low thermal expansion characteristics of ceramic materials assures the long term stability of optical fiber to laser chip alignments. A tungsten copper (90/10) heat sink was brazed on the bottom of the fired alumina package to support the thermoelectric heat pump mounted on BeO base. A kovar lid was welded to a kovar seal ring which was brazed to the ceramic to complete the hermetic package structure. In this paper, we shall analyze the material requirements for the optical packages. Properties of metals and ceramics will be reviewed and evaluated as candidates for optical package

440

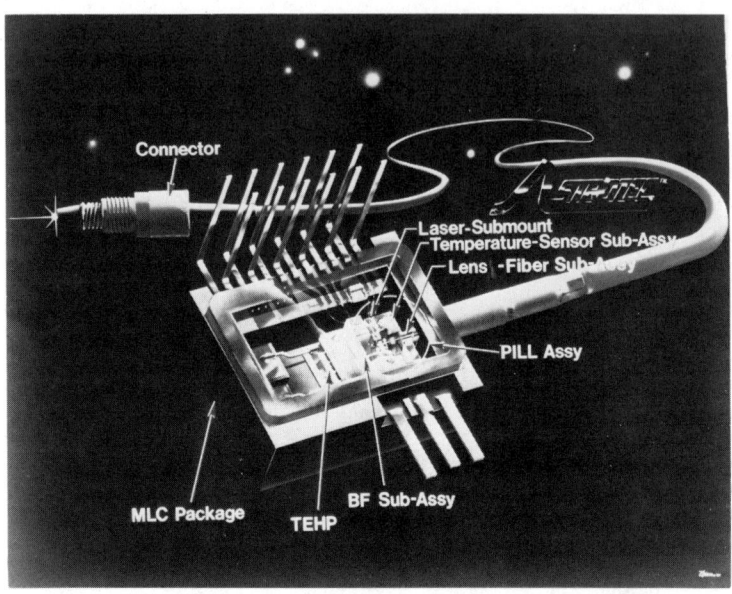

ASTROTEC® INTEGRATED TRANSMITTER

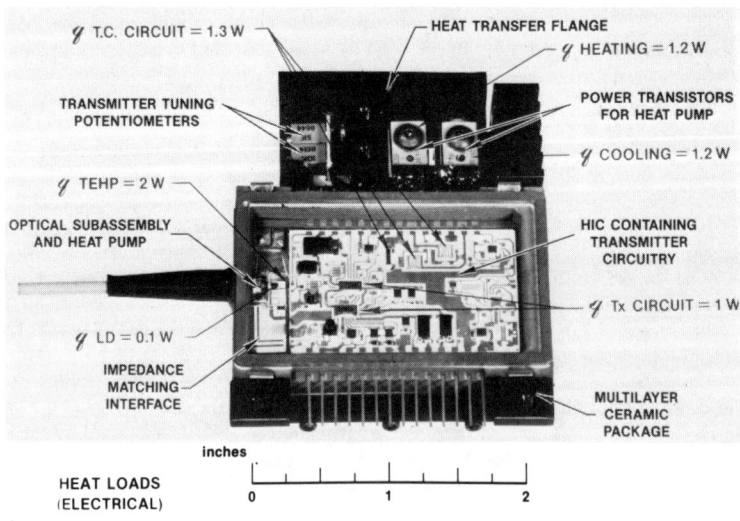

Figure 1   Astrotec Lightwave Transmitter Package developed by Lightwave Device Packaging Department at AT&T Bell Laboratories.[1-3]

applications.

MATERIAL REQUIREMENTS IN CERAMICS

The desirable attributes of ceramic materials for optical package applications are summarized in Table I and these attributes are discussed in the following sections.

High Hermeticity: The most important function of a ceramic package for the light sources and electronic components is to provide a hermetic housing. Virtually all dense ceramics are impervious to moisture, water vapor and air.[4] In glasses, the intrinsic permeability rate of gases is thermally activated and increases rapidly with temperature following an Arrhenius relation. Furthermore, the activation energy increases with the atomic or molecular radius of the diffusing gas. Among the gaseous species, helium atoms have the highest permeability in glasses and their permeability is 10 and 10^7 times higher than those of hydrogen and oxygen molecules respectively. At 200°C, probably the maximum operating temperature of an optical package, the He permeability rates are 2.28×10^{-13}, 7.6×10^{-15} and 7.6×10^{-19} cm^3 He/cm^2 sec under a pressure gradient of 1 Pa/cm in fused silica, borosilicate and lead borate glasses respectively.[4] For example, when a hollow vacuum cube $(1 \times 1 \times 1$ cm$^3)$, enclosed with 1 mm thick glass walls, becomes submerged in He at the atmospheric pressure $(10^5$ Pa), the hollow cube will be completely filled with He after 21 months, 52.8 yrs, 5.3×10^5 years for glass walls made of fused silica, borosilicate and lead borate respectively. The relatively high permeability of fused silica is due to the large openings in the glass networks. The modifying cations in the borosilicate and lead borate glasses block the openings in the glass network and decrease the gas permeability. The gas permeability is even smaller in crystalline oxides.[4] Thus, any lack of hermeticity in a ceramic package is probably due to structural defects, e.g. cracks and porosity in either the ceramic or the metal vias.

High Thermal Conductivity: The heat generated by the light sources and electronic components may raise the temperature of these devices well above their maximum operating temperatures if the ceramic package has a limited thermal conductivity. In one design of optical packages, a thermoelectric heat pump is installed inside the optical package to transfer heat from an operating laser.[1-3] However, the heat must be dissipated to the ambient through conduction. In a current design in our laboratory, a metallic bottom plate is used for thermal conduction.[1-3] However, the use of a metallic plate may impose additional constraints on electrical insulation and interconnection. Furthermore, the metal to ceramic seal in this design may compromise the hermeticity. Thus, in the more advanced designs a monolithic ceramic package is preferred.

Low Dielectric Constants: In the present design of optical packages, the dielectric constant, ϵ, of the ceramic does not have a large effect on the package performance. However, in the more advanced designs, many electronic components will be housed in the

TABLE I

Requirements on Ceramics for Optical Package Applications

1. High Hermeticity
2. High Thermal Conductivity
3. Low Dielectric Constant
4. Low Sintering Temperature
5. High Mechanical Strength
6. Good Thermal Expansion Match with other Components

optical package and an efficient interconnection among these packaged electronic components will be a critical consideration. It has been known that in the design of ceramic packages for electronic applications,[5] the propagation delay, τ_d, of signal transmission is given as

$$\tau_d = \frac{\ell\sqrt{\epsilon}}{c} \tag{1}$$

where ℓ is the distance the signal travels and c is the speed of light. For transmission over a distance of 1 m in a packaging medium, e.g. Al_2O_3, with a dielectric constant of 9.5, the propagation delay is $\simeq 10$ ns.

Low Sintering Temperature: Ceramic materials with a lower sintering temperature may permit the use of less expensive and electrically and thermally more conductive electrodes. A lower sintering temperature usually leads to a lower production cost of the ceramic packages.

High Mechanical Strength: A high mechanical strength is always desirable for high reliability package applications. However, the mechanical strength of a package is affected by the processing conditions, package design and the compatibility with co-fired electrodes as well as the intrinsic strength of the ceramic material itself.

Good Thermal Expansion Matching with other Components: It is desirable for the ceramic material to have a good thermal expansion match with the optical and electronic components mounted in an optical package. A good match in their thermal expansion coefficients will reduce the stress on the interfaces between the package and the components. However, the optical and electronic components are usually fabricated from materials with different thermal expansion coefficients and it is unlikely that such dissimilar materials can achieve a good thermal match. Thus, in the more advanced designs, a package may consist of different sections fabricated from different appropriate ceramic materials to accommodate various components. While a thermal expansion mismatch may still exist between different ceramic sections, the thermal stresses would act on the ceramics instead of the more critical components.

Material Requirements in Metals

The desirable attributes of metallic materials for optical package applications are summarized in Table II and these attributes are discussed in the following sections.

Low Electrical Resistivity: In the present design of electronic packages the relatively large cross-section (125-250 μm diameter) of strip lines,[6,7] even with the use of relatively resistive metals such as Mo or W in the ceramic package, has resulted in small resistive loss. However, the large dimension of strip lines prevents a further miniaturization and leads to large capacitance between signal lines. In the more advanced designs, a 20-fold increase in wire density has been proposed by the use of lossy but distortion-free

TABLE II

Requirements on Metals for Optical Package Applications

1. Low Electrical Resistivity
2. Good Compatibility with Ceramic During Co-firing

TABLE III

Material Properties of Ceramics

Materials	ϵ	tan δ	ρ at 25°C (Ω cm)	Flexural Strength (MPa)	Thermal Expansion Coefficient at 25°C 10^{-6}/K	Thermal Conductivity W/mK	Sintering Temperature (°C)
SiO_2	3.78			30-100	0.6	1.5	1000-1200
96 SiO_2 + 4B_2O_3 (vycor)	3.85	8×10^{-4}	$>10^{17}$		0.75		1000-1200
73.2 SiO_2 + 26.8 B_2O_3	4.05						1000
Borosilicate (Corning 7740)	4.60	2.6×10^{-2}	10^{15}		3.25	1.7	1000
KCl	4.75	1×10^{-4}		4	36.6	7	500-700
Soda-borosilicate (Pyrex-Corning 7060)	4.84				2.75		1000
KBr	4.90	2×10^{-4}		5	38.5	4.9	500-700
2 MgO + 2 Al_2O_3 + 5 SiO_2 (cordierite)	5.0			245	1.5	4.3	1400-1500
Li_2O − ZnO − SiO_2 (5% ZnO) glass ceramic	5.0	2.3×10^{-3}	10^{15}		8	2.8	1000-1200
70 P_2O_5 + 30(Al_2O_3 + ZnO)	5.25						
Li_2O − ZnO − SiO_2 (30% ZnO) glass ceramic	5.3	6.3×10^{-3}	10^{14}		8	2.2	
Li_2O − MgO − SiO_2 glass ceramic	5.4	2.2×10^{-3}				1.7-3.8	
Li_2O − ZnO − PbO − SiO_2 glass ceramic	5.8	3×10^{-4}	$>10^{14}$			1.7-3.8	
NaCl	5.9	2×10^{-4}	10^{15}	10	41	6.2	500-700
Si_3N_4	6.0			580	0.8	33.5	2000
Li_2O − ZnO − SiO_2 (10% ZnO) glass ceramic	6.0	1.3×10^{-3}					
BeO	6.7	1×10^{-4}	10^{14}	490	6.3	250	1500-1600
3 Al_2O_3 + 2 SiO_2 (mullite)	6.6			270	5.0	6.7	1500
Li_2O − Al_2O_3 − SiO_2 (glass ceramic)	6.6	1.8×10^{-3}	10^{14}		5.0	5.4	
92% Al_2O_3	8.5	3×10^{-4}	$>10^{13}$	331	6.5	16.7	1600
AlN	8.8	5×10^{-4}	5×10^{13}	490	4.5	320	1800
LiF	9.0	2×10^{-4}			34.2	5	500-600
96% Al_2O_3	9.3		10^{14}	317	6.4	25.1	1600
Al_2O_3	9.5	3×10^{-4}	10^{14}	350	5.5	30	1600
SiC (BeO doped)	40	5×10^{-3}	10^{13}	450	3.7	270	2100

transmission lines. The proposed lossy transmission lines may be fabricated from Cu strip lines with a cross-section of 5×9 μm.[6,7] However, the maximum length, ℓ_{max}, of a strip line for low distortion transmission is limited by the resistivity, ρ, of the electrode material as

$$\ell_{max} = \frac{2Z_0A}{\rho} \qquad (2)$$

where A is the cross-sectional area of the transmission line with a characteristic impedance $Z_0 = \sqrt{L/C}$. Thus, for a given geometry, it is desirable to decrease the resistivity in order to achieve the distortion-free transmission in a lossy strip line.

Good Compatibility with Ceramics During Co–firing: Since metal and ceramic are co-fired during the preparation of packages it is important that both materials can be sintered at the same temperature and atmosphere. The shrinkage during sintering of the metal should be close to that of the ceramic to avoid deformation and stresses. It is also important to prevent cracks and voids in the electrode vias and the ceramic body in order to preserve the package hermeticity. In principle, a macroscopic void space would not remain in the interface between metal and ceramic if both materials have the same relative densities before firing and both were sintered to their theoretical densities. While such processing conditions are difficult to obtain in practice, it is sometimes sufficient to adjust the green densities and sintering conditions such that any remaining cracks and voids would not provide a continuous path for air leakage. Densification behavior and stress field developed during sintering of a metal-ceramic multilayer composite have been analyzed by Bordia and Raj.[8] They showed that the tensile stress generated by the densification process is usually not large enough to cause brittle cracking in ceramic. Instead, diffusional pore growth mechanism is more likely to produce defects; and this pore growth mechanism can be minimized by increasing the spacing between pores.[8]

REVIEW OF MATERIAL PROPERTIES OF CERAMICS

Ceramic properties pertinent to optical package applications will be reviewed to provide a useful data base for the package design.

Table III lists the material properties of selected ceramics with a potential application in optical packages. In particular, we list the dielectric constant, dissipation factor (tan δ), electrical resistivity, mechanical strength, thermal expansion coefficient, thermal conductivity and typical sintering temperature of the ceramic materials reviewed in this study. In the following sections, we classify the ceramic materials according to their ranking in a given material property.

Dielectric Constant: General trend of dielectric constants, ϵ, reported in different classes of ceramics is shown in Table IV. Silica-based glasses probably have the lowest dielectric constant among ceramics. Alumina based ceramics have a dielectric constant twice that of silica glasses. The delay times of electronic signals propagating along the strip lines imbedded in different ceramic media were calculated and normalized with respect to that in SiO_2 medium. It is interesting to observe that the delay time is reduced by $\sim$32% by the use of silica glass instead of alumina based ceramics since the delay time is proportional to $\sqrt{\epsilon}$ as shown in Eq. (1).

In a composite with different ceramic constituents, it is desirable to estimate the effect on the dielectric constant of the composite. Several mixing rules for the dielectric constant of a composite have been reported in the literature.[4,5] In a simple mixing rule for the constituents in series, the dielectric constant, ϵ_s, of a composite is given as

$$\frac{1}{\epsilon_s} = \sum_i \frac{V_i}{\epsilon_i} \tag{3}$$

where V_i and ϵ_i are the volume fraction and dielectric constant of the individual phase, i, in the composite. In a composite with the constituents in parallel, the dielectric constant, ϵ_p, is given as

$$\epsilon_p = \sum_i V_i \epsilon_i. \tag{4}$$

These two mixing rules can be generalized as

$$\epsilon^n = \sum_i V_i \epsilon_i^n \tag{5}$$

where $n = 1$ for the constituents in a series arrangement and $n = -1$ for a parallel arrangement. In an intermediate case such that $n \simeq 0$ it can be shown that $\epsilon_i^n \simeq 1 + n \ln \epsilon_i$ and a logarithmic mixing rule can be obtained as

$$\ln \epsilon_\ell = \sum_i V_i \ln \epsilon_i . \tag{6}$$

In a more complex structure[4] with a dispersion of spherical second phase particles imbedded in a continuous matrix phase, the resultant dielectric constant, ϵ_c, is

$$\epsilon_{\hat{c}} = \frac{V_{\hat{m}} \epsilon_{\hat{m}} (2 + \epsilon_{\hat{d}}/\epsilon_{\hat{m}}) + 3V_{\hat{d}} \epsilon_{\hat{d}}}{V_{\hat{m}} (2 + \epsilon_{\hat{d}}/\epsilon_{\hat{m}}) + 3V_d} \tag{7}$$

where V_d and V_m are the volume fractions of the dispersed and matrix phases with dielectric constants of ϵ_d and ϵ_m respectively.

Curves of dielectric constant versus the volume fraction of second phase as predicted by these mixing rules are given by Kingery et al.[4] Furthermore, dielectric measurements of TiO_2 powder mixed with various matrix materials, e.g. ZrO_2, kaolin clay $(Al_2(Si_2O_5)(OH)_4)$ and polystyrene, have shown a quantitative agreement with these mixing rules in different ranges of TiO_2 volume fraction.

TABLE IV

General Trend of Dielectric Constants and Delay Times over 1 m Ceramic Media

	Materials	Range of Dielectric Constant	Delay Time (ns)
a.	Silica based glass	3.8-4.8	6.5-7.3
b.	Halides	4.8-6	7.3-8.2
c.	Glass ceramics	5-7	7.5-8.8
d.	Cordierite/Mullite	5-7	7.5-8.8
e.	Si_3N_4	6	8.1
f.	BeO	6.7	8.6
g.	Al_2O_3 based ceramics	8.5-9.5	9.7-10.3

TABLE V

Typical Sintering Temperatures of Different Classes of Ceramics

Materials	Sintering Temperature (°C)
1) PbO based glasses	400-700
2) Cordierite (amorphous)	800-900
3) Silica based glasses	1000-1200
4) Glass Ceramics	<1200
5) Mullite, cordierite (crystalline)	1400-1500
6) Al_2O_3 based	1600
7) AlN	1800
8) Si_3N_4	2000
9) SiC	2100

Sintering Temperature: Table V lists the typical sintering temperatures of different classes of ceramics. PbO-based glasses can be sintered at a rather low temperature of 400-700°C while SiO_2-based glasses require a higher sintering temperature of 1000-1200°C.[10] Amorphous cordierite powder can be sintered at an even lower temperature range of 800−900°C than the silica glasses.[11] However, crystallized powders with the mullite and cordierite phases require a substantially higher sintering temperature of 1400-1500°C which is close to that of alumina-based ceramic.[12] Most covalent carbides and nitrides require an even higher temperature during pressureless sintering. For example, AlN,[13] Si_3N_4[14] and SiC[15] have been sintered to >90% of their theoretical densities at 1800, 2000 an 2100°C respectively.

The sintering temperature of a given ceramic material can be affected by the characteristic of the starting powder. In general, a lower sintering temperature can be achieved by a smaller particle size, a uniform particle size distribution, an equiaxed particle shape and a uniform powder packaging.

Selected chemical dopants have been introduced to ceramic to increase the sintering rate. Certain roles of these dopants have been identified. For example, dopants may dissolve in the lattice to create point defects which are necessary for the material transport during densification. Alternatively, dopants may segregate at grain boundaries to provide a high diffusivity path to remove pores from the interior. However, some chemical dopants are also known to affect the thermal conductivity. In fact, several dopants and impurities have been reported to reduce the thermal conductivity of AlN[13] and SiC.[16]

Mechanical Strength: Mechanical properties of ceramics are often critically dependent on their microstructures. In particular, the flexural strength can be increased by reducing the flaw size. For example, a microstructure with a uniform distribution of small grain and pore sizes in a well densified body with a minimal amount of porosity can often lead to the maximum flexural strength. In fact, the flexural strength, σ, is related to the flaw size, ç, by

$$\sigma = K_{Ic} / \sqrt{c} \qquad (6)$$

where K_{Ic} is the fracture toughness and it is usually a material property and independent of the microstructures. Most metals have a larger toughness value than ceramics. For example, most alloy steels have $K_{Ic} \simeq 45 − 50$ MPa · $m^{1/2}$ in contrast with the $K_{Ic} = 0.9$ MPa · $m^{1/2}$ for cordierite glass;[11] 4 MPa · $m^{1/2}$ for Al_2O_3[11] and SiC,[17] 5 MPa · $m^{1/2}$ for Si_3N_4[17] and $15 − 17$ MPa · $m^{1/2}$ for transformation toughened ZrO_2 ceramics.[18] The fracture toughness can be increased by modifying the material properties near a crack tip. Recently, it was discovered that in ZrO_2 containing ceramics the phase transformation from the metastable tetragonal to monoclinic phase may be initiated by the stress field of a crack tip. The volume increase due to the phase transformation imposes a compressive stress on the crack tip and prevents crack propagation.[19] In some other ceramics and glasses, surface flaws often lead to fracture under a tensile stress. It has been shown that the surface finish of Al_2O_3,[20] SiC[21] and Si_3N_4[22] has a significant effect on the flexural strength. The flexural stress measurements of these materials showed that the strength is extrinsically controlled by flaws induced by surface machining in ceramics with a small grain size, whereas the strength is controlled by the intrinsic flaws within the largest grains in ceramics with a coarse grain size. Compressive layers are thus intentionally fabricated on specimen surfaces, e.g. by ion exchange of the smaller matrix ions with larger ions, to increase the fracture toughness. In many glasses, fiber reinforcement and controlled devitrification are useful methods to increase flexural toughness.

Table VI lists the mechanical strengths of selected ceramic materials. Among all ceramic materials, the transformation-toughed ZrO_2 containing ceramics probably have the highest strength of ~2000 MPa exceeding even some commercial steels.[18] Covalent nitrides and carbides, e.g. Si_3N_4,[23] AlN,[13] and SiC[16,21], usually have a much higher strength than most oxides, e.g. Al_2O_3,[24] mullite[25] and cordierite.[26] Ordinary commercial silica glasses[27] have a flexural strength about 5-15 times smaller than most covalent carbides and nitrides.

TABLE VI

Mechanical Flexural Strength of Ceramic Materials

Materials	Flexural Strength (MPa)	References
$ZrO_2–Al_2O_3$[*]	2000	(18)
Si_3N_4	580	(20)
SiC[†]	450	(16,21)
AlN	490	(13)
Al_2O_3	350	(24)
Mullite	270	(25)
Cordierite	245	(26)
SiO_2[‡‡]	30-100	(27)

* A composite with 20 wt.% Al_2O_3 and 80 wt.% tetragonal phase ZrO_2 stabilized with 3 mole % Y_2O_3 dopant.
† SiC with ~2 wt.% BeO dopant to achieve an electrically insulating ($\rho \geqslant 4 \times 10^{13}$ Ω cm) property.
‡ ‡ While the theoretical strengths of silica glass is ~1.8×10^4 MPa, the flexural strengths cited above are much smaller due to surface flaws in ordinary commercial glasses.[27]

Thermal Conductivity: Diamond holds the honor of being the material with the highest thermal conductivity (2000 W/mK) at room temperature.[28,29] Graphite also has a similarly high thermal conductivity for conduction perpendicular to the[29,30] c-axis. Table VII lists the thermal conductivity of ceramics at 25°C.

The thermal conductivity of a dielectric material usually varies with temperature.[4] The conductivity, k, starts from a zero value at 0°K at a rate of k ~ exp $(-\theta/\alpha T)$ and it reaches a maximum value near the Debye temperature, θ. At a higher temperature, the conductivity decreases with temperature with k ~ 1/T and at a sufficiently high temperature the conductivity reaches a constant value. Near room temperature the thermal conductivity usually decreases with temperature in most ceramic materials. In order for a nonmetallic crystal to possess high thermal conductivity it must have low average atomic mass, strong interatomic bonding, a simple crystal structure and low anharmonicity.[29] For example, SiC[16] and BeO have a higher thermal conductivity than Al metal. In most oxides and carbides, the thermal conductivity decreases with an increase in the atomic weights of cations.[4] The difference in the atomic weights of the constituents reduces the energy transfer by lattice vibration and thus leads to a lower thermal conductivity. Furthermore, ceramics with complex crystalline structures tend to have more phonon scattering and also a lower thermal conductivity. For example, Al_2O_3[31] has a relatively small conductivity of about 13% of Al. Complex oxide crystals with heavy cations have even lower thermal conductivities, e.g. ZrO_2[31] has a conductivity at about 0.6% of Al. Solutes, second phase inclusions and porosity tend to reduce the mean free path of phonons and thus reduce the thermal conductivity. An amorphous material usually has a much lower thermal conductivity than a crystalline material because the phonon mean free path in a glass is limited to a few interatomic spacings by the random amorphous structure. Thus, fused silicon, borosilicate and soda-lime glasses[32] have a very small thermal conductivity of about 0.4-0.6% of Al.

In a composite structure it is desirable to evaluate the equivalent thermal conductivity. In a simple structure with laminated slabs, the equivalent conductivity, K_p, for heat conduction parallel to the plane of slabs is given as

$$K_p = V_1 k_1 + V_2 k_2 \qquad (5)$$

where V_1 and V_2 are the volume fractions of each component with conductivities of k_1 and k_2 respectively.

However, for heat conduction normal to the plane of slabs, the equivalent heat conductivity, K_n, is given as

$$\frac{1}{K_n} = \frac{V_1}{k_1} + \frac{V_2}{k_2} . \qquad (6)$$

In a more complex structure[4] with a dispersed phase imbedded with in a continuous phase, the resultant conductivity, K_e, of the composite is given by

$$K_e = \frac{A}{B} k_c \qquad (7a)$$

where
$$A = 1 + \frac{2V_d(1 - k_c/k_d)}{(2K_c/K_d) + 1} \quad \text{and} \qquad (7b)$$

$$B = 1 - \frac{V_d(1 - k_c/k_d)}{(1 + k_c/k_d)} \qquad (7c)$$

In the above equations k_c and k_d refer to the conductivity of the continuous and dispersed phases respectively, and V_d is the volume fraction of the dispersed phase. When $k_c > k_d$, the resultant conductivity is

$$K_e = k_c(1 - V_d)/(1 + V_d) \ . \tag{8}$$

However, if $k_d > k_c$, then

$$K_e = k_c(1 + 2V_d)/(1 - V_d) \ . \tag{9}$$

TABLE VII

Thermal Conductivity of Ceramics at 25°C

Materials	Thermal Conductivity (W/mK)	K/K_{Al} * (%)	References
Diamond	2000	844	28,29
Graphite (⊥ c-axis)	2000	844	29,30
AlN	320	34	13,29
SiC	270	114	16
BeO	250	105	29
Al_2O_3	30	13	31
$3Al_2O_3 - 2SiO_2$ (Mullite)	6.7	2.8	31
$Li_2O - Al_2O_3 - SiO_2$ (glass ceramic)	5.4	2.3	32
$2MgO - 2Al_2O_3 - 5SiO_2$ (Cordierite)	4.3	1.8	31
SiO_2 (fused silica)	1.5	0.6	32
Borosilicate glass	1.7	0.4	32
Soda-lime-silica	1.5-1.7	0.6-0.7	32
ZrO_2	1.5	0.6	31

* Thermal conductivity is normalized with respect to that of Al metal which has a conductivity of 237 W/mK at 25°C.

Thermal Expansion Coefficients: Table VIII lists the thermal linear expansion coefficients of selected ceramic materials and materials for the fabrication of optical and electronic components. We should note that the thermal expansion coefficient of a given material usually varies with temperature. The coefficient generally increases with temperature.[4] For the package applications, the relevant values are those near 25°C and they are shown in Table VIII. Al_2O_3[33] has a relatively good match in thermal expansion coefficient with GaAs laser material. For InP laser application, mullite appears to be a better choice in terms of thermal matching. However, SiC and borosilicate glass substrates have a near perfect match with silicon for integrated circuits applications.

Review of Material Properties of Metals

Table IX lists the material properties of metal electrodes[5] with a potential application in optical packages. In this table, metals are listed according to an ascending order of their melting point. Among the metals reviewed there are four groups of metals with different ranges of electrical resistivity: Ag, Cu and Au with a rather low resistivity in the range of $(1.6-2.2) \times 10^{-6} \Omega$ —cm; Mo, W and Ni have a resistivity about 3-4 times that of Ag; Palladium and Pt have resistivity values 6-7 times that of Ag; and Chromium has a resistivity 12-13 times that of Ag.

Metals usually have a much larger thermal expansion coefficient than a ceramic or semiconductor. For example, Ni, Au, Cu and Ag have an expansion coefficient about 2.4-3.6 times that of Al_2O_3 and about 4-5.5 times that of Si. However Mo, W and Cr

TABLE VIII

Thermal Linear Expansion Coefficients at 25°C

Materials	Thermal Expansion Coefficient $(10^{-6}/°K)$	References
BeO[*]	6.3	33
(GaAs)	5.8	33
Al_2O_3[*]	5.5	33
$3Al_2O_3-2SiO_2$ (Mullite)	5.0	34
(InP)	4.3	33
SiC	3.7	16
(Si)	2.6	33
Borosilicate glasses[‡]	3.2-5.2	35
$2MgO-2Al_2O_3-5SiO_2$ (Cordierite)	1.5	33
SiO_2 (fused silica)	0.6	4

[*] Polycrystalline Samples
[‡] For example, Corning glasses 7740, 7050, 7056 have thermal linear expansion coefficients of 3.25, 4.6 and 5.15 ppm/K respectively between 0 and 300°C.[35]

have a thermal expansion coefficient close to the typical ceramic and semiconductor materials. For example, Molybdenum and W have a near perfect thermal match with mullite, Al_2O_3 and InP. Chromium has a good thermal match with BeO and GaAs.

Most metals have a higher thermal conductivity than ceramics because electrons are generally more effective than phonons in the thermal conduction process. For example, Au, Cu and Ag have a thermal conductivity 1.1-1.5 times those of SiC and BeO which have the highest thermal conductivities among polycrystalline ceramics except diamond. Silver has the highest thermal conductivity among the metals investigated with a conductivity 2-3 times that of W or Mo and nearly 6 times that of Pd or Pt. While Cr, Pd and Pt probably have the lowest thermal conductivity among the metals investigated, they are still 2-2.5 times more conductive than alumina ceramics.

DISCUSSION AND SUMMARY

In this paper, we reviewed the requirements on materials for optical package applications. The basic need for an optical package is to provide a hermetic and protective environment for the optical and electronic components. Thus, the design, materials and processing must provide a finished product to satisfy a set of minimum requirements on hermeticity and mechanical strength.

After these basic requirements are satisfied thermal management will be considered such that the critical components within the package do not exceed a certain maximum temperature at the operating condition. It is evident that the thermal conductivities of materials used in a package are important consideration for thermal management.

In the more advanced designs, integrated circuits will also be housed in an optical package. The interconnection among these circuits and their communication with external

TABLE IX

Material Properties of Metal Electrodes[6]

Metals	Melting point (°C)	Electrical resistivity (10^{-6} Ω-cm)	Coefficient of thermal expansion (10^{-6}/°C)	Thermal conductivity (W/m·K)
Ag	960	1.6	19.7	418
Au	1063	2.2	14.2	297
Cu	1083	17	1.7	393
Ni	1455	6.8	13.3	92
Pd	1552	10.8	11	71
Pt	1774	10.6	9	71
Cr	1900	20	6.3	67
Mo	2625	5.2	5	146
W	3415	5.5	4.5	201

systems could be an important consideration. However, the topological and geometrical design of the interconnection wiring cannot be optimized without inputs from the dielectric properties of the ceramics and conductivity properties of the metal electrodes.

The different materials used in an optical package must be compatible during processing. Thus, during co-firing of metals and ceramics in an optical package, these materials must have similar densification rates. Furthermore, a lower production cost could be achieved by the use of low temperature and ambient atmosphere sintering. With this objective, ceramics with a low sintering temperature and nonprecious metals with an oxidation resistant capability will offer a competitive edge.

The thermal expansion mismatch among different materials at operating temperatures of an optical package could also increase its failure rate due to mechanical stress and material fatigue from thermal cycling. This thermal consideration also imposes an additional constraint on the choice of materials.

Among the materials reviewed, it is evident that none can satisfy all the material requirements and a certain compromise will be necessary. Thus, this review on material properties will serve a useful purpose to provide system and package designers with the necessary data base to make an informed decision on their designs.

A composite material usually can combine the favorable attributes of its constituents. New composite materials may lead to a new degree of flexibility in the designs of optical packages. The mixing rules reviewed in this paper provide useful guidelines to estimate the resultant dielectric constant and thermal conductivity of a composite. Furthermore, it has been reported that composite materials can have a substantial improvement in their mechanical properties.

REFERENCES

1. P. D. Smeltz, "Ceramic Applications in Laser Packaging", Proceedings of International Symposium on Ceramic Substrates and Packages, Denver, CO, October 18-21 (1987).
2. P. D. Smeltz and W. R. Holbrook, "Design of the Multilayer Ceramic Package for AT&T/8 Astrotec Lightwave Transmitter", ibid.
3. J. Thomson, R. Buck, P. Smeltz and Holbrook, "Manufacture of Multilayer Ceramic Packages for Astrotec Lightwave Transmitter", ibid.
4. W. D. Kingery, H. K. Bowen and D. R. Uhlmann, "Introduction to Ceramics", John Wiley & Sons, N.Y. (1976).
5. B. Schwartz, "Review of Multilayer Ceramics for Microelectronic Packaging", J. Phys. Chem. Solids, Vol. 45 [10] 1051-1068 (1984).
6. A. J. Blodgett and D. R. Barbour, "Thermal Conduction Module: A High Performance Multilayer Ceramic Package", IBM J. Res. Develop. 26 [1] 30-36 (1982).
7. C. W. Ho, D. A. Chance, C. H. Bajorek and R. E. Acosta", Thin Thin-Film Module as a High-Performance Semiconductor Package", IBM J. Res. Develop. 26 [3] 286-296 (1982).
8. R. K. Bordia and R. Raj, "Sintering Behavior of Ceramic Films Constrained by a Rigid Substrate", J. Am. Ceram. Soc. 68 [6] 287-92 (1985).
9. A. E. Paladino, "Temperature-Compensated $MgTi_2O_5 - TiO_2$ Dielectrics", J. Am. Ceram. 54 [3] 168-69 (1971).
10. M. F. Yan, J. B. Mac Chesney, S. R. Nagel and W. W. Rhodes, "Sintering of Optical Wave-Guide Glasses", J. Mat. Sci. 15 1371-1378 (1980).
11. E. A. Geiss, J. P. Fletcher and L. W. Herron, "Isothermal Sintering of Cordierite-Type Glass Powders", J. Am. Ceram. Soc. 67 [8] 549-552 (1984).
12. B. H. Mussler and M. W. Shafer, "Preparation and Properties of Mullite-Cordierite Composites", Ceram. Bull. 63 [5] 705-714 (1984).
13. Y. Kurokawa, K. Utsumi, H. Takamizawa, T. Kamata and S. Noguchi, "AlN substrates with High Thermal Conductivity", IEEE Trans. CHMT-8 [2], 6 (1985).
14. S. Prochazka and C. Greskovich, "Effect of Some Impurities on Sintering Si_3N_4", pp. 489-502 in "Proc. Int. Symposium on Factors in Densification and Sintering of Oxide and Non-oxide Ceramics". Edited by S. Somiya and S. Saito, Assoc. for Sci. Doc. Information, Tokyo Inst. Technology, Ookayama, Meguro, Tokyo, Japan (1979).
15. S. Prochazka, C. A. Johnson and R. A. Giddings, "Atmosphere Effects in Sintering of Silicon Carbide", pp. 361-381, ibid.
16. K. Maeda, T. Miyoshi, Y. Takeda, K. Nakamura, S. Ogihara and M. Ura, "Grain Boundary Effect in Highly Resistive SiC Ceramics with High Thermal Conductivity", pp. 260-268. Advances in Ceramics vol. 7. Edited by M. F. Yan and A. H. Heuer, Am. Ceram. Soc., Columbus, Ohio (1983).
17. A. G. Evans and E. A. Charles, "Fracture Toughness Determinations by Indentation", J. Am. Ceram. Soc. 59 [7-8] 371-372 (1976).
18. M. V. Swain and L. R. F. Rose, "Strength Limitations of Transformation-Toughened Zirconia Alloys", J. Am. Ceram. Soc. 69 [7] 511-18 (1986).
19. R. C. Garvie, R. H. Hannink and R. T. Pascoe, "Ceramic Steels ?", Nature (London) 258, 703 (1975).
20. R. E. Tressler, R. A. Langensiepen and R. C. Bradt, "Surface-Finish Effects on the Strength-vs-Grain Size Relation in Polycrystalline Al_2O_3", J. Am. Ceram. Soc. 57 [5] 226-27 (1974).

21. D. C. Cranmer, R. E. Tressler and R. C. Bradt, "Surface Finish Effects and the Strength-Grain Size Relation in SiC", J. Am. Ceram. Soc. *60* [5-6] 230-32 (1977).

22. M. Kawai, H. Abe and J. Nakayama, "The Effect of Surface Roughness on the Strength of Silicon Nitride", pp. 545-556 in "Proc. Int. Symposium on Factors in Densification and Sintering of Oxide and Non-oxide Ceramics", edited by S. Somiya and S. Saito, Assoc. for Sci. Dco. Information, Tokyo Inst. Technology, Ookayama, Megino, Tokyo, Japan (1979).

23. G. R. Terwilliger, "Properties of Sintered Si_3N_4", J. Am. Ceram. Soc. *57* [1] 48-9 (1974).

24. Alumina as a Ceramic Material; edited by W. H. Gitzen, The American Ceramic Society, Columbus, OH (1970).

25. K. S. Mazdiyasni and L. M. Brown, "Synthesis and Mechanical Properties of Stoichiometric
Am. Ceram. Soc. *55* [11] 548-52 (1972).

26. R. J. Pletka and S. M. Wiederhorn, "A Comparison of Failure Predictions by Strength and Fracture Mechanics Techniques", J. Mat. Sci. *17* [5] 1247-68 (1982).

27. R. H. Doremus pp. 281-295, "Glass Science", John Wiley & Sons, N.Y. (1973).

28. R. Berman, pp. 384, "Physical Properties of Diamond", Oxford University Press, Oxford (1965).

29. G. A. Slack, "Nonmetallic crystals with High Thermal Conductivity", J. Phys. Chem. Solids *34* 321-35 (1973).

30. C. Y. Ho, R. W. Powell and P. E. Liley, U.S. Nat. Bureau Standards, Rept. NSRDS-NBS16 (1968).

31. Y. S. Touloukian, R. W. Powell, C. Y. Ho and P. G. Klemens, "Thermophysical Properties of Matters — Thermal Conductivity", Vols. 1 and 2. IFI/Plenum, N.Y. (1970).

32. P. W. McMillan, "Glass Ceramics", Second Edition, Academic Press (1979).

33. Y. S. Touloukian, R. K. Kirby, R. E. Taylor and T. Y. R. Lee, "Thermophysical Properties of Matter — Thermal Expansion", Vol. 13, IFI/Plenum, N.Y. (1977).

34. B. L. Metcalfe and J. H. Sant, "The Synthesis, Microstructure and Physical Properties of High Purity Mullite", Trans. Br. Ceram. Soc. *74* [6] 193-201 (1975).

35. Data Sheet by Corning, "Properties of Glasses and Glass-Ceramics".

36. D. L. Evans, G. R. Fischer, J. E. Geiger and F. W. Martin, "Thermal Expansions and Chemical Modifications of Cordierite", J. Am. Ceram. Soc. *63* [11-12] 629-34 (1980).

Nb CONTACTS TO GaAs: THERMAL STABILITY AND PHASE FORMATION

KEVIN J. SCHULZ, XIANG-YUN ZHENG, and Y. AUSTIN CHANG
University of Wisconsin-Madison, Department of Metallurgical and Mineral
Engineering, 1509 University Avenue, Madison, WI 53706
* This work was sponsored by the U.S. Department of Energy under contract
No. DE-FG02-86ER452754.

ABSTRACT

The applicability of Nb as a Schottky barrier on GaAs depends to a large
extent on the thermal stability of the contacts. In this study, bulk diffusion couple and phase diagram studies in addition to thin film studies were
completed to understand the stability of and the reactions at the Nb/GaAs
interface. Nb thin films were deposited onto GaAs substrates by dc magnetron
sputtering and were annealed in the temperature range 300 to 1000°C. Analysis was done using plan-TEM and XTEM. The Nb/GaAs interface was found to
break down into a series of binary compounds above 500°C. Bulk diffusion
couples annealed at 600°C were analyzed using an electron microprobe. The
stable sequence of phases formed in the couple, i.e., the diffusion path, was
determined and was used to rationalize the observed compound formation in the
thin film contact system.

INTRODUCTION

A critical issue for the development of Schottky and ohmic contacts to
III-V semiconductors is the thermal stability at the metal/semiconductor
interface. For device applications, a metallization must be able to undergo
high temperature annealing (in excess of 800°C) to remove ion implantation
damage from the semiconductor substrate. Most metal contacts react with GaAs
resulting in degradation of the contact electrical properties. Refractory
metals such as niobium are of interest for use as a Schottky contact because
of their high thermal stability. Earlier work by Wu et al. [1], however, has
shown that rapid thermal annealing (RTA) for 10 seconds above 750°C results
in degradation of contact electrical properties. A recent study by Ding [2]
has shown that the binary compounds Nb_4As_3 and Nb_5Ga_3 form during RTA at
700°C for 10 seconds. The determination of the stable or equilibrium configuration of reaction products is of practical interest. The Nb-Ga or Nb-As
binary compound which is stable in contact with GaAs is a good candidate for
use as a contact material. This material could be directly deposited onto
GaAs by sputter deposition or co-evaporation as has been demonstrated by
Lince et al. [3] for $AuGa_2$ on GaSb.
The sequence of phases formed in thin film reactions can be very complicated. The reaction products which initially form at the interface may be
thermodynamically or kinetically unstable. A series of phases may form and
decompose before the final stable configuration is reached. This stable configuration is termed the diffusion path. The diffusion path is determined by
the thermodynamics and kinetics of the system and is unique for a given temperature and pressure. Thus, a knowledge of the relevant phase diagram
information is essential for an clear understanding of interfacial reactions.
The importance of understanding thermodynamics when considering metal reactions with III-V compounds has been emphasized in recent works by Beyers et
al. [4], Tsai and Williams [5], and McGilp [6]. Recently, Lin et al. [7]
have demonstrated the merits of using bulk diffusion couples to determine the
Pd/GaAs diffusion path and have used it to rationalize thin-film Pd contact
reactions with GaAs. A similar approach employing phase diagram samples,
bulk diffusion couples, and thin film couples has been used in this study to
understand the Nb/GaAs system.

EXPERIMENTAL METHODS

Phase diagram study

A series of seven Nb–Ga–As compositions along the Nb/GaAs tie line were annealed at 600°C for 30 days and analyzed using x-ray diffraction. Nb and GaAs powders were intimately mixed, pressed into pellets, and sealed in quartz tubes evacuated to 10^{-4} torr for annealing. X-ray diffraction was done using a Picker diffractometer.

Bulk diffusion study

Single crystal GaAs and fine-grained Nb were cut into 5x5x3 mm pieces. The contact surfaces were mechanically polished and cleaned. The pieces were placed in a quartz tube and were held tightly together by quartz rods which were sealed into the tube. These were evacuated to a pressure of 10^{-4} torr. The diffusion couple was annealed at 600°C for seven days and then quenched in water. The sample was cross sectioned and analyzed with an electron microprobe.

Thin film study

Undoped GaAs (100) oriented wafers were prepared for deposition by degreasing and etching in a 1:1 solution of HCl:deionized water. Niobium films with a thickness of 30 nm were deposited by dc magnetron sputtering in a vacuum chamber with a base pressure of 10^{-7} torr. Samples were encapsulated in 5 mm quartz tubes evacuated to 10^{-4} torr for annealing. Samples were annealed at temperatures between 300 and 1000°C in 100°C intervals for 1 hour.

Plan-TEM specimens were prepared by back-thinning the GaAs substrate with a 5% bromine in methanol solution. Cross-sectional specimens were thinned using an ion mill. The TEM specimens were examined with a JEOL 100B TEM and a JEOL 200CX TEM/STEM equipped with an energy dispersive x-ray spectrometer (EDS).

EXPERIMENTAL RESULTS

Phase diagram

The Nb–Ga–As phase diagram with the 7 sample compositions studied is shown in Fig. 1. X-ray results support thermodynamic calculations of the phase diagram using Miedema's model reported by Schmid-Fetzer [8]. No evidence of ternary compound formation was observed. There was some difficulty in identifying Nb_3As and certain Nb–Ga binary compounds (Nb_2Ga_3, Nb_4Ga_5, and Nb_5Ga_4) because of the limited number of samples studied and the overlap of major x-ray peaks. Formation of these compounds was observed to be very slow at 600°C. Regions of the phase diagram not unambiguously determined are shown with dashed lines. The major feature to be noted in the diagram is the high relative stability of NbAs.

Bulk Diffusion Couple

The diffusion couple held at 600°C for 7 days consisted of the layer sequence $GaAs/NbAs/NbGa_3/Nb$. Nominal atomic concentrations obtained with the microprobe for NbAs were: 54%Nb-44%As-2% Ga and for $NbGa_3$: 27%Nb-72%Ga-1%As. The composition profile across the couple is shown in Fig. 2.

Fig. 1. Nb–Ga–As phase diagram. Dashed lines indicate tie-lines not substantiated experimentally. Dotted lines indicate GaAs/Nb thin-film diffusion
path. Numbers indicate location of phase diagram sample compositions.

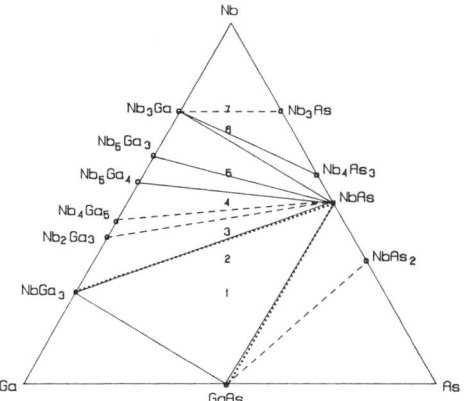

Fig. 2. Photomicrograph of bulk diffusion couple annealed for 7 days at
600 °C and accompanying composition profile. Phase regions are (a) GaAs, (b)
NbAs, (c) NbGa₃, and (d) Nb.

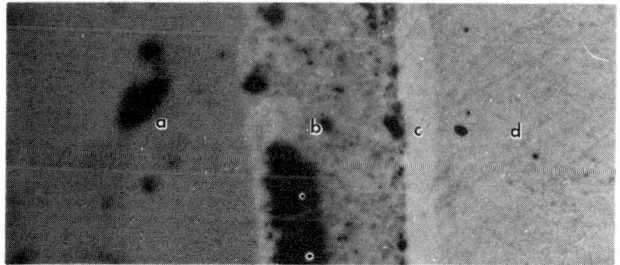

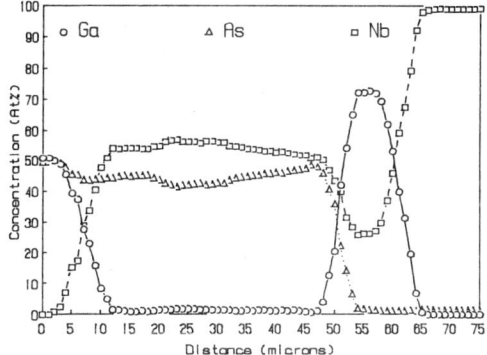

Thin Film Couples

Niobium films appeared to be stable on GaAs when annealed for 1 hour up to 400°C. At 500°C interdiffusion was observed by XTEM as shown in Fig. 2a. The stable configuration GaAs/NbAs/NbGa₃ was observed with XTEM after annealing 1 hour at 800°C and is shown in Fig. 2b. A series of plan-view micrographs and diffraction patterns are illustrated in Fig. 3. Nb_4Ga_3 was observed to form first at the interface. At 600°C, Nb_4As_3 and $NbGa_3$ are the major phases with some evidence of Nb_5Ga_3, Nb_5Ga_4, and the unreacted Nb film still apparent with electron diffraction. Annealing at 700°C and above for 1 hour results in complete reaction of the Nb to form NbAs and $NbGa_3$. The interface between the reacted layers and GaAs is very irregular. All phases observed showed little, if any, texture.

Fig. 2. XTEM images of 30 nm Nb/GaAs annealed for 1 hour at (a) 500°C, and (b) 800°C.

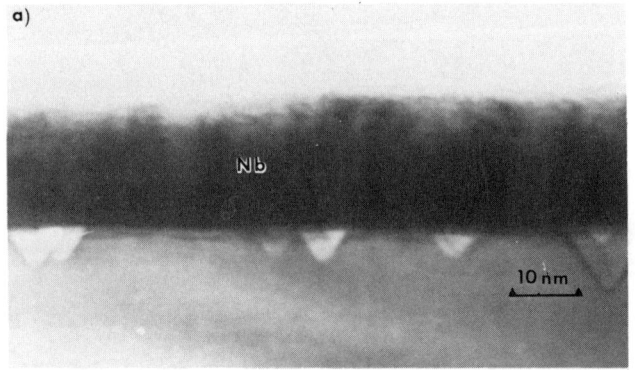

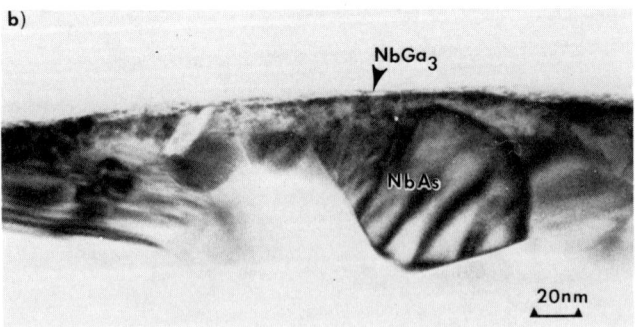

Fig. 3. Plan–TEM images and diffraction patterns for 30 nm Nb/GaAs
annealed for 1 hour at (a) 600°C and (b) 700°C.

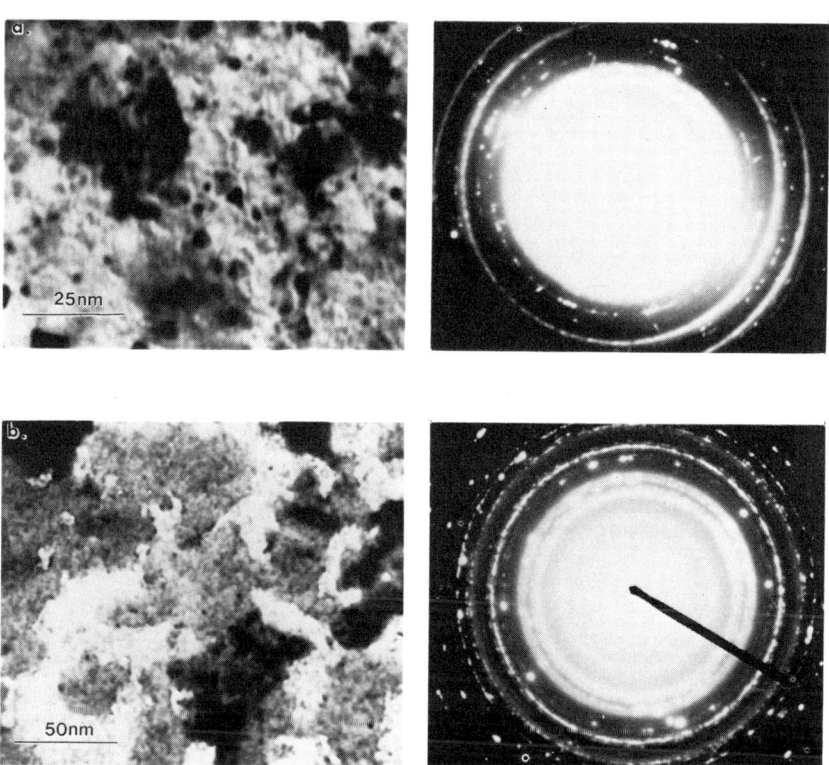

DISCUSSION

Elemental niobium in contact with GaAs is thermodynamically unstable. Interdiffusion of Ga and As into the Nb matrix results in compound formation at the interface. The native oxides present on the surface of GaAs are likely reduced by the Nb [9]. There appear to be no ternary phases or phases with extended solubility which preferentially form first at the GaAs/Nb interface; therefore, at least two binary compounds must form to conserve the equal amounts of Ga and As from the GaAs matrix [10]. In systems with similar phase diagrams, e.g. Pt and Ir [11], Ga diffuses most rapidly into the metal leaving excess As at the GaAs interface. Consequently, Nb_4As_3 forms first in contact with GaAs. During this initial transition period, Nb_5Ga_3, Nb_5Ga_4, and $NbGa_3$ also form, although $NbGa_3$ appears to be the predominant Nb–Ga phase. From the phase diagram, it is apparent that the configuration of phases present during this transition period will not be in local equilibrium and, thus, will not be stable. NbAs, which is stable in contact with GaAs, nucleates and consumes the parent Nb_4As_3. In the thin film case, complete consumption of elemental Nb results in a shift in the overall composition of the couple toward the GaAs side of the phase diagram. Thus, the final configuration observed in the thin-film diffusion couples, GaAs/NbAs/NbGa₃, is consistent with the phase diagram and is the thin film diffusion path. The observed configuration in the bulk couple, GaAs/NbAs/NbGa₃/Nb, is consistent with the thin film results; however, this configuration does not represent the final diffusion path because NbGa₃ and Nb cannot be in local equilibrium. It is believed that the other Nb-Ga compounds predicted by the phase diagram would form in the diffusion couple if given sufficient time for all interfaces to reach local equilibrium. Considerable difficulty was also observed in forming the remaining Nb-Ga binary compounds at 600°C in the phase diagram study.

The irregular contact interface observed in the thin-film reactions would certainly degrade the electrical contact properties. Therefore, Nb is unsuitable as a contact to GaAs. The non-uniform growth of phases at the GaAs interface may be the result of a non-uniform oxide layer at the interface causing regions of faster diffusion and dissolution. The pyramidal shape of the Nb₄As₃ grains shown in Fig. 2a may also suggest preferential growth into the (100) GaAs; however, it is interesting to note that the growth of the product phases was found to be, at most, weakly textured. In Fig. 2b it is apparent that regions of the film are layered while in other areas the NbAs and NbGa₃ phases are both in contact with the GaAs substrate. This lateral distribution of phases would be expected for films under 100 nm where grain size and lateral diffusion effects can be important.

Although the interfaces in the thin-film couples were irregular, those observed in the bulk diffusion couple were planar. Considering the criteria for interfacial stability discussed by Wagner [12] and Lin et al. [10], planar interfaces would be stable if the growth fronts of the NbAs and NbGa₃ phases were at the GaAs and Nb interfaces respectively. This situation would be expected if Ga was the predominant diffusing species and As the slowest. The origin of the voids observed in the NbAs phase of the bulk diffusion couple is unclear.

CONCLUSION

The phase formation sequence in GaAs/Nb bulk diffusion couples has been rationalized with the aid of the phase diagram and diffusion path concept. NbAs was found to be the stable phase in contact with GaAs. Niobium appears to be unsuitable as a contact to GaAs due to interfacial reaction at temperatures as low as 500°C.

REFERENCES

1. X.W. Wu, L.C. Zhang, P. Bradley, D.K. Chin, and T. Van Duzer, Appl. Phys. Lett. 50, 288 (1987).
2. J. Ding (private communication).
3. J.R. Lince, C.T. Tsai, and R.S. Williams, J. Mater. Res. 1, 537 (1986).
4. R. Beyers, K.B. Kim, and R. Sinclair, J. Appl. Phys. 61, 2195 (1987).
5. C.T. Tsai and R.S. Williams, J. Mater. Res. 1, 352 (1986).
6. J.F. McGilp, J. Mater. Res. 2, 516 (1987).
7. J.-C. Lin, K.-C. Hsieh, K.J. Schulz, and Y.A. Chang, J. Mater. Res. (in press).
8. R. Schmid-Fetzer, J. Electron. Mater. (in press).
9. M.R. Yu, F.R. Zhu, X. Wang, B.Q. Wang, K. Zao, P.S. Pu, and C.L. Lei, Chin. J. Semicond. 6, 55 (1985).
10. J.C. Lin, K.J. Schulz, K.-C. Hsieh, and Y.A. Chang, J. Electrochem. Soc. (in press).
11. T. Sands, V.G. Keramidas, K.M. Yu, J. Washburn, and K. Krishnan, J. Appl. Phys. (in press).
12. C. Wagner, J. Electrochem. Soc. 103, 571 (1956).

PRODUCTION CONTACT VIA ETCHING FOR GaAs MMICS.

P.J. Astell-Burt*, G.A. Ditmer*, V.B. Kadakia**, B.C. Cochran***, D-R Webb***.
*P.T. Technology Inc., 145 Sidney St., Cambridge, MA 02139.
**U. of Lowell, Lowell, MA 01854.
***Adams Russell Electronics Co., Burlington, MA 01803.

ABSTRACT.

Research has been undertaken in several major research laboratories to achieve a commercially successful GaAs via hole etching process. This work presents the achievement of results with the desired characteristics for through-wafer GaAs interconnects or vias, for production of MMICs.

We report vias acceptable for use in the production of microwave GaAs field effect transistors fabricated by a reactive ion etching process. Using conventional resist masks and non toxic gases, aspect ratios of up to 10:1 have been achieved with via sizes as small as 40μm x 40μm. Typical wafer thicknesses were greater than 100μm and the process showed high selectivity to the frontside metallization. The etching conditions had to be varied as a function of overall substrate area and exposed GaAs. To determine optimum process parameters, the RF power and resultant DC bias; gas flow rates and ratios; and system pressure were systematically varied. The effects of changes in area of wafer to be etched and of the composition of the electrode material were also studied. Details are presented of the optimum etching parameters for use in 2" and 3" GaAs MMIC production.

INTRODUCTION.

The intrinsic high speed properties of III-V compounds such as GaAs set these materials apart from other semiconductors and make them suitable for special electronic and optical applications. One application of GaAs is in very high frequency devices in which the performance is becoming increasingly dependent on the physical layout of the various components.

One of the fundamental speed restrictions is the nature of the connections to ground which contributes towards the LC delay time. The use of low inductance connections through the wafer to a ground plane on the backside of the chip eliminates problems associated with long thin wires to ground[1].

This type of connection can be accomplished by metallizing a via hole that has been etched from the backside to gold pads on the frontside of the wafer as depicted in Figure 1.

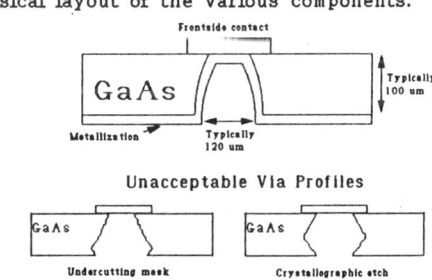

Unacceptable Via Profiles

Undercutting mask   Crystallographic etch

Figure 1. Via hole features.

The essential structure of a via hole is such that it allows for a metallization without discontinuities or voids which can potentially lead to reliability problems. An undercutting etch or one exhibiting crystallographic characteristics are not acceptable.

Typical via holes are in the range of 25 - 200μm in diameter with either square or circular geometries. Etch depth requirements range from 400μm in bulk wafers to more usually 100μm through a lapped wafer.

Production level processing of GaAs IC's contributes additional process requirements for via hole etching. The process must be repeatable and uniform for maximum yield. The etch rate must be fast enough for through-put requirements. Also, the masking material must be both resistant to etching and easily removable following the etch process. Wafer handling must also be a consideration to minimize breakage.

Traditionally, wet etching has been used to make via holes but this suffers from several disadvantages. Firstly, it is prone to the lack of reproducibility typical of wet etching processes. Secondly, as wet etching is essentially isotropic and crystallographically dependent, an unacceptable amount of GaAs is consumed.

Dry processing, on the other hand, presents a satisfactory solution to these problems enabling the process to be tailored to specific device requirements[2]. The process for production contact via etching must meet the following requirements:

1. No overhang, requiring minimal crystallographic etching.
2. The least GaAs "real estate" consumption.
3. High etch rate.
4. High etch selectivity of GaAs over mask material.
5. Run to run reproducibility.
6. Uniformity across a 2" (or 3") wafer.
7. Polymer and mask free surfaces following etch.

In this paper a process is presented which utilizes the advantages of dry processing to meet the above requirements for production via etching.

EXPERIMENTAL.

All the work described herein has been carried out on a Plasma Technology Plasmalab reactive ion etcher, schematic shown in Figure 2. Similar results have been achieved on the larger RIE 800 machine used on MMIC production lines.

To remove operator dependence the system was always run in fully automatic mode. A 13.56MHz discharge was generated inside a plasma chamber consisting of two parallel anodised aluminium electrodes enclosed in a pyrex vacuum vessel. RF power was supplied to the lower electrode, and the top electrode plate was grounded.

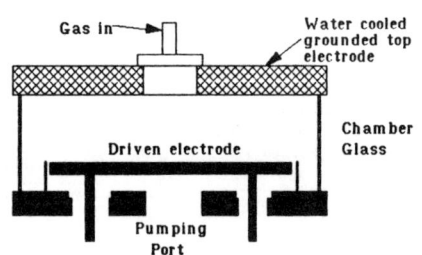

Figure 2. Basic layout of etching system.

Both electrodes were connected to a heater/chiller unit to maintain the chamber at a desired temperature between -5 and 95°C, which was normally 40°C. The vacuum system comprised a rootes / rotary combination. An extensive pump and purge cycle was employed to evacuate the chamber of any contaminants, especially water vapour. As an alternative, a turbomolecular pump is available on the system which would also ensure a good base pressure.

The gas inlet system was of the conventional shower head design, and the gases were all regulated by mass flow controllers. The vias were etched using a Freon-12 (CCl_2F_2) and O_2 gas mixture. High conductance pipe work was used, along with an auto pressure controller, to ensure total gas flow and process pressure could be controlled as two independent variables. An essential part of good process reproducibility is chamber cleanliness and this is maintained by a rigorous clean of the chamber prior to every etch cycle using an O_2 plasma.

The reported data was obtained from via holes etched in <100> semi-insulating GaAs. The wafers to be etched were mounted on ceramic carriers to minimize breakage due to handling 100μm thick wafers. 75μm circular holes were masked with 5.5μm of positive photo resist. Other mask materials such as Ni and UV hardened photo resist have also been used with some success, but removal of these masks after the etch process is difficult.

Process parameters were systematically varied in order to determine their effects on etch rate and aspect ratio (vertical etch rate : lateral etch rate). The parameters which were studied are :

1. Total system pressure
2. Applied RF power
3. Oxygen gas flow

The vias were etched, from the wafer backside to 2μm thick Au frontside metallization. Following the via etch, the mask was stripped, then the wafer was dismounted and cleaned using a positive photoresist stripper and an oxygen plasma. This sample clean is done to prepare the backside surface for metallization. The deposited backside metal consisted of sputtered TiW and Au, 500Å and 10,000Å thick, respectively.

For all samples systematically examined, the mask size was a 75μm diameter hole, and the etch time was 45 minutes. A temperature controller kept the top and bottom electrodes at 40°C. The wafer temperature on the surface is estimated to reach a maximum of 90-110°C. The effect of temperature variations remains to be studied.

RESULTS AND DISCUSSION.

The first set of samples was processed varying the system pressure from 40 - 100mT.

Process conditions for these runs were 250 Watts with 8% oxygen in the etchant gas stream. These results are shown in Figure 3. A maximum etch rate is found at 60mT. The optimal flow of reactants to the surface and reaction products away from the surface has been achieved at 60 - 80mT. The aspect ratio of the etch is also maximized in this pressure regime.

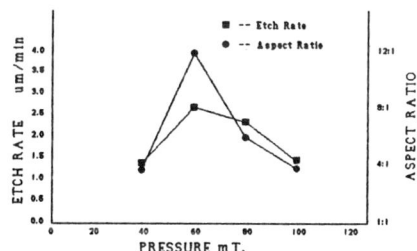

Figure 3. GaAs Etch rate and aspect ratio vs. System pressure.

Utilizing a system pressure of 60mT and the same oxygen percentage of 8%, the applied RF power was varied between 50 and 300 watts.
These values translate to power densities of .11 to .44 W/cm^2 for a 240mm (9.4") electrode. The DC bias varied with applied RF power as indicated in Figure 4. However, RF power was the monitored parameter during wafer etching.

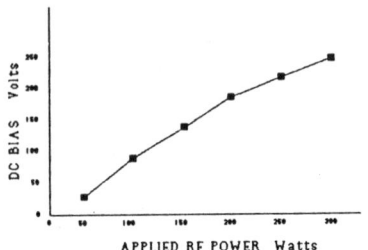

Figure 4. DC Bias vs. Applied RF power.

The etch rate was found to be neglible at 50 Watts, increasing to 1.1μm/min at 100Watts. The relationship of etch rate to incident power is shown in Figure 5. The etch rate increases with power up to 250 watts, then drops off sharply as crystallographic etching becomes more prominent. A similar maximum was observed by C.B. Cooper et al.[3] at DC biases greater than 175 V while etching GaAs with SiCl$_4$/Cl$_2$.

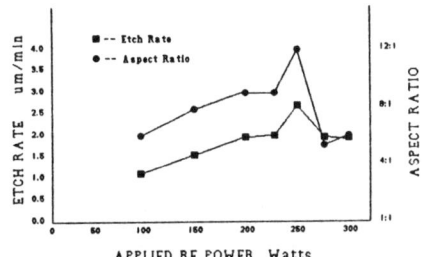

Figure 5. GaAs Etch rate and aspect ratio vs. Applied RF power.

It is proposed that high ion energies sputtered some adsorbed reactants off the surface prior to reaction completion reducing the etch rate.
The final parameter to be examined was the percentage of oxygen in the CCl$_2$F$_2$/O$_2$ etchant gas mixture.

The oxygen percent was varied from 0% up to 15%. Figure 6. shows results from this series of runs. The etch rate was observed to increase with additional oxygen up to approximately 10% of total gas flow, then diminish at higher percentages. The etch also began to undercut and proceed with a more crystallographic nature at higher than 6% O$_2$.

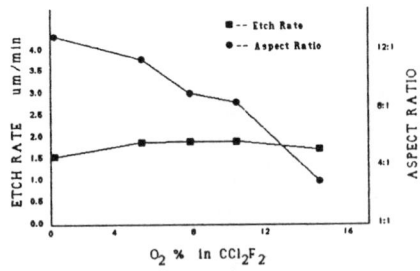

Figure 6. GaAs Etch rate and aspect ratio vs. Oxygen %.

Mounted wafer samples containg functional passive devices for RF characterization were then processed utilizing the optimum conditions developed through this work. The process parameters used were 60mT with 6% oxygen under incident RF power of 230 watts.

The resultant RIE vias gave an inductance of 80 picohenrys (pH) associated with the through to ground connection at RF frequencies of 0 - 20 GHz. An HP 8510 Network Analyzer was used for RF characterization. On wafer standards were used to calibrate the parameter measurements. The imaginary part of the inductance parameter Z_{11} equals 2π times the frequency times the inductance of the via.

The inductance can therefore be derived from the slope of Z_{11} plotted versus frequency as shown in Figure 7. The RIE via data shown is an average of measurements made on a 110μm thick 2" diameter GaAs wafer with 120±10 micron diameter via holes. For comparison, the data is normalized to the inductance associated with a 100μm thick wafer. The normalized inductance is measured at 78.0 pH. The normalized inductance of a wafer processed identically except with wet chemical etched vias is 69.1 pH. The RIE via and wet etched via inductances are comparable.

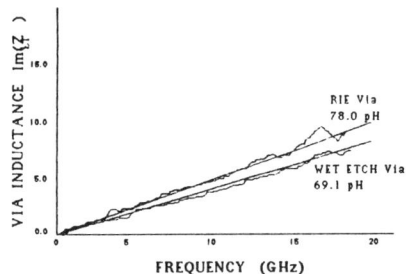

Figure 7. Via hole inductance vs. Frequency

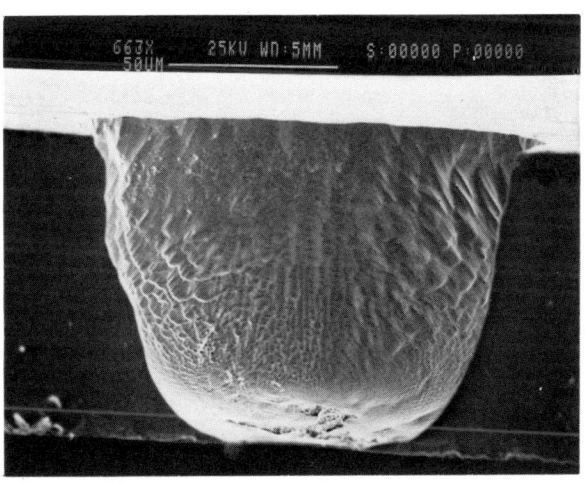

Figure 8. SEM Micrograph of RIE via

The lower value for the wet chemical etched via is due to the fact that it is 250±50μm in diameter, more than twice that for the via on the RIE sample. The larger diameter reduces the packing density of components with vias by greater than 100 %, consuming considerable GaAs 'real estate'. Large via holes also contribute to reduced yield due to wafer breakage. The RF inductance for the RIE vias is stable across the 0 - 20 GHz measurement range.

CONCLUSIONS.

A process was developed for reactive ion etching vias for GaAs MMIC production. Masking with conventional photoresists and etching with non-toxic freon - oxygen gas mixtures contribute to compatibility of the process with a production environment. Optimized etching conditions of 6% Oxygen in Freon-12 at 60mT with 230 Watts applied power were used to etch 120±10μm diameter via connections through 100±10μm thick 2" GaAs wafers. Resultant vias are shown in Figure 8. The RIE vias provide good low inductance (80pH) contacts to ground for GaAs MMIC's up to 20 GHz.

ACKNOWLEDGMENTS.

This work was a collaboration between P.T.Technology Inc., University of Lowell, and Adams Russell Electronics Co. We are grateful to Ed DiMattia, and the staff of Adams-Russell semiconductor division, particularly Margret Cole, Dave Danzillio, Brian Murphy, and Maura Tully for processing assistance. RF measurements taken by Scott Gatley and Dave Wandrei were also appreciated.

REFERENCES.

[1] L.A. D'Asaro, A.D. Butherus, J.V. DiLorenzo, D.E. Iglesias, and S.H. Wemple, Inst.Phys.Conf. 56, 267 (1980).

[2] K.P. Hilton and J. Woodward, Electronics Letters, 21 (21), 962 (1985).

[3] C.B. Cooper, M.E. Day, C. Yuen, and M. Salimian, J.Electrochem.Soc., 134 (10), 2533 (1987).

HIGH FREQUENCY DIELECTRIC RESPONSE OF CALCIUM ALUMINATE CEMENTS: POTENTIAL PACKAGING SUBSTRATE MATERIALS

PAUL SLIVA, M. LEFFLER, M. BLISS, L.E. CROSS AND B.E. SCHEETZ
The Pennsylvania State University, Materials Research Laboratory, University Park, PA 16802

ABSTRACT

Dielectric permittivity and loss tangent were measured on: 1) pressed disks of four different calcium aluminate cements shear mixed with water and poly(vinyl alcohol) and cured at room temperature and 2) laminates of SECAR 71 cement sintered at 1450°C. Post resonance and perturbation methods were used for microwave frequency measurements. Far infrared dielectric response of the cements was determined by FTIR in the diffuse reflectance mode. The real and imaginary parts of the relative dielectric permittivity were obtained via a conventional Kramers-Kronig analysis.

INTRODUCTION

Recent advancements in the development of GaAs integrated circuits have successively pushed gate densities into the 5000-6000 range and switching delays to as low as 52 ps at 0.55 mW power dissipation [1,2]. However, to fully exploit these and other advantages of GaAs devices, a different approach to GaAs packaging must be realized as compared with present silicon VLSI. Included in this alternative approach is the evolution of low dielectric permittivity substrate materials needed to meet the more stringent requirements of GaAs packages.

One of the more critical properties of a substrate material is its relative dielectric permittivity. Several important performance characteristics are directly related to the dielectric permittivity including: signal propagation delay, and crosstalk between and impedance of signal traces [3,4,5]. Signal transmission velocity is inversely proportional to the square root of the dielectric permittivity. Consequently, using a substrate with a dielectric permittivity as low as possible will result in an increase in signal propagation. In addition, lowering the dielectric permittivity effectively permits a higher signal trace density without increasing crosstalk through mutual inductance and allows thinner substrate layers to maintain the same impedance of the signal trace.

There are a large number of materials, both organic and inorganic, which possess a relatively low dielectric permittivity [6,7,8]. The organics, mostly polymeric materials, are not stable in radiation environments. Inorganic materials have been restricted mostly to polycrystalline ceramics and glasses. Several studies have been undertaken at this laboratory to develop new and often novel materials for low dielectric permittivity applications [9,10,11,12]. The approach of the present study is to ascertain the feasibility of utilizing commercial calcium aluminate cement as a low relative dielectric permittivity substrate material.

Calcium aluminate cements are hydraulic cements in that, when mixed with water, they are capable of setting and hardening at room temperature. In addition, these cements will form a refractory bond at elevated temperatures. Most cements, including the calcium aluminates, are electrical insulators. Previous work has shown that several calcium aluminate cements, when prepared at room temperature, exhibit a dielectric permittivity ranging from 14 to 4 at frequencies of 100Hz to 1MHz with SECAR 71 having the lowest permittivity [13]. Losses were reported as low as 0.02 at 1MHz for samples still containing adsorbed water and poly (vinyl alcohol), a plasticizer used during mixing. In an investigation of the properties of sintered laminates prepared from SECAR 71, it was shown that the average value for the dielectric constant up to 1MHz was 10.67 with an average dissipation factor of 0.0038 [14]. The present study extended dielectric measurements of the room temperature cured cements and the sintered laminates of SECAR 71 into the microwave and far infrared frequencies.

Mat. Res. Soc. Symp. Proc. Vol. 108. ©1988 Materials Research Society

EXPERIMENTAL PROCEDURE

Sample Preparation

Four different commercial calcium aluminate cements, under the trade name SECAR[TM] (LaFarge Calcium Aluminates, Inc., Chesapeake, VA), were used in the study. For each of the cements, the chemical composition and phases detected by x-ray diffraction are summarized in Table I. The number associated with the cement trade name reflects the approximate bulk Al_2O_3 content. Other than subtle differences in the minor constituents, the cements are chemically comprised primarily of Al_2O_3 and CaO (as calcium aluminate and/or calcium aluminosilicate phases). A comprehensive analysis of these cements was reported previously [13].

Samples were prepared for room temperature curing by using a shear mix process that involves the formation of a plastic cement dough by combining deionized water, calcium aluminate cement powder and 80% hydrolyzed poly(vinyl alcohol) (PVA). A typical mix consists of 22.22g of a 10 wt.% PVA solution and 100.00g of cement resulting in a water/cement of 0.20 and a PVA content of 1.8 wt.%. The materials are homogenized by high shear mixing (C.W. Brabender Preparation Mill, S. Hackensack, NJ) under vacuum. Circular disks are formed by uniaxially pressing the mix at ~138MPa in a 1.905cm diameter steel die coated with teflon mold release. Each disk is removed from the die and placed into a dessicator over drierite at room temperature for curing. For microwave dielectric measurements, the surfaces were polished to ~30μm fineness. Samples for infrared measurements were polished further to 1μm fineness.

TABLE I. Chemical[a] and Phase Analysis of SECAR Calcium Aluminate Cements

	SECAR 51	SECAR 60	SECAR 71	SECAR 80
SiO_2	3.43	23.02	0.52	0.31
Al_2O_3	50.8	59.1	71.2	79.8
TiO_2	2.70	0.58	0.013	0.008
Fe_2O_3	2.37	0.38	0.11	0.07
MgO	0.46	0.19	0.38	0.20
CaO	35.9	13.3	27.5	16.7
MnO	0.001	<0.002	<0.002	<0.002
SrO	0.05	0.01	0.01	0.01
Na_2O	0.07	0.22	0.37	0.74
K_2O	0.28	0.10	0.04	0.02
P_2O_5	0.27	0.05	0.02	0.13
SO_3	0.10	0.05	0.05	0.02
LOI(900)[b]	2.62	3.60	0.36	1.55
TOTALS	99.20	100.60	100.57	99.56
PHASES[c]	CA	CA	CA	CA
	(C_2AS,CT)	(CA_2,S)	(CA_2)	(CA_2,αA)

a) weight percent oxide
b) includes CO_2
c) standard abbreviations: C= CaO, A= Al_2O_3, S= SiO_2, T= TiO_2, minor
   phases in parentheses

Sintered laminates of SECAR 71 were prepared from tapes cast in an acetate-based binder system (Metaramic Sciences, Carlesbad, CA) [14]. Each laminate is ~6.45 square centimeters and consists of a stack of 12 tapes sintered at 1450°C in air. For both microwave dielectric and infrared measurements, the laminates were polished to 1μm fineness.

Microwave Measurements

At microwave frequencies, two methods were used to measure ε' and ε'': the post resonance method [15] and the cavity perturbation technique [16]. All sample measurements were taken at room temperature and several samples (minimum of 4 for shear mixed and 3 for sintered laminates) of each type cement were measured to ensure reproducibility.

The post resonance method is an exact resonance technique that uses samples in the form of a circular disk. Varying the thickness of the sample disks for a given material allows measurements of ε' and ε'' to be taken at different frequencies. Cured cement paste samples prepared for this study were typically 1.90cm in diameter while varying in thickness from 0.4 to 1.0cm, corresponding to a measured frequency range of ~8GHz to ~17GHz.

In the post resonance technique, the sample is placed between two large parallel brass plates, which simulate mathematically infinite conducting plates. Two probes are placed directly on either side of the sample disk. A signal is introduced through one of the probes using a sweep oscillator (HP Model 83508). The other probe detects the signal as it passes out of the sample. The signal is then fed to an oscilliscope also linked to a CRT display (HP Model 182T), a microwave frequency counter (HP Model 5343A) and a storage normalizer (HP Model 8750A). The $TE_{01\delta}$ mode is located on the display, and its resonant frequency and the width of the resonant peak at the 3dB point, $f_{\Delta 3dB}$, are measured. Taking into consideration the sample geometry, ε' and ε'' are calculated using a computer program which uses an iterative technique for solving the Bessel functions for electric fields inside and outside the sample.

The cavity perturbation method was used to measure ε' and ε'' of sintered laminates of SECAR 71 cement. The technique makes use of a rectangular cross section wave-guide cavity, constructed of brass, coupled to conducting plates with small apertures (2.5mm wide) on either side of the waveguide to provide inductive coupling to the waveguide cavity. Five resonant peaks (TE modes) are generated in the cavity with a network analyzer (HP Model 8510T). A sample, in the form of a thin rectangular plate (<0.5mm in thickness) is introduced into the cavity through a slot cut in the top, causing the resonant peaks to shift in frequency. From the peak shifts and change in the width of the 3dB points of the five resonant peaks, ε' and ε'' can be calculated.

FTIR Measurements

Samples of both cured cement pastes and sintered SECAR 71 were prepared for analysis of infrared properties. Specular reflectance from the sample surface was measured in the frequency range of 7.5×10^{11} Hz to 1.2×10^{14} Hz and normalized to specular reflectance from a front surfaced aluminum mirror in the same frequency range (IBM IR98 Fourier Transform Infrared Spectrometer). Because the samples are polycrystalline and, in some cases polyphasic, the maximum possible reflectance intensities are not obtained (the absolute intensities for dielectric single crystals are true material parameters). The measurements obtained were adequate for comparison between the samples [17]. A Kramers-Kronig analysis of the reflectance spectrum in conjuction with application of the Fresnel equation makes it possible to extract the dispersion behavior of the real and imaginary parts of the dielectric constant [18,19].

RESULTS AND DISCUSSION

Microwave Measurements

Samples prepared by shear mixing showed very little hydration as indicated by the absence of any hydrated phase in the x-ray diffraction pattern. Sample disks were within 92% of theoretical bulk density, suggesting the presence of some porosity.

Microwave frequency dielectric response of each of the cements is given in Table II. The results are considered accurate to within 5%. The variation in dielectric response within each sample set may be due to damped harmonic oscillations in the material changing with frequency. However, within experimental error, all of the cements exhibit a similar response. Variations in phase composition were observed to have little effect on the dielectric response. The calcium titanate in the SECAR 51 is not present in a high enough concentration to contribute significantly to the dielectric constant, although the presence of Ti and Fe does contribute to the losses [20]. It is apparent from the low values obtained for tan δ that potential loss contributions from adsorbed water and PVA have been clamped.

The resultant sintered laminates of SECAR 71 were translucent and comprised of CA and CA_2, the precursor phases in the as-received cement. Most of the porosity was very fine ($<1\mu m$) and found at grain boundaries and triple points. There was little evidence of larger pores.

The relative dielectric permittivity and tan δ of the sintered laminates at the five resonant frequencies are given in Table III. The measurements for the cements are accurate to within 5%. Variation in dielectric constant with frequency is due to damped harmonic oscillations,

TABLE II. Relative Dielectric Permittivity (K) and Loss (tanδ) of Calcium Aluminate Cements Cured at Room Temperature

CEMENT	FREQ. (GHz)	K	TAN δ
SECAR 51	16.446±0.228	6.222±0.478	0.0177±0.0035
	12.130±0.368	6.582±0.258	0.0079±0.0024
	9.175±0.134	6.194±0.064	0.0095±0.0030
SECAR 60	16.921±0.595	6.251±0.133	0.0059±0.0010
	11.830±0.506	6.842±0.224	0.0064±0.0026
	9.275±0.615	5.652±0.529	0.0065±0.0014
	8.680±0.089	5.863±0.109	0.0084±0.0012
SECAR 71	15.650±0.384	6.370±0.247	0.0044±0.0001
	11.112±0.551	6.566±0.144	0.0058±0.0007
	9.604±0.364	6.005±0.202	0.0056±0.0010
	8.884±0.146	5.529±0.072	0.0072±0.0016
SECAR 80	16.496±0.309	6.061±0.080	0.0095±0.0053
	11.770±0.269	6.254±0.452	0.0078±0.0025
	9.099±0.194	5.739±0.185	0.0075±0.0012

TABLE III. Relative Dielectric Permittivity (K) and Loss (tanδ) of Tape Cast SECAR 71 Laminates

FREQ. (GHz)	K	TAN δ
8.350±0.012	4.086±0.307	0.0020±0.0008
9.467±0.356	4.398±0.379	0.0028±0.0008
9.892±0.008	3.314±0.164	0.0024±0.0022
10.637±0.022	5.171±0.433	0.0039±0.0009
11.567±0.023	4.038±0.519	0.0035±0.0016

variation in the chemical composition and measurement error. The dielectric constant reported at 9.892 GHz appears to be anomalously low, even though it is reproducible. The overall results are however, comparable with data obtained by the post resonance method on sintered disks of SECAR 71 (K:5.466, tan δ:0.0027 at 18.452 GHz) [14]. When compared with alumina measured by the same technique (K:9.35-9.47 at 8.542 to 11.846 GHz), the SECAR 71 has a lower dielectric constant across the frequency range [21]. The SECAR 71 however, has a much higher loss when compared with an average value of 0.0006 for aluminum, presumably from the extensive chemical composition.

Infrared Measurements

The behavior of K' and K'' in the infrared is strongly dependent on the atomic structure and bond character of the material. The number of peaks observed is determined, for crystalline materials, by symmetry and the number of atoms per unit cell. Poorly crystalline regions, such as grain boundaries and amorphous regions can have many more resonance peaks because of their reduced symmetry. The calcium aluminate cements are a mixture consisting primarily of two monoclinic phases which would be expected to have many more resonant vibrations than a higher symmetry monophasic material such as alumina.

The peaks in K' and K'' are caused by mechanical resonances of atomic units usually only involving nearest neighbor interactions. The peak locations are determined by the masses of the atoms, the strength of the bonds involved and the geometry of the atom displacements during the vibrations. Peak widths are related to the mechanical damping of the vibration and peak amplitudes and areas are determined by the number of atomic units with that resonant frequency and the magnitude of the dipole moment caused by the vibration.

Because the spectra are collected as reflectance from a surface, the data can only be used in a comparative fashion between samples with equivalent surface qualities. SECAR 51 and 80 gave adequate reflectance spectra for the Kramers-Kronig analysis. SECAR 71 had reflectance values of about half that of SECAR 51 and 80; the spectra was too noisy for Kramers-Kronig analysis. The lower reflectance values can be attributed to a poorer surface quality or K' and K'' are substantially lower in SECAR 71. SECAR 60 did not yield a reflectance spectrum at all, which can be attributed to the inability to get a good surface finish on this particular sample.

The real and imaginary parts of the dielectric constant, as derived from the Kramers-Kronig analysis, for SECAR 51 and 80, are shown in Figure 1. The SECAR 51 sample shows a much greater number of oscillators than SECAR 80 indicating a much more complicated phase mixture in the SECAR 51. There are however, peaks in common between the samples, indicating some similarities in composition. Both cements have a K' of < 9, but SECAR 80 has a more stable K' at the higher frequencies.

The sintered laminate of SECAR 71 was compared to sintered alumina, the present industry standard (Figure 2). The SECAR 71 shows a much greater number of oscillators than

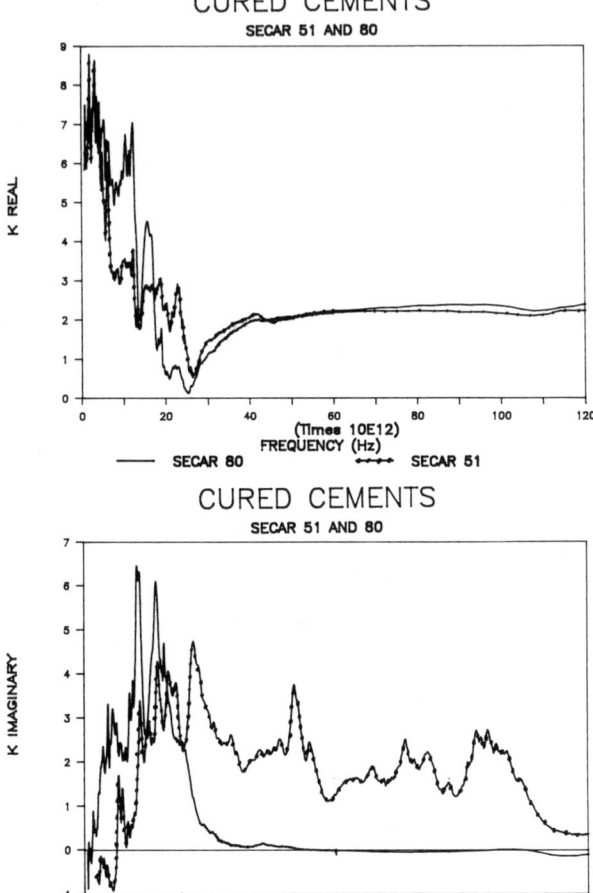

Fig. 1. Real and imaginary parts of the dielectric constant, as determined by Kramers-Kronig analysis, of SECAR 51 and 80 calcium aluminate cements cured at room temperature.

the alumina which is characteristic of the differences in crystal structures. The SECAR 71 shows an overall flatter response indicating that although there are more vibrating units, the oscillators are not as strong as seen in the alumina sample. There are fewer oscillators generating each peak and/or the induced dipole moments are much smaller in the sintered SECAR 71 than the alumina.

CONCLUSIONS

The dielectric response of four different calcium aluminate cements cured at room temperature and SECAR 71 laminates sintered at 1450°C has been determined at both microwave and far infrared frequencies. The results of the investigation indicate the following:

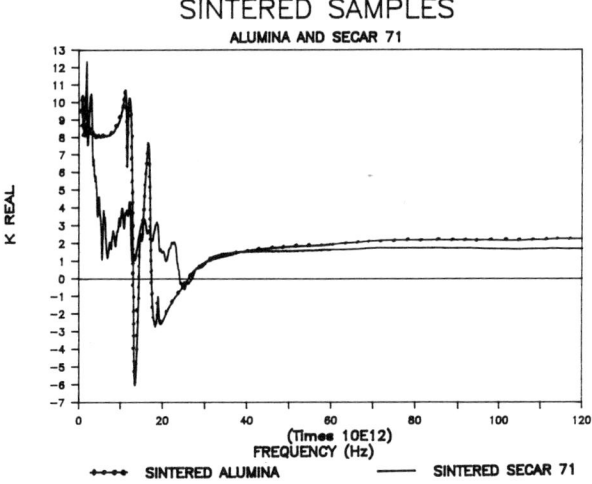

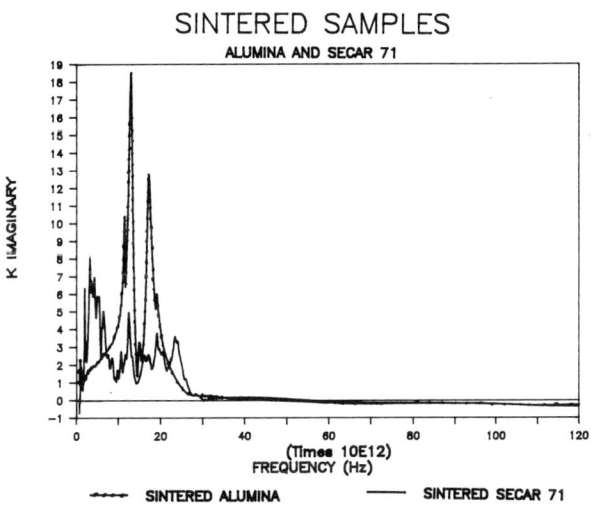

Fig. 2. Real and imaginary parts of the dielectric constant, as determined by Kramers-Kronig analysis, of SECAR 71 calcium aluminate cement laminates sintered at 1450°C compared with sintered alumina.

1) Within experimental error, all of the cured cements exhibit a similar response of the dielectric constant and tanδ (~8 to 17 GHz).

2) Sintered laminates of SECAR 71 have a lower dielectric constant than alumina but higher losses (~8 to 12 GHz).

3) For the two cured cement samples (SECAR 51 and 80) that ran through the Kramers-Kronig analysis, both have a dielectric constant below 9, but the losses in SECAR 80 are much more stable above ~30 THz.

4) Sintered SECAR 71 shows an overall flatter response of both $K^{'}$ and $K^{''}$ than alumina.

The low dielectric constant and loss obtained for these cements, coupled with the low temperature processing (in the case of the cured cements) and low cost make calcium aluminate cements a possible alternative to alumina for GaAs IC packaging applications.

ACKNOWLEDGEMENTS

The authors would like to thank LaFarge Calcium Aluminates for supplying the SECAR cements. A special thanks goes to Ms Vicki Zimmerman for typing the manuscript. The work was supported by the Defense Advanced Research Projects Agency (ONR Contract N00014-84-K-0721).

REFERENCES

1. B.K. Gilbert, B.A. Naused, D.J. Schwab and R.L. Thompson, Computer $\underline{19}$(10), 29-43 (1986).
2. A. Rhoads, T. Flegel and G. LaRue in IEEE Gallium Arsenide IC Symposium, IEEE Publication #83CH1876-2, 178-181 (1983).
3. B.K. Gilbert, D.J. Schwab and M.L. Samson in VHSIC Packaging Workshop, Naval Surface Weapons Center, Silver Spring, MD, 103-118 (1985).
4. S. Matsushita and T. Moto-Oka, IEEE Trans. on Computers $\underline{11}$, C-23, 1122-1132 (1974).
5. C.R. Paul, IEEE Trans. on EMC, EMC-25, 182-192 (1984).
6. J.M. Herbert, Ceramic Dielectrics and Capacitors, (Gordon and Breach Science Publishers, New York, 1985), pp. 95-124.
7. A.J. Beuhler, J.A. Wrezel, J.L. Maas and R.C. Sundahl, Jr., in Electronic Packaging Materials Science II, edited by K.A. Jackson, R.C. Pohanka, D.R. Uhlmann and D.R. Ulrich (Mater. Res. Soc. Proc. $\underline{72}$, Pittsburgh, PA 1986) pp. 223-234.
8. B. Schwartz, ibid., p. 58.
9. J. Yamamoto, MS Thesis, The Pennsylvania State University, 1986.
10. A. Das, R. Messier, T.R. Gururaja and L.E. Cross in Electronic Packaging Materials Science II, edited by K.A. Jackson, R.C. Pohanka, D.R. Uhlmann and D.R. Ulrich (Mater. Res. Soc. Proc. $\underline{72}$, Pittsburgh, PA 1986) pp. 27-34.
11. L.E. Cross and T.R. Gururaja, ibid., pp. 53-66.
12. M.J. Leap and W. Huebner, presented at the 1987 International Symposium on Ceramic Substrates and Packages, Denver, CO, 1987 (submitted).
13. P. Sliva, L.E. Cross, T.R. Gururaja and B.E. Scheetz, Mat. Let. $\underline{4}$(10), 409-413 (1986).
14. P. Sliva, G.O. Dayton, L.E. Cross and B.E. Scheetz, presented at the 1987 International Symposium on Ceramic Substrates and Packages, Denver, CO, 1987 (submitted).
15. B.W. Hakki and P.D. Coleman, IRE Trans. on Microwave Theory and Tech. MTT-8, 402-410 (1960).
16. M. Lanagan, Ph.D. Thesis, The Pennsylvania State University, 1987.
17. F. Gervais and J. Lecomte, Sol. State Comm. $\underline{53}$(8), 711-713 (1985).
18. G. Andermann, A. Caron and D.W. Dows, J. of the Optical Soc. of Am. $\underline{55}$(10), 1210-1216 (1965).
19. H.C. Chen, Theory of Electromagnetic Waves, A Coordinate-Free Approach, McGraw-Hill Book Company, New York, 1983).
20. L.E. Cross (private communication).
21. D.C. Dube, M.T. Lanagan, J.H. Kim and S.J. Jang, J. of Appl. Phys. (submitted).

Author Index

Subject Index